Electrical and Electronic Devices, Circuits and Materials

Electrical and Electronic Devices, Circuits and Materials

Design and Applications

Edited by
Suman Lata Tripathi, Parvej Ahmad Alvi, and
Umashankar Subramaniam

CRC Press is an imprint of the
Taylor & Francis Group, an **informa** business

First edition published 2021
by CRC Press
6000 Broken Sound Parkway NW, Suite 300, Boca Raton, FL 33487-2742

and by CRC Press
2 Park Square, Milton Park, Abingdon, Oxon, OX14 4RN

CRC Press is an imprint of Taylor & Francis Group, an Informa business

Library of Congress Cataloging-in-Publication Data
Names: Tripathi, Suman Lata, editor. | Alvi, Parvej Ahmad, editor. |
Subramaniam, Umashankar, editor.
Title: Electrical and electronic devices, circuits and materials : design and
applications / edited by Suman Lata Tripathi, Parvej Ahmad Alvi, Umashankar Subramaniam.
Description: First edition. | Boca Raton, FL : CRC Press/Taylor & Francis
Group, LLC, 2021. | Includes bibliographical references and index.
Identifiers: LCCN 2020037199 (print) | LCCN 2020037200 (ebook) |
ISBN 9780367564261 (hardback) | ISBN 9781003097723 (ebook)
Subjects: LCSH: Electronic apparatus and appliances. | Electric apparatus
and appliances. | Electronic circuits. | Electronics—Materials.
Classification: LCC TK7870 .E155 2021 (print) | LCC TK7870 (ebook) |
DDC 621.3—dc23
LC record available at https://lccn.loc.gov/2020037199
LC ebook record available at https://lccn.loc.gov/2020037200

ISBN 13: 978-0-367-56426-1 (hbk)
ISBN 13: 978-1-003-09772-3 (ebk)

Typeset in Times
by codeMantra

Contents

Preface

PREFACE TO THE FIRST EDITION

The extensive use of electronic devices and components in domestic and industrial purposes is increasing their importance by the time, and lead designers and researchers explore new electronic devices and circuits that can perform several tasks efficiently with low IC area and low power consumption. In the present era, smart and portable systems also intensify the researches to design sensor elements, an efficient storage cell and large-capacity memory elements. The objective of this edition is to provide a broad view of advanced electronic device and circuit design and challenges in a concise way for fast and easy understanding. It also explores device fabrication and characterization with new materials. This book provides information regarding almost all the aspects to make it highly beneficial for all the students, researchers and teachers of this field. Fundamental concepts of electrical and electronic device and circuit design are illustrated herein in a clear and detailed manner with an explanatory diagram wherever necessary. All the chapters are organized and described in a very simple manner to facilitate readability of the chapters.

Chapter Organization

This book is organized into 23 chapters in total.

Chapter 1 intends to study the basic MOSFET concepts, their characteristics, optimization methods and their evolution into novel device architectures for low-power applications.

Chapter 2 presents the temperature effect on RF/analog figure of merits (FOMs) for step-FinFET through Sentaurus 3D Technology Computer Aided Design (TCAD) simulator.

Chapter 3 focuses on the strategies to reduce the operating voltage of SRAM designed using FinFET technology and enhance battery life for ULP IoT devices.

Chapter 4 deals with various design issues of SRAMs from CMOS to various nanoscale devices from their advantages to limitations with their designs and applications.

Chapter 5 delivers the study approaching an emerging GaN-based HEMTs device technology appropriate for radio frequency and high-level power applications.

Chapter 6 explores the novel materials and architectures in modern electronic devices for high-speed and high-frequency electronic systems.

Chapter 7 intends to serve a fundamental understanding of the energy storage devices and how they can fit in for e-mobility and power grid by a careful technological design.

Chapter 8 describes the history of solar cell right from the time of inception till the present day. The prevalent technologies of today were discussed in detail with their pros and cons.

Chapter 9 addresses the problem of model order reduction and controller synthesis together of an automatic voltage regulator (AVR).
Chapter 10 presents the model reduction and developing the controller scheme of a single-machine infinite bus (SMIB) system using a heuristic approach, namely, the grey wolf optimizer (GWO).
Chapter 11 provides challenges and development of IoT in energy systems, in general, and in the context of smart grids particularly.
Chapter 12 attempts to review the state-of-the-art development of automobile industry and its battery management system to create a platform for smart electric vehicles.
Chapter 13 presents the reduced-order modelling (ROM) and controller design one after another for load frequency control of dual area network. Model order reduction is performed by employing a composite method.
Chapter 14 predicts the electricity price, 24 hours in advance, on the day-head market operation with effectively forecasting energy prices such as maximizing energy storage, allowing building flexibility on the demand side and enabling them to reduce consumption in costly times.
Chapter 15 deals with the role of thin films used in the field of micro-electromechanical system or MEMS, which manufactures different sensors that include physical and chemical sensors, actuators and microelectronics.
Chapter 16 mainly deals with undoped and Ba-doped $SrSnO_3$ samples that were synthesized using sol–gel method followed by calcination and sintering for analyzing lattice parameters, volume and sample density.
Chapter 17 describes the PbS powder and its films on flexible fabric-based substrate prepared via chemical root and characterized by XRD, FTIR, SEM, optical and electrical methods.
Chapter 18 explores a direct relation between stiffness of materials used in rectangular spade cantilever and electrical sensitivity piezoresistive MEMS force sensor.
Chapter 19 gives an insight into environment-friendly successive ionic layer adsorption and reaction (SILAR) chemical synthesis route that was adopted to deposit ZnS-nanoparticle/ZnO-nanoflower structure.
Chapter 20 is related to a non-conventional e-beam approach utilized for the deposition of polycrystalline thin film which has replaced the conventional one like reactive ion etching (RIE) and low chemical vapor deposition (LPCVD).
Chapter 21 embraces the most recent accomplishments in the related field and affords an impending into the aqueous electrolytes development.
Chapter 22 emphasizes graphene-based flexible and stretchable electronic devices and their mechanical properties with graphene-growth techniques, subsequent printing and transfer technologies.
Chapter 23 discusses a broad information on flexible microfluidics sensor, microstructure of paper with selected properties and making of paper as a substrate for the fabrication of flexible microfluidic devices.

Editors

Dr. Suman Lata Tripathi is currently working as professor, in the School of Electronics and Electrical Engineering, Lovely Professional University, India. She has more than 17 years of experience in academics and has published more than 45 research papers in refereed journals and conferences. She has organized a number of workshops, summer internships and expert lectures for students. She has worked as a session chair, conference steering committee member, editorial board member and reviewer in international/national IEEE journal and conferences. She has received the Research excellence award in 2019 at Lovely Professional University. She had received the best paper at IEEE ICICS-2018. She has published an edited book titled *Recent advancement in Electronic Device, Circuit and Materials* by **Nova Science Publishers**. She has also published edited books *Advanced VLSI Design and Testability Issues* and *Electronic Devices and Circuit Design Challenges for IoT application* in **CRC Taylor & Francis and Apple Academic Press**. Her area of expertise includes microelectronics device modeling and characterization, low-power VLSI circuit design, VLSI design of testing and advance FET design for IOT, embedded system design and biomedical applications.

Dr. Parvej Ahmad Alvi is currently working as an associate professor, in the Department of Physics, Banasthali University, Rajasthan, India. His major area of research is MEMS/NEMS technology (nanotechnology). He has more than 14 years of teaching and research experience in the area of modern physics, semiconductor physics, nanotechnology, etc. He has worked as an editorial board member and reviewer in several international/national IEEE/Elsevier journals and conferences. He has published more than 100 research papers in refereed international journals and conferences. He has also published more than six national/international text and reference books in the area of physics. He has guided around 20 Ph.D. scholars.

Dr. Umashankar Subramaniam is an associate professor in Renewable Energy Lab, College of Engineering, Prince Sultan University, Saudi Arabia. He has more than 15 years of teaching, research and industrial experience. Previously, he worked as associate professor and head, VIT Vellore, as well as senior R&D and senior application engineer in the field of power electronics, renewable energy and electrical drives. He is an associate editor in IEEE Access. He is a senior member in IEEE, PES, IAS, PSES, YP and ISTE. He has taken charge as Vice Chair – IEEE Madras Section and Chair – IEEE Student Activities from 2018 to 2020. He was an executive member (2014–2016) and a Vice Chair of IEEE MAS Young Professional from 2017 to 2019 by IEEE Madras Section. He has published more than 250+ research papers in national and international journals and conferences. He has also authored/co-authored/contributed 12 books/chapters and 12 technical articles on power electronics applications in renewable energy and allied areas. He is an editor of Heliyon, an Elsevier journal and various other reputed journals. He received Danfoss Innovator Award-Mentor during 2014–2015 and 2017–2018, and Research Award from VIT University during 2013–2018. Also, he received the INAE Summer Research Fellowship for the year 2014.

Contributors

Shini Agarwal
Department of Electrical and Instrumentation Engineering
Thapar Institute of Engineering and Technology
Patiala, Punjab, India

Sayeed Ahmad
Zakir Husain College of Engineering & Technology
Aligarh Muslim University
Aligarh, Uttar Pradesh, India

Jamil Akhtar
Flex MEMS Research Centre, Department of Electronics and Communication Engineering
Manipal University Jaipur
Jaipur, Rajasthan, India

Naushad Alam
Zakir Husain College of Engineering & Technology
Aligarh Muslim University
Aligarh, Uttar Pradesh, India

Parvej Ahmad Alvi
Department of Physics
Banasthali Vidyapith
Banasthali, Rajasthan, India

K. Arya
Department of Electrical and Electronics Engineering
Amrita School of Engineering
Amrita Vishwa Vidyapeetham
Coimbatore, Tamil Nadu, India

Srimanta Baishya
Electronics and Communication Engineering Department
National Institute of Technology Silchar
Silchar, Assam, India

S. Bhattacherjee
Department of Physics
JISCE
Kolkata

Brinda Bhowmick
Electronics and Communication Engineering Department
National Institute of Technology Silchar
Silchar, Assam, India

Shilpi Birla
Department of Electronics and Communication Engineering
Manipal University Jaipur
Jaipur, Rajasthan, India

Prakash Chand
Department of Physics
National Institute of Technology
Kurukshetra, Haryana, India

K.R.M. Vijaya Chandrakala
Department of Electrical and Electronics Engineering
Amrita School of Engineering
Amrita Vishwa Vidyapeetham
Coimbatore, Tamil Nadu, India

Himanshu Chaudhary
Department of Electronics and Communication Engineering
Manipal University Jaipur
Jaipur, Rajasthan, India

T. Chinnadurai
Department of Instrumentation and control Engineering
Sri Krishna College of Technology
Coimbatore, Tamil Nadu, India

Rohan S. Deshmukh
Department of Mechanical Engineering
SKN Sinhgad College of Engineering
Pandharpur, Maharashtra, India

Sampat G. Deshmukh
Department of Applied Physics
S. V. National Institute of Technology
Surat, Gujarat, India
and
Department of Engineering Physics
SKN Sinhgad College of Engineering
Pandharpur, Maharashtra, India

Anup Dey
Department of Electronics and Telecommunication Engineering
Jadavpur University
Kolkata

J. Dhanaselvam
Department of Instrumentation and Control Engineering
Sri Krishna College of Technology
Coimbatore, Tamil Nadu, India

S. Dwivedi
Department of Physics
S.S. Jain Subodh P.G. (Autonomous) College
Jaipur, Rajasthan, India

Souvik Ganguli
Department of Electrical and Instrumentation Engineering
Thapar Institute of Engineering and Technology
Patiala, Punjab, India

Bijoy Goswami
Department of Electronics and Telecommunication Engineering
Jadavpur University
Kolkata

Saumyadip Hazra
Department of Electrical and Instrumentation Engineering
Thapar Institute of Engineering and Technology
Patiala, Punjab, India

Bushra Khan
Centre of Material Sciences
University of Allahabad
Allahabad, Uttar Pradesh, India

Vipul Kheraj
Department of Applied Physics
S. V. National Institute of Technology
Surat, Gujarat, India

Abhimanyu Kumar
Department of Electrical and Instrumentation Engineering
Thapar Institute of Engineering and Technology
Patiala, Punjab, India

Aditya Kumar
Centre of Material Sciences
University of Allahabad
Allahabad, Uttar Pradesh, India

Arun Kumar
Panipat Institute of Engineering and Technology
Samalkha, Haryana, India

Mahesh Kumar
Flex MEMS Research Centre, Department of Electronics and Communication Engineering
Manipal University Jaipur
Jaipur, Rajasthan, India

Upendra Kumar
Department of Physics
Banasthali Vidyapith
Banasthali, Rajasthan, India

Monica Lamba
Flex MEMS Research Centre, Department of Electronics and Communication Engineering
Manipal University Jaipur
Jaipur, Rajasthan, India

Parag Nijhawan
Department of Electrical and Instrumentation Engineering
Thapar Institute of Engineering and Technology
Patiala, Punjab, India

Ashish K. Panchal
Department of Electrical Engineering
S. V. National Institute of Technology
Surat, Gujarat, India

Sweety Panchal
Department of Physics
Uka Tarsadia University
Surat, Gujarat, India

M. Karthigai Pandian
Department of Instrumentation and control Engineering
Sri Krishna College of Technology
Coimbatore, Tamil Nadu, India

Kinjal Patel
Department of Physics
Uka Tarsadia University
Surat, Gujarat, India

Himanshu Priyadarshi
School of Electrical Electronics and Communication
Manipal University Jaipur
Jaipur, Rajasthan, India

Jaymin Ray
Department of Physics
Uka Tarsadia University
Surat, Gujarat, India

Shasanka Sekhar Rout
GIET University
Gunupur, Odisha, India

Rajesh Saha
Electronics and Communication Engineering Department
Malaviya National Institute of Technology Jaipur
Jaipur, Rajasthan, India

Samridhi
Department of Physics
Banasthali Vidyapith
Banasthali, Rajasthan, India

T. S. Arun Samuel
Department of Electronics and Communication Engineering
National Engineering College
Kovilpatti, Tamil Nadu, India

Subir Kumar Sarakar
Department of Electronics and Telecommunication Engineering
Jadavpur University
Kolkata

K. Saravanakumar
Department of Instrumentation and Control Engineering
Sri Krishna College of Technology
Coimbatore, Tamil Nadu, India

Dipanjan Sen
Department of Electronics and Telecommunication Engineering
Jadavpur University
Kolkata

Sauhardh Sethi
Department of Electrical and Instrumentation Engineering
Thapar Institute of Engineering and Technology
Patiala, Punjab, India

Adeeba Sharif
Zakir Husain College of Engineering & Technology
Aligarh Muslim University
Aligarh, Uttar Pradesh, India

Niti Nipun Sharma
Department of Mechanical Engineering
Manipal University Jaipur
Jaipur, Rajasthan, India`

Sharad Sharma
Maharishi Markandeshwar (Deemed to be University)
Mullana, Haryana, India

Ashish Shrivastava
School of Electrical Electronics and Communication
Manipal University Jaipur
Jaipur, Rajasthan, India

N. K. Shukla
Department of Electrical Engineering
King Khalid University
Abha, Saudi Arabia

Kulwant Singh
Flex MEMS Research Centre, School of Electrical Electronics and Communication
Manipal University Jaipur
Jaipur, Rajasthan, India

Manoj K. Singh
Centre of Material Sciences
University of Allahabad
Allahabad, Uttar Pradesh, India

Neha Singh
Department of Electronics and Communication Engineering
Manipal University Jaipur
Jaipur, Rajasthan, India

Rishabh Singhal
Department of Electrical Engineering
Roorkee Institute of Technology
Roorkee, Uttarakhand, India

Manish Kumar Singla
Department of Electrical and Instrumentation Engineering
Thapar Institute of Engineering and Technology
Patiala, Punjab, India

Yashonidhi Srivastava
Department of Electrical and Instrumentation Engineering
Thapar Institute of Engineering and Technology
Patiala, Punjab, India

Ashish Tiwary
GIET University
Gunupur, Odisha, India

P. Vimala
Department of Electronics and Communication Engineering
Dayananda Sagar College of Engineering
Bangalore, Karnataka, India

Supriya Yadav
Department of Biosciences
Manipal University Jaipur
Jaipur, Rajasthan, India

1 MOSFET Design and Its Optimization for Low-Power Applications

P. Vimala
Dayananda Sagar College of Engineering

M. Karthigai Pandian
Sri Krishna College of Technology

T. S. Arun Samuel
National Engineering College

CONTENTS

1.1 INTRODUCTION

Metal-Oxide Semiconductor Field Effect Transistors (MOSFETs) find a major role in digital circuits because of their better efficiency in terms of power consumption and reduced silicon-area usage compared to bipolar digital technologies. In the current scenario, they are the basic building blocks of integrated circuits (ICs) and microprocessors. Continuous development of CMOS technology is guided and improved by CMOS scaling. As the size of the devices is reduced to greater proportions, their switching capabilities tend to get reduced. Decreasing the size of the device also reduces the distance between the source and drain, resulting in a phenomenon called Short-Channel Effects (SCEs). In a continuous effort to overcome the problems posed by SCEs, the classical, planar, single-gate MOSFET device has evolved into a three-dimensional device with structural modifications. Increasing the effective number of gates around a channel will enhance the electrostatic control of the channel by the gates and thus reduce the SCEs. In recent times, scaling of MOSFETs has reached the physical limitation of their size (5 nm), and this has raised the need for novel device architectures to replace the MOSFET technology for semiconductor industries.

1.2 VLSI DESIGN HIERARCHY

Very-Large-Scale Integration (VLSI) is the method of generating an IC by merging transistors, resistors, or other electronic components in a single chip. VLSI design is used to minimize the size of circuits and cost of devices. It can be achieved by increasing the number of transistors used in ICs.

Figure 1.1 shows the VLSI design hierarchy with different design levels. The VLSI design can be divided into five levels of hierarchy based on the top-down approach. The first level is considered the system level or IC level and is also known as the final outcome of the VLSI design. The second level is known as the module level. IC design is generally divided into different modules. Based on the complexity,

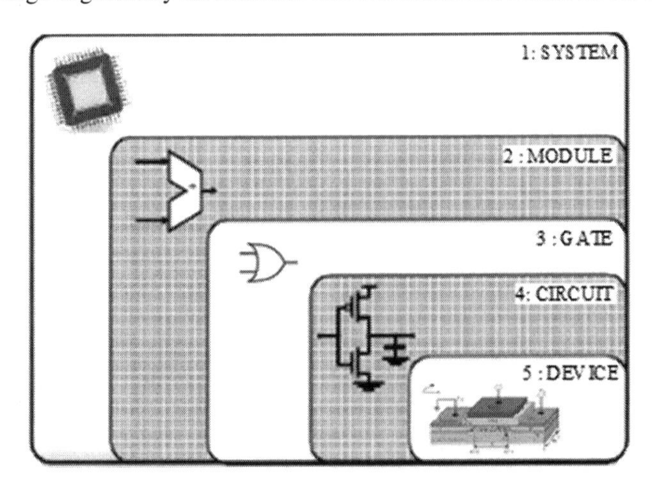

FIGURE 1.1 VLSI design hierarchy levels.

the module can again be divided into submodules. This can be carried on till the complexity could be reduced. Below the module level is the gate level. In this hierarchy level, each module or submodule is designed using the basic logic gates. The fourth level of hierarchy occurs at the circuit level. Each logic gate in the gate level is implemented by using transistors at this point of design. The final level is the device level.

The size of transistors used in VLSI design is constantly reduced to provide more functionality, minimize cost, and to speed up the design process. If there is a change in the device level as the size of the transistor gets reduced, obviously the circuit level also changes as per the transistor size. The impact on the circuit level leads to changes in the gate level. The gate level changes have an influence on the module level, and finally it is all reflected in the system level. Thus, the device level plays a highly significant role in VLSI design. The commonly used device for VLSI design is a Metal-Oxide Semiconductor Field Effect Transistor (MOSFET).

1.3 MOSFET BASICS

A MOSFET is a frequently used type of transistor with a "Metal Oxide" gate, and this part of the device is electrically sheathed from the channel of the semiconductor. And due to this phenomenon, we can say "NO current flows into the gate". MOSFETs can generally be classified into two types – depletion mode and enhancement mode, and each of these two types has an n/p channel type.

When no voltage is applied at the source and drain junction, a depletion mode transistor acts much like a switch that is "Normally Closed". A depletion mode n-channel MOSFET structure is shown in Figure 1.2. A "positive" voltage applied to the gate expands the channel for an n-channel MOSFET, increasing the flow of the current in the drain. Similarly, when the gate is supplied with a "negative" voltage, the drain current is highly reduced as the channel shrinks in size. The same concept is applicable for the devices made of p-channels, too.

Enhancement devices are normally preferred to depletion mode devices in practical applications. A device is normally in an "ON" condition and a gate source

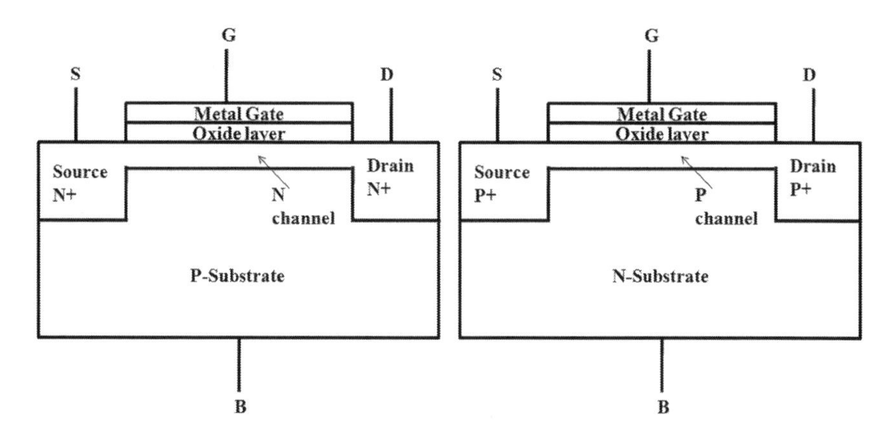

FIGURE 1.2 Structure of an n-channel and p-channel depletion MOSFET.

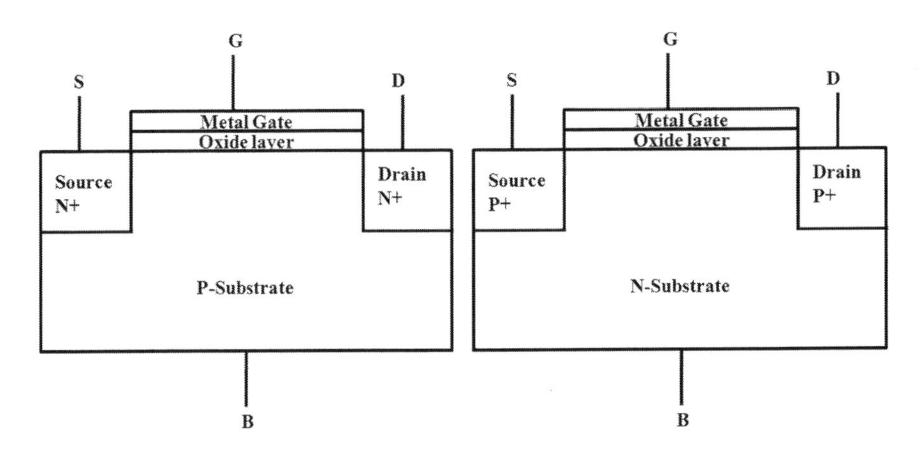

FIGURE 1.3 Structure of an n-channel and p-channel enhancement MOSFET.

TABLE 1.1
ON and OFF States of MOSFET Types

	Gate Source Voltage (V_{gs})		
Device Type	**Positive**	**Zero**	**Negative**
P-Depletion MOSFET	OFF	ON	ON
P-Enhancement MOSFET	OFF	OFF	ON
N-Depletion MOSFET	ON	ON	OFF
N-Enhancement MOSFET	ON	OFF	OFF

voltage (V_{gs}) is required to turn the device "OFF". An enhancement mode n-channel MOSFET structure is shown in Figure 1.3. An electrical field is produced within a channel by applying a positive voltage that decreases the resistance of the channel and allows the electrons to get attracted towards the oxide layer, resulting in channel conduction. As the positive voltage applied to the gate is gradually increased, the drain current is also increased. This concept is acceptable for p-channel enhancement forms too. MOSFETs can generally be used as electronic switches or amplifiers due to their very low power consumption. The four types of ON and OFF states of a MOSFET switch are summarized in Table.1.1.

1.3.1 NMOS Enhancement Mode MOSFET Operation

An N-channel Metal Oxide Semiconductor (NMOS) structure is a better candidate to easily understand the methodology and rules of design and provide a basic introduction to VLSI design. To build a channel for conduction, the minimum voltage known as the threshold voltage must be defined. There are three sets of conditions with respect to drain source voltage (V_{ds}), gate source voltage (V_{gs}), and threshold voltage (V_t) to understand the operation of an NMOS transistor operating in enhancement mode.

- When $V_{gs} > V_t$ and $V_{ds} = 0$

  A channel is formed, but there is no current between the source and drain regions as the drain source voltage is zero.

$$I_d = 0 \tag{1.1}$$

- When $V_{gs} > V_t$ and $V_{ds} < V_{gs} - V_t$

  When applying V_{ds}, current flows in the channel. Hence, a potential is developed in the region between the gate and the channel, and this potential is found to be varying with the distance across the channel at the source end. Here the maximum voltage is V_{gs}. The device operates in a non-saturated region of operation, and the current rises linearly. A constraining situation occurs when $V_{ds} = V_{gs} - V_t$. The saturation drain current is given as,

$$I_{ds} = C_{ox}\mu_n \frac{W}{L}\left[\left(V_{gs} - V_t\right)V_{ds} - \frac{V_{ds^2}}{2}\right] \tag{1.2}$$

  where C_{ox} is the oxide capacitance, μ_n is the electron mobility, L is the channel length, and W is the channel width. The oxide capacitance per unit area is given as,

$$C_{ox} = \frac{\varepsilon_{ox}}{t_{ox}} \tag{1.3}$$

  where ε_{ox} is the gate dielectric constant and t_{ox} is the gate dielectric thickness.

- When $V_{gs} > V_t$ and $V_{ds} > V_{gs} - V_t$

  A feeble electric field inversion layer cannot give rise to an inversion layer and the concept of pinch-off occurs in the channel. In this case, diffusion current completes the path from the source to drain, resulting in a high resistance in the channel and it starts behaving like a constant current source. This area of the channel is called the saturation region. The saturation current is given as,

$$I_{ds} = \frac{C_{ox}\mu_n}{2}\frac{W}{L}\left(V_{gs} - V_t\right)^2 \tag{1.4}$$

1.4 COMPACT MODELS FOR MOSFET

In compact models, terminal voltages are used to define the terminal properties of a device. A compact model is normally implemented inside a circuit simulation engine. Compact model bridges technology/process development with IC design. The Pao-Sah model was used initially based on the gradual channel approximation (GCA) for checking the accuracy of compact models, but it was declared too numerically complicated to be used as the heart of compact models. When a charge sheet is applied to the GCA of the incremental path, potential surface models are developed [1,2]. Generally, Pao-Sah and charge sheet model formulations have been considered too difficult to compute. The solution finally adopted for compact models was based on a threshold voltage based MOSFET model formulation. Advanced MOSFET models under investigation in the modern era are generally charge-based and surface potential-based models.

1.5 CMOS TECHNOLOGY AND SCALING

Complementary Metal-Oxide Semiconductor (CMOS) technology is the prime technology used to build ICs. The CMOS technology [3] is a method compatible with VLSI for the fabrication of ICs. Both NMOS-type and P-channel Metal Oxide Semiconductor (PMOS)-type transistors are paired in a CMOS circuit in a symmetric push–pull arrangement and are used to realize various digital, analog, and mixed mode designs. Due to its characteristics like low power dissipation, great resistance to noise, large input impedance, and state-of-the-art processing technology, CMOS technology is found to have greater advantages over all currently available IC technologies. Continuous developments can be made in CMOS technology and these developments are controlled and maintained by continuous CMOS scaling or miniaturization widely referred to as Moore's law. CMOS scaling is the method of decreasing the size of field effect transistors and interconnects without disturbing the functionality of the particular IC. Figure 1.4 shows the recent logic technologies used by Intel company due to scaling down of devices from 2010 to 2020. For future generations of CMOS, research on quantum nanoelectronics or nanomagnetics beyond CMOS is the core area that aims to identify new innovations in IC technology, which offer improved performance and power.

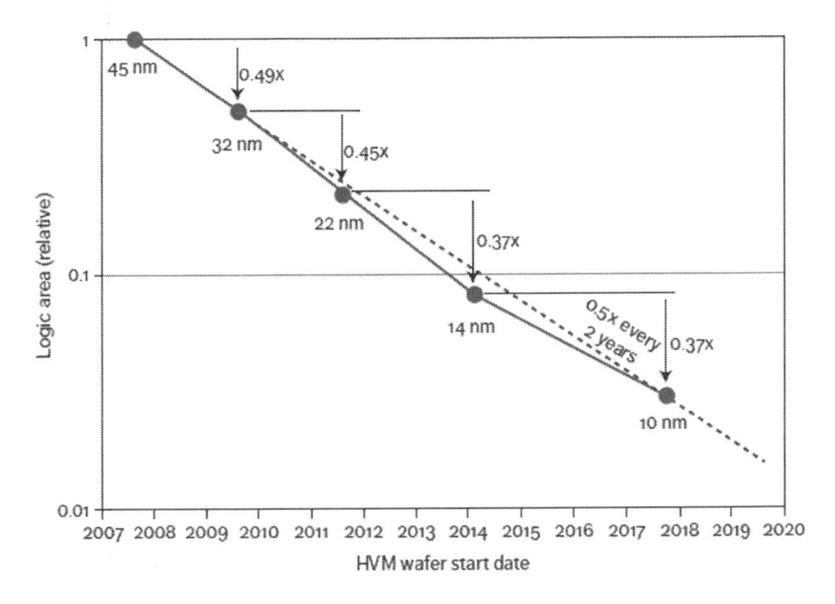

FIGURE 1.4 Intel's trend over the past five generations towards scaling down the logic circuit area.

1.6 SHORT CHANNEL EFFECTS

Departure from long-channel behavior can occur as the length of the channel is decreased. Such deviations, normally called SCEs, occur in the channel region due to the impact of two-dimensional potential distribution and strong electric fields across the channel. When the channel length of a conventional MOSFET is decreased, the gate loses its complete control over the potential developed across the channel. The threshold voltage starts to diminish with the rise in the drain bias due to Drain-Induced Barrier Lowering (DIBL). Another associated effect that hampers the performance of short channel devices is the roll-off that occurs in the threshold voltage. The electric field rises as the channel length decreases, and this results in carrier scattering near the surface. Hence, the electron mobility [4,5] at the surface decreases. Subthreshold current is another SCE which occurs before strong inversion. The channel is primarily influenced by the gate for long-channel MOSFETs. However, in terms of short-channel devices, source and drain regions play a major role in supporting the channel charge and hence the threshold voltage tends to decrease gradually. So, in this region, a very feasible amount of charge is only supported by the gate leading to a further decrease of the threshold voltage, and this impact is called the charge sharing effect. The transport of energetic electrons over the barrier into the oxide is another effect known as the hot-electron effect. This leads to the trapping of electrons in the oxide layer and degradation of I–V characteristics and the threshold voltage of the device.

Therefore, MOSFET scaling is becoming increasingly difficult and new transistor designs are required, which offer better scalability. Several alternative MOS structures are proposed to resolve various SCEs of low-dimensional MOSFETs, such as SOI technology, multiple gate structures, etc.

1.7 SILICON-ON-INSULATOR (SOI) MOSFET

In a conventional IC, the active components are implemented as a thin layer of surface and a depletion layer of a p–n junction as shown in Figure 1.5a separates them from the silicon body. Temperature changes tend to increase the junction leakage current, and this reduces the reliable nature of the device. Increased leakage current and huge dissipation of power in the form of surface heat due to increased temperatures will critically restrict the performance of microcircuits. The silicon-on-insulator (SOI)

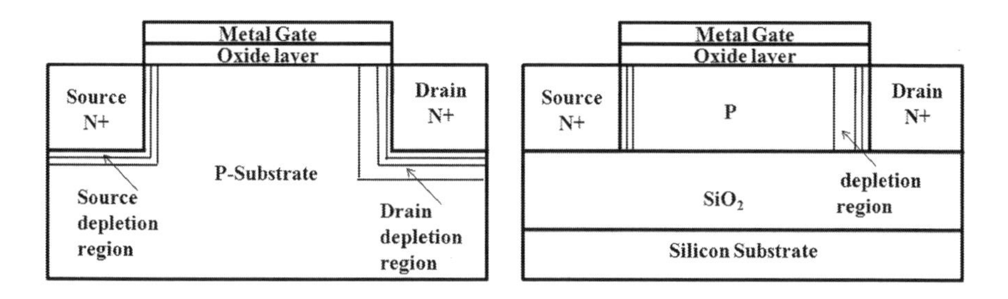

FIGURE 1.5 Cross-sectional view of a (a) bulk MOSFET and (b) SOI MOSFET.

technology has become a very attractive alternative to replace the bulk CMOS technology since the traditional bulk MOSFET transistor downscaling process exceeded the limits of miniaturization and manufacturing of devices. Because of their cost relative to other semiconductors, silicon wafers are still retained as the starting point in the fabrication of ICs. SOI technology uses a thin layer of semiconductor silicon, which is separated from a sheet of silicon substrate by a slightly thicker SiO_2 layer, as shown in Figure 1.5b. Dielectric isolation of components can be achieved using the SOI methodology and in combination with lateral isolation, it can get rid of latch-up failures and parasitic capacitances are also avoided.

1.8 MULTI-GATE MOSFETs

To enhance the device current characteristics and suppress the SCEs, planar transistor devices with a solitary gate have evolved into advanced three-dimensional device structures such as the SOI technology, multiple-gate transistors, High Electron Mobility Transistors (HEMTs), and Tunnel Field Effect Transistors (TFETs). Different devices have been studied extensively as the ultimate solution for extremely scaled devices.

In conventional MOSFET devices, the channel is controlled by the gate on any one side only. Conversely, in the case of multiple-gate structures, the channel is controlled by gates from many surfaces, and this results in highly improved device performance and decreased SCEs. The term multi gate does not actually refer to the number of individualistic electrodes used in the device, but it is actually a reference to the presence of gate electrodes on more than two sides of the semiconductor structure. The family of multi-gate devices generally includes Double Gate (DG) MOSFETs, Dual Material Double Gate (DMDG) MOSFETs, Surrounding Gate (SG) MOSFETs, DM SG MOSFETs, and Triple Gate (TG) MOSFETs [6].

Sekigawa and Hayashi (1984) are the scientists who have the honor of publishing the first article on DG devices [7]. They explained that SCEs occurring in a conventional device can be significantly reduced by squeezing a fully depleted SOI structure in the middle of two gate electrodes that are interconnected to each other. Since the cross-section of this new device looked like a Greek letter, this device was called an XMOS device. And also, the effect that the drain electric field had on the channel was considerably decreased in this device structure. In general, DG MOSFETs shown in Figure 1.6 have two gates that control the charge in the thin silicon body layer, and the current flow through both channels is enhanced. Two gate electrodes are employed in a DG MOS structure, which improve the device's control over the channel, leading to enhanced electron mobility and very good protection against SCEs. Linking of the gate electrodes in a DG device leads to volume inversion in the silicon film, and hence, a DG MOSFET is found to have a subthreshold slope, which is very much suitable for applications with low doping concentrations [8]. As a metal gate work function is used to control the variations of the threshold voltage, performance degradation of the device structure is highly reduced.

Another modification of a DG device is the Multiple Independent Gate FET (MIGFET) [9]. Here there is no interconnection between the two gate electrodes. Hence, two independent voltages are used for biasing the electrodes of this device.

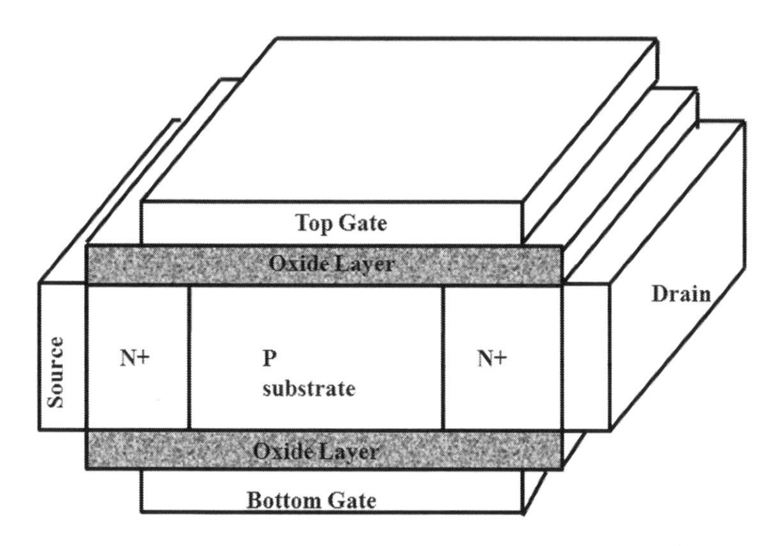

FIGURE 1.6 Schematic representation of a double gate MOSFET.

The most important aspect of a MIGFET is that the bias voltage applied to any one gate of the device can be used to modulate the threshold voltage of the other gate, akin to the body effect in fully depleted SOI devices. These types of devices are used in the field of signal modulation.

TG MOSFETs, shown in Figure 1.7, are also known as tri-gate MOSFETs in the semiconductor industry. They are very promising candidates for modern ICs with highly enhanced performance characteristics and low-power applications in 45 nm CMOS technology [10]. They are uneven transistors where the silicon body is surrounded by gates on three surfaces. The presence of gates on three sides of the device allows for continuous downscaling, and reverse channel doping gradient is also reduced [11]. The two channels on the lateral side and also the top horizontal

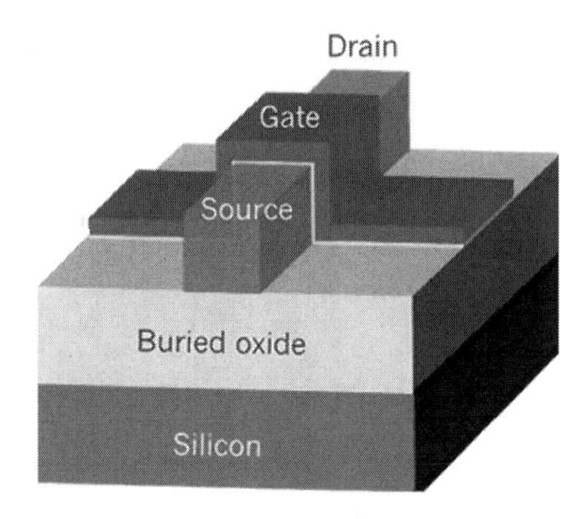

FIGURE 1.7 Schematic diagram of a triple gate MOSFET.

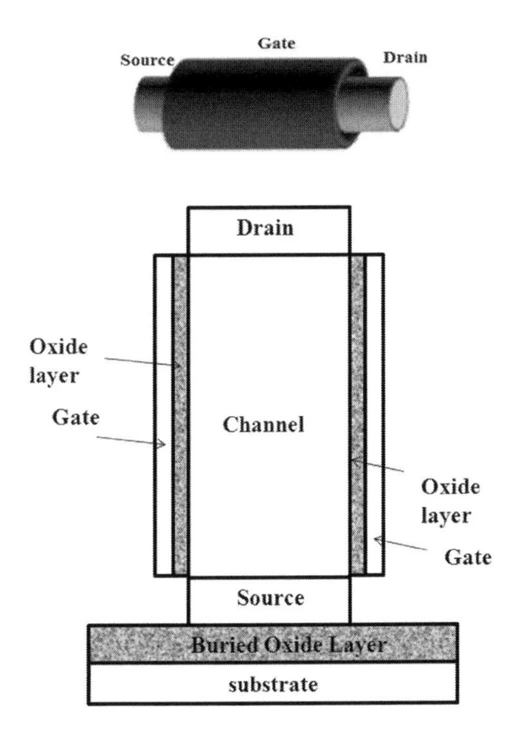

FIGURE 1.8 (a) Schematic diagram and (b) cross-section view of a surrounding gate MOSFET.

channel of the semiconductor body are controlled by the three gates simultaneously. This device represents a thin film consisting of a narrow silicon island with gates on its three sides. The lower side of the channel is totally under the control of the electric fields generated by the lateral gates. The drain current characteristics are improved and SCEs are highly reduced in TG devices compared to their DG counterparts.

Among the various available MOSFET devices, surrounding gate MOSFETs are touted to be some of the reliable structures to combat SCEs [12]. Figure 1.8 shows the cross-section view and schematic diagram of SG MOSFETs. SG devices are very much like TG MOSFETs except for the condition that the channel region is surrounded by the gate structure on all of its sides. Steep subthreshold characteristics, high packing density, and excellent drain current are the salient features of surrounding gate MOSFETs, which make them a good choice for semiconductor device design [13].

Two types of novel devices used in low-power applications for the semiconductor industry are silicon nanowire transistors and TFETs.

1.9 SILICON NANOWIRE TRANSISTORS

A nanowire transistor is a promising candidate with great potential and ability to resolve SCEs in SOI MOSFETs and has received substantial interest from developers of both circuits and devices in recent years [14]. A nanowire is a one-dimensional entity where the width is less than a few tens of nanometers and the length to width ratio is much more than 10. Multi-gate nanowire transistors show excellent current drive, reduction in SCEs due to their enhanced gate strength, and are also consistent with traditional CMOS technology. Hence, they find applications in nanoelectronics and nano-optoelectronics as interconnects.

Nanowire transistors are expected to play an important role in the forthcoming decade in providing a solution for the scaling problem faced by the semiconductor industry [15]. A number of advantages possessed by nanowire transistors have made them the primary and feasible candidates for replacing conventional MOSFETs. First, a huge quantity of nanowire devices can be produced for ULSI applications with the electronic properties required to meet the expectations. Second, the limits of lithography are not a great constraint for these nanowire transistors; great control over the size of the device is offered by nanowire materials that use the "bottom-up" approach compared with the nanofabricated structures employing the "top-down" approach [16]. Scattering in these devices is highly reduced due to their even surfaces, crystalline nature, and the capability to fabricate complicated heterostructures. And also, the carrier mobility concentration is also vastly improved compared to other nanodevices. Finally, the nanowire diameter, i.e., the thickness of the nanowire body can be efficiently controlled even below the size of 10 nm. This leads to a condition where even if the length of the channel is scaled to very low values, the electrical properties of the devices are not disturbed or diminished. This is a phenomenon that planar MOSFETs normally could not achieve.

Contrary to planar MOSFETs, source and drain contacts are made of metal in silicon nanowire FETs. The concept is that normal contacts made out of degenerately doped semiconductors are effectively replaced using metal contacts. Because of this, the performance of the device is greatly impacted by the physical properties of the Schottky barriers formed at the device interfaces. To avoid this problem, ohmic contacts are formed in the surfaces using annealing, which in effect increase the device's ON current and further improve the carrier mobility too.

Silicon nanowire transistors are identified to have great control over channel conduction and hence improved drain current characteristics can be observed in the intrinsic channels. Variations in the structures of nanowire devices are extensively studied to determine their performance characteristics and resistance to SCEs. The most common structures under investigation are double, triple, and surrounding gate nanowire transistors. Of all these devices, it is understood that in a SG nanowire device, the threshold voltage is not affected by the biasing voltage applied at the substrate level as the channel body is completely insulated from the electrostatic interferences.

1.9.1 Rectangular Surrounding Gate Silicon Nanowire Transistors

Fabrication of ultra-thin body SOI MOSFETs can be carried out using numerous variations such as deploying a single or double gate on the sides of the channel, and the shape of the gates can either be cylindrical or rectangular. A simple SOI single-gate transistor can be converted into a rectangular surrounding gate device by adding three gates at the bottom, forefront, and rear sides that are in alignment with the front gate [17]. Based on the gate voltages applied, they can be termed either symmetric or asymmetric type of transistor. As the name suggests, in a symmetric rectangular gate-all-around device, all the gates surrounding the channel are bound to exhibit similar metal work functions, the oxide thickness also remains the same for all gates, and each gate is applied the same biasing voltage uniformly. On the other hand, in an asymmetric rectangular gate-all-around device, the input voltage applied to the four gates can either be the same or different, or the gates are designed in such a fashion that they have different work functions.

1.9.2 Junctionless Cylindrical Surrounding Gate Nanowire Transistors

Formation of precise source and drain junctions is always a very big challenge in very short channel devices. This is supposed to have adverse effects on the thermal budget and the doping methods employed in the devices. A junctionless (JL) nanowire transistor is a reliable solution to overcome these problems. These devices can also be called gated resistors. In this novel architecture, a uniform concentration of doping is applied all over the source, drain, and channel. Colinge et al. are the pioneers to propose these transistors as the replacement for complementary MOSFETs [18]. A heavily doped nanowire will perform the function of a channel in these multi-gate devices, and the device can be turned off by the complete depletion of the nanowire channel. As there are no junctions available in the device, it can be easily fabricated. Variations of the device characteristics are greatly reduced and electrical properties tend to improve in junctionless devices. The major difference between a junctionless MOSFET and a conventional bulk device is that the doping across the channel in a junctionless device is very high and in the OFF state it is fully depleted. Figure 1.9 shows the schematic diagram of a SOI junctionless Nanowire Field Effect Transistor (NWFET). To initiate the source–drain conduction, a biasing voltage has to be applied at the gate to retrieve the channel from depletion. But the restriction is

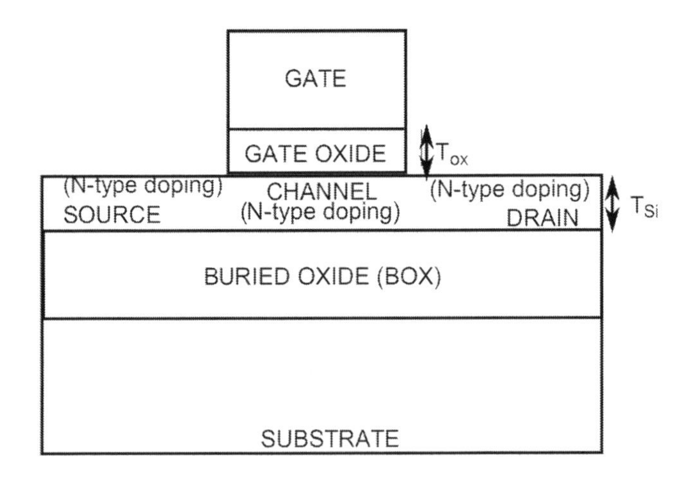

FIGURE 1.9 Schematic diagram of a SOI junctionless NWFET.

that the doping concentration should be very high to initiate this conduction process. And also, a semiconductor with heavy doping is bound to have a very little depletion width, and thus in a junctionless device, an ultra-thin silicon body is employed to meet the expected performance levels. In the n-channel operation of the device, an n-type impurity is used in the doping of the source, channel, and drain, and a gate with p-type work function is employed. The doping concentration of the channel region and the gate/channel work function difference are the two major factors in deciding the depletion across the channel. Once the channel is fully depleted, the charge carrier transmission between the source and the drain is greatly reduced. In the absence of an applied gate voltage, a block is created along the source and the channel of the device, on account of shortage of charge carriers along the channel. When a positive biasing voltage is applied at the gate, depletion is wiped off and the block between the source and channel gradually reduces. As a result, a very high conduction is experienced by the channel, leading to a surge in its drain current. When the gate bias voltage exceeds the gate/channel work function difference, flat band condition is attained and the transistor is switched on. Nevertheless, the basic requirement for this mode of operation is that the thickness of the channel remains small than its depletion width; or else, zero gate bias will not turn the device OFF. Hence, for a junctionless device to perform satisfactorily, we need to ensure that the semiconductor thickness always remains very low.

Among the various parameters available to gauge the performance of nanowire transistors, the scaling factor plays a major role [19]. The effect of the scaling factor upon the threshold voltage characteristics of a junctionless cylindrical surrounding gate nanowire transistor is demonstrated in Figure 1.10. The simulation is carried out for three varying channel lengths of the device, i.e., 6, 8, and 10 nm, respectively. Degradation in the threshold voltage is found to be less when the channel length is equal to 10 nm. Further reduction of the channel lengths to 8 nm and then to 6 nm shows that there is a serious variation in terms of their threshold voltage roll-off.

Apart from silicon nanowire transistors, the other devices that have a huge role to play in low-power applications are TFETs.

1.10 TUNNEL FIELD EFFECT TRANSISTORS

As a result of device scaling, multi-gate MOS transistor structures are predominantly confined by SCEs such as leakage current, surface scattering, DIBL, and non-scalability of subthreshold swing (SS). These limitations can be resolved with the aid of TFETs – promising novel device structures. In semiconductor devices, electron tunneling is the aspect that the thin barriers could not expel, which brings strength to TFETs. Also, the current transport mechanism of TFETs is totally different from that of MOSFETs. In MOS transistors, the current transport from the source to drain relies on thermionic emission, whereas the current transport in TFETs focuses on the tunneling of electrons from the valence band to the conduction band named Band-to-Band Tunneling (BTBT). TFETs have several supreme attributes, such as reduced leakage current in the femto–pico ampere range [20], capability of reducing the SS to less than 60 mV/decade [21], lower temperature dependence [22], and limited gate length. From the above, reduced leakage current and SS help

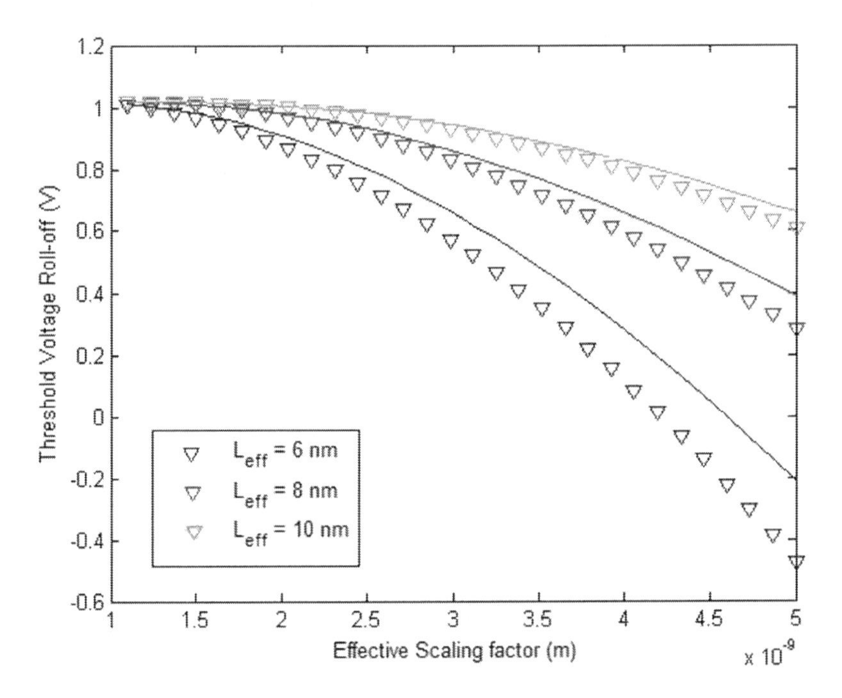

FIGURE 1.10 Threshold voltage roll-off versus effective scaling factor for different channel lengths.

circuit designers to design a circuit that has low power consumption. In addition, the structural similarities between TFETs and MOSFETs facilitate the use of typical MOSFET fabrication steps.

The fundamental structure of TFETs is distinctive from that of MOS transistors. In MOSFETs, the source and drain are comprised of the same sort of materials/doping, whereas in TFETs, the source and drain are comprised of different types of materials/doping. Figure 1.11 depicts a basic n-type TFET device where the source (P+) is doped with trivalent impurities and the drain (N+) is doped with pentavalent impurities. The channel or substrate has a p-type or n-type semiconductor, either intrinsic or lightly doped. A TFET operates under reverse biased conditions. As the gate is applied with a positive voltage, the energy gap between the source and channel is sufficiently narrow to cause electron tunneling when it exceeds the threshold voltage (Figure 1.12a). The narrow potential barrier enables the electrons to tunnel from the source region of the valence band to the channel region conduction band. This phenomenon is known as BTBT and also the ON state of an n-type TFET. Similarly, when the threshold voltage is above the gate voltage, the energy gap between the source and channel region is wider and thus there is no tunneling. Such a state is called the OFF state (Figure 1.12b) of an n-type TFET. However, there is a possibility of very low leakage current on the order of femto–pico amperes.

Although a TFET is considered an appropriate device for CMOS technology, it also faces certain limitations such as ON current reduction and ambipolar behavior.

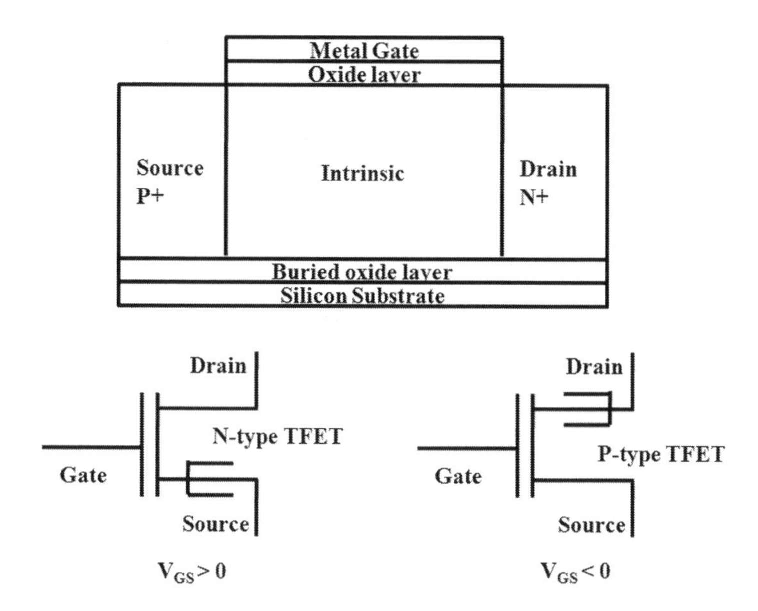

FIGURE 1.11 Basic schematic and symbolic representation of a TFET.

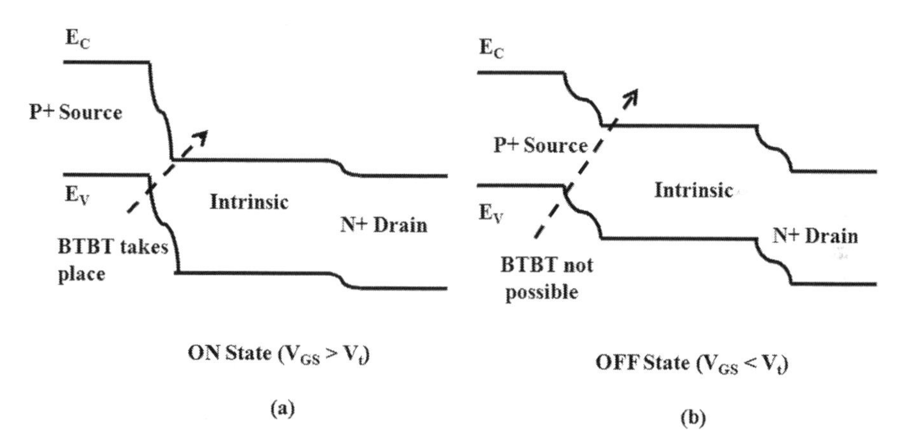

FIGURE 1.12 Energy band diagram of an n-type TFET representing the (a) OFF state of a TFET and (b) ON state (not scaled).

The basic principle of carrier tunneling and silicon material's large indirect band gap in a TFET result in far less ON current than the requirements of the International Technology Roadmap for Semiconductors (ITRS) [23]. In order to use a TFET as an alternative to CMOS technology, the ON current must be within the range of micro amperes to milli amperes and the OFF current must be within the range of femto–pico amperes. The ambipolar behavior is a further demerit and a unique property of TFETs [24]. TFETs are found to be less effective for complementary circuits due to their ambipolar conduction, and this limits their functionality in the design of digital circuits. Therefore, while designing TFETs for CMOS applications, it is significant to reduce the ambipolar current as far as possible.

Many optimization methods have been proposed [25] to overcome the constraints of reduced ON current such as DG TFETs, DMDG TFETs, SG TFETs, high-k gate dielectric TFETs, strained-SiGe TFETs, and heterojunction TFETs. Together with silicon, germanium and III–V semiconductor materials are often considered as valid materials for the manufacture of TFET devices due to the lower value of the direct energy band gap. For all these methods, exponentially increasing ON current has been achieved using the BTBT electron generation rate, which depends on the energy band gap of the source and channel material, width of the tunneling barrier, gate oxide thickness, and effective mass of the charge carrier. Similarly, the simple way of reducing the ambipolar behavior of TFETs is to maintain source doping higher than drain doping. However, different optimization methods have been proposed to overcome the ambipolar behavior of TFETs such as heterogate dielectric TFETs, gate drain over/under lap structures, and asymmetric device structures.

The device physics of a basic single gate SOI TFET and its analytical model have been reported [26], and also the ON current and OFF current are obtained as 10^{-10} and 10^{-17} A, respectively. However, the output current performance is not equivalent to CMOS technology standards. Hence, a Dual Material Gate (DMG) TFET [27] using SOI technology is proposed. Figure 1.13 shows the schematic diagram of a

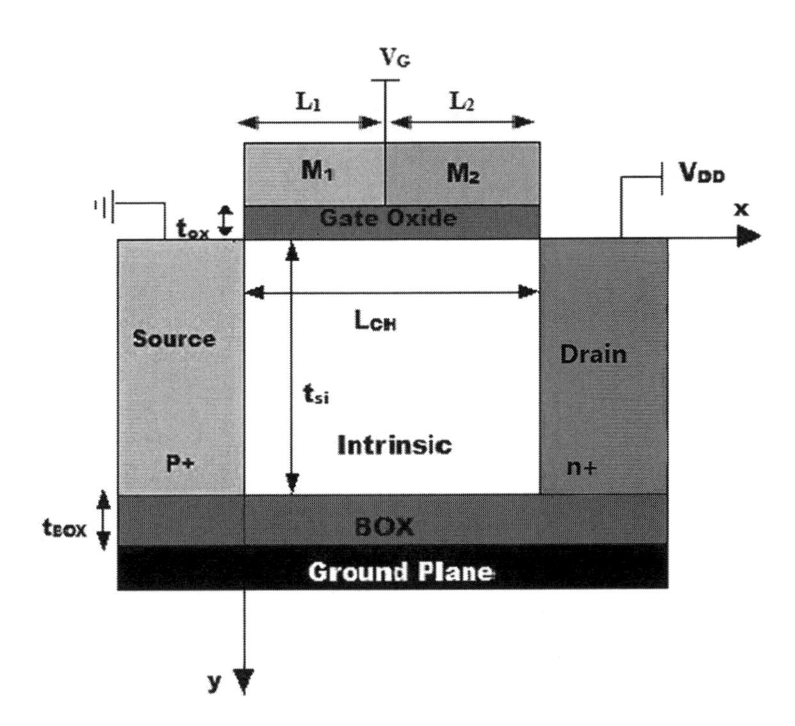

FIGURE 1.13 Schematic diagram of a dual material gate SOI TFET.

DMG SOI TFET, where two different metal gates with two different work functions are connected to the same external terminal called gate. Metal gate M1 has a lower work function than metal gate M2 gate for a DMG TFET, which results in lower OFF current and higher ON current. There is a slight improvement in the DMG TFET structure compared to a single gate SOI TFET.

1.10.1 Multi-Gate Tunnel FETs

The main objective of TFET optimization is ON current enhancement, a SS less than 60 mV/dec, and OFF current reduction as much as possible. Multi-gate TFET structures exploit many gates around the channel region for better control over the channel region in short-channel devices. More than one gate structure is used to form channels under all the gates, thus increasing the overall electron tunneling area. Multi-gate device structures include DG TFETs, DMDG TFETs, SG TFETs, etc.

1.10.1.1 Double Gate TFETs

Several research studies have been carried out to resolve the shortcomings of TFETs. DG TFETs with a high-k gate dielectric [28] have attracted researchers to achieve the ON and OFF current requirements as mentioned in the ITRS. The schematic diagram of a DG TFET is shown in Figure 1.14, where gate 1 is taken from the top surface and gate 2 is taken from the bottom surface. Moreover, the same gate bias is introduced to gate 1 and gate 2. High-k dielectric is used to distinguish the intrinsic silicon layer from a metal gate work function of 4.5 eV. The source doping ($10^{20}\,cm^{-3}$) is held higher than the drain doping ($5 \times 10^{18}\,cm^{-3}$) to suppress ambipolar current of the DG TFET. The device has been designed with an optimum silicon thickness value of 7–8 nm, the dielectric thickness is 3 nm, and the channel length has been chosen as 50 nm. This DG TFET structure provides an I_{ON} current of 0.23 mA and an I_{OFF} current which is less than 1 fA by applying a high-k gate dielectric.

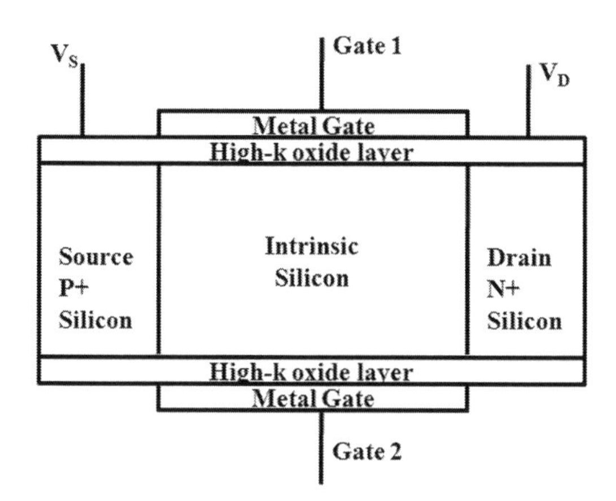

FIGURE 1.14 Schematic diagram of a double gate TFET.

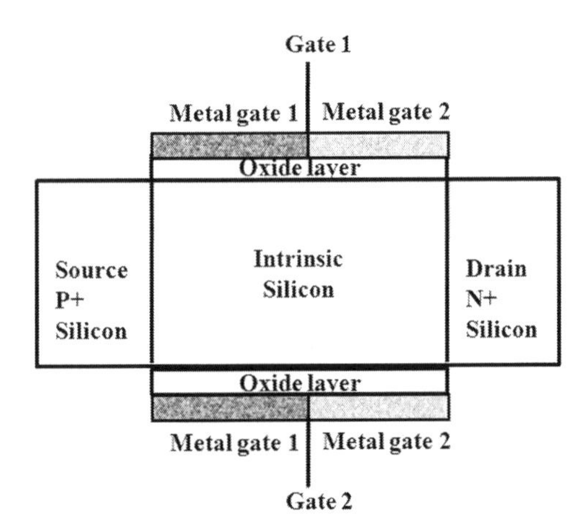

FIGURE 1.15 Schematic diagram of a dual material double gate TFET.

1.10.1.2 Dual Material Double Gate TFETs

Another TFET device structure proposed to meet the requirements of ITRS is a DMDG TFET. Figure 1.15 illustrates the schematic diagram of a DMDG TFET [29] wherein the two gates are made up of two different materials with two separate work functions. The length of the channel is set as 50 nm – the length of metal gate 1 (tunnel gate) is selected as 20 nm by which I_{ON} remains constant and I_{OFF} declines. However, the length of metal gate 2 (auxiliary gate) is 30 nm, gate oxide thickness is 3 nm, and silicon body thickness is 10 nm. The I_{ON}/I_{OFF} ratio acquired for these parameters is 1×10^{10}, which greatly limits the SS_{avg} compared to that of SMGDG TFETs. The SS_{avg} rises from 73 mV/dec for a SMGDG TFET to 58 mV/dec for a DMG-DG TFET. This indicates that the work functions of the gates determine the optimal SS. The work function of the tunnel gate is lower than that of the auxiliary gate to accomplish a lower V_{Dsat} (drain voltage at saturation) and higher I_{ON}. Thus, it increases the device's actual performance. A DMG-DG TFET is also found to be more resistant to DIBL effects. It has been shown that the DMG technique can be used in DG TFETs to obtain good ON current, OFF current, and threshold voltage exchanges, as well as enhancements in the average SS, to improve the quality of device characteristics and to reduce the effects of DIBL.

1.10.1.3 DG and DMDG TFETs with SiO_2/High-k Stacked Gate-Oxide Structure

Introducing a HfO_2 dielectric layer over the SiO_2 dielectric layer in a DG TFET structure forms a novel device called the stacked dielectric DG TFET [30]. If the HfO_2 dielectric layer is directly placed over a silicon layer, there is a possibility of large lattice mismatch. Thus, the HfO_2 layer must be placed over a SiO_2 dielectric layer. This device structure provides both reduced ambipolar current and improved SS even below 50 mV/dec. The stacked dielectric DG TFET has been subtly changed and Kumar et al. [31] proposed a new device structure called the stacked SiO_2/HfO_2 DMDG TFET. This structure provides better results in terms of the ambipolar effect,

I_{ON}/I_{OFF} ratio, and SS. However, the ON current of the DMDG TFET with stacked dielectric needs to be improved to the mA range of current.

1.10.1.4 Tri-Gate SOI TFETs

Tri-gate SOI TFETs [32] are another type of TFET device structure suggested to comply with the requirements of the ITRS. The top surface and walls of the silicon body are surrounded by tri-gate electrodes. Figure 1.16 illustrates the schematic diagram of tri-gate SOI TFETs. The performance of a TMG TFET is compared with that of SMG and DMG TFETs. SiO_2 is used as a dielectric material for SMG and DMG TFETs, whereas silicon dioxide, silicon nitride, sapphire, hafnium dioxide, tantalum pentoxide, and titanium dioxide are the dielectric materials used in TMTG TFETs. However, titanium dioxide as a dielectric material displays outstanding efficiency compared to all other dielectric materials and decreases SCEs. A heavily doped source region is employed in the device structure in order to reduce the ambipolar behavior and a prominent doping profile is used for an enhancement of drain current. For $V_{gs} = 1.2\,V$ and $V_{ds} = 0.1\,V$, the TMG tri-gate TFET ON current is increased by 45%, that is 7.25 μA, whereas the drain currents of SMG and DMG TFETs are 5 and 6.75 μA as shown in Figure 1.17.

1.10.1.5 Heterojunction Triple Material DG TFETs

Figure 1.18 depicts the schematic diagram of a heterojunction triple material (TM) DG TFET. The heterojunction TM DG TFET [33] is implemented with germanium and silicon materials as a heterojunction along the source and channel interface, and silicon dioxide (SiO_2) and hafnium dioxide (HfO_2) as a hetero-dielectric gate stack. This demonstrates that the higher dielectric constant remarkably achieves an optimal performance in drain current.

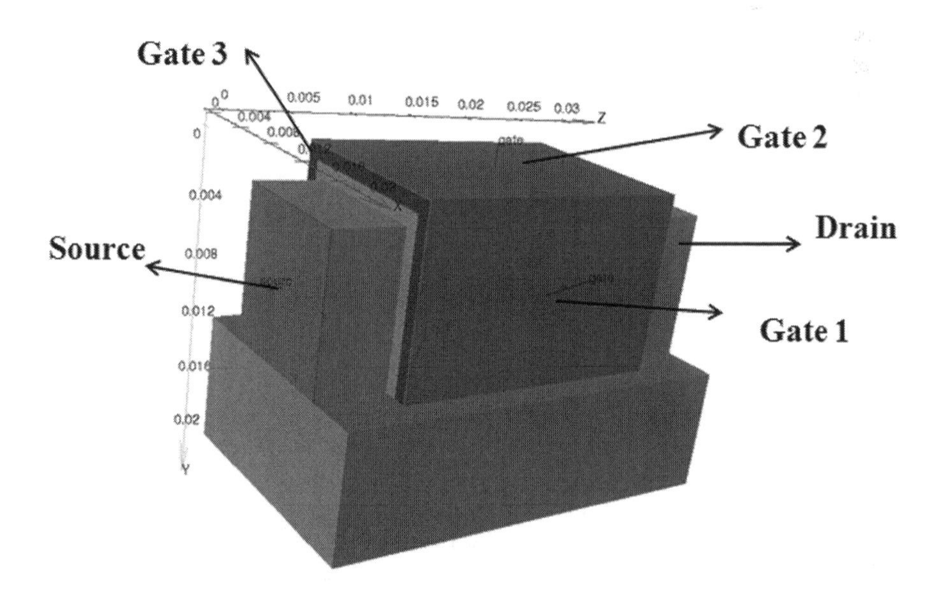

FIGURE 1.16 Three-dimensional (3D) diagram of a tri-gate SOI TFET.

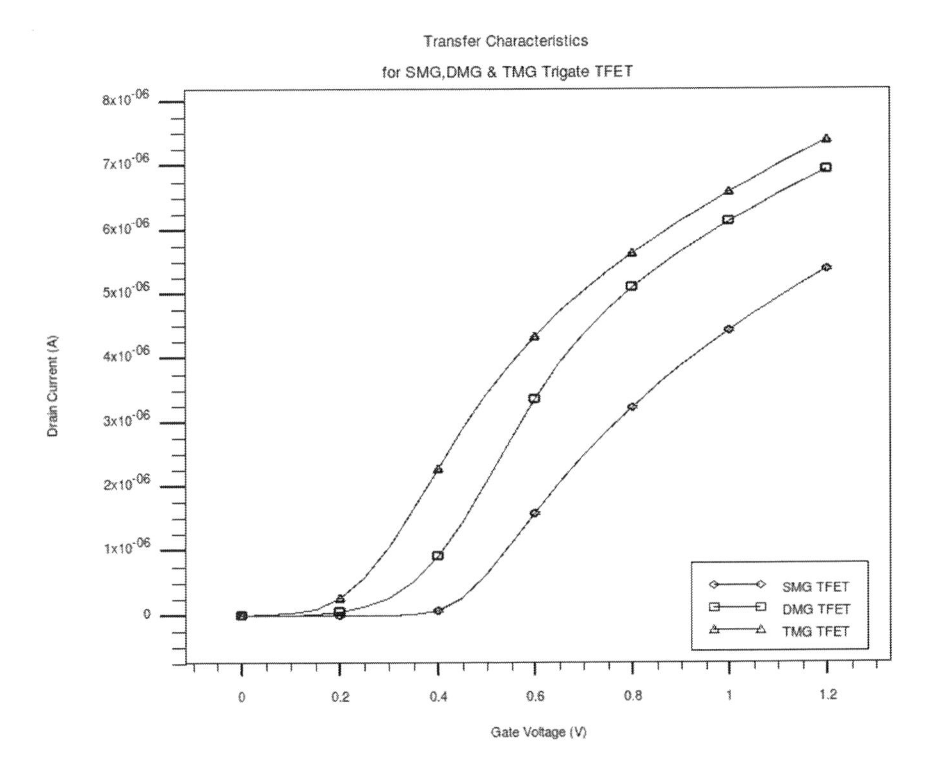

FIGURE 1.17 Transfer characteristics of SMG, DMG and TMG tri-gate TFETs for $V_{ds} = 0.1\,\text{V}$.

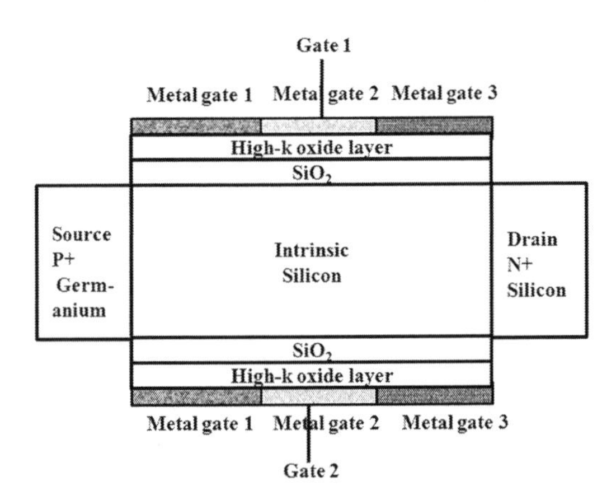

FIGURE 1.18 Schematic diagram of a heterojunction triple material DG TFET.

An integrated gate length ratio of 3:1:1 is chosen to repress the I_{OFF} without worsening I_{ON}, and I_{OFF} significantly improved for a work function of 4.8 eV. This seems to result in ambipolar current reduction and an increased I_{ON}/I_{OFF} ratio. Figure 1.19 shows the drain current comparison of a SMG TFET, DMG TFET,

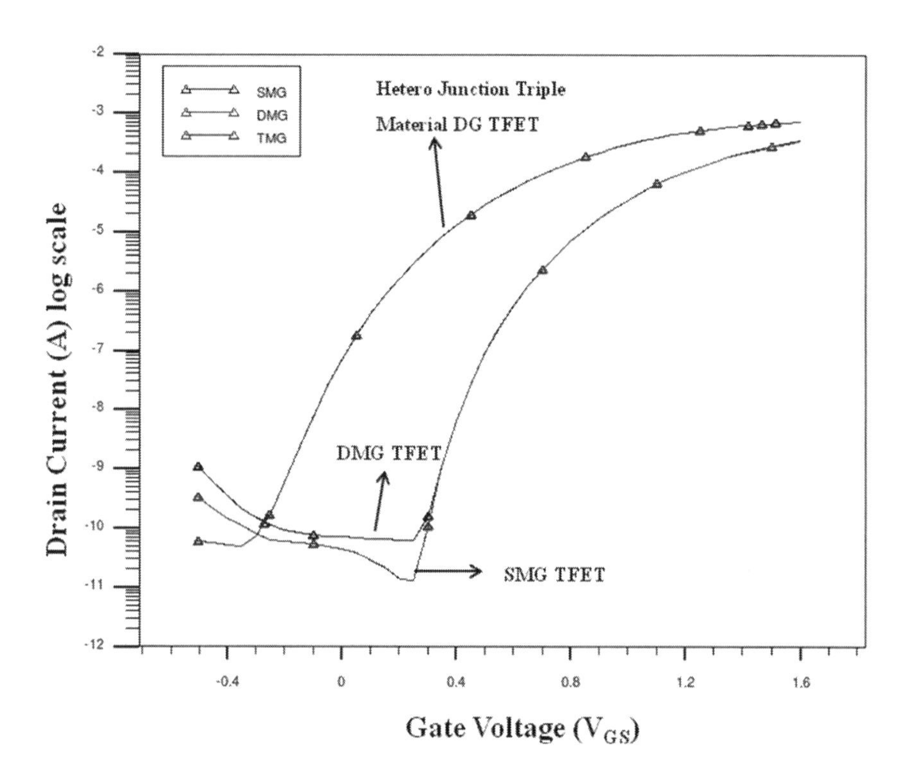

FIGURE 1.19 Comparison of drain current vs. gate voltage plot for single material, double material and triple material double gate TFETs.

and TM heterojunction TFET. In addition, the effect of the germanium/silicon heterojunction also limits the exact depiction of the tunneling barrier width, and the ON current of the heterojunction TM DG TFET (10^{-3} A) is strengthened at the CMOS transistor level.

1.11 SUMMARY

A brief description of the MOSFET, its types, and device level characteristics have been presented in this chapter. Practical problems caused by scaling and SCEs leading to the evolution of one-dimensional planar semiconductor devices into multi-gate devices are addressed. Silicon nanowire transistors and TFETs are two types of devices that are found to have excellent properties for use in low-power applications. Various modifications in the structure of these devices and their applications are also presented.

REFERENCES

1. Baccarani G, Rudan M, and Spadini G. 1978. Analytical i.g.f.e.t. model including drift and diffusion currents, *IEE Journal on Solid-State and Electron Devices*, 2(2): 62–68.
2. Brews JR. 1978. A charge sheet model for the MOSFET, *Solid-State Electronics*, 21:345–355.

3. Kittl JA, Lauwers A, and Dal Mv. 2006. Ni, Pt and Yb based fully silicided (FUSI) gates for scaled CMOS technologies, *ECS Transactions*, 3(2):233–246.
4. Caughey DM and Thomas RE. 1967. Carrier mobilities in silicon empirically related to doping and field, *Proceedings of the IEEE*, 55(12): 2192–2193.
5. Darwish MN, Lentz JL, and Pinto MR. 1997. An improved electron and hole mobility model for general purpose device simulation, *IEEE Transactions on Electron Devices*, 44(9):1529–1538.
6. Colinge JP. 2007. Multi-gate SOI MOSFETs, *Microelectronic Engineering*, 84: 2071–2076.
7. Sekigawa T, Hayashiya Y, and Ishii K. 1993. Feasibility of very-short-channel mos transistors with double-gate structure, *Electronics and Communication in Japan (Part II: Electronics)*, 76(10): 39–48.
8. Ortiz-Conde A, Garzia-Sanchez FJ, Muci J, Malobabic S, and Liou JJ. 2007. Review of core compacts models for undoped double-gate SOI MOSFETs, *IEEE Transaction on Electron Devices*, 54(1): 131–140.
9. Langen M, Thean AV-Y, Vendooran A. 2004. CMOS vertical multiple independent gate field effect transistor (MIGHT). *IEEE International SOI Conference Proceedings*. 187–189.
10. Nandi S and Sarkar M. 2013. Analytical modeling of triple-gate MOSFETs, *International Journal of Semiconductor Science and Technology*, 3(4): 1–10.
11. Max Christian L, Mollenhauer T, and Henschel W. 2004. Sub-threshold behaviour of triple-gate MOSFETs on SOI material, *Solid State Electronics*, 48(4): 529–534.
12. Bo Y, Wei-Yuan L, Huaxin L, and Yuan T. 2007. Analytic charge model for surrounding gate MOSFETs, *IEEE Transactions on Electron Devices*, 54(3): 492–496.
13. Benjamin I, David J, Jaume R, Hamdy A, Hamid L, Marsal F, and Joseph, P. 2005. Explicit continuous model for long-channel undoped surrounding gate MOSFETs, *IEEE Transactions on Electron Devices*, 44 (8): 1868–1873.
14. Cui Y, Duan X, Hu Y, and Lieber CM. 2000. Doping and electrical transport in silicon nanowires, *Journal of Physical Chemistry B*, 104 (22): 5213–5216.
15. Cui Y, Zhong Z, Wang D, Wang WU, and Lieber CM. 2003. High performance silicon nanowire field effect transistors, *Nano Letters*, 3: 149–152.
16. Wu Y et al. 2004. Controlled growth and structures of molecular-scale silicon nanowires, *Nano Letters*, 4: 433–436.
17. Karthigai Pandian M and Balamurugan NB. 2014. Analytical threshold voltage modelling of surrounding gate silicon nanowire transistors with different geometries, *Journal of Electrical Engineering & Technology*, 9(6): 2079–2088.
18. Colinge JP, Balestra F, Raskin JP, Gamiz F, and Lysenko V. (eds). *Semiconductor-On Insulator Materials for Nanoelectronics Applications. Engineering Materials*. Springer, Berlin, Heidelberg. 187–200.
19. Karthigai Pandian M, Balamurugan NB, and Manikandan S. 2014. Analytical modelling of junctionless surrounding gate silicon nanowire transistors, *Journal of Nanoelectronics and Optoelectronics*, 9(4): 468–473.
20. Arun Samuel TS, Balamurugan NB, Bhuvaneswari S, Sharmila D and Padmapriya K. 2013. Analytical modelling and simulation of single-gate SOI TFET for low-power applications, *International Journal of Electronics*, 101: 779–788.
21. Narimani KS, Glass P, Bernardy N, Von Den Driesch QT, Zhao S, and Mant L. 2018. Silicon tunnel FET with average subthreshold slope of 55 mV/dec at low drain currents, *Solid-State Electronics*, 143: 62–68.
22. Nirschl T, Peng-Fei W, Hansch W, and Schmitt-Landsiedel D. 2004. The tunnelling field effect transistors (TFET): the temperature dependence, the simulation model, and its application. *IEEE International Symposium on Circuits and Systems (IEEE Cat. No.04CH37512)*, Vancouver. BC: III-713.

23. International Technology Roadmap for Semiconductor. Available: http://www.itrs.net/.
24. Hraziia A, Vladimirescu C, Amara A, and Anghel, C. 2012. An analysis on the ambipolar current in Si double-gate tunnel FETs, *Solid-State Electron*, 70: 67–72.
25. Usha C and Vimala P. 2015. A tunneling FET exploiting in various structures and different models: A review. *2015 International Conference on Innovations in Information, Embedded and Communication Systems (ICIIECS)*: 1–6 doi: 10.1109/ICIIECS.2015.7192878.
26. Lee MJ and Choi WY. 2011. Analytical model of single-gate silicon-on insulator (SOI) tunneling field-effect transistors (TFETs), *Solid-State Electronics*, 63(1): 110–114.
27. Arun Samuel TS, Balamurugan NB, Sibitha S, Saranya R, and Vanisri D. 2013. Analytical modeling and simulation of dual material gate tunnel field effect transistors, *Journal of Electrical Engineering & Technology*, 8(6): 1481–1486.
28. Boucart K and Ionescu AM. 2007. Double-gate tunnel FET with high-κ gate dielectric, *IEEE Transactions on Electron Devices*, 54: 1725–1733.
29. Saurabh S and Kumar MJ. 2011. Novel attributes of a dual material gate nanoscale tunnel field-effect transistor, *IEEE Transactions on Electron Devices*, 58(2): 404–410.
30. Kumar S et al. 2017. 2-D analytical modeling of the electrical characteristics of dual-material double-gate TFETs with a SiO_2/HfO_2 stacked gate-oxide structure. *IEEE Transactions on Electron Devices*, 64(3): 960–968.
31. Kumar S, Goel E, Singh K, Singh B, Kumar M, and Jit S. 2016. A compact 2-D analytical model for electrical characteristics of double-gate tunnel field-effect transistors with a SiO_2/high-k stacked gate-oxide structure, *IEEE Transactions on Electron Devices*, 63(8): 3291–3299.
32. Vimala P, Arun Samuel TS, and Karthigai Pandian M. 2019. Performance investigation of gate engineered tri-gate SOI TFETs with different high-k dielectric materials for low power applications, *Silicon*, doi:10.1007/s12633-019-00283-6.
33. Vimala P, Arun Samuel TS, Nirmal D and Ajit Kumar P. 2019. Performance enhancement of triple material double gate TFET with heterojunction and heterodielectric, *Solid State Electronics Letters*, 1: 64–72.

2 RF/Analog and Linearity Performance Evaluation of a Step-FinFET under Variation in Temperature

Rajesh Saha
Malaviya National Institute of Technology Jaipur

Brinda Bhowmick and Srimanta Baishya
National Institute of Technology Silchar

CONTENTS

2.1 INTRODUCTION

Scaling of MOSFETs provides an enormous improvement in switching speed, packing density, and cost of microprocessors, which leads to increased number of transistors and hence increased power density [1]. However, to fulfil the industrial demands, scaling of MOSFETs leads to several unwanted issues such as huge subthreshold swing (SS), large amount of drain-induced barrier lowering (DIBL), large leakage current, etc. and these need to be addressed [2,3]. To address these issues, various engineering techniques such as material change, gate structure change, modification of work function, modification of device architecture, use of spacers, change of gate oxide material, etc. have emerged over time [4–6]. Combination of these techniques has led to emerging devices that can replace MOSFETs. In this regard, FinFETs present themselves as challengers capable of overcoming the adverse short-channel effects

(SCEs), and here, the conducting channel is wrapped by all three sides of the gate [7,8]. They are categorized as multi-gate MOSFETs, where the vertical fin is placed between source and drain regions. A large number of articles on analytical modelling and simulation of electrical parameters in FinFETs are reported in the literature. The performance of FinFETs is also improved by various modified FinFET architectures such as cylindrical FinFETs [9], FinFETs with a gate structure like pie [10], omega structured FinFETs [11], FinFETs with a gate made of two different gate materials named dual material gate (DMG) FinFETs [12], likewise triple material gate (TMG) FinFETs [13], and many more. We have reported a step-like fin structure named step-FinFET, and its electrical performance for different fin dimensions is analysed [14]. Further, circuit realizations of devices are extremely essential, and therefore, study on circuit parameters in FinFETs is important. A number of studies are available in the literature on the analog/RF performance of MOS devices [15–17], a quadruple gate-all-around (GAA) MOSFET [18], a junctionless FinFET [19], and DMG/TMG FinFETs [20]. Experimental demonstration of the analog/RF parameters of FinFETs exists in the literature [21]. On the other hand, temperature is one major factor that has a significant impact on device performance, and therefore, it is very important that the performance of a device sustains under variation in temperature. A simulation study of RF/analog performance of DMG FinFETs is reported with variation in temperature [22]. The study also investigated the linearity characteristics of DMG FinFETs. Therefore, analysis of temperature effect on RF/analog as well as linearity parameters of other non-conventional FinFET architectures is one of the attractive research areas.

This chapter presents the temperature effects on RF/analog figures of merit (FOMs): transconductance (g_m), total gate capacitance (C_{gg}), output conductance (g_d), transconductance generation factor (TGF = g_m/I_D), cut-off frequency (f_t), intrinsic gain (g_m/g_d), and gain transconductance frequency product (GTFP) for a step-FinFET through a Sentaurus 3D Technology Computer Aided Design (TCAD) simulator. Linearity parameters such as higher order harmonics (g_{m2} and g_{m3}), voltage intercept points (VIP2 and VIP3), third order power intercept point (IIP_3), and third order intermodulation distortion (IMD_3) are discussed by changing the temperature from 250 to 450 K with a step of 50 K. We also present the transfer characteristics and the current ratio of a step-FinFET by varying the temperature.

2.2 LITERATURE SURVEY

Raskin et al. reported a fair investigation of the RF/analog performance by changing the number of gates, that is, single gate (SG), double gate (DG), and triple gate (TG) MOSFETs, and modifying the gate structure, such as pie and omega gate MOSFETs, through 3D simulations for wideband applications [16]. It is reported that as the number of gate increases better RF/analog performance is obtained than for a SG device, and this is due to a large amount of early voltage, which leads to improved intrinsic gain. Subramanian et al. investigated the comparative analysis of RF performance between FinFETs and bulk MOSFETs [17]. They have summarized that at low-frequencies FinFETs show good RF performance, whereas at high-frequencies MOSFETs have improved RF characteristics.

An analytical model for various analog/RF FOMs such as intrinsic gain, capacitance, transconductance, output resistance, maximum frequency, and oscillation frequency of a quad gate GAA MOSFET has been developed by Sharma and Vishvakarma. They have found a close agreement of analytical data with simulation results obtained from an ATLAS TCAD simulator [18].

Jegadheesan et al. investigated the stability of RF performance with variation in temperature, fin width, fin height, and channel doping in Silicon on Insulator (SOI) junctionless FinFETs [19]. With increased temperature, maximum frequency and the stability of junctionless FinFETs reduce. However, as channel doping and fin dimensions increase, the maximum cut-off frequency increases, whereas the stability of junctionless FinFETs degrades. Lederer et al. showed that the RF performance of FinFETs varies by changing the gate length and fin width [21]. Results reveal that as the fin width reduces, the cut-off frequency decreases, and this is because of an increase in parasitic resistance with decreased fin width. Kumar reported a comparative study of RF/analog parameters among TG FinFETs, cylindrical GAA (CY-GAA) FinFETs, and rectangular GAA (RE-GAA) FinFETs [23] obtained through simulation. They concluded that CY-GAA FinFETs are suitable for RF applications as they have maximum capacitance, whereas RE-GAA FinFETs are suitable for analog applications as they have the highest transconductance.

Analytical design guidelines for RF/analog application of FinFETs were suggested by Sohn et al. [24], who reported that the RF performance can be improved by decreasing the ratio of fin-spacing to fin height. This is because cut-off and maximum frequencies are more influenced by reducing parasitic resistance than by increasing series resistance. Tinoco et al. discussed the RF behaviour of FinFETs due to variation in extrinsic gate capacitances [25]. They have reported that the RF performance is enhanced significantly by decreasing the fin spacing, modifying the ratio of fin height to width of the fin, and optimizing the ratio of fin spacing to extension source/drain regions of the fin. The RF stability performance at particular bias and device dimensions of FinFETs exists in the literature [26]. They have reported the optimized spacer length of the gate region, fin height, and fin width as well as have chosen appropriate work function for the gate material and bias settings to obtain better stability performance of FinFETs in the RF range.

Krivec et al. investigated the RF performance of bulk and SOI FinFETs with a gate length of 20 nm for both doped and undoped channels [27]. The cut-off frequency is more for bulk-substrate based FinFETs than for SOI FinFETs, whereas the maximum oscillation frequency is higher in SOI substrate FinFETs than in bulk substrate FinFETs. Mohapatra et al. presented a study on the impact of fin dimensions (fin width and height) on various RF/analog parameters of FinFETs [28] and depending on the feature ratio (fin width to height ratio), they act as FinFETs and trigate and planar MOSFETs. They systematically presented both low-frequency and high-frequency performance parameters of these devices.

A comparative study of RF/analog/linearity performance among conventional FinFETs (conv. FinFETs), DMG FinFETs, and TMG FinFET is realized through a TCAD simulator. Analysis reveals that DMG and TMG FinFETs have improved high-frequency and analog performances compared to conv. FinFETs [20], which is due to improved gate control with an increase in the number of gate materials.

Also, DMG and TMG FinFETs exhibit enhanced linearity performance compared to conv. FinFETs. Saha et al. reported that RF/analog and linearity characteristics of DMG FinFETs are a function of temperature. RF parameters degrade, whereas an improvement in linearity performance is visualized with the rise in temperature of DMG FinFETs [22]. For the variation of metal gate work function (ϕ_M), high-frequency and analog performances of a multifin-FinFET are reported. It is seen that f_t and the intrinsic gain of the multifin-FinFET improve with an increase in the work function of the gate material [29].

The RF performance of an NC-FinFET is reported in the literature, and it is found that the RF performance is a function of the ferroelectric layer thickness [30]. As the thickness of the ferroelectric layer increases, the RF performance such as gain and maximum frequency of the NC-FinFET degrade. Also, at lower values of supply voltage, the RF/analog performance of the NC-FinFET is similar to that of conv. FinFETs.

2.3 DEVICE ARCHITECTURE

The 3D cross-section of a step-FinFET is shown in Figure 2.1a and the corresponding 2D cross-section along with different materials is depicted in Figure 2.1b. The fin of the step-FinFET is divided into two sections, namely upper and lower fins, and this fin looks like a step. The performance improvement of the step FinFET in comparison with conv. FinFETs is highlighted in our earlier work [14]. The fin of the step-FinFET is designed with silicon material, whereas HfO_2 having relative electrical permittivity (=22) is chosen as the gate oxide material. The gate material used in this work is aluminium and 4.5 eV is considered as ϕ_M. Source/drain portions are highly doped, and they have a doping concentration of $10^{20}cm^{-3}$. However, the channel portion is lightly doped with a doping concentration of $10^{16}cm^{-3}$. The dimensions used for the step-FinFET are the following: gate length (L_g) = 30 nm, width of the lower fin (W_{fin1}) = 10 nm, height of the lower fin (H_{fin1}) = 20 nm, width of the upper fin (W_{fin2}) = 8 nm, height of the upper fin (H_{fin2}) = 10 nm, gate dielectric thickness of the lower fin (t_{ox1}) = 1.5 nm, gate dielectric thickness of the upper fin (t_{ox2}) = 2.5 nm, height of the buried oxide (t_{box}) = 40 nm, and thickness of the buried oxide (T_b) = 17 nm.

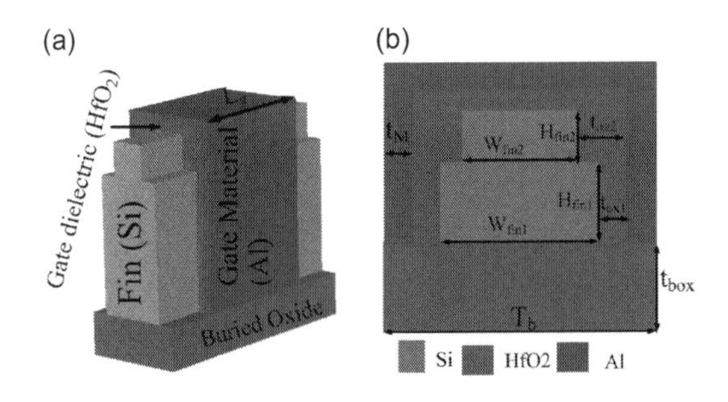

FIGURE 2.1 Schematics views of the step-FinFET: (a) 3D cross-section and (b) 2D cross-section.

2.4 TCAD CALIBRATION

TCAD is a powerful tool for research and development of new technologies with reduced design cost. It is better to use cheap and time effective computer simulation to develop new technologies instead of costly and time-consuming fabrication processes. The TCAD tool can accurately measure the physical characteristics of oxide, doping distributions, etc. with their accurate physics model. Thus, TCAD can be used to analyse the electrical characteristics of various novel devices.

Our work is based on TCAD simulation, and therefore, it is very necessary to calibrate our TCAD physics model with the existing experimental data. We have extracted fabricated data from Ref. [31] and calibrated the physics model with the extracted data, so that accurate simulation results can be achieved. The presence of highly doped source and drain regions in the step-FinFET leads to activation of the Fermi-Dirac Statistics in the TCAD simulator [32]. To account for the recombination rate, the SRH [32] model is adopted in the simulator. The Masetti Mobility model, which is doping dependent, is enabled in the simulator to consider the effect of doping concentration on charge carriers. As the fin width dimensions are below 10 nm, the quantum confinement effect may be present in the simulator, and in this regard, the quantum density gradient model is adopted in the simulator. All these physics models are calibrated with the fabricated data as shown in Figure 2.2 for two different drain biases. During calibration, the various mobility parameters are adjusted to match the simulation data with experimental results, and the tuned parameters considered in the simulator are the following: $\mu_{min1} = 140\,cm^2/Vs$, $\mu_{min2} = 25\,cm^2/Vs$, $\mu_1 = 25\,cm^2/Vs$, $C_r = 25.7 \times 10^8 cm^{-3}$, and $\beta = 5$. An exact agreement is acquired for the experimental data with simulated results as summarized in Figure 2.2.

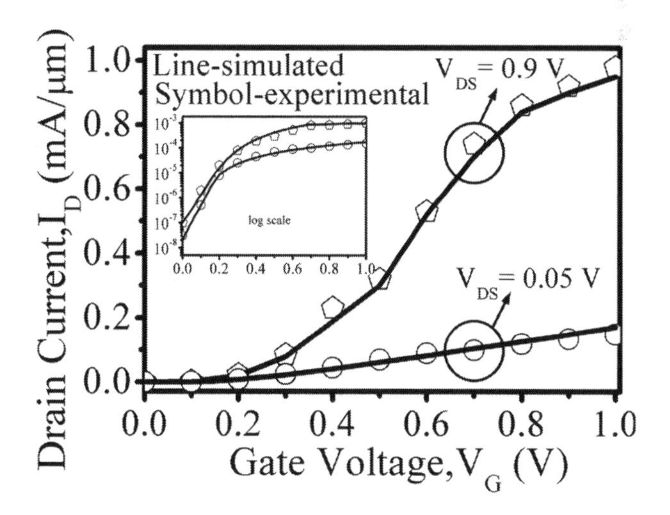

FIGURE 2.2 Calibrated transfer characteristics in the TCAD tool.

2.5 RESULTS AND DISCUSSION

This division describes the transfer characteristics and RF/analog performance of the step-FinFET under extensive change of temperature. We have also discussed about the linearity performance of the step-FinFET with varying temperature. In this work, a drain–source bias (V_{DS}) value of 0.5 V is considered.

2.5.1 Temperature Effect on Transfer Characteristics

The transfer characteristics with change in temperature in the linear and log scale are presented in Figure 2.3a and b, respectively. It is well known that both mobility of charge carriers and energy bandgap are functions of temperature. The mobility is expressed in terms of temperature as below [33]:

$$\mu_{eff} = \mu_{eff0}\left(\frac{T}{T_0}\right)^{-2} \tag{2.1}$$

where μ_{eff} and μ_{eff0} are the mobilities at any temperature T and at ambient temperature T_0, respectively. The relationship between temperature and energy bandgap is given by [34]:

$$E_g(T) = E_g(300) - \frac{\alpha T^2}{T+\beta} \tag{2.2}$$

The parameter values in equation (2.2) are the following: $\alpha = 4.73\times10^{-4}$ eV/K, $\beta = 636$ K, and E_g (300) = 1.16 eV for silicon. Equations (2.1) and (2.2) describe that both mobility and energy bandgap decrease with increased temperature. It is seen from Figure 2.3 that the input characteristic degrades both in subthreshold and superthreshold regions with the rise in temperature. For a lower value of voltage at the gate terminal, the effect of decrease in energy gap is more prominent than mobility degradation, which in turn increases the OFF current as temperature changes from 250 to 450 K. However, at the strong inversion region, the decrease in mobility values leads to degradation in ON current with the rise in temperature. The effect of

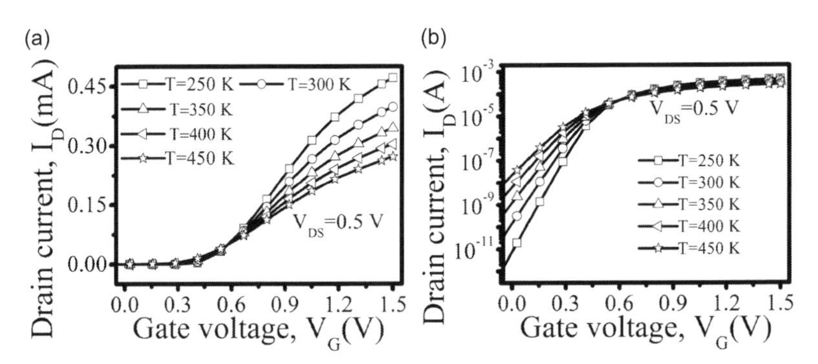

FIGURE 2.3 Input characteristics of the step-FinFET with variation in temperature in the (a) linear and (b) log scale.

temperatures on drain current is negligible at gate bias (V_G) = 0.58 V, and this voltage represents the zero temperature coefficient (ZTC) voltage. Below V_G = 0.58 V, the device shows positive temperature coefficient (PTC) because of the increase in current with the rise of temperature, and beyond the ZTC, the step-FinFET exhibits a negative temperature coefficient (NTC) because of the reduction in drain current with increased temperature. It is also noted from Figure 2.3 that with a rise in temperature, the threshold voltage falls, which is primarily due to the increased drain current below the threshold point.

The temperature effect on the ON to OFF current ratio (I_{ON}/I_{OFF}) taking temperature as a parameter is presented in Figure 2.4. As discussed, the rise in temperature degrades ON as well as OFF current, and this behaviour leads to the decrease of the I_{ON}/I_{OFF} ratio. Therefore, there is a significant amount of degradation in the I_{ON}/I_{OFF} ratio as temperature increases from 250 to 450 K.

2.5.2 Temperature Effect on RF/Analog Characteristics

One of the foremost analog parameters is transconductance $\left(g_m = \partial I_D / \partial V_{GS}\right)$ and it can be used to find the gain of any circuit. A greater peak value of g_m indicates improvement in circuit gain, and Figure 2.5a reports that transconductance is a function of temperature. It is understood that the peak value of g_m increases with the decrease in temperature, which indicates better analog performance as temperature decreases from 450 to 250 K. At the strong inversion region, the g_m falls due to the degradation of carrier mobility.

Figure 2.5b presents the total gate capacitance (C_{gg}) with temperature variation, and it is perceived that the amount of C_{gg} increases as the temperature changes from 250 to 450 K. It has already been discussed that the energy bandgap decreases with the rise in temperature, which increases the amount of gate charge, and to follow the charge balance equation, the quantity of charge in the channel section also increases. This increased charge indicates an increase in total gate capacitance as the temperature rises from 250 to 450 K. It is also observed that C_{gg} increases

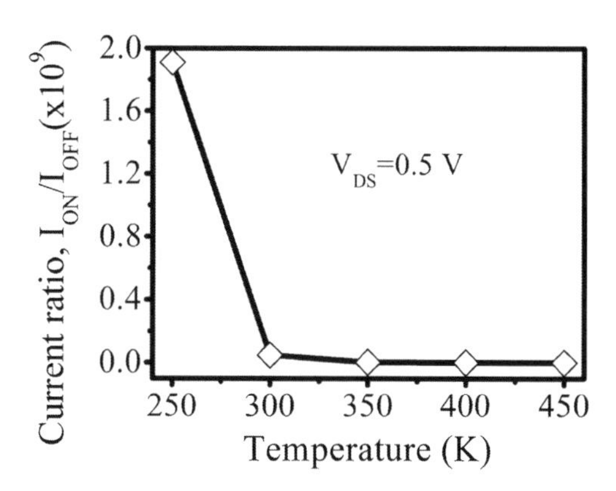

FIGURE 2.4 ON to OFF current ratio of the step-FinFET with variation in temperature.

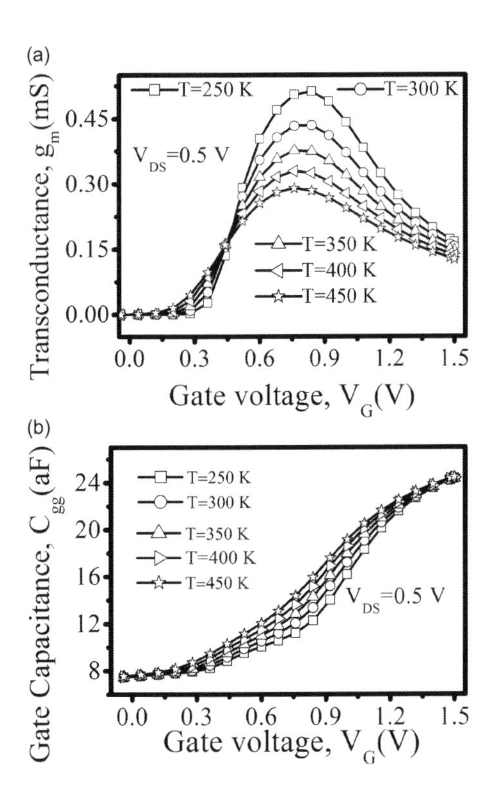

FIGURE 2.5 (a) Transconductance and (b) gate capacitance of the step-FinFET with variation in temperature.

continuously with gate voltage at different temperatures, which is because of an increase in the amount of charge with gate bias.

The temperature effect on g_d is shown in Figure 2.6a, and the lower the value of g_d, the greater the amount of intrinsic gain. As temperature decreases, the driving capability of the step-FinFET improves, and this is because of the reduction in g_d as the temperature falls from 450 to 250 K.

The intrinsic delay $\left(\tau = \left(\left(C_{gg}V_{DD}\right)/I_{on}\right)\right)$ taking temperature as a changing parameter of the step-FinFET is depicted in Figure 2.6b, and it states the speed of the FinFET [35]. A lower value of τ implies that the step-FinFET is quicker, and τ is directly and inversely related to C_{gg} and I_{ON}, respectively, at fixed V_{DD}. As C_{gg} increases and I_{ON} decreases with the rise in temperature at $V_{DD} = 0.5$ V, there is degradation in intrinsic delay as the temperature changed from 250 to 450 K. Therefore, the device is faster at lower values of temperature.

Like g_m and g_d, the influence of temperature on g_m/g_d vs. gate bias is depicted in Figure 2.7a. It has already been discussed that g_m decreases, whereas g_d increases as temperature changes from 250 to 450 K, which signifies degradation in gain within the weak to moderate inversion region. At high gate bias, there is insignificant variation in g_m and g_d, which leads to a negligible influence of temperature on g_m/g_d.

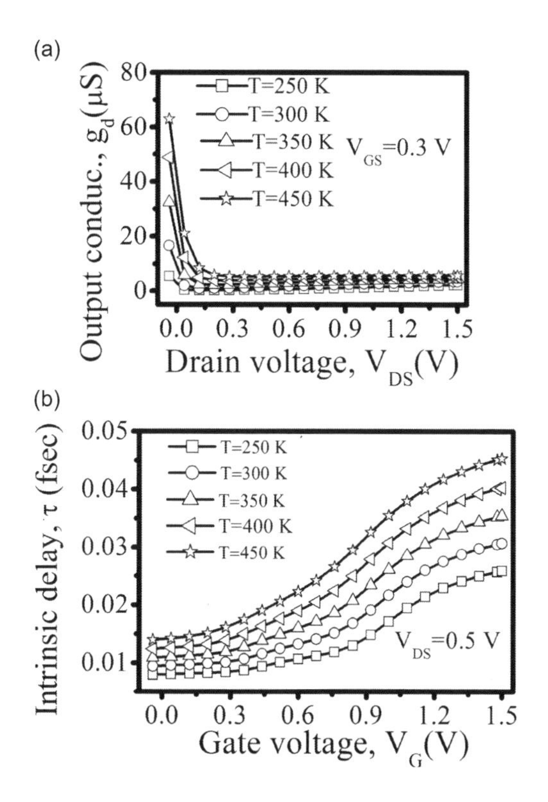

FIGURE 2.6 (a) Output conductance (g_d) and (b) intrinsic delay of the step-FinFET with variation in temperature.

For high-frequency application, the cut-off frequency $\left(f_t = g_m / 2\pi C_{gg}\right)$ plays a very major role and the temperature influence on f_t is portrayed in Figure 2.7b. A larger peak value of f_t implies superior RF performance of the step-FinFET. As already pointed out, g_m increases, whereas C_{gg} decreases with a rise in temperature, which implies degradation in f_t as the temperature increases from 250 to 450 K. However, at high gate bias, f_t falls, which is because of roll-off of g_m and insignificant deviation in C_{gg} with the variation of temperature.

The ratio of g_m to I_D is popularly recognized as TGF (g_m/I_D), and the temperature influence on TGF is portrayed in Figure 2.8a. As temperature decreases, the reduction of drain current (I_D) is more significant than the decrease in g_m, which leads to an increase in TGF within the weak to moderate inversion regions as summarized in Figure 2.8a. However, at high gate bias, an insignificant variation of TGF is observed, and this is because the values of g_m and I_D are on the order of 10^{-3} with the variation in temperature.

The parameter that defines the overall RF/analog performance is named GTFP $\left(=\left(g_m/g_d\right)\times\left(g_m/I_D\right)\times f_t\right)$, and a higher value of GTFP signifies better overall

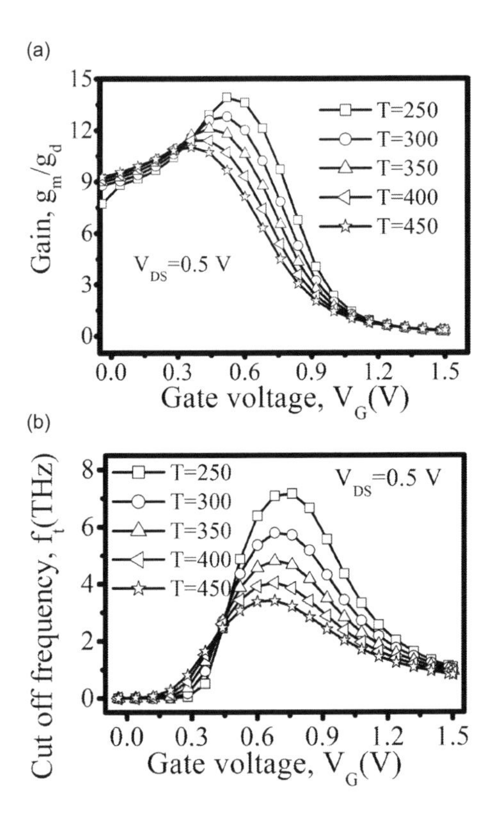

FIGURE 2.7 (a) Intrinsic gain (g_m/g_d) and (b) cut-off frequency (f_t) of the step-FinFET with variation in temperature.

RF/analog performance. The GTFP for wide temperature variation is shown in Figure 2.8b, and as g_m/g_d, g_m/I_D, and f_t increase with a decrease in temperature, the GTFP also improves as the temperature drops from 450 to 250 K. Like other parameters at the strong inversion region, GTFP decreases and this is due to mobility degradation of the charge carriers.

2.5.3 Temperature Effect on Linearity Characteristics

Performance evaluation of linearity parameters is particularly important for RF, 3G mobile communication applications [36,37]. The device non-linearity would be predicted from the higher order derivate of input characteristics, and a lower value of higher order parameters indicates improvement in linearity performance. Such higher order harmonics are g_{m2} and g_{m3}: g_{m2} is the double derivative of input characteristics with gate bias, and g_{m3} is the triple derivative of input characteristics with gate voltage, at a fixed drain bias [38]. In this chapter, the various linearity parameters of the step-FinFET such as g_{m2}, $g_{m3,}$ VIP2, VIP3, IIP_3, and IMD_3 are presented under the variation of temperature [39]. Initially, we will present the harmonics of transconductance such as g_{m2} and g_{m3} of the step-FinFET, and they represent the

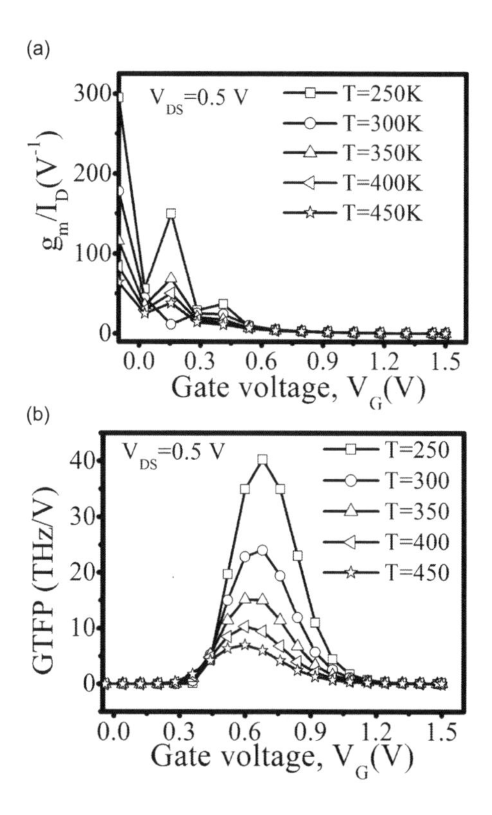

FIGURE 2.8 (a) Transconductance generation factor (TGF= g_m/I_D) and (b) gain transconductance frequency product (GTFP) of the step-FinFET with variation in temperature.

amount of disturbance in any communication system. They are expressed as the following [38–40]:

$$g_{m2} = \frac{\partial^2 I_D}{\partial V_G^2} \tag{2.3}$$

$$g_{m3} = \frac{\partial^3 I_D}{\partial V_G^3} \tag{2.4}$$

It is well known that the disturbance in the presence of g_{m2} can be lessened by using a balanced modulator circuit, and thus, g_{m3} is the leading factor whose presence produces disturbance in the circuit.

The temperature effects on g_{m2} and g_{m3} are shown in Figure 2.9a and b, respectively, and for superior linearity, the value of higher order harmonics must be minimized. It is visualized that with a rise in temperature, the values of g_{m2} and g_{m3} reduce, which indicates improved linearity performance as temperature rises from 250 to 450 K. It is visualized from Figure 2.5a that the g_m curve becomes more flat with increased temperature and this signifies better linearity characteristics. The value of g_{m2} is

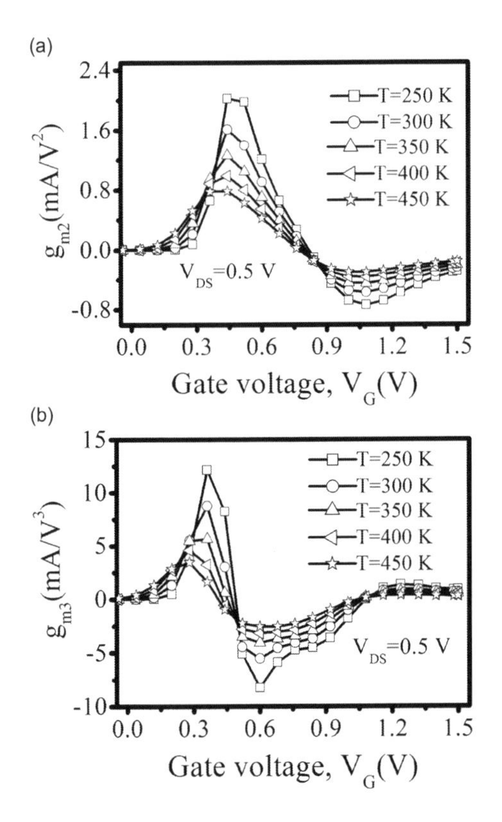

FIGURE 2.9 Higher order harmonics (a) g_{m2} and (b) g_{m3} of the step-FinFET with variation in temperature.

negative after a certain gate voltage and this is because of the roll-off of g_m at high gate bias. Also, the singularity in g_{m3} is observed at those gate biases where the value of g_{m2} is zero.

The induced bias at which fundamental and second order harmonics are equal is termed second voltage intercept point (VIP2). Similarly, the extrapolated bias where the fundamental harmonic is equivalent to the third order harmonic is defined as VIP3. The formulae for VIP2 and VIP3 are given by the following:

$$\text{VIP2} = 4\left(g_m / g_{m2} \right) \tag{2.5}$$

$$\text{VIP3} = \sqrt{24\left(g_m / g_{m3} \right)} \tag{2.6}$$

Figure 2.10a and b depict, respectively, VIP2 and VIP3 vs. gate voltage, considering temperature as a parameter. Greater values of these parameters imply enhanced linearity performance. Figure 2.10 presents that the maximum values of voltage intercept points increase as temperature changes from 250 to 450 K, and this is

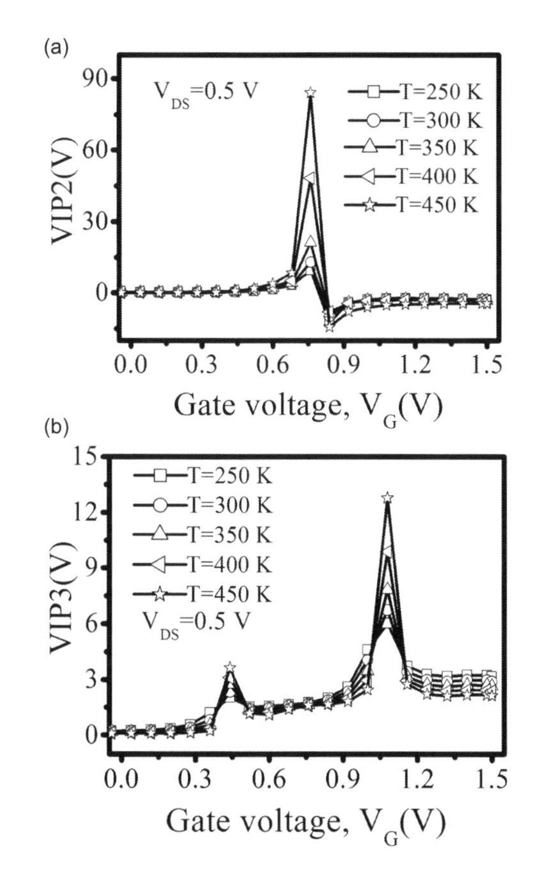

FIGURE 2.10 (a) VIP2 and (b) VIP3 of the step-FinFET with variation in temperature.

due to the decrease in g_{m2} and g_{m3}, respectively. This behaviour of VIP2 and VIP3 with temperature implies that the linearity performance improves with increased temperature.

IIP_3 is the extrapolated power where the first order harmonic power is the same as the third order harmonic power. IMD_3 corresponds to the intermodulation power where the first order intermodulation power is equal to the third order intermodulation power. IIP_3 and IMD_3 are stated as [40] the following:

$$\mathrm{IIP}_3 = \left(\tfrac{2}{3}\right)\left(g_m / \left(g_{m3} R_{se}\right)\right) \tag{2.7}$$

$$\mathrm{IMD}_3 = \left[\frac{9}{2}(\mathrm{VIP3})^2 \left(g_{m3}\right)\right]^2 R_{se} \tag{2.8}$$

where R_{se} is taken as 50 Ω for RF application.

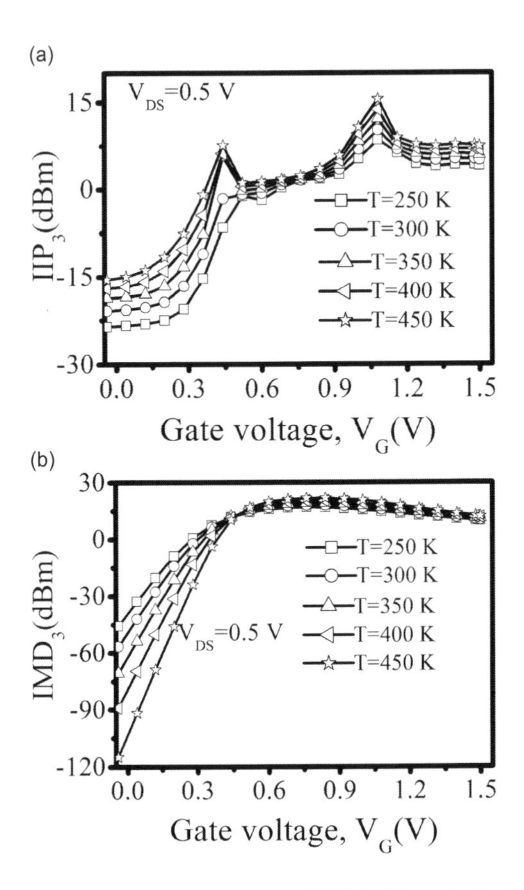

FIGURE 2.11 (a) IIP_3 and (b) IMD_3 of the step-FinFET with variation in temperature.

The IIP_3 and IMD_3 for a wide variation of temperature are presented in Figure 2.11a and b, respectively. A greater amount of IIP_3 indicates superior linearity characteristics. Figure 2.11a shows that the value of IIP_3 increases as the temperature changes from 250 to 450 K, which in turn improves the linearity performance. The temperature effect on IMD_3 is summarized in Figure 2.11b, and a lower value of intermodulation distortion signifies improved linearity characteristics. As the temperature increases, the value of IMD_3 reduces at lower gate bias, which summarizes better SCEs and reduced effects of hot carriers. Also, the decrease of IMD_3 value with an increase in temperature improves the linearity characteristics.

2.6 SUMMARY

In this chapter, the performance of a step-FinFET is evaluated for a wide variation in temperature. Results reveal that the current ratio I_{ON}/I_{OFF} of the step-FinFET degrades on the order of 10^4 when the temperature rises from 250 to 450 K. Improvements in various RF/analog properties such as g_m, gain (g_m/g_d), and f_t are obtained with decreased temperature. Also, the GTFP which gives the overall RF/analog performance of the step-FinFET is improved as the temperature falls from 450 to 250 K. This work will be beneficial for RF applications of the step-FinFET at different environmental temperatures. It is also seen that for high speed applications, the temperature must be maintained as low as possible, so that the intrinsic delay of the device is minimized. Furthermore, higher order harmonics g_{m2} and g_{m3} decrease with an increase in temperature, which indicates superior linearity performance. Observable enhancements in linearity performances such as VIP2 and VIP3; IIP_3, and IMD_3 are attained when the temperature rises from 250 to 450 K. RF/analog and linearity parameters are functions of temperature and it has a very significant impact on these parameters. Thus, the RF/analog performance of the step-FinFET is suppressed, and the linearity characteristics are improved with an increase in temperature. Therefore, this work provides valuable information to design any device for wireless communication and low-power applications.

ACKNOWLEDGMENT

The authors acknowledge the funding by the Science & Engineering Research Board (SERB), Govt. of India (sanction order no. SRG/2019/000628).

REFERENCES

1. Haensch, W., Nowak, E. J., Dennard, R. H., et al. 2006. Silicon CMOS devices beyond scaling. *IBM J. Res. Dev.* 50, 6: 339–361. doi: 10.1147/rd.504.0339.
2. Tsividis, Y. 2011. Small-Dimension Effects in Operation and Modeling of the MOS. *Transistor*, Rachael Zimmermann (ed.), 2nd ed.: New York, Oxford University Press.
3. Vandamme, E. P., Jansen, P., Deferm, L. 1997. Modeling the subthreshold swing in MOSFET's. *IEEE Electron Device Lett.* 18, 8 (Aug.): 369–371. doi: 10.1109/55.605442.
4. Rai, M. K., Narendar, V., Mishra, R. A. 2014. Significance of variation in various parameters on electrical characteristics of FinFET devices, *Students Conference on Engineering and Systems*, Allahabad, (August): 1–6. doi: 10.1109/SCES.2014.6880096.
5. Kumar, M. J., Chaudhry, A. 2004. Two-dimensional analytical modeling of fully depleted DMG SOI MOSFET and evidence for diminished SCEs. *IEEE Trans. Electron Devices* 51, 4 (April): 569–574. doi: 10.1109/TED.2004.823803.
6. Mizuno, T., Takagi, S., Sugiyama, N., et al. 2000. Electron and hole mobility enhancement in strained-Si MOSFET's on SiGe-on-insulator substrates fabricated by SIMOX technology. *IEEE Electron Device Lett.* 21, 5 (May): 230–232. doi: 10.1109/55.841305.
7. Bhattacharya, D., Jha, N. K. 2014. FinFETs: From devices to architectures. *Adv. Electron.* 21 pages. doi: 10.1155/2014/365689.
8. Pei, G., Kedzierski, J., Oldiges, P., Ieong, M., Kan, E. C.C. 2002. FinFET design considerations based on 3-D simulation and analytical modeling. *IEEE Trans. Electron Devices* 49, 8 (August): 1411–1419. doi: 10.1109/TED.2002.801263.
9. Mehrad, M., Orouji, A. A. 2010. Partially cylindrical fin field-effect transistor: A novel device for nanoscale applications. *IEEE Trans. Device Mater. Reliab.* 10, 2 (March): 271–275. doi: 10.1109/TDMR.2010.2046663.
10. Ritzenthaler, R., Lime, F., Iniguez, B., Faynot, O., Cristoloveanu, S. 2010. 3D analytical modelling of subthreshold characteristics in Pi-gate FinFET transistors. *Solid-State Device Research Conference* (ESSDERC), (November): 448–451. doi: 10.1109/ESSDERC.2010.5618179.

11. Yeh, M. S., Wu, Y. C., Hung, M. F., et al. 2013. Fabrication, characterization and simulation of Ω-gate twin poly-Si FinFET nonvolatile memory. *Nanoscale Res. Lett.* 8, 1 (July): 331. doi: 10.1186/1556-276X-8-331.
12. Saha, R., Baishya, S., Bhowmick, B. 2017. 3D Analytical modeling of surface potential, threshold voltage, and subthreshold swing in dual material gate (DMG) SOI FinFET. *J. Comput. Electron.* 17, 1 (October): 153–162. doi: 10.1007/s10825-017-1072-x.
13. Li, C., Zhuang, Y., Zhang, L. 2012. Simulation study on FinFET with tri-material gate. *2012 IEEE International Conference on Electron Devices and Solid State Circuit (EDSSC),* (March): 1–3. doi: 10.1109/EDSSC.2012.6482802.
14. Saha, R., Bhowmick, B., Baishya, S. 2017. Si and Ge step-FinFETs: work function variability, optimization and electrical parameters. *Superlattices Microstruct.* 107, (July): 5–16. doi: 10.1016/j.spmi.2017.04.001.
15. Kranti, A., Armstrong, G. A. 2007. Comparative analysis of nanoscale MOS device architectures for RF applications. *Semicond. Sci. Technol.* 22, 5 (March): 481–491. doi: 10.1088/0268-1242/22/5/005.
16. Raskin, J. P., Chung, T. M., Kilchytska, V., Lederer, D., Flandre, D. 2006. Analog/RF performance of multiple gate SOI devices: wideband simulations and characterization. *IEEE Trans. Electron Devices* 53, 5 (May): 1088–1095. doi: 10.1109/TED.2006.871876.
17. Subramanian, V., Parvais, B., Borremans, J., et al. 2006. Planar bulk MOSFETs versus FinFETs: An analog/RF perspective. *IEEE Trans. Electron Devices* 53, 12 (December): 3071–3079. doi: 10.1109/TED.2006.885649.
18. Sharma, D., Vishvakarma, S. K. 2015. Analyses of DC and analog/RF performances for short channel quadruple-gate gate-all-around MOSFET. *Microelectron. J.* 46, 8 (August): 731–739. doi: 10.1016/j.mejo.2015.05.008.
19. Jegadheesan, V., Sivasankaran, K. 2017. RF stability performance of SOI junctionless FinFET and impact of process variation. *Microelectron. J.* 59, (January): 15–21. doi: 10.1016/j.mejo.2016.11.004.
20. Saha, R., Bhowmick, B. 2018. Comparative Analysis among SMG, DMG, and TMG FinFETs: RF/analog and digital inverter performance. *J. Nanoelectron. Optoelectron.* 13, 6 (June): 803–811. doi: 10.1166/jno.2018.2336.
21. Lederer, D., Parvais, B., Mercha, A. et al. 2006. Dependence of FinFET RF performance on fin width. Digest of Papers. *2006 Topical Meeting on Silicon Monolithic Integrated Circuits in RF Systems*, San Diego, CA. doi: 10.1109/SMIC.2005.1587887.
22. Saha, R., Bhowmick, B. 2018. Temperature effect on RF/analog and linearity parameters in DMG FinFET. *Appl. Phys. A: Mater. Sci. Process.* 124, 9 (September): 642. doi: 10.1007/s00339-018-2068-5.
23. Kumar, A. 2016. Analog and RF performance of a multigate FinFET at nano scale. *Superlattices Microstruct.* 100, (December): 1–8. doi: 10.1016/j.spmi.2016.10.073.
24. Sohn, C. W., Kang, C. Y., Baek, R. H., et al. 2012. Device design guidelines for nanoscale FinFETs in RF/analog applications. *IEEE Electron Device Lett.* 33, 9 (September): 1234–1236. doi: 10.1109/LED.2012.2204853.
25. Tinoco, J. C., Rodriguez, S. S., Martinez-Lopez, A. G., Alvarado, J., Raskin, J. P. 2013. Impact of extrinsic capacitances on FinFET RF performance. *IEEE Trans. Microwave Theory Tech.* 61, 2 (Feburary): 833–840. doi: 10.1109/SiRF.2012.6160141.
26. Sivasankaran, K., Mallick, P. S. 2014. Bias and geometry optimization of FinFET for RF stability performance. *J. Comput. Electron.* 13, (September): 250–256. doi: 10.1007/s10825-013-0507-2.
27. Krivec, S., Prgić, H., Poljak, M., Suligoj, T. 2014. Comparison of RF performance between 20 nm-gate bulk and SOI FinFET. *37th International Convention on Information and Communication Technology, Electronics and Microelectronics (MIPRO),* Opatija, (July): 45–50. doi: 10.1109/MIPRO.2014.6859530.

28. Mohapatra, S. K., Pradhan, K. P., Singh, D., Sahu, P. K. 2015. The role of geometry parameters and fin aspect ratio of sub-20nm SOI-FinFET: An analysis towards analog and RF circuit design. *IEEE Trans. Nanotechnol.* 14, 3 (May): 546–554. doi: 10.1109/TNANO.2015.2415555.
29. Hirpara, Y., Saha, R. 2020. Analysis on DC and RF/analog performance in Multifin-FinFET for wide variation in work function of metal gate. *Silicon*, (February), doi: 10.1007/s12633-020-00408-2.
30. Singh, R., Aditya, K., Parihar, S. S. et al., 2018. Evaluation of 10-nm bulk FinFET RF performance-conventional versus NC-FinFET. *IEEE Electron Device Lett.* 39, 8 (August): 1246–1249. doi: 10.1109/LED.2018.2846026.
31. Rios, R., Cappellani, A., Armstrong, M., et al. 2011. Comparison of junctionless and conventional trigate transistors with down to 26 nm. *IEEE Electron Device Lett.* 32, 9 (September): 1170–1172. doi: 10.1109/LED.2011.2158978.
32. *Sentaurus Device User Guide*. Synopsys, Inc., Mountain View, CA, University of California, Berkeley, 2011.
33. Tenbroek, B. M., Lee, M. S. L., White, W. R. et al. 1996. Self-heating effects in SOI MOSFETs and their measurement by small signal conductance techniques. *IEEE Trans. Electron Devices* 43, 12 (December): 2240–2248. doi: 10.1109/16.544417.
34. Narang, R., Saxena, M., Gupta, R. S., Gupta, M. 2013. Impact of temperature variations on the device and circuit performance of tunnel FET: A simulation study. *IEEE Trans. Electron Nanotechnol.* 12, 6 (November): 951–957. doi: 10.1109/TNANO.2013.2276401.
35. Rai, S., Sahu, J., Dattatray, W., Mishra, R. A., Tiwari, S. 2012. Modelling, design, and performance comparison of triple gate cylindrical and partially cylindrical FinFETs for low-power applications. *ISRN Electron.* 2012, Article ID 827452: 7 pages. doi: 10.5402/2012/827452.
36. Adan, A. O., Yoshimasu, T., Shitara, S., Tanba, N., Fukurni, M. 2002. Linearity and low-noise performance of SOI MOSFETs for RF applications. *IEEE Trans. Electron Devices* 49, 5 (May): 881–888. doi: 10.1109/16.998598.
37. Ma, W., Kaya, S., Asenov, A. 2003. Study of RF linearity in sub-50 nm MOSFETs using simulations. *J. Comput. Electron.* 2, (December): 347–352. doi: 10.1023/B:JCEL.0000011450.37111.9d.
38. Kumar, S. P., Agrawal, A., Chaujar, R., Gupta, R. S., Gupta, M. 2011. Device linearity and intermodulation distortion comparison of dual material gate and conventional AlGaN/GaN high electron mobility transistor. *Microelectron. Reliab.* 51, 3 (March): 587–596. doi: 10.1016/j.microrel.2010.09.033.
39. Ghosh, P., Haldar, S., Gupta, R. S., Gupta, M. 2012. An investigation of linearity performance and intermodulation distortion of GME CGT MOSFET for RFIC design. *IEEE Trans. Electron Devices* 59, 12 (December): 3263–3268. doi: 10.1109/TED.2012.2219537.
40. Razavi, B. 1998. *RF Microelectronics*. Castleton, NY: Prentice Hall.

3 Low-Power Memory Design for IoT-Enabled Systems

Part 1

Adeeba Sharif, Sayeed Ahmad, and Naushad Alam
Aligarh Muslim University

CONTENTS

3.1 INTRODUCTION

The emergence of the Internet of Things (IoT) based electronics industry necessitates the design of ultra-low-energy sensor nodes. Sustainable IoT nodes demand an extremely energy efficient operation of cyber physical systems (CPS) so as to enhance battery life or to operate with harvested energy [1]. Large embedded memories (primarily SRAM) are used in IoT devices for the storage of collected data as well as instructions [2]. Consequently, a significant fraction of the overall system energy is consumed by these embedded memories. It has been reported that SRAM is the major contributor to the overall power dissipation in typical systems-on-chip (SoCs) [3]. Therefore, designing ultra-low-power (ULP) and robust SRAM for sustainable growth of IoT applications is a new challenge.

SRAM applications are primarily spread over two distinct domains – high speed SRAM and ULP SRAM. Both these application domains have their own set of features and applications. Traditionally, products where SRAMs are deployed need them either for their high speed or their ultra-low power consumption, but rarely for both. However, recently there has been an increasing demand for SRAM that offers high speed while consuming ultra-low power to perform computations while running on batteries or with energy harvested from the environment. This demand is driven by a plethora of new applications of IoT-based products in personal use, healthcare systems, consumer electronics, communication systems and industrial controllers [4]. There will be a huge impact of the IoT in the field of smart wearables and on the automation of industrial/commercial operations extending from individual houses to large factories and entire cities [5]. While smart wearables demand SRAMs having a small area and ultra-low power consumption, the latter require ULP SRAMs along with high performance (HP). Therefore, to be suitable for ULP applications such as IoT designs, SRAM needs to be designed in a way that it simultaneously satisfies the performance and power requirements.

Planar Complementary Metal Oxide Semiconductor (CMOS) has been the work horse for the semiconductor industry for more than four decades. However, with technology scaling, the power dissipation of planar CMOS-based SRAM cells increased significantly due to increased leakage current, V_T roll-off, Drain Induced Barrier Lowering (DIBL) and process variations [6]. Operating SRAMs with near-/sub-threshold supply voltage can significantly reduce the energy consumption. However, smaller drive current and threshold voltage (V_T) variations pose major challenges for the robust operation of SRAM cells at scaled operating voltages [7]. Consequently, many of the recent studies have investigated various techniques such as power gating and multi-V_T for leakage reduction while sacrificing the cell area [8–10]. Nevertheless, degraded subthreshold slope (SS) at room temperature is the real constraint for supply voltage scaling in a deep submicron planar CMOS, and therefore, it is challenging to suppress the leakage power in SRAM cells designed using a planar CMOS [11]. Therefore, circuit designers switched to 3-dimensional FinFET structures after the 22 nm technology node and FinFETs completely replaced planar CMOS in the state-of-the art HP circuits [12].

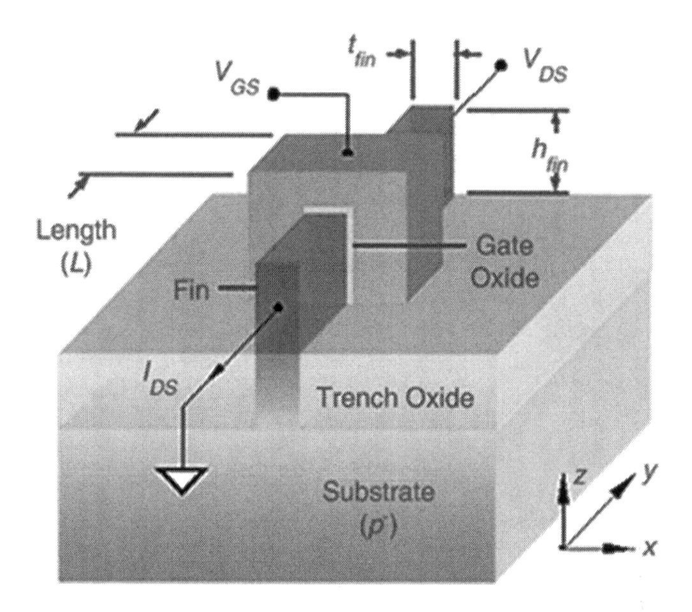

FIGURE 3.1 Cartoon of a FinFET.

3.2 OVERVIEW OF FinFETs

FinFETs have a 3-D structure [13] as shown in Figure 3.1, which is different from that of conventional planar Metal Oxide Semiconductor Field Effect Transistor (MOSFETs). Here the 'Fin' resembles a fish fin and rises above the substrate. A rectangular source and drain are connected by the fin-shaped channel portion. The gate surrounds the fin from three sides, left, top and right, to achieve better control over the channel. This form of gate structure facilitates tight electrical control of the channel conduction, and it enables leakage current suppression through reduced short-channel effects. A gate on three sides of the channel results in very strong electrostatic integrity and suppresses leakage current significantly when the device is in the OFF state [14]. Because of the strong electrostatic integrity, FinFET technology allows the use of lower threshold voltages, which offers advantages in terms of better performance and lower power dissipation [15].

3.3 LOW POWER SRAM DESIGN CONSIDERATIONS

Data collected by IoT devices need to be stored before they are transmitted/processed. Therefore, SRAM as a data storage element is an essential component of such systems [16]. While the memory occupies a significant area on modern SoCs, the activity factor of memory remains very low [17]. This implies that at any point of time only a small fraction of the overall memory circuit performs read/write operations. The major fraction of the memory remains in idle mode. The SRAM cells that perform the read/write operation contribute towards dynamic power dissipation and the major fraction of SRAM cells in idle mode contribute towards static/leakage power dissipation [18]. The total leakage power of idle SRAM cells is much larger than the dynamic power of the small fraction of active SRAM cells. Consequently, the overall power consumption of memory is dominated by static/leakage power [19]. The breakup of total power into leakage power and dynamic power of SRAM with technology scaling is shown in Figure 3.2a.

Scaling the supply voltage is very effective in containing the power consumption of digital circuits [20]. However, there exists power–performance trade-off for supply voltage scaling. Voltage scaling in logic circuits leads to degraded performance and may be acceptable for low speed circuits. This acceptability is also because of the fact that in digital circuits, voltage scaling will rarely lead to functional failure. However, voltage scaling in SRAM may lead to functional failure and will never be acceptable. Therefore, voltage scaling in memory circuits is more challenging than in logic circuits [21]. This is also the reason why, on a state-of-the-art SoC, the logic and memory circuits are operated with different supply voltages to strike a balance between power and performance. To be more precise, memory circuits are operated at a higher supply voltage than the logic circuits [22]. Figure 3.2b shows the power–performance along with the supply voltage for an Intel near-threshold voltage Pentium core processor. This processor designed using 32 nm technology

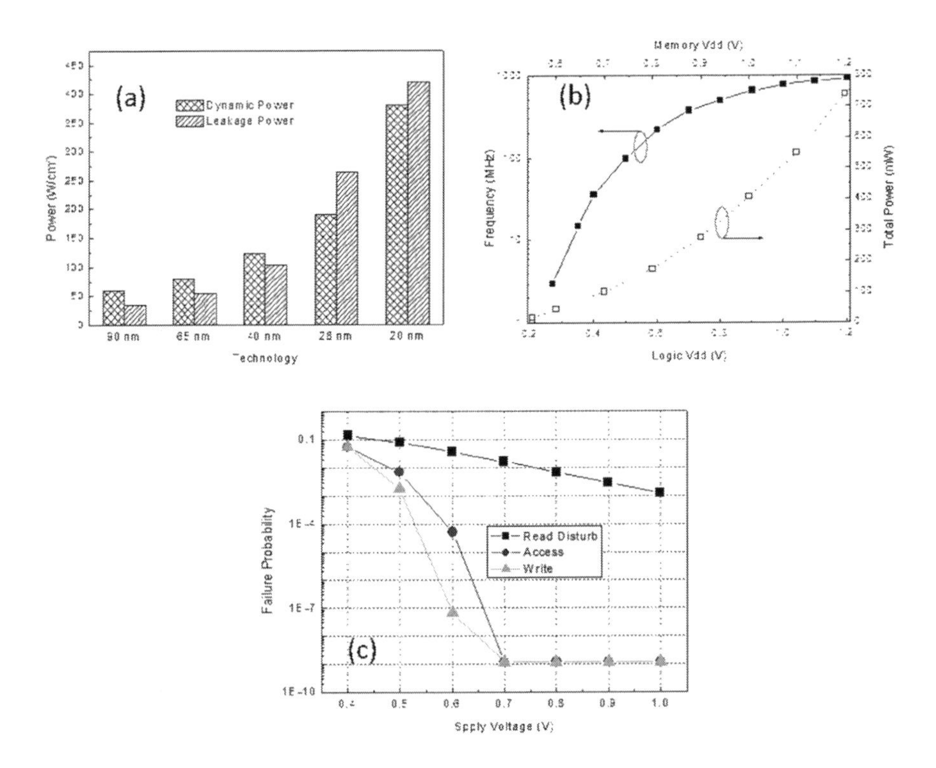

FIGURE 3.2 (a) Dynamic and leakage power components in SRAM, (b) power–performance plot of Intel's near-threshold voltage processor and (c) failure probability vs supply voltage scaling.

accommodates nearly 6 million transistors in a 2 mm^2 die area. This processor uses two separate voltage domains: one for logic and another for cache memory. While the cache needs a minimum supply of 0.55 V, the logic can operate with only 0.28 V power supply. It can be observed from Figure 3.2b that at 3 MHz, 62% of the power dissipation is attributed to the cache memory.

The fact that a lower supply voltage will lead to lower overall power dissipation in SRAM cells has led to aggressive voltage scaling in SRAM cells. However, the minimum supply voltage required for a cell operation depends upon various stability metrics of SRAM cells [23]. These stability metrics are read static noise margin (RSNM), write margin (WM) and hold static noise margin (HSNM). RSNM puts a constraint on the minimum supply voltage ($V_{DD,\text{min}}$) at which the data can be read from an SRAM cell without causing data upset in the cell [24]. Typically, it should be more than 26 mV to avoid read upset in an SRAM cell. The WM puts a constraint on the $V_{DD,\text{min}}$ at which the data can be successfully written into an SRAM cell. Typically, it should be positive to ensure successful write operation. Lastly, the HSNM puts a constraint on the minimum supply voltage at which an SRAM cell can retain data in the steady state. Typically, it should be larger than 26 mV. It has been widely reported that while an SRAM cell can retain data at a very small supply voltage, it needs a little larger voltage to perform read/write operations. Therefore, it is the requirement of RSNM/WM that puts a limit on the minimum supply voltage for a conventional 6T SRAM cell [25]. It can be observed from Figure 3.2c that the functional failure probability of an SRAM cell increases with supply voltage scaling. It should be noted that the failure probability due to read disturb is significantly higher than the other failure mechanisms related to memory access.

This implies that it is the small number of active cells that necessitates the major fraction of memory to be operated at a voltage higher than what is necessary for retaining the data, thereby ending up with increased power dissipation. To do away with this unnecessary wastage of power, various techniques have been suggested by researchers. These techniques are referred to as read assist and write assist techniques. The read assist technique decouples/isolates the read path from the storage nodes of an SRAM cell, thereby avoiding any chance of read upset [26]. This is achieved by using a separate read buffer in the cell. On the other hand, in an SRAM cell operating at scaled supply voltage, the write assist ensures sufficient write current for successful write operation. This is achieved by techniques such as word line boosting, negative bitline, lowering V_{DD}, raising V_{SS}, etc. [27]. However, these assist techniques involve power and performance penalties, which are often acceptable for achieving stable operation while dissipating minimum power. Therefore, these assist techniques are an integral part of ULP SRAM design [28].

3.3.1 Read Assist Techniques

Figure 3.3 shows that the '0' storing node 'Q' is raised to a non-zero value during read operation in a 6T cell, which may cause read failure. To avoid this, one of the easiest techniques, which has been extensively used, is decoupling the read operation by utilizing a separate read buffer. The read operation is performed by an extra wordline (WL) and bitline, generally referred to as read wordline (RWL) and read

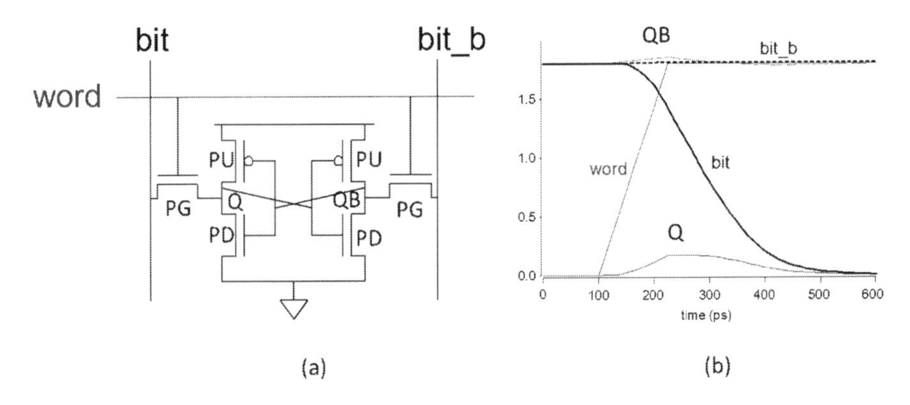

FIGURE 3.3 (a) Schematic of a 6T SRAM cell and (b) node voltages during read operation.

bitline (RBL). The internal storage nodes are isolated from the bitlines during read operation, and hence, the read disturbance is not a concern. The read stability of a cell is then the same as the hold stability. Obviously, such an advantage is obtained by sacrificing cell area. Various read buffers reported in the literature are discussed here. These read buffers necessitate a separate RBL. Therefore, in such cells, a single-ended scheme is deployed that senses the voltage drop developed on RBLs.

3.3.1.1 Read Buffers

An 8T cell proposed by Liang et al. [29] uses the simplest read port (Read Buffer-A) along with a conventional 6T SRAM cell as shown in Figure 3.4. This eliminates the worst case stability condition without disturbing the internal nodes of the cell. The read buffer/port comprises two transistors named NR1 and NR2. The gate of NR2 is connected to a separate row-based word line named RWL, whereas the gate of NR1 is controlled by node QB of the SRAM cell. The write operation is identical to that in

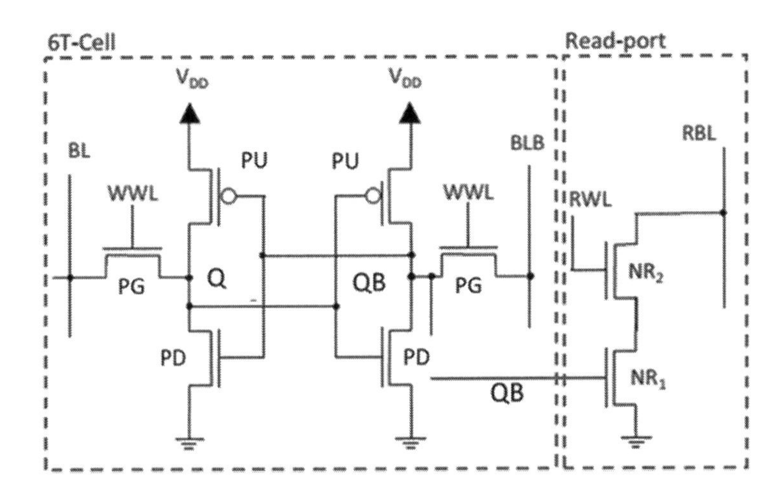

FIGURE 3.4 Schematic of an 8T SRAM cell using Read Buffer-A.

the conventional 6T cell. However, the read operation is performed by disabling WL and activating RWL with a pre-charged RBL. The RBL may or may not discharge depending on the data, 0 or 1, stored in the cell, and the voltage swing on the RBL is sensed by a single-ended sense amplifier.

Calhoun and Chandrakasan proposed a 10T cell [30] that adds a 4T read buffer (Read Buffer-B), shown in Figure 3.5, to a conventional 6T SRAM cell. This read buffer helps in reducing the RBL leakage current. When the RWL is disabled, R3 is in the OFF state, thereby avoiding the discharge of node QBB to ‘0’ even though QB = ‘1’. As the node QBB floats near V_{DD} because of P-type Metal Oxide Semiconductor (PMOS) transistor R2, it limits the subthreshold leakage current through R1. Moreover, the leakage through R3 and R4 is reduced because of the stacking effect. Kim et al. proposed a 10T SRAM cell [31] to combat the effect of data dependent bitline leakage. The modified read buffer (Read Buffer-C) circuit

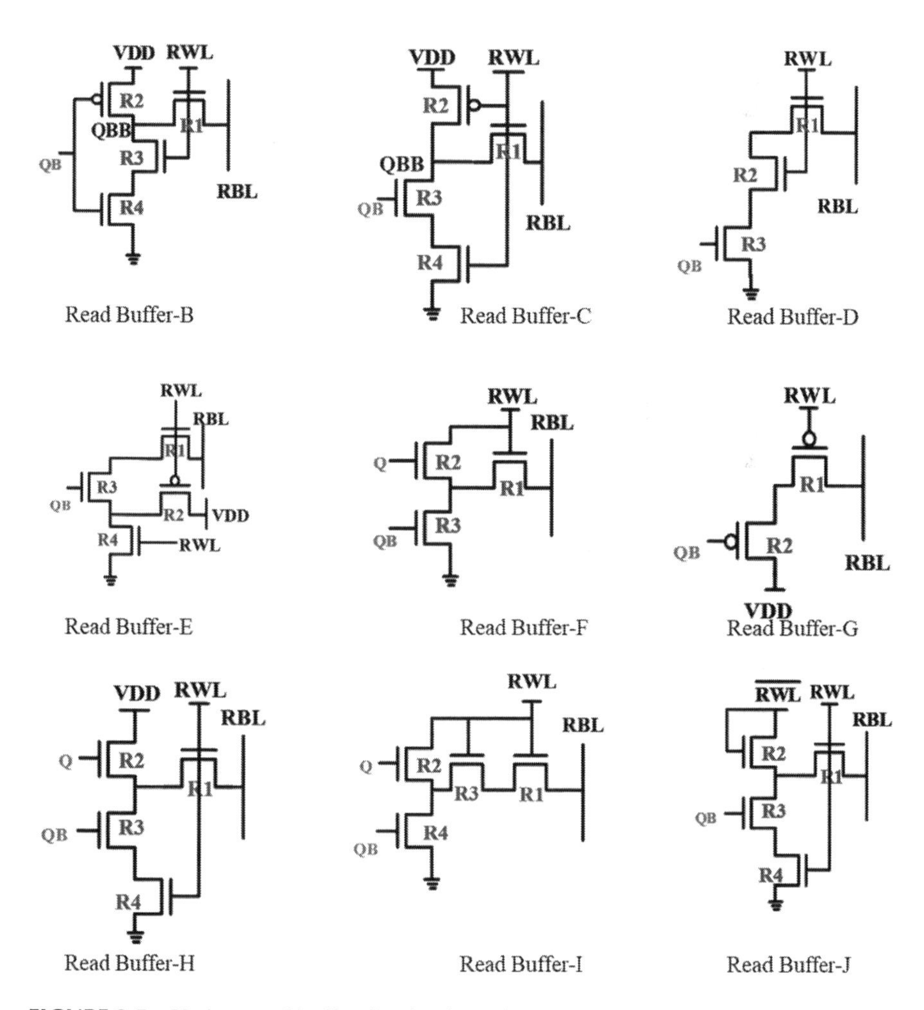

FIGURE 3.5 Various read buffer circuit schematics.

is shown in Figure 3.5. For unselected rows, when RWL = 0, node QBB is at V_{DD} through R2, hence subthreshold leakage current through R1 is reduced irrespective of the data stored in the SRAM cell. Thus the I_{ON}/I_{OFF} ratio of the cell is significantly improved enabling more cells to be allowed per RBL. Lin et al. proposed a 9T SRAM cell [32] that employs a 3T read buffer (Read Buffer-D) as shown in Figure 3.5. It consists of two stacked transistors R1 and R2 driven by RWL. Due to the stack effect, the bitline leakage is significantly reduced. It is possible to have more SRAM cells on a bitline in a column because of reduced bitline leakage. Therefore, it is suitable for high-density SRAM designs. Pasandi and Fakhraie proposed another read buffer (Read Buffer-E) circuit [33] to improve the I_{ON}/I_{OFF} ratio for facilitating a larger number of cells to share a common bitline in a column. The circuit schematic of the 4T read buffer is shown in Figure 3.5. In this read buffer, PMOS R2 is in the ON state for unselected cells and the source voltage of R3 will be clamped to V_{DD}. This reduces the leakage current through R3. To mitigate the data-dependent bitline leakage issue, bitline equalization was proposed in Ref. [34]. The read buffer (Read Buffer-F) shown in Figure 3.5 consists of three transistors to realize equal bitline leakage for unselected cells. When the data stored is 0, i.e. $Q = 0$ and QB = 1, the bitline leakage current will flow through R1 and R3 to ground. In the same manner, when $Q = 1$, the bitline leakage current flows to RWL (which is grounded for unselected cells) through transistors R1 and R2. Hence, one of the R2 and R3 transistors is always ON irrespective of stored data. Therefore, leakage strength is the same for both leakage paths, and hence, a constant bitline leakage is formed independent of the stored data in an SRAM cell. Read Buffer-G comprising two PMOS transistors to reduce the bitline leakage is shown in Figure 3.5. In this circuit, a pre-discharge read mechanism (i.e. charging RBL through R1 and R2 for QB = 0) is employed instead of the conventional pre-charge mechanism. However, PMOS has weak drive capability than N-type Metal Oxide Semiconductor (NMOS) and it deteriorates the read performance. The worst situation arises in this circuit when for the selected cell QB = 1 and for the unselected cells QB = 0. The problem of drive-ability gap between NMOS and PMOS transistors becomes worse with the scaling of supply voltage. Hence, adding an inverter to this cell for sensing improves the performance.

Gupta et al. proposed a read buffer (Read Buffer-H) [36] as shown in Figure 3.5. The read buffer shown in Figure 3.5 reduces the bitline leakage by employing R2 that charges the source of R1 to a voltage $V_{DD} - V_{TH}$ for a cell storing 1. Therefore, the leakage current through R1 is significantly reduced. The leakage current is also reduced due to the stacking effect of R1, R3 and R4. Therefore, this read buffer achieves sufficient I_{ON}/I_{OFF} ratio and enables a larger number of cells sharing a bitline in a column. The authors in Ref. [36] proposed another read buffer (Read Buffer-I) as shown in Figure 3.5. This circuit employs two stacked transistors R1 and R3 to exploit stacking induced leakage reduction. The circuit also maintains data-independent I_{ON}/I_{OFF} ratio by the use of R2 and R3. At any point of time for any cell, either R2 or R4 is ON and, therefore, provides a symmetrical leakage path. Therefore, the magnitude of the leakage current is maintained equal for both the read '0' and read '1' cases. This facilitates the maintenance of the necessary difference in the magnitudes of current between accessed and un-accessed cells. Yet another read buffer (Read Buffer-J) circuit was reported by the authors in Ref. [36], which is shown in Figure 3.5. All three read buffers consist of four NMOS transistors. As claimed by the authors, Read Buffer-H improves data-dependent RBL leakage and also achieves high performance. However, this cell still has some undetermined region for read '0' and '1' cases. Read Buffer-I and Read Buffer-J provide data-independent RBL leakage and are suitable for low-power and low-area applications, respectively.

3.3.2 Write Assist Techniques

When wordlines are activated, the low-going bitline discharges the cell node storing '1'. For successfully writing the data, the internal nodes of the cell must flip and reach the pre-defined V_{DD} level. Write failure will occur if the internal nodes do not flip. WM is defined as the measure of writeability of an SRAM cell. With technology scaling, even at nominal supply voltages, it is becoming difficult to write data into SRAM, and this issue becomes more severe at lower supply voltages as shown in Figure 3.2c. During the write operation, the techniques employed to aid the bit cell to change the state are known as write assist techniques, and nowadays, these techniques are widely employed to design low-power SRAM. Several write assist techniques have been suggested in the literature to improve the WM of an SRAM cell [37].

3.3.2.1 Wordline Boost Assist

This technique helps to improve the WM of an SRAM cell by increasing the overdrive voltage of access transistors, thereby increasing its driving strength. The boosted voltage can be generated initially by a charge pump [38], or by capacitor coupling [39] or can be routed as a separate power supply. This technique works on a row of SRAM array. Therefore, chances of read upset in all the half selected cells in a row increase due to the reduction in their dynamic read noise margins. Hence, it hurts the stability of the unselected cells. However, this technique is easy to implement and also the power consumption penalty is smaller.

3.3.2.2 Negative Bitline Write Assist

One of the several techniques to provide write assist is pulling down the bitline below ground (GND) during write '0'. Consequently, the increased V_{GS} of the access transistor results in a significant increase in write current, and hence the bits can be flipped easily [40]. This approach is suited for both single port and dual port memories. Power consumption issues also occur because all selected bitlines need to be brought to low voltage.

3.3.2.3 Lowering of V_{DD}

In this technique, V_{DD} is reduced by discharging the V_{DD} supply, which weakens the pull-up (PU) device with respect to the access transistor in an SRAM cell. It is easier to write new data into a cell once the PU device is weakened. Many techniques are there to temporarily lower the supply voltage of an SRAM cell. Some of the techniques include a second external lower supply which is connected through a multiplexer to write selected columns [41] and another approach is to use on-chip voltage regulators [42]. The main challenge posed by this technique is to ensure that the lowered column voltage remains higher than the data retention voltage of unselected cells in that column. However, to solve this issue, during the write operation, the supply voltage of the whole array is lowered. However, the dynamic read noise margins of half selected cells are compromised in this technique.

3.3.2.4 Raising of V_{SS}

The idea behind this technique is to weaken the strength of pull-down (PD) transistors by raising the source voltage of PD NMOS transistors, thereby improving the writeability [42]. This extra ground voltage can be routed as a separated ground or can be generated internally using a regulator. However, the effect of raised V_{SS} saturates at a certain fraction of the power supply to the cell. This is because the access transistor has to pull the '0' storing internal nodes to write '1', but has limits up to $V_{DD} - V_{TH}$ due to the V_{TH} drop across NMOS access transistors. On the other hand, the V_{DD} lowering technique is free from such limitations.

3.4 SIMULATION SETUP AND MEASUREMENT TECHNIQUES

Simulations for analysis and comparison were conducted using HSPICE [43] with the 16 nm predictive technology model for the FinFET process [44]. The device parameters and simulation conditions for this analysis are given in Table 3.1. In FinFETs, the width of the device is defined by the height of the fin. The equivalent width of a FinFET is given as $W_{Eq} = n\ (2H_{fin} + T_{fin})$, where H_{fin} is the fin height, n is the number of fins and T_{fin} is the thickness/width of the fin. Both configurations, low standby power (LSTP) and HP of FinFET have been used for the simulation and analysis.

We simulated an SRAM cell, as shown in Figure 3.6a, with two different sizes. In one case, the transistors are sized such that the number of fins in PU, pass gate (PG) and PD devices are 1, 2 and 3, respectively, i.e. PU:PG:PD = 1:2:3. In another case, we used only one finger for all the transistors of the SRAM cell, i.e. PU:PG:PD = 1:1:1. For comprehensive analysis, we implemented the above two configurations of SRAM

TABLE 3.1
Device Parameters and Simulation Conditions

Parameter	Values
L_g	20 nm
T_{fin}	12 nm
H_{fin}	26 nm
V_{DD} variation	0.85 to 0.35 V
Temperature variation	25°C–100°C
SRAM bitline capacitance	50 fF

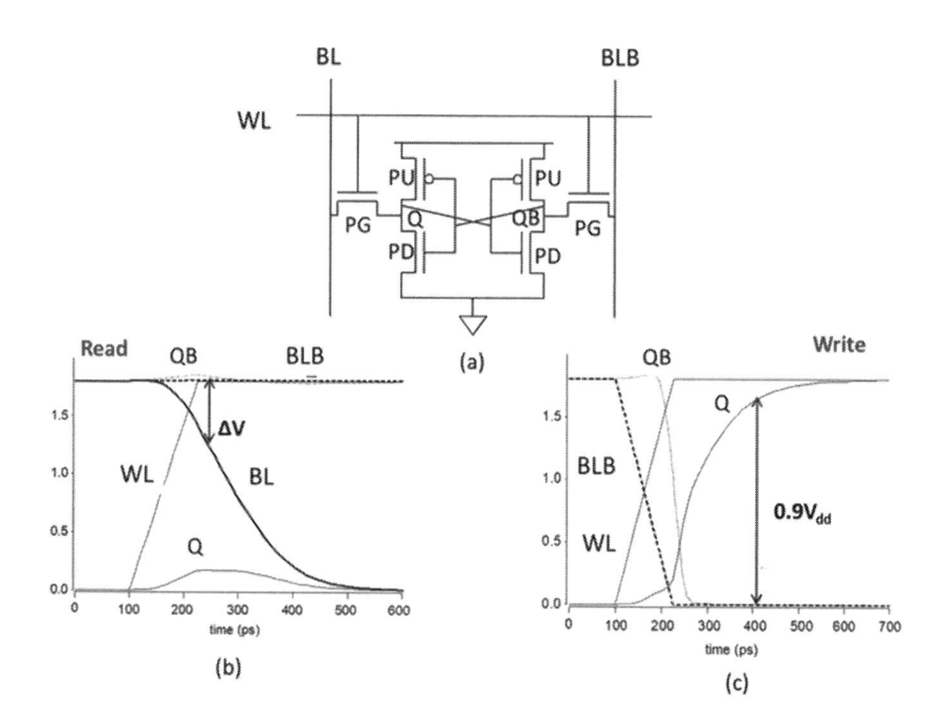

FIGURE 3.6 (a) Schematic of a 6T SRAM cell, (b) voltage transients for read operation (assuming the cell stores '0') and (c) voltage transients for write operation (assuming write '1').

cells with both the devices i.e. high V_{TH} low power (LSTP) and low V_{TH} HP FinFET devices [44].

3.4.1 SRAM Cell Operation

The 6T based SRAM cell that we implemented with FinFET devices is shown in Figure 3.6a. It comprises two cross-coupled inverters (PU1, PD1 and PU2, PD2) and two access transistors (PG1 and PG2). The inverter pair forms a simple latch circuit that stores information at its storage nodes Q and QB. Two bitlines, bit and bit_b, are connected to the storage nodes, Q and QB, through access transistors PG1 and PG2. These access transistors are controlled by WL. In the case of read operation, as shown in the Figure 3.6b, I_{READ} flows through PG1 and PD1 and the voltage level at node Q (storing '0') is slightly raised, and if this raised voltage is greater than the switching threshold of the right inverter (consisting of PU2 and PD2), then this may switch PD2 ON and the stored value gets altered. This is known as 'read upset'. To avoid this, PD transistors (PD1 and PD2) should be stronger than access transistors (PG1 and PG2). During a successful read operation, the bitline discharges and creates a voltage difference ΔV between the bitlines. Subsequently, this voltage difference is sensed by a sense amplifier and the read operation is completed. For a successful write operation, I_{WRITE} through PG2 should overcome the I_{PULLUP} of

PU2. This is achieved by making access transistors (PG1 and PG2) stronger than PU transistors (PU1 and PU2). The write operation is considered as complete when the charging node Q has reached 90% of V as shown in Figure 3.6c.

3.4.2 Measurement of Performance Metrics of an SRAM Cell

SRAM cells are assessed on the basis of various performance metrics. SRAM design for targeted applications uses trade-off among these metrics. In this section, various figures of merit (FoM) of SRAM cells along with their measurement methodologies are discussed.

3.4.2.1 Read Access Time

The read access time (read delay) is measured when a cell performs the read operation. It is estimated as the time difference between the activation of wordline and the discharge of Bitline/Bitline_bar (BL/BLB) by ΔV (nearly 50 mV) from its initial high value as shown in Figure 3.6b. This 50 mV voltage difference is sufficient for detection by a sense amplifier. Clearly, a speedy discharge of a bitline can be achieved by increasing the current of the discharge path. However, such improvements incur area penalty attributed to the increased size of transistors, which is not recommended for high density SRAMs.

3.4.2.2 Write Access Time

The write access time (write delay) is measured when a cell performs the write operation. It is estimated from the time when a wordline is activated to the time when the charging node Q or QB has reached 0.9 V_{DD} as shown in Figure 3.6c. Once the charging node has reached 90% of the supply voltage, the write operation is considered complete. Clearly, a faster charging of the storage node can be achieved by reducing the resistance in the charging path, i.e. by having strong access transistors and weak PD transistors. However, transistor size requirement for write operation has a conflict with the size required for read operation.

3.4.2.3 Hold Static Noise Margin (HSNM)

HSNM determines the stability of an SRAM cell in the standby mode. The HSNM is determined by the length of a side of the largest square that can be inscribed inside the smaller of two lobes as shown in Figure 3.7a. These two lobes are actually the voltage transfer characteristics of the two inverters of an SRAM cell.

3.4.2.4 Read Static Noise Margin (RSNM)

The read stability of an SRAM cell is defined in terms of read SNM (RSNM). The RSNM is graphically measured as the length of a side of the largest square that can be inscribed inside the smaller lobe of the butterfly curve [45]. The RSNM is measured following the same technique as discussed for HSNM, except that the WL is kept at V_{DD} for tracing the butterfly curve. It has been discussed earlier that during the read operation, the I_{READ} raises the voltage level of the '0' storing node. Therefore, the Voltage Transfer Characteristics (VTC) of the inverter does not reach ground, rather shifts upward, thereby reducing the size of the butterfly lobes

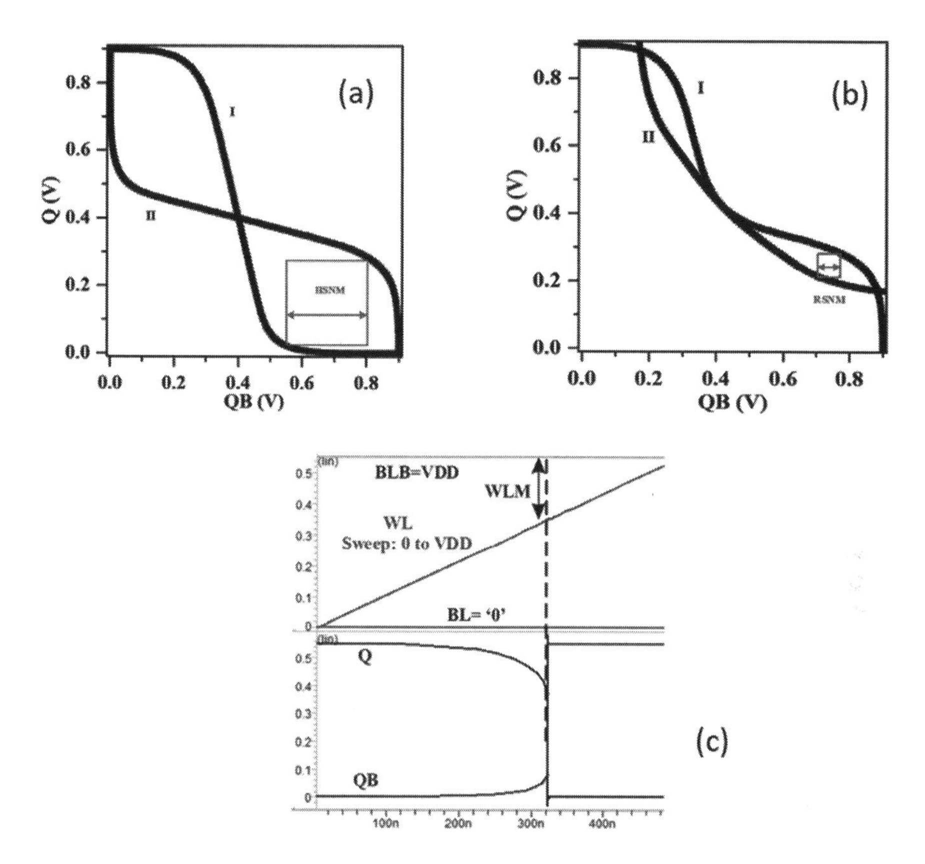

FIGURE 3.7 Butterfly curve used for quantifying (a) HSNM, (b) RSNM and (c) voltage transients depicting the write margin measurement for an SRAM cell.

(shown in Figure 3.7b). Hence, RSNM is reduced. RSNM is, therefore, more critical than HSNM and dictates the stability of an SRAM cell.

3.4.2.5 Write Margin (WM)

Wordline margin (WLM) [46] is a very useful and easy to measure writeability metric. Figure 3.7c illustrates the signal waveform for WM measurement for the write '0' case. The data to be written are placed on bitlines, and then the WL is swept from '0' to V_{DD}. The WM is defined as the difference between V_{DD} and WL voltages at which the nodes Q and QB flip. The lower the value of WM is, the harder it is to write data into the cell, implying a poor WM. The WLM method replicates a real write operation since data are placed on bitlines and then the WL is enabled during SRAM write.

3.4.2.6 Read Current to Leakage Current Ratio (I_{READ}/I_{LEAK} or I_{ON}/I_{OFF})

I_{READ}/I_{LEAK} is one of the important FoM of an SRAM cell as it dictates the maximum number of cells that can be connected to a bitline in an array. Leakage current decreases the SRAM density by reducing the number of cells that can be connected to a column. Therefore, a high I_{ON}/I_{OFF} ratio is required to acquire high density of SRAM. In the case of read operation, a BL is discharged by I_{READ} of the selected cell. Assuming that the BLB (Q storing '0') is at V_{DD}, i.e. not discharged, the combined leakage current may also discharge the BLB if the number of cells in the column is too large and may cause false read. Even if this does not happen, at least read delay increases.

3.4.2.7 Power Consumption

To minimize the overall power consumption of an SRAM cell, identification of such components is a critical issue. The power consumption of an SRAM cell comprises two major components: static or leakage power and dynamic power. Static power is the power that is drawn from the supply by the cell when it is in the steady state. Unlike nonvolatile memory devices, SRAM is a volatile memory and demands that the power supply be ON for retaining the data. Though the power that a cell consumes for retaining the data is relatively small, in an array, there are a large number of cells and the total static power consumption becomes significant. The dynamic power consumption during read/write operation of an SRAM cell can be measured by adding up the dynamic power drawn by different capacitive loads that are charged and discharged during the read and write operations. Clearly, the total dynamic power consumption is mainly dominated by the long interconnects which impose a large capacitive load to the signal paths in the SRAM unit. Better sense amplifiers with a smaller sense margin can be used to achieve lower power consumption.

3.5 SIMULATION-BASED ANALYSIS AND DISCUSSION

3.5.1 Simulation of a Conventional 6T SRAM Cell

First of all, we simulated a 6T SRAM cell and measured various performance metrics. Two transistor sizes of SRAM cells and two FinFET device types, i.e. low V_{TH} or HP and high V_{TH} or LSTP, result in four configurations of SRAM cells, namely LSTP with PU = PG = PD = 1 fin; LSTP with PU = 1 fin, PG = 2 fins and PD = 3 fins; HP with PU = PG = PD = 1 fin and HP with PU = 1 fin, PG = 2 fins and PD = 3 fins. Figure 3.8a shows the plot of the HSNM of these four SRAM configurations. We observe that LSTP configurations offer better HSNM. This is attributed to better VTC of inverters implemented using high V_{TH} transistors.

Figure 3.8b shows the plot of the read SNM of the SRAM cell configurations. It can be observed that the voltage scaling results in a sharp reduction in RSNM, which poses stability problem at reduced supply voltages. Figure 3.8c shows the plot of the write margin of SRAM cell configurations with changing supply voltage. It can be observed that LSTP configurations offer a poor WM due to a reduced drive strength of access transistors. The HP configuration with PU = 1 fin, PG = 2 fins and PD = 3 fins offers best WM as the access transistors are much stronger than the PU transistors.

Figure 3.9a shows the plot of the read delay of SRAM cell configurations. Supply voltage scaling deteriorates the read performance. Among the four configurations, LSTP configurations deteriorate fast and the LSTP with only 1 fin for PU,

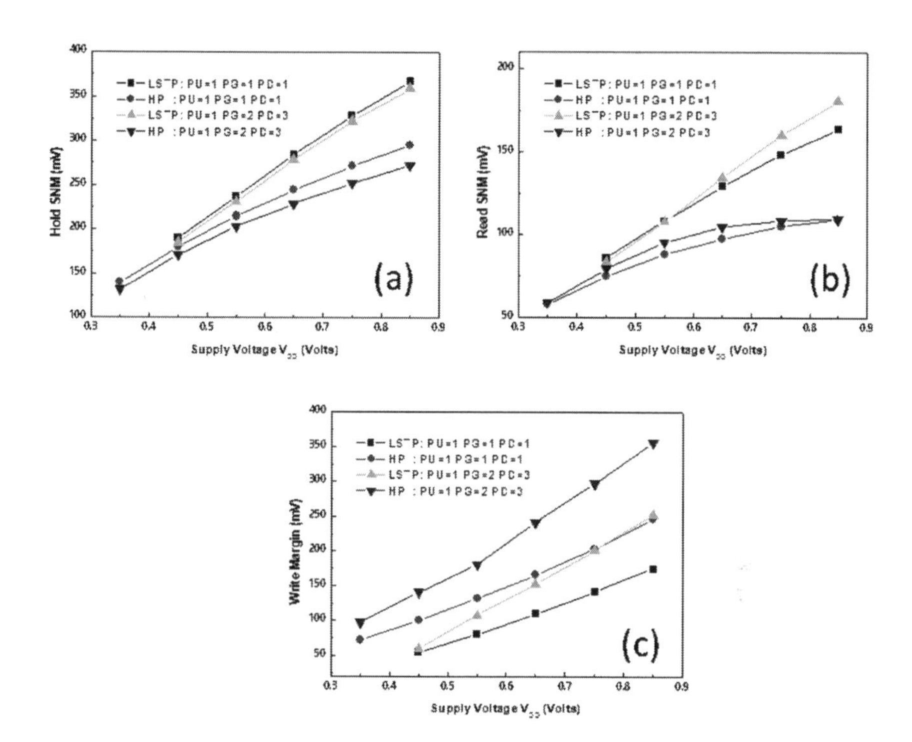

FIGURE 3.8 (a) Hold SNM, (b) read SNM and (c) write margin of SRAM cells vs supply voltage.

PG and PD transistors performs worst due to the reduction of read current through PG and PD transistors. For this configuration, scaling beyond 0.45 V is not possible due to the unacceptably high read delay. Figure 3.9b shows the plot of the write delay of SRAM cell configurations. It can be observed that the write delay of HP configurations increases gradually with voltage scaling. However, the write delay of LSTP configurations increases sharply with supply voltage scaling. This is because reduction in voltage brings LSTP devices to the near-threshold operation region, where the write current is significantly reduced. Figure 3.9c shows the plot of the leakage power of SRAM cell configurations. We observe that the leakage powers of LSTP configurations are of the order of pico watts, which is three orders smaller than the leakage powers of HP configurations that are of the order of nano watts. Among these configurations, the HP configuration with PU = 1 fin, PG = 2 fins and PD = 3 fins consumes the maximum leakage power per bit.

3.5.2 Simulation of Read Buffers

Subsequently, we simulated and analyzed the performance of read buffers implemented using LSTP and HP configurations of FinFETs. Figure 3.10a shows the plot of the LSTP configuration of read buffers with varying supply voltage. It can be observed that Read Buffer-F offers minimum delay due to a smaller number of transistors in

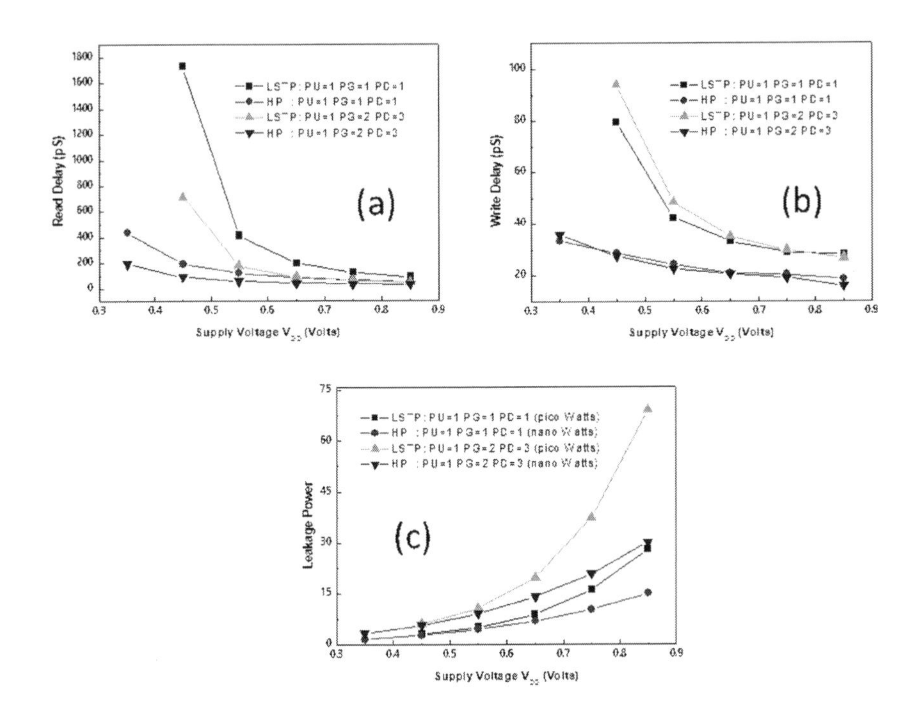

FIGURE 3.9 (a) Read delay, (b) write delay and (c) leakage power of SRAM cells vs supply voltage.

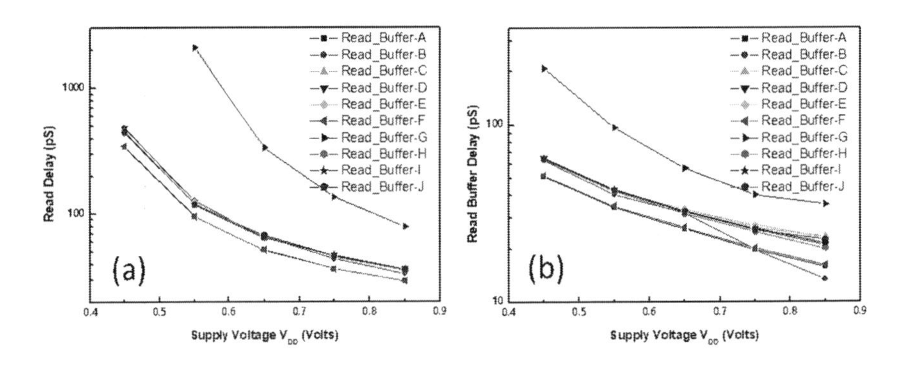

FIGURE 3.10 (a) LSTP read buffer delay and (b) HP read buffer delay vs supply voltage.

the read path. On the other hand, Read Buffer-G incurs maximum delay due to the use of p-type transistors in the read path. This trend more or less remains the same for HP configurations of read buffers, as shown in Figure 3.10b. However, the overall delay of HP configurations is much smaller than that of the LSTP configurations. This is attributed to the larger read current in low-V_{TH} transistors.

Figure 3.11a and b show the plot of the read powers of the two configurations of read buffers. It can be observed that the LSTP configurations offer a smaller read power than HP configurations; however, it is at the cost of higher read delay.

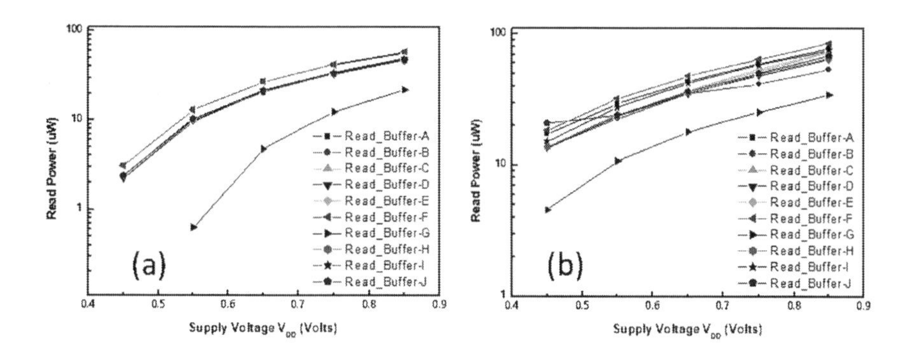

FIGURE 3.11 (a) LSTP buffer read power and (b) HP buffer read power vs supply voltage.

3.5.3 Simulation of Write Assist Techniques

Finally, we analyzed the performance of various write assist techniques. Figure 3.12a shows the plot of the write delay of an LSTP SRAM cell. It is observed that a 6T cell with an Negative Bitline (NBL) assist offers the minimum write delay, whereas the V_{DD} lowering technique shows maximum write delay, even larger than that of a 6T cell without any assist. However, the smallest write power compared to other write assists makes V_{DD} lowering a perfect choice for low-power applications as shown in Figure 3.12b. It can be observed from Figure 3.12b that the write power reduces for

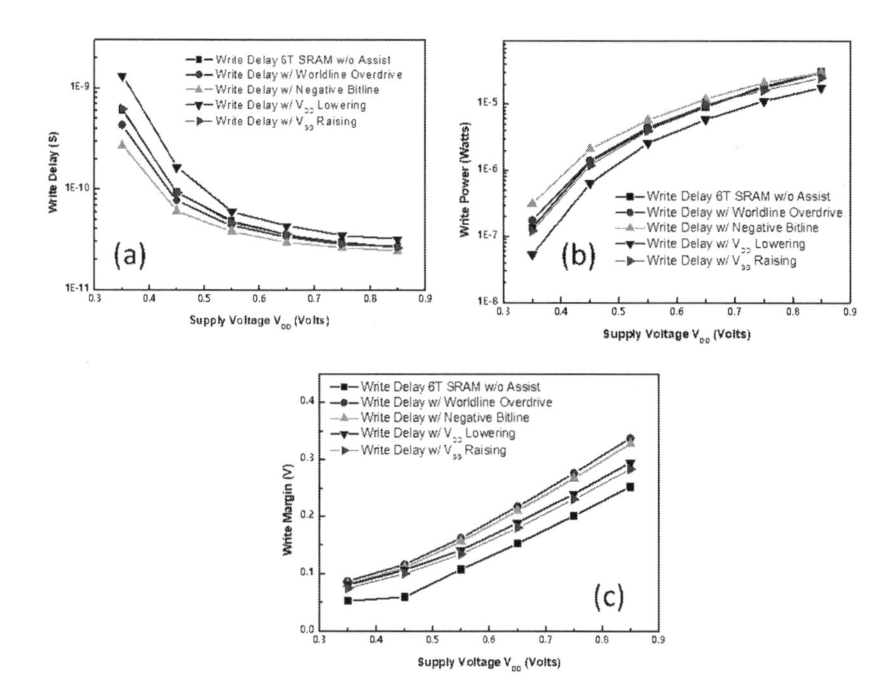

FIGURE 3.12 (a) Write delay, (b) write power and (c) write margin of an LSTP SRAM cell with assist techniques.

all write assist techniques, except for the WL overdrive technique, for which the write power is comparable to that of a 6T cell without any assist. It can also be concluded that LSTP SRAM consumes low write power, however, write delay is penalized. Figure 3.12c shows the comparison of the WMs of various write assist techniques for LSTP SRAM. It can be observed that the WM significantly improves for all the write assist techniques. The V_{SS} raising, V_{DD} lowering, negative bitline and WL overdrive assist techniques show an improvement of 17%, 20%, 32% and 35% (at $V_{DD} = 0.55$ V), respectively, in the WM compared to the 6T cell without any write assist.

3.6 CONCLUSION

In this chapter, various low-power design considerations for SRAM have been thoroughly studied. It is observed that supply voltage scaling is very effective in containing power dissipation and achieving ULP operation of SRAM cells. However, design of robust low-power SRAM cells becomes very challenging with technology scaling down to the deep nanometer range. Challenges arise due to a significant increase in read, write and data retention failures. Nowadays, read and write assist techniques are becoming an integral part of SRAM design to have robust operation with improved performance at scaled supply voltages. In this chapter, we studied various write and read assist techniques and their impact on the writeability, readability and stability of SRAM cells. We presented simulation-based analysis of various assist techniques and their toll on the performance metrics of SRAM cells. Finally, we conclude that by using these assist circuits, WM and read margin of SRAM cells are improved even at scaled supply voltages. Therefore, employing assist techniques facilitates the design of low-power memory for IoT-enabled systems.

REFERENCES

1. Y. Liu et al., "Data backup optimization for nonvolatile SRAM in energy harvesting sensor nodes," *IEEE Transactions on Computer-Aided Design of Integrated Circuits and Systems*, vol. 36, no. 10, pp. 1660–1673, 2017.
2. H. N. Patel et al., "Optimizing SRAM bitcell reliability and energy for IoT applications," *17th International Symposium on Quality Electronic Design (ISQED)*, pp. 12–17, 2016.
3. Y. Lee et al., "Circuit and system design guidelines for ultra-low power sensor nodes," *IPSJ Transactions on System LSI Design Methodology*, vol. 6, no. 2, pp. 17–26, 2013.
4. J. P. Kulkarni et al., "*Energy Efficient Volatile Memory Circuits for the IoT Era*," In Enabling the Internet of Things, Switzerland: Springer International Publishing, pp. 149–170, 2017.
5. I. Lee, K. Lee, "The Internet of Things (IoT): Applications, investments, and challenges for enterprises," *Business Horizons*, vol. 58, no. 4, pp. 431–440, 2015.
6. M. Belleville et al., "Designing digital circuits with nano-scale devices: Challenges and Opportunities," *Solid-State Electronics*, vol. 84, pp. 38–45, 2013.
7. S. Kurude et al., "Statistical variability analysis of SRAM cell for emerging transistor technologies," *IEEE Transactions on Electron Devices*, vol. 63, pp. 3514–3520, 2016.
8. H. Pilo et al. "A 64Mb SRAM in 22nm SOI technology featuring fine-granularity power gating and low-energy power-supply-partition techniques for 37% leakage reduction," *IEEE International Solid-State Circuits Conference Digest of Technical Papers*, pp. 322–323, 2013.

9. S. Lutkemeier et al. "A 65 nm 32 b subthreshold processor with 9T multi-Vt SRAM and adaptive supply voltage control," *IEEE Journal of Solid-State Circuits*, vol. 48, no. 1, pp. 8–19, 2012.
10. C. T. Chuang et al. "High-performance SRAM in nanoscale CMOS: Design challenges and techniques," *IEEE International Workshop on Memory Technology, Design and Testing*, pp. 4–12, 2007.
11. S. Ramey et al., "Intrinsic transistor reliability improvements from 22 nm tri-gate technology," *IEEE International Reliability Physics Symposium (IRPS)*, pp. 1–6, 2013.
12. H. J. Lee et al. "Intel 22nm FinFET (22FFL) process technology for RF and mm wave applications and circuit design optimization for FinFET technology," *International Electron Devices Meeting (IEDM)*, pp. 14-1, 2018.
13. M. Poljak et al., "Technological constrains of bulk FinFET structure in comparison with SOI FinFET," *International Semiconductor Device Research Symposium*, IEEE, pp. 1–2, 2007.
14. X. Wang et al., "Statistical variability and reliability in nanoscale FinFETs," *International Electron Devices Meeting*, IEEE, pp. 5-4, 2011.
15. R. V. Joshi et al. "FinFET SRAM for high-performance low-power applications," *Proceedings of the 30th European Solid-State Circuits Conference*, IEEE, pp. 69–72, 2004.
16. T. Hiramoto et al., "Ultra-low power and ultra-low voltage devices and circuits for IoT applications," *IEEE Silicon Nanoelectronics Workshop (SNW)*, pp. 146–147, 2016.
17. G. Chen et al., "Yield-driven near-threshold SRAM design," *IEEE Transactions on Very Large Scale Integration (VLSI) Systems*, vol. 18, no.11, pp. 1590–1598, 2009.
18. E. Morifuji et al., "Power optimization for SRAM and its scaling," *IEEE Transactions on Electron Devices*, vol. 54, no. 4, pp. 715–722, 2007.
19. B.-Y. Nguen et al., "A path to energy efficiency and reliability for ICs," *IEEE Solid-State Circuits Magazine*, pp. 24–33, Fall 2018.
20. R. Gonzalez et al., "Supply and threshold voltage scaling for low power CMOS," *IEEE Journal of Solid-State Circuits*, vol. 32, no. 8, pp. 1210–1216, 1997.
21. K. Itoh et al., "Low-voltage scaling limitations for nano-scale CMOS LSIs," *Solid-State Electronics*, vol. 53, no. 4, pp. 402–410, 2009.
22. S. Vangal et al., "Near-threshold voltage design techniques for heterogenous manycore system-on-chip," *Journal of Low Power Electronics and Applications*, vol. 10, no. 2, pp. 1–23, 2020.
23. S. Ahmad et al., "Single-ended schmitt-trigger based robust low power SRAM cell," *IEEE Transactions on VLSI Systems*, vol. 24, no. 8, pp. 2634–2642, 2016.
24. K. Cheng et al., "ETSOI CMOS for system-on-chip applications featuring 22nm gate length, sub-100nm gate pitch, and 0.08μm2 SRAM cell," *Symposium on VLSI Technology-Digest of Technical Papers*, IEEE, pp. 128–129, 2011.
25. S. Ahmad et al., "Low leakage single bitline 9T (SB9T) static random access memory," *Microelectronics Journal*, vol. 62, pp. 1–11, 2017.
26. S. Ahmad et al., "Pseudo differential multi-cell upset immune robust SRAM cell for ultra-low power applications," *International Journal of Electronics and Communications*, vol. 83, pp. 366–375, 2018.
27. V. Chandra et al., "On the efficacy of write-assist techniques in low voltage nanoscale SRAMs," *Design, Automation & Test in Europe Conference & Exhibition*, IEEE, pp. 345–350, 2010.
28. B. Zimmer et al., "SRAM assist techniques for operation in a wide voltage range in 28-nm CMOS," *IEEE Transactions on Circuits and Systems II: Express Briefs*, vol. 59, no. 12, pp. 853–857, 2012.
29. L. Chang et al., "Stable SRAM cell design for the 32 nm node and beyond," *Digest of Technical Papers-Symposium on VLSI Technology*, pp. 128–129, 2005.

30. B. H. Calhoun and A. Chandrakasan, "A 256kb Sub-threshold SRAM in 65nm CMOS," *IEEE International Solid State Circuits Conference-Digest of Technical Papers*, pp. 2592–2601, 2006.
31. T. Kim et al., "A 0.2V, 480kb subthreshold SRAM with 1k cells per bitline for ultra-low-voltage computing," *IEEE Journal of Solid-State Circuits*, vol. 43, pp. 518–529, 2008.
32. S. Lin et al., "Design and analysis of a 32nm PVT tolerant CMOS SRAM cell for low leakage and high stability," *Integration, the VLSI Journal*, vol. 43, pp. 176–187, 2010.
33. G. Pasandi and S. M. Fakhraie, "A 256-kb 9T near-threshold SRAM with 1k cells per bitline and enhanced write and read operations," *IEEE Transactions on Very Large Scale Integration (VLSI) Systems*, vol. 23, pp. 2438–2446, 2015.
34. B. Wang et al., "Design of an ultra-low voltage 9T SRAM with equalized bitline leakage and CAM-assisted energy efficiency improvement," *IEEE Transactions on Circuits and Systems I: Regular Papers*, vol. 62, pp. 441–448, 2015.
35. J. Cai et al., "A PMOS read-port 8T SRAM cell with optimized leakage power and enhanced performance," *IEICE Electronics Express*, vol. 14, no. 3, pp. 20161188–20161188, 2017.
36. S. Gupta et al., "Low-power near-threshold 10T SRAM bit cells with enhanced data-independent read port leakage for array augmentation in 32-nm CMOS," *IEEE Transactions on Circuits and Systems I: Regular Papers*, vol. 66, pp. 978–988, 2019.
37. F. B. Yahya et al., "Combined SRAM read/write assist techniques for near/sub-threshold voltage operation," *6th Asia Symposium on Quality Electronic Design (ASQED)*, IEEE, pp. 1–6, 2015.
38. C. C. Wang et al., "A boosted wordline voltage generator for low-voltage memories," *10th IEEE International Conference on Electronics, Circuits and Systems (ICECS)*, vol. 2, pp. 806–809, 2003.
39. M. Iijima et al., "Low power SRAM with boost driver generating pulsed word line voltage for sub-1V operation," *JCP*, vol. 3, no. 5, pp. 34–40, 2008.
40. N. Shibata et al., "A 0.5-V 25-MHz 1-mW 256-kb MTCMOS/SOI SRAM for solar-power-operated portable personal digital equipment-sure write operation by using step-down negatively overdriven bitline scheme," *IEEE Journal of Solid-State Circuits*, vol. 41, no.3, pp. 728–742, 2006.
41. K. Zhang et al. "A 3-GHz 70-Mb SRAM in 65-nm CMOS technology with integrated column-based dynamic power supply," *IEEE Journal of Solid-State Circuits*, vol. 41, no.1, pp. 146–151, 2005.
42. H. Pilo et al., "A 64 Mb SRAM in 32nm high-k metal-gate SOI technology with 0.7V operation enabled by stability, write-ability and read-ability enhancements," *IEEE Journal of Solid-State Circuits*, vol. 47, no.1, pp. 97–106, 2011.
43. Synopsys, HSPICE® Reference Manual: Commands and Control Options. E-2010.12. Copyright Notice and Proprietary Information. Copyright © 2010 Synopsys, Inc.
44. Predictive Technology Model, available online at http://ptm.asu.edu/. Accessed in April 2020.
45. E. Seevinck et al., "Static-noise margin analysis of MOS SRAM cells," *IEEE Journal of Solid-State Circuits*, vol. 22, pp. 748–754, 1987.
46. H. Qiu et al., "Comparison and statistical analysis of four write stability metrics in bulk CMOS SRAM cells," *Japanese Journal of Applied Physics*, vol. 09, no. 04DC09, pp. 1–4, 2015.

4 Low-Power Memory Design for IoT-Enabled Systems
Part 2

Shilpi Birla and Neha Singh
Manipal University Jaipur

N. K. Shukla
King Khalid University

CONTENTS

4.1 INTRODUCTION

The quick advancement of the Internet of Things (IoT) has led to various challenges and opportunities for researchers to innovate and design memories that are low cost, small size, and have efficient power for smart handheld devices, as the basic requirement for these gadgets is memory.

From conventional RAM-based and flash memory to further developed, chip-based memory arrangements, there are a lot of alternatives to explore. Because of the remarkable development of IoT gadgets around the world, an advancement in new innovation has been initiated, with a specific focus on memory. This chapter discusses and focuses on the design and role of different memories in IoT applications.

4.2 BACKGROUND

IoT has become one of the most fascinating and captivating words of this century. The word IoT is an acronym for Internet of Things. It can be described as the juncture of three things, i.e., internet, things, and information. IoT can be summed as the idea of connecting a device to the internet and to other connected devices, and the connecting devices must have an on/off switch. The other definition of IoT is as follows: "Sensors and actuators installed in physical items are connected through wired and remote systems". Basically, IoT is a huge network connecting things and people. The common thing is that all of them collect and share data. All of them gather and offer information about the environment and the way in which they are utilized. IoT is a mere perception of connecting any device with an on and off switch to the internet or with each other. We can see many applications of IoT in our day to day life such as cellphones, coffee makers, washing cars, headphones, lamps, wearable devices and so on. As per business and specialized technical experts, the applications of IoT will be tremendous.

IoT is a measure that lets devices to exchange data with other devices. Thus, overall the key role is of data which keep on changing from one point of connection to the other. Thus, in IoT networks, data will be transported at all times. The conclusion is that it is an overall system that associates different items anytime and anywhere through the web. The fundamental objective of IoT is to screen and control things from anyplace on the planet. The control or the administration of IoT systems can be done by a machine or by anybody, such as home automation, which can be done using a mobile phone. Security is the major concern in using this system as the major work is data transmission, which can be accessed easily by hackers if security is not strong.

The internet is already a vast section of the IoT, began as a feature of DARPA (Defense Advanced Research Projects Agency) in 1962 and developed into ARPANET in 1969. Numerous business specialists started supporting open utilization of ARPANET, permitting it to develop into our advanced internet during the 1980s. In 1993, the Global Positioning System (GPS) became a reality with satellites and landlines offering the essential infrastructure to a great part of the IoT [1].

To enhance the growth of RFID, the term Internet of Things was invented in 1999. Though this concept was with us long before 1970, the word "Internet of Things" was gifted to us by Kevin Ashton who was working with Procter & Gamble in 1999. The word has evolved, since in 1999 internet itself was a sizzling word. This is how the IoT started its journey.

The overall spending on IoT till 2019 was around $745 billion, an expansion of 15.4% from the $646 billion spent in 2018, as indicated by another update to the

International Data Corporation (IDC) Worldwide Semiannual Internet of Things Spending Guide in January 2019. The IDC expects that the overall IoT spending will have a double-digit yearly development rate all through 2017–2022 and outperform the $1 trillion imprint in 2022. So we can see this graph of growth and can analyze how it is going to dominate us in every application of life whether it is transportation, medical, automation or others. [2].

4.3 IoT ARCHITECTURE

The IoT architecture was started with three layers, which include the application, network, and perception layer, but the present architecture of IoT has five layers as shown in Figure 4.1.

Let us review the role of each layer of the IoT architecture. The first layer which is the application layer is used in the three-layer architecture also. It is also the top layer of the three-layer architecture. This layer offers customized assistance depending on the client's needs. The primary obligation of this layer is to interface the wide gap between clients and applications, which include many applications such as health monitoring, disaster management, transportation, and other automated smart applications. The network layer works like the brain of the system, and its main job is secure transmission of data between the application and the perception layer [3]. The lowest layer of the three-layer IoT architecture is the perception layer, which is responsible for gathering information and then transmitting it to another environment such as real time applications, WSN, etc.

The object layer is the lowest layer in the five-layer architecture, whose job is to gather information/data from various devices and then digitize them after processing, and it also transmits the data processed by it to other upper layers. After the object layer, the object abstraction layer comes, which is similar to the network layer and acts as a facilitator between the object layer and the upper service management layer. It assists in processing information and controlling other tasks. The task of the application layer is to provide customers with high-end features according to their requirements. So this concludes the brief description about the IoT architecture.

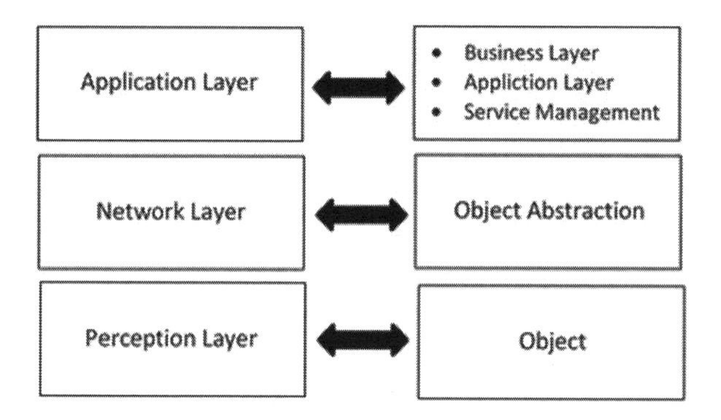

FIGURE 4.1 IoT architecture with five layers.

4.4 ROLE OF MEMORIES IN IoT

In the above section we have discussed about the IoT and its architecture. In this section, we will discuss about intelligent IoT and its memory technology. A prediction was made by Gartner that by 2020 there will be almost 26 billion IoT gadgets, and in another estimation given by ABI Research, 30 billion gadgets will be present by the same time. This indicates a substantial open door for any organization that turns into an innovator in the IoT field, one that can comprehend and use progress in the availability and system data transfer capacity to create innovative gadgets and new applications.

Generally, human information is not required in every IoT system and only in particular examples it is required, otherwise most of them depend on M2M communication, i.e.., machine to machine. The point here is that even when human mediation is not required, information is continuously gathered and transmitted to other servers. Just as internet has changed our lives and its uses are endless, IoT has changed our lives and the applications are evolving every day starting from smart cities to home automation, transportation, medical aids, agriculture, and industries, and in almost every sector of our day to day life, IoT has entered, which is why people started calling it intelligent IoT. A system that can transmit, collect, and analyze data with the ability to intrude from the analysis is IIoT [3]. Now the issue is that a huge collection of data in IoT systems requires storage, which becomes one of the key components in IoT devices. This storage in IoT is nothing but the memory, which can be of any type such as volatile or non-volatile memory (NVM) depending on the application.

A quickly advancing IoT application environment additionally implies fast developing memory supplies. The demand for storing data is increasing. This increase in demand for memories allows researchers to think about various aspects to improve the performance, such that they can be fit into an IoT system. Various different approaches for designing efficient memory for small sized IoT devices have been proposed. There are various factors such as fast memories for data sharing, small sized memories that can be embedded into a system, and the most important is low-power memory. So advancement in memory is also one of the important factors in developing new IoT devices.

As the demand for IoT is increasing, the IoT processors are required to be produced in bulk and at a low cost, and so the memories need to be manufactured in volume. The demand for IoT applications directly raised the demand for memories. Conventional embedded memories such as flash memory and EEPROM are already in use, but new configurations have also been developed to meet the requirements of IoT applications as there is a requirement for flexible circuits for wearable devices. To reduce the cost of memories, researchers are focusing on other non-traditional memories such as ferroelectric RAM, and charge trapping- and resistance-based memories which are compatible with the CMOS logic family. The required features of these memories are speed, low-power operation, and low cost.

IIoT is a word which is used for smart systems having a smart processing network that can quickly and efficiently analyze data along with the response according to the conclusions, and so the IIoT requires a storage that is fast, accurate, low power, low cost, and small in size. Since they are required to store data which a processor will analyze further, these memories can be volatile or non-volatile as per the requirement of the applications and the systems. In the next section, we will discuss various memories which are the possible solutions for IIoT processors.

4.5 MEMORIES FOR IoT-ENABLED SYSTEMS

The memory hierarchy consists of various levels; the lower level consists of mass storage, which consists of the flash and HDD memory, the next is the storage class memory, and then come the DRAM, SRAM and registers. The access time will be slow towards the bottom of the memory hierarchy.

For example, if a processor runs at 1 ns, the speed of the SRAM will be 10 ns and the DRAM memory will try to run at 100 ns. Macronix, NTHU, NCTU, and IBM gave phase change memory (PCM), which was considered to be a possible storage class memory, but with high latency [4]. Hard drive storage is now increased up to 2 Tb, which is provided by vertically stacked chips.

There are different levels of cache where embedded SRAM is generally used in large servers because of its high speed. Low operating SRAMs are also significant. Magnetic RAM is also another NVM with fast switching and can be used as the last level cache. Since in standby there is no power consumption, it makes possible the reduction of overall power compared to SRAM, when configured as a "4-way SET Associative cache" with a 362 K-bit SRAM TAG [5]. This cache minimized the total energy by 93% compared to a conventional SRAM LLC, and the standby energy also reduced by 77%.

If we see the memory hierarchy, the main memory is below the cache, with the majority having high density DRAM. DRAM as we know is slower than SRAM but faster than other storage such as high-density flash memory which is non-volatile. DRAM chips have a high capacity e.g. 64 Gb. Samsung made high bandwidth memory with 8 GB capacity [6]. This used the "GDDR5 high bandwidth memory (HBM)" interface. This high bandwidth memory has 160 GB/s memory bandwidth, but consumes up to 70% less energy per bit with respect to existing technologies.

4.5.1 SRAM IN IoT SYSTEMS

The basic IoT architecture has various sub-systems such as applications, gateways, processors, and sensors. These sub-systems require low power and high speed memories such as SRAM and ROM. The applications of IoT define the type of memory, i.e., either high speed or low power. Since low-power devices are more in demand, ultra-low-power SRAMs are required for portable and handheld devices. At the same time, the performance of SRAM should not be degraded. Wearable devices are nowadays the latest trend, so these devices require ultra-low-power SRAMs. The need for memory in IoT systems depends on the application. For example, in applications where huge data storage and handling is required, the memory requirement switches to DRAM and flash memories. The applications that require high data transfer speed need high speed SRAM memory. High speed data transfer is vital for communication among IoT devices. So, SRAM is chosen as a cache memory because of its quicker response.

Integration of standalone SRAMs into processors was the decision of Intel around the mid-1990s, which entirely changed the market for SRAMs. The external SRAMs consist of either 4T or 6T per cell, which is difficult to minimize with respect to other memories. The IoT processors are small as the process technology has been scaled from 180 to 10 nm, which has led to the miniaturization of processors and the same is required for SRAMs also [7]. So people thought that the end of SRAM is near, as already for IoT processors low power, low cost, and scaled memories are required. So, changes are also required in SRAM. Already SRAMs consume 70% of the power of a chip, but the power required for IoT systems is very less, so there are many challenges such as cost and power with regard to SRAM for IoT. So, embedding SRAM into a processor was one of the feasible solutions to save SRAM as it has its own potential advantages.

Since access time is one of the key parameters for all the cache memories, embedded SRAMs perform much better in terms of access time. As all the processors have enhanced their performance in terms of size, area, and power, the memories used in these processes should also be enhanced and improved. But the limitation with SRAMs is their size as mentioned, that is, minimum 4T or 6T transistors are required so high-density cache memories require typically a larger area than smaller processors. Though the process technology for CMOS has been scaled from 130 to 10 nm, it has also increased many other factors such as soft error rates. Also, various defects have arisen due to process variations in nanotechnologies. Power issue is also one of the dominant factors, as scaling of the technology increases the leakage current, which results in an increment of the overall standby power consumption, which is not favored in high-end processors. So these are the serious concerns about embedding SRAMs.

In the present day scenario, there is a lot of talk about handheld devices, and recently wearable devices are in huge demand. Also we can observe that in such devices, size and power are the two main attractive USPs of the products. The microcontrollers (MCUs) present in these devices are of very small size and run on low power supply. In today's world, MCUs are available in a wide range of devices. Moreover, every day, new features are being added to these wearable devices which need more memory and it is possible that the on-chip memory may fall short leading to the requirement of external memory, so SRAM can be the best option as an external memory. The reason for this is that the improved SRAM has low power consumption compared to DRAM and other flash memories, and the other reason is low access time [7].

In the last few years, fast and low-power circuits are in demand and so is the SRAM memory. The SRAMs are used for two different types of applications: one where speed is required and the other where low power consumption is required because achieving both is difficult. People require high performance and low-power devices for complex functionalities. Such types of applications are medical devices, handheld devices, wearable devices, communication systems, and all IoT applications. The thrust areas of SRAMs are as follows: smaller size, lower pin count, high density, soft error correction, and lastly high performance and low power. Presently, many of the SRAMs have a parallel interface. The serial SRAMs available in the market have low density. The present need is to produce high-density serial SRAMs. Some of the advanced features of SRAMs which make them compatible for IoT are as follows: (i) Chip scale packaging: it is one of the prevailing techniques, which reduces the size of the chip. The chip scale must have an area not more than one and half times the area of the packaged die and the linear dimensions should not be more than 1.2 times the dimensions of the die, whereas in the standard package die, the chip area can be 10 times more than the die area. This reduction is possible by using scaled technology and also by removing the first level packaging, which includes the lead frame, wire bonds, etc. The pinout used is similar to the Ball grid Array (BGA). Many companies have started offering CSO SRAM, yet it is not in bulk production as it is mostly required in specific applications such as wearable devices. (ii) Lower pin count: a parallel interface is one of the key problems in memory expansion compared to other memories, although it consumes less power than other memories. The use of a parallel interface will increase the speed, i.e. fast read and write, but the problem is that many input/output pins are required for interfacing, which is a problem in small sized devices such as wearable devices, so the interfacing should be changed to make it compatible. Since serial memory is a big success, SRAM with serial interfacing requires fewer pins which in turn results in a simpler interfacing. The other advantage is that as density increases, the number of pin count does not increase, and so people are working on improving serial SRAMs. As the requirement for new add-on-features is increasing day by day, the density of SRAMs will also be high in the near future, so in conclusion, CSP and serial interfacing are some of the possible solutions which would really make SRAMs a prominent memory for IoT systems [7].

The design aspects for designing any SRAM are the same: The leakage power should be minimized, the supply voltage should be low for low-power applications, and high stability, which includes the read, write, and hold stability. The cell should be of minimum area. Though the CMOS technology is useful in designing SRAM memories efficiently and also compatible in reducing the size, the limitation is process variation, which increases the leakage power and affects the stability. So many researchers have started proposing other devices such as FinFET and TFET, which can remove the limitations of CMOS SRAM memories.

So many SRAM types have been researched, such as in Ref. [8], an 8T SRAM bit cell has been proposed with low leakage and low energy, and the chip has been tested at 130nm. In Ref. [9], a 9T TFET SRAM bit cell is proposed for IoT sensor nodes, and this cell removes the half-select disturb issue, which makes it suitable for the bit-interleaving architecture. In Ref. [10], a 10T SRAM cell is proposed, which is fabricated in 28nm CMOS with minimum standby for IoT applications. The authors in Ref. [11] proposed a 10T SRAM which is an ultra-low-power cell particularly designed for IoT applications, and this cell works at very low voltages with high stability in terms of noise margin. They have used a power gated transistor, which is a PMOS that reduces the leakage power, which makes it suitable for all IoT applications. A 10T SRAM cell is proposed in Ref. [12] at 28nm for low-voltage SRAM for IoT applications, and they have used the mixed Vth technique for reducing the power and improving the stability. The new device trends in Ref. [13] proposed an ultra-low-leakage and ultra-low-power SRAM cell for IoT applications.

A TFET based on the tunneling mechanism-based SRAM is proposed in Ref. [14] for giving an option beyond CMOS. It is a reliable and effective memory circuit due to its key sensing and data processing unit and can be used for IoT applications. The 7 nm CMOS technology for mobile applications is proposed in Ref. [15] for the first time. In this fourth generation, FinFET transistors are optimized and used to achieve the low power and high performance design requirements. The multi-threshold technique is also used. A new 3T TFET SRAM cell is used for ultra-low-power applications such as IoT [16]. Voltage scaling has been used to allow low-power operation without reducing the data stability. The authors in Ref. [17] proposed a transmission gate-based low-power 9T SRAM for IoT applications in 16 nm CMOS technology.

Since high speed and power efficient SRAMs are required in IoT devices, various new ultra-low-power subthreshold SRAM memories and their architectures have been proposed. With CMOS technology, scaling down the supply voltage was the most popular method to improve the overall power consumption. However, scaling of the supply voltages in scaled devices has various limitations such as an increase in leakage current, and scaling the device beyond the nanometer regime increases the variation in threshold voltages, which leads to other problems. These proccss variations affect the performance of SRAMs, particularly the read and write stability, which degrades the performance of SRAMs. Most of the SRAM cells fail during read in ultra-low-power applications. Various methods such as the butterfly curve and N curves are used to find the stability of SRAM cells [18]. Researchers have started working on the subthreshold regime to make ultra-low-power SRAM cells where power and stability are the main concerns. To address these issues, several architectures of SRAM cells have been proposed using CMOS technology. The electronics market started considering the CMOS SRAM to be the stakeholder in the memory market. Due to the scaling quality of CMOS, SRAM was having its grip on the market for the last few decades. In the last few years, CMOS scaling is confined to a limit which raised various issues such as short-channel effects (SCEs) and threshold voltage variations. The CMOS limitations have led to alternative devices such as FinFET and other nano-scaled devices such as CNTFETs and TFETs.

FinFETs are evolving as some of the most suitable choices for CMOS memory circuits. In a small time frame FinFETs became promising devices and in some areas a better choice than MOSFETs, since they have some properties such as reduction in leakage current due to better gate control and also because of smaller size. FinFETs are considered to be superior to CMOS because of various reasons, e.g., FinFETs have superb control over the gate channel, which results in reduction of the drain and source leakage currents, and SCEs can be reduced in FinFETs [19]. FinFETs are also known as multi-gate transistors, due to which various gate control results in excellent electrostatic properties. FinFET-based SRAMs are successful in various applications and used nowadays in most of the mobile phones and many IoT devices. Similarly, CNTFET- and TFET-based SRAMs have also been proposed, and in future we may expect their production.

4.5.2 Embedded Flash and EEPROM Memory in IoT Systems

Embedded flash and EEPROM memories are NVMs, and they must have different characteristics from standalone memories, they should be cost efficient, and their volume production should be possible like the conventional CMOS fabrication. The challenge with this flash memory is that they need to be made well-matched with the conventional CMOS process, which can be easily accessible in different foundries. Fabrication of CMOS memories with new materials will increase the cost and may raise reliability issues. So, we should bring some variations in conventional memories to bring the embedded NVM in the practical use of IoT processors.

4.5.3 Floating Gate Flash Memory

In 1985, Philips made a single polysilicon memory in CMOS. It was made for the purpose of microprocessors and other logic circuits [20]. It was fabricated in 12.5 μm technology with 2 kb capacity. It was designed for a power supply of 5 V and with a 13 V programming voltage. It is shown in Figure 4.2.

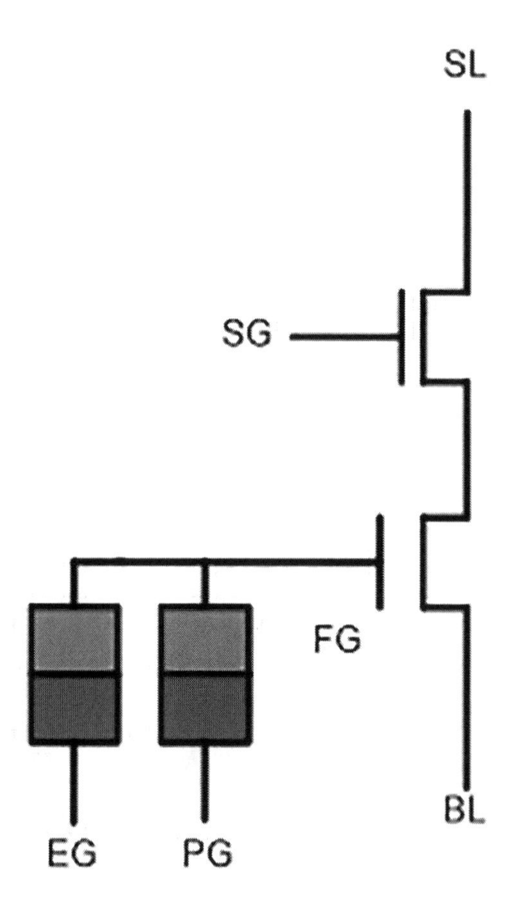

FIGURE 4.2 Single-poly memory.

The programming is done by reducing the injector oxide thickness to 8 nm and along with it increasing the capacitive coupling between the floating gate and a control gate. This is done at the cost of increasing the cell area by 30%. The injector oxide is responsible for data retention because the retention time is directly related to the area of the oxide thickness to the floating gate at the cost of 30% increment in the cell.

Toshiba developed 256 K-bit EEPROM in 1985 using a single polysilicon cell for a MCU chip in a 1.2 μm process [21]. A single polysilicon cell shown in Figure 4.3. EEPROM similar to that of Philips was also used where the cell area was reduced by applying a “bird’s beak” isolation technology. IBM in 1994 discussed another single-poly EEPROM cell produced in a conventional CMOS fabrication process. In 2000, Lucent Technologies and Bell Labs also proposed and discussed CMOS-compatible embedded flash memory [22]. It is a three-transistor memory cell similar to the IBM cell discussed in Ref. [23] and is presented in Figure 4.4, but in this, FN (Fowler–Nordheim) tunneling is used to change the control of gate

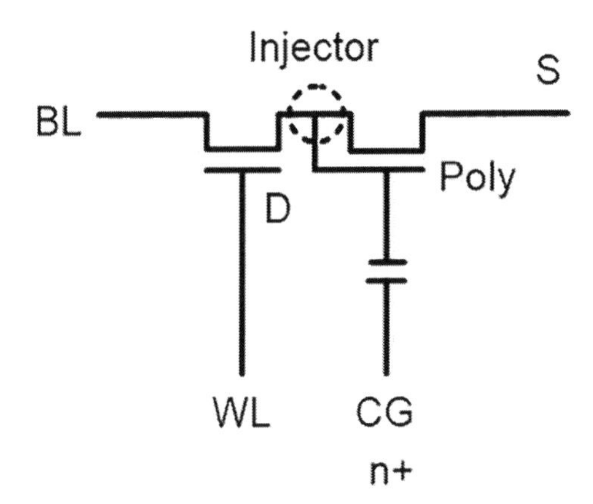

FIGURE 4.3 A single-poly embedded flash cell by IBM.

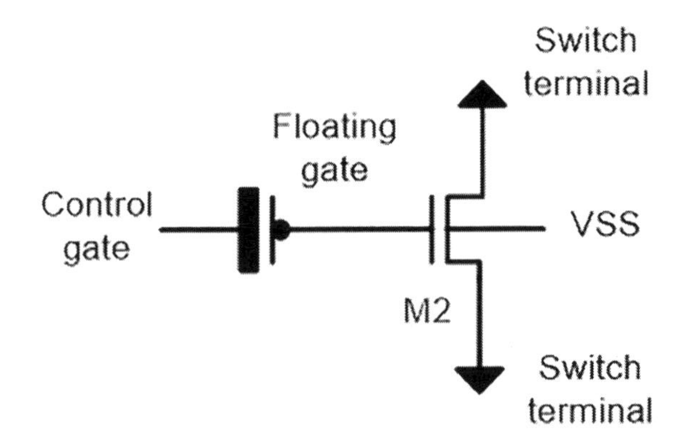

FIGURE 4.4 A single-poly embedded flash cell by Bell Labs.

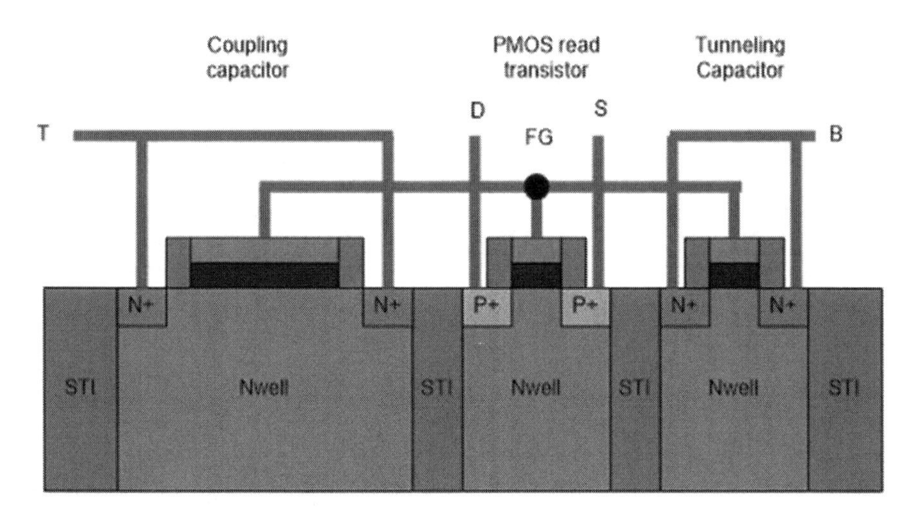

FIGURE 4.5 A 3T single-poly embedded flash cell.

design and erase gate, which lowers the threshold voltage, so it can have low power supply because of low threshold voltage. This cell required a low gate voltage for programming with more data retention time.

10 V was the erase voltage of the cell, and the cell was fabricated in 0.25 μm technology with 2.5 V as the CMOS logic process. The gate control voltage was 6.5 V, source was 0 V, and the drain was kept at 5.5 V for programing, while for the read, drain was at 1.5 V and gate voltage was 2.25–2.75 V.

Another CMOS embedded flash memory was suggested by Synopsis (earlier Virage Logic). It was proposed at 130 nm and is known as non-volatile electrically alterable memory [24]. It was actually proposed for the encryption of security code. The embedding of the memory on the logic circuit is based on security concerns. It was fabricated with the CMOS logic standard process without the need for any extra fabrication steps. The cell has a PMOS which is a read transistor along with a tunneling and a coupling capacitor, and these all are connected collectively with a single floating gate as shown in Figure 4.5.

A 7 nm oxide layer is grown all over the active area, and this was 10 nm for the double polysilicon embedded flash memory. One bit is made using two memories connected in a parallel fashion, and they are programmed with the opposite data. A password is used for security. It has a differential voltage of less than 100 mV and a high voltage NMOS having an NWell source and drain, which is comparable with the standard CMOS. A high-voltage generator is used to provide a programming voltage of 8 V.

4.5.4 Programmable Single-Poly Embedded NVM

TSMC in 2013 talked about embedded NVM, and they discussed about the data retention method for multiple-time programmable (MTP) applications. TSMC explained two ways of data retention in MTP applications [25]. Embedded flash is

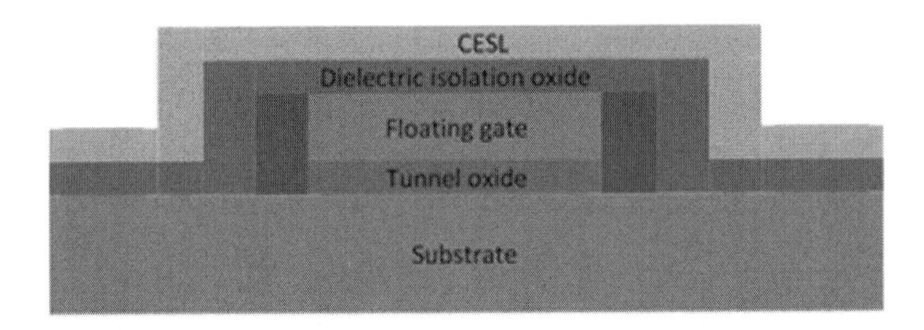

FIGURE 4.6 Cross-section of an embedded NVM cell.

replaced by the logic embedded NVMs because of the high speed, low cost, and logic compatibility. Also, data retention in embedded NVMs is not a matter of concern since a thick tunnel oxide is present for a 5 V device in BJT, MOS, and high voltage technologies. To understand the mechanism of data retention in these memories, a cross-sectional view of the embedded NVM logic is depicted in Figure 4.6.

Along the cover of the floating gate, "an etch layer known as contact etch stop liner covers the isolation oxide which is a dielectric". The schematic of the MTP bit cell memory is presented in Figure 4.7. This will be operated by a 5 V logic device having a tunnel oxide of 12 nm thickness. PG and EG are the capacitors for controlling program and erase, respectively. Electrons are ejected to the EG during erase, and electrons are injected during programming to the PG. A model has been made for data.

For a bit cell, the effect of bake temperature on the data retention degradation is observed, which lies between 25°C and 250°C implying that at a high bake temperature or if the duration of the baking is more, the degradation is more. The effect of temperature shows that if the bake temperature is increased higher than 125°C, the degradation can be recovered in later baking, but for temperatures less than this, there is no recovery; it is based on the phenomenon that for high bake temperature, trans-conductance Gm recovers faster above a temperature of 125°C–250°C, as low temperature will not have sufficient thermal energy to activate electron detrapping, and due to this, there is no possibility of recovery. The data programming code and the electron detrapping due to thermal energy are responsible for data retention degradation.

ST Microelectronics with the University of Brescia in 2013 discussed another MTP cell which is fabricated using 130 nm CMOS conventional technology, and they called it "Half Cell" structure. The application of this NVM is in RFID chips, IP security, etc. These applications need low cost NVM, which may be programmable.

4.5.5 Single-Poly Fully CMOS Embedded EEPROM Devices

Flash silicon proposed an NVM device which can be fabricated using CMOS scalable logic in 2014 at technology nodes of 110, 55 and 40 nm. NOR flash arrays were designed according to the process design rules of the CMOS technology ranging from 100 to 40 nm. This EEPROM used three MOSFETs with single-poly floating

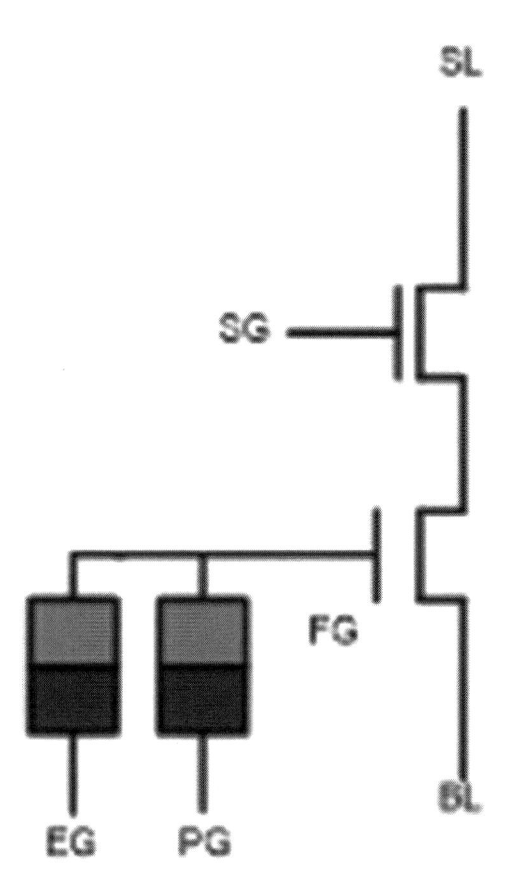

FIGURE 4.7 Circuit diagram of an embedded NVM cell.

gates processed using a CMOS technology. Early EEPROMs used NMOS as the channel and PMOS as the controlling gate, and the charge storage floating gate is the CMOS logic gate itself. The advanced EEPROM later used an N-type control gate embedded in a P-type silicon substrate. In this, the gate length of the logic device forms the floating gate which stores charge. The oxide layer having a thickness of 6.5–8 nm stores and holds the charge for the retention time.

This logic compatible EEROM was integrated into a 6T conventional SRAM which makes it a non-volatile SRAM. They have integrated it into a non-volatile FPGA and to a non-volatile register, and the foundry was from UMC from 40 to 55 nm and from 1 to 8 Mb of size. Another embedded EEPROM with a gate of tungsten was proposed by National Tsing Hua University (NTHU) [26]. Since it is a single-poly with a tungsten gate, it has smaller Drain-Induced Barrier Lowering (DIBL), low resistance, and no parasitic depletion, and this structure has a lower cost and fewer defects as it was completely fabricated with CMOS process technology. It can further integrate an advanced system-on-chip due to its scaling property, modest design, and a better isolation between the floating gate and the control gate.

4.5.6 Embedded Flash Cells Using CMOS Logic

Embedded flash cells have been fabricated using CMOS logic transistors with CMOS process technology without any variation in the existing process rules. The University of Minnesota discussed a five-transistor embedded flash memory, which uses only standard MOS transistors, in June 2012 [27]. The cell is represented in Figure 4.8.

This 5T cell has a selected row, which is a refreshing scheme to enhance the endurance [27]. It is fabricated in a standard process of low power with a tunnel oxide thickness of 5 nm. It can be used as a secure on-chip NVM for the conventional CMOS logic, and all transistors are fabricated with 2.5 V and 5 nm oxide thickness. M1 is kept eight times bigger than M2 or M3 to get improved programming and erasing. This cell has two PMOS transistors and the rest is NMOS. This is useful for a system with zero standby power systems as it can save critical data in power down mode. The selective word line (WL) is used to refresh and improve the threshold voltage of the cell. The new cell is at 65 nm with 1.9 V as the threshold window.

4.5.7 Novel Embedded Flash Memory

A*STAR and the University of Singapore introduced a novel embedded flash memory in 2013 [28]. It is a "onetime programmable (OTP) antifuse non-volatile memory (NVM)" designed with a "TaN microbeam movable arm". It needs one extra mask

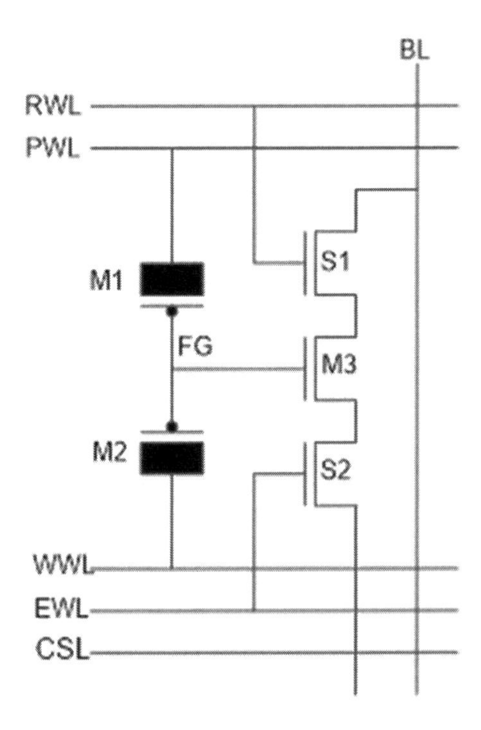

FIGURE 4.8 Circuit diagram of a 5T single-poly embedded flash cell.

than conventional CMOS and can be integrated. This memory cell includes one transistor and one microbeam per bit in the memory array. In this, the open state is "0" and the closed/fused state is 1. It operates at 4 V, which is useful for low-power applications, and it can be switched on and reading can be done by the transistor.

4.5.8 Thin Film Polymer and Flexible Memories

Processors require low-power and low-cost embedded memories, but if we consider wearable devices, flexible circuits are required. So, people started moving away from the conventional semiconductor processing which can reduce the cost. This is the reason why researchers started exploring polymer circuits. At present, many embedded memories such as Resistive Random Access Memory (RRAM), ferroelectrics, and charge trapping memories can be made using printing methods. This can be useful as it can produce a flexible circuit that is low cost as well as low power but with the limitation of low performance. Polymer RRAMs and nanocrystal memories are being developed.

These memories are useful for low-cost chips such as RFID chips as they do not need high performance but are required to be of low cost. But for various IoT applications such as wearable devices and medical systems, the performance of the polymer circuits is low. For such types of applications, high performance is required. So, silicon chips can be needed, and such technologies have evolved for transferring silicon chips on flexible substrates. Integration of silicon chips with flexible substrates has also been made possible. Using silicon-on-insulator base wafers, the silicon chips can be transferred to flexible substrates.

4.5.9 Memories for the Intelligent IoT

A 6T SRAM which is made from organic transistors is represented in the circuit in Figure 4.9. For writing, the WL is kept at VDD and the bitline (BL) is set to VDD and bitline (BL') is made zero. The node 2 is then charged to VDD. When the WL

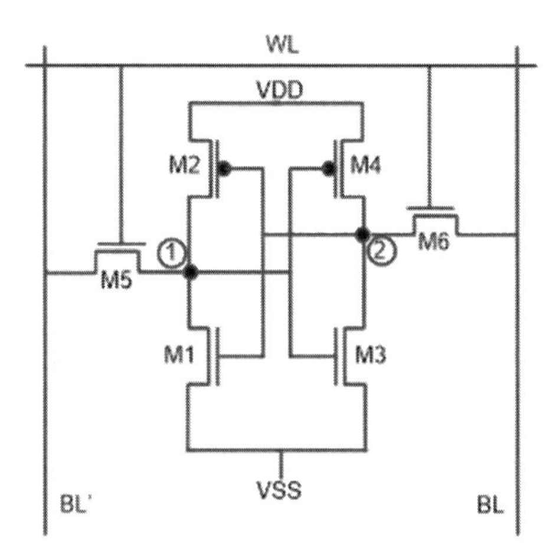

FIGURE 4.9 Circuit diagram of a 6T SRAM cell.

is at zero voltage, the data is stored. This can be used in RFID tags [29]. Xerox and Thinfilm thought to introduce printed memory to the electronics market in 2015 [30]. Though it does not require a voltage source or a battery, it needs a reading device to read the data. Some of the application areas are smart devices and protection for expensive fashion items. TFTs were also used to design memories in 2016 by Hewlett Packard and Hong Kong University.

Xerox propelled printed memory items proposed to battle falsifying of marked and administrative items in 2015 [31]. Printed memory with cryptographic security features was also developed by XEROX PARC, which can confirm the integrity of the product, and Xerox planned to set up "1.3 billion smart printed memories" [32].

4.5.10 Polymer Resistive RAMs with Flexible Substrate

RRAMs are prepared with low price materials and simple manufacturing methods. RRAM can be used for many IoT applications, but it is limited with respect to performance and yield. Many inorganic RRAMs can be integrated on flexible substrates with low voltage. IZO and IGZO RRAMs have high performance as well as high density. IGZO TFT embedded NVM has high density with small size, An amorphous IGZO TFT can be a possible platform for embedded NVM which can be low cost, high density, and light weight [33]. An RRAM designed using a-IGZO has features such as low cost, high density along with flexible memory characteristics which make it appropriate for SoP applications.

4.6 CONCLUSION

IoT is a new area which involves several applications that require an enormous amount of data transfer. Data transfer means there is a requirement for memory. Several volatile and non-volatile memories have been proposed by researchers from time to time. In this chapter, we have covered some volatile and non-volatile memories which are used in IoT applications. In this chapter, the role of memories in IoT has been explored to explain the use of memories in IoT. Several memories starting from SRAM, embedded flash, EEPROM, RRAM, polymer memories, and NVM have been discussed. Various memories designed using new devices and materials are discussed in this chapter.

REFERENCES

1. Foote, K. D. (2016). A Brief History of the Internet of Things in the World Wide Web. https://www.dataversity.net/brief-history-internet-things/.
2. Framingham, M. (2019). IDC Forecasts Worldwide Spending on the Internet of Things in the World Wide Web. https://www.idc.com/getdoc.jsp?containerId=prUS44596319.
3. Prince, B., & Prince, D. (2018). *Memories for the Intelligent Internet of Things*. California: John Wiley & Sons.
4. Khwa, W. S., Chang, M. F., Wu, J. Y., Lee, M. H., Su, T. H., Yang, K. H., & Kim, S. (2016). A resistance drift compensation scheme to reduce MLC PCM raw BER by over $100\times $ for storage class memory applications. *IEEE Journal of Solid-State Circuits*, *52*(1), 218–228.

5. Noguchi, H., Ikegami, K., Takaya, S., Arima, E., Kushida, K., Kawasumi, A., ... & Fujita, S. (2016). 7.2 4MB STT-MRAM-based cache with memory-access-aware power optimization and write-verify-write/read-modify-write scheme. In *2016 IEEE International Solid-State Circuits Conference (ISSCC)* (pp. 132–133).
6. Sohn, K., Yun, W. J., Oh, R., Oh, C. S., Seo, S. Y., Park, M. S., & Yu, H. S. (2016). A 1.2V 20nm 307 GB/s HBM DRAM with at-speed wafer-level IO test scheme and adaptive refresh considering temperature distribution. *IEEE Journal of Solid-State Circuits, 52*(1), 250–260.
7. Jacob Morgan. (2004). A Simple Explanation of 'The Internet of Things' in the World Wide Web. https://www.forbes.com/sites/jacobmorgan/2014/05/13/simple-explanation-internet-things-that-anyone-can-understand/#430142871d09.
8. Patel, H. N., Yahya, F. B., & Calhoun, B. H. (2016). Optimizing SRAM bitcell reliability and energy for IoT applications. In *2016 17th International Symposium on Quality Electronic Design (ISQED)* (pp. 12–17).
9. Ahmad, S., Alam, N., & Hasan, M. (2018). Robust TFET SRAM cell for ultra-low power IoT applications. *AEU-International Journal of Electronics and Communications, 89*, 70–76.
10. Chien, Y. C., & Wang, J. S. (2018). A 0.2V 32-Kb 10T SRAM with 41 nW standby power for IoT applications. *IEEE Transactions on Circuits and Systems I: Regular Papers, 65*(8), 2443–2454.
11. Singh, P., & Vishvakarma, S. K. (2017). Ultra-low power, process-tolerant 10T (PT10T) SRAM with improved read/write ability for internet of things (IoT) applications. *Journal of Low Power Electronics and Applications, 7*(3), 24.
12. Fujiwara, H., Chen, Y. H., Lin, C. Y., Wu, W. C., Sun, D., Wu, S. R., & Chang, J. (2016, November). A 64-Kb 0.37V 28nm 10T-SRAM with mixed-Vth read-port and boosted WL scheme for IoT applications. In *2016 IEEE Asian Solid-State Circuits Conference (A-SSCC)* (pp. 185–188).
13. Hiramoto, T., Takeuchi, K., Mizutani, T., Ueda, A., Saraya, T., Kobayashi, M., & Kamohara, S. (2016, June). Ultra-low power and ultra-low voltage devices and circuits for IoT applications. In *2016 IEEE Silicon Nanoelectronics Workshop (SNW)* (pp. 146–147).
14. Pandey, S., Yadav, S., Nigam, K., Sharma, D., & Kondekar, P. N. (2018). Realization of junctionless TFET-based power efficient 6T SRAM memory cell for internet of things applications. In *Proceedings of First International Conference on Smart System, Innovations and Computing* (pp. 515–523). Springer, Singapore.
15. Wu, S. Y., Lin, C. Y., Chiang, M. C., Liaw, J. J., Cheng, J. Y., Yang, S. H., ..., & Chang, V. S. (2016). A 7nm CMOS platform technology featuring 4th generation FinFET transistors with a 0.027 um 2 high density 6-T SRAM cell for mobile SoC applications. In *2016 IEEE International Electron Devices Meeting (IEDM)* (pp. 2–6).
16. Gupta, N., Makosiej, A., Vladimirescu, A., Amara, A., & Anghel, C. (2016, March). 3T-TFET bitcell based TFET-CMOS hybrid SRAM design for ultra-low power applications. In *2016 Design, Automation & Test in Europe Conference & Exhibition (DATE)* (pp. 361–366).
17. Pal, S., Gupta, V., Ki, W. H., & Islam, A. (2019). Transmission gate-based 9T SRAM cell for variation resilient low power and reliable Internet of things applications. *IET Circuits, Devices & Systems, 13*(5), 584–595.
18. Birla, S., Shukla, N. K., Pattanaik, M., & Singh, R. K. (2010). Device and circuit design challenges for low leakage SRAM for ultra low power applications. *Canadian Journal on Electrical & Electronics Engineering, 1*(7), 156–167.
19. Birla, S. (2019). Variability aware FinFET SRAM cell with improved stability and power for low power applications. *Circuit World, 45*, 1–12.

20. Cuppens, R., Hartgring, C., Verwey, J., & Peek, H. (1984). An EEPROM for microprocessors and custom logic. In *1984 IEEE International Solid-State Circuits Conference. Digest of Technical Papers* (Vol. 27, pp. 268–269).
21. Matsukawa, N., Morita, S., Shinada, K., Miyamoto, J., Tsujimoto, J., Iizuka, T., & Nozawa, H. (1985). A high density single-poly Si structure EEPROM with LB (lowered barrier height) oxide for VLSI's. In *1985 Symposium on VLSI Technology. Digest of Technical Papers* (pp. 100–101).
22. McPartland, R. J., & Singh, R. (2000). 1.25 volt, low cost, embedded flash memory for low density applications. In *2000 Symposium on VLSI Circuits. Digest of Technical Papers (Cat. No. 00CH37103)* (pp. 158–161).
23. Ohsaki, K., Asamoto, N., & Takagaki, S. (1994). A single poly EEPROM cell structure for use in standard CMOS processes. *IEEE Journal of Solid-State Circuits, 29*(3), 311–316.
24. Raszka, J., Advani, M., Tiwari, V., Varisco, L., Hacobian, N. D., Mittal, A., & Shubat, A. (2004). Embedded flash memory for security applications in a 0.13/spl mu/m CMOS logic process. In *2004 IEEE International Solid-State Circuits Conference (IEEE Cat. No. 04CH37519)* (pp. 46–512).
25. Liao, Y. Y., Tsai, L. Y., Leu, L. Y., Lee, Y. H., Wang, W., & Wu, K. (2013). Investigation of data retention window closure on logic embedded non-volatile memory. In *2013 IEEE International Reliability Physics Symposium (IRPS)* (pp. MY–7).
26. Chung, C. P., & Chang-Liao, K. S. (2015). A highly scalable single poly-silicon embedded electrically erasable programmable read only memory with tungsten control gate by full CMOS process. *IEEE Electron Device Letters, 36*(4), 336–338.
27. Song, S. H., Chun, K. C., & Kim, C. H. (2012). A logic-compatible embedded flash memory featuring a multi-story high voltage switch and a selective refresh scheme. In *2012 Symposium on VLSI Circuits (VLSIC)* (pp. 130–131).
28. Singh, P., Li, C. G., Pitchappa, P., & Lee, C. (2013). Tantalum-nitride antifuse electromechanical OTP for embedded memory applications. *IEEE Electron Device Letters, 34*(8), 987–989.
29. Guerin, M., Bergeret, E., Bènevent, E., Daami, A., Pannier, P., & Coppard, R. (2013). Organic complementary logic circuits and volatile memories integrated on plastic foils. *IEEE Transactions on Electron Devices, 60*(6), 2045–2051.
30. Savastano. (2015). Xerox, Thinfilm Look to Bring Xerox Printed Memory to Market in the World Wide Web. www.printedelectronicsnow.com.
31. Xerox. Xerox Launches Printed Memory Products to Combat Counterfeiting, Xerox Press in the World Wide Web. https://www.news.xerox.com/news/Xerox-Launches-Printed-Memory-to-Combat-Counterfeiting, Release, September 15, 2015.
32. Xerox. Xerox to set up 1.3 billion capacity Thin film smart label print line, Plastic Electronics in the World Wide Web. https//www.plusplasticelectronics.com, January 21, 2016.
33. Hou, T. H., Wu, S. C., Yu, M. J., Liu, P. S., & Chi, L. J. (2013). Low-cost embedded RRAM technology for system-on-plastic integration using a-IGZO TFTs. In *2013 Twentieth International Workshop on Active-Matrix Flatpanel Displays and Devices (AM-FPD)* (pp. 71–74).

5 Performance Evaluation of a Novel Channel Engineered Junctionless Double-Gate MOSFET for Radiation Sensing and Low-Power Circuit Application

Dipanjan Sen, Bijoy Goswami,
Anup Dey, and Subir Kumar Sarakar
Jadavpur University

CONTENTS

5.1 INTRODUCTION

In previously reported articles, a JLDG MOSFET has been demonstrated as one of the MOS devices with impressive electrostatic control, and it provides better immunity in terms of short-channel effects than conventional MOSFET [1,2] devices. Besides, conventional bulk MOSFETs exhibit major short-channel characteristics [3] in terms of threshold voltage roll-off, high drain-induced barrier lowering (DIBL) and subthreshold slope degradation as the device dimension is scaled down to the nanoscale regime. However, to overcome the issues related to short-channel effects, various novel MOSFET structures have been proposed such as asymmetric junctionless double-gate silicon-on-nothing MOSFETs [4]. In addition to this, the ease of the fabrication procedure, reduction of the detrimental effects related to impurity doping and sharp formation of doping concentration make junctionless devices [5] a better option for several high-end applications. However, the above-mentioned research studies on these modified MOSFET structures have not considered the effect of the interface trapped charges on the device electrical behaviour. The fixed charges at the SiO_2/Si interface change the flat band voltage of the device, thereby an impact can be observed on the complete device characteristics and hence this needs specific attention in the case of nanoscale devices. In this case, a channel engineered JLDG MOSFET shows promising results as far as the short-channel effects are concerned, with improved radiation sensitivity [6] and excellent circuit performance (power, delay, noise margin and power–delay product (PDP)). Radiation sensors are the backbone of nuclear industries, radiotherapy, space etc. Besides, the proposed device shows low power consumption and high reliability, which are much desired in both radiation sensing and digital circuit application. Threshold voltage shift is used as the measurement methodology of sensitivity due to irradiation induced damage in the form of trap charges.

Here, incorporation of SiO_2 barriers within the channel shows lower surface charge leakage resulting in minimum off-state current. The proposed device provides an increment in the channel length (50 nm) as the effective length increases. Therefore, the proposed JLDG MOSFET contributes to the suppression of the I_{OFF} and improves the subthreshold slope (SS) due to the faster switching of the drain current from the off-state to the on-state with gate voltage. Thus, the proposed MOSFET structure can be termed an alternative in providing ultra-low power dissipation [7], minimum delay, good noise immunity and lower amount of PDP.

5.2 DEVICE DESCRIPTION & PERFORMANCE

Figure 5.1 demonstrates the schematic representation of the novel junctionless double-gate MOSFET and the device parameters are depicted in Table 5.1. Additionally, an oxide region within the channel has been included in the form of a trench. The channel length of the device is represented as L_g (nm), whereas the silicon or body thickness and gate oxide thickness are represented as T_{si} and T_{ox} (nm). SiO_2 has been considered as the gate oxide material, which eventually traps the localized

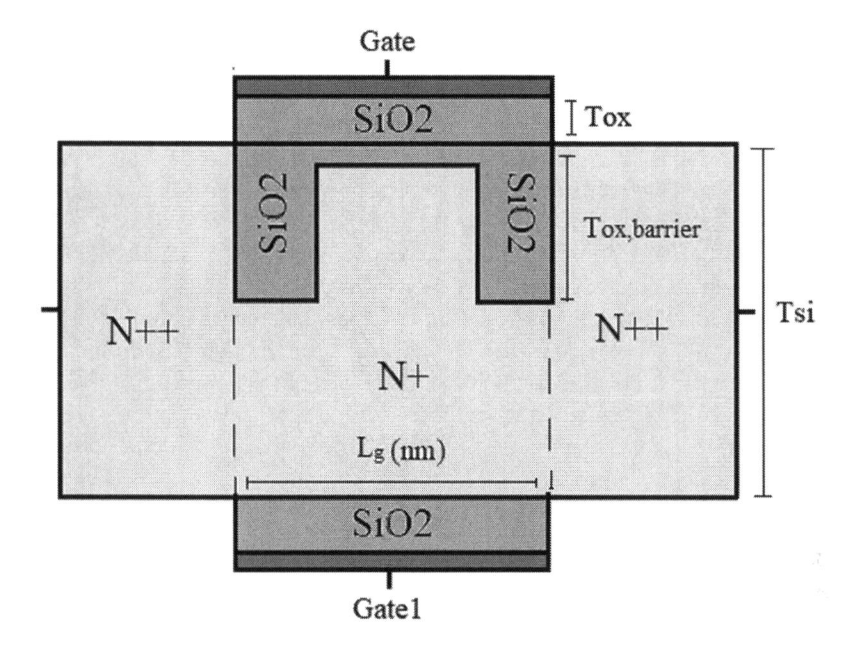

FIGURE 5.1 2D cross-sectional image of the proposed channel engineered junctionless DG-MOSFET.

TABLE 5.1
Device Parameter Chart of the Proposed Channel Engineered Junctionless DG-MOSFET

Device Parameters	Values
Channel length or L_g (nm)	50
Silicon thickness or T_{si} (nm)	10
Gate oxide thickness (nm)	2
Oxide barrier thickness (within channel) (nm)	6, 1
Gate metal work-function (eV)	4.8
Source/drain doping, channel doping concentration (/cm^3)	10^{20}, 10^{17}

charges at the SiO_2/Si interface. The trench oxide barriers are denoted as $T_{ox,\text{barrier}}$ (nm) within the channel region of the proposed MOSFET. Besides, the heavily doped source and drain regions are denoted as N++, and the channel is denoted as N+. In this work, the electrical characteristics of the MOSFET have been shown by using a SILVACO ATLAS [8] 2D simulator. Here, the Shockley–Read–Hall (SRH) generation and recombination model [8] is included for the life-time of the minority carriers. The CVT or Lombardi model [8] has been added to show the dependency of the mobility of the carriers on temperature. The BGN or band-gap narrowing model [9] has also been added. The FLDMOB or field dependent mobility model [10] is taken into consideration to accommodate the effects of saturation velocity caused by high electric fields. The CONMOB or concentration dependent mobility model [10] has also been included. In addition to this, numerical methods such as Gummel, Newton etc. [11] have been included for the purpose of calculating the carrier transport mechanism at each step of simulation. The mixed-mode simulation method has been used to perform the circuit analysis by using the proposed MOSFET device.

However, the fabrication feasibility of a junctionless double-gate MOSFET is high compared to other emerging MOSFET structures. Here, a 10 nm thick and 100 nm long un-doped silicon thin film is taken into consideration for design. Besides, the un-doped silicon is then doped with boron through an ion-implantation process. Then, an annealing method has been used at a temperature of 1,050°C. Additionally, the ion implantation method is again used to form the source, channel and drain regions of the n-type MOSFET, where arsenic at a concentration of $1 \times 10^{18}\,cm^{-3}$ is doped with a dose of $1 \times 10^{12}\,cm^{-2}$ at an energy of 10 keV. The annealing method for activating the dopants is performed at a temperature of 900°C for 5 min. Now, a 2 nm thin silicon oxide layer is grown by the dry oxidation process, and the deposition of the gate metal is performed over the whole surface by an isotropic method. Also, the trench regions are created by selective etching on the silicon surface for the purpose of depositing the oxide layer. The unwanted parts of the gate oxide layer have been etched out from the silicon surface by using an anisotropic etching method. Also, the gate metal is etched out from the unwanted parts using anisotropic etching.

Table 5.2 depicts the short-channel performance (SS, DIBL and I_{ON}/I_{OFF}) of the proposed MOSFET at $L_g = 50$ nm, where the V_{GS} is 2 V. The proposed MOSFET shows a subthreshold slope of 64.7 mV/decade, which is almost close to the ideal one (60 mV/decade), and also a DIBL value of 16.65 mV/V at V_{DS} (low) = 0.1 V and V_{DS} (high) = 1 V. Besides, a low value of DIBL justifies the better short-channel performance [12] of the proposed device at $L_g = 50$ nm. Also, the switching ratio (I_{ON}/I_{OFF}) is about 10^{10} at $N_f = 0$. Therefore, higher switching ratio results in faster operation of circuits designed using the proposed device. Figure 5.2a and b depict the electron concentration ($10^{19}\,cm^{-3}$) and conduction current density contour plots for the complete silicon region, which justify that the carrier transport in the channel region is quite evident and the conduction current density (6.7 A/cm^2 in the log scale) is uniform throughout the channel region. Therefore, the contour plot shows that the proposed device exhibits excellent electrical characteristics at $N_f = 0$.

TABLE 5.2
Short-Channel Performance Chart of the Proposed Channel Engineered Junctionless DG-MOSFET at $N_f = 0$

SS (mV/decade)	DIBL (mV/V) (V_{DS}: 1.0 & 0.1V)	I_{ON}/I_{OFF}
64.7	16.65	10^{10}

(a)

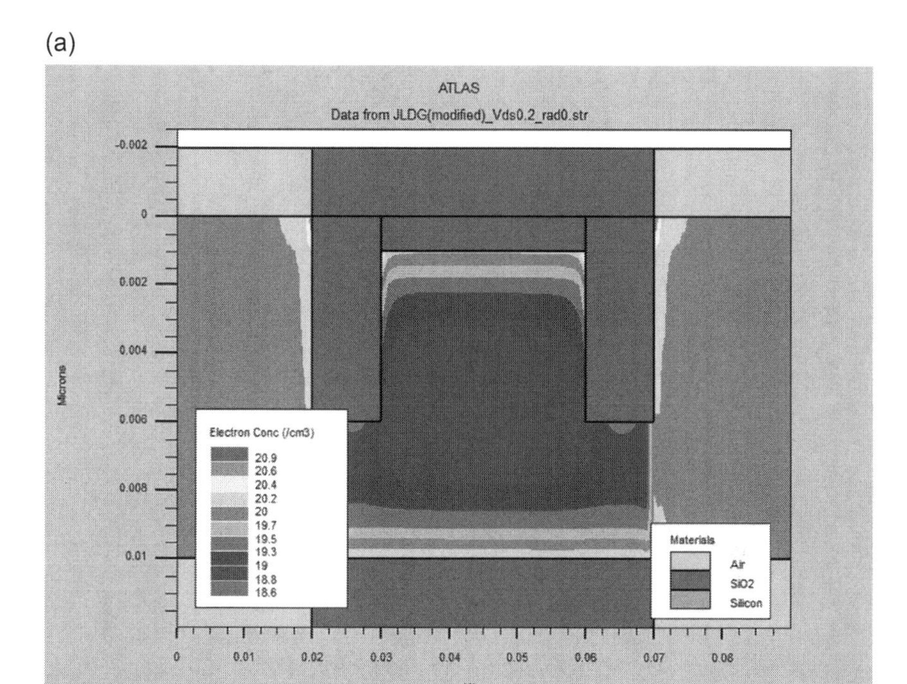

(b)

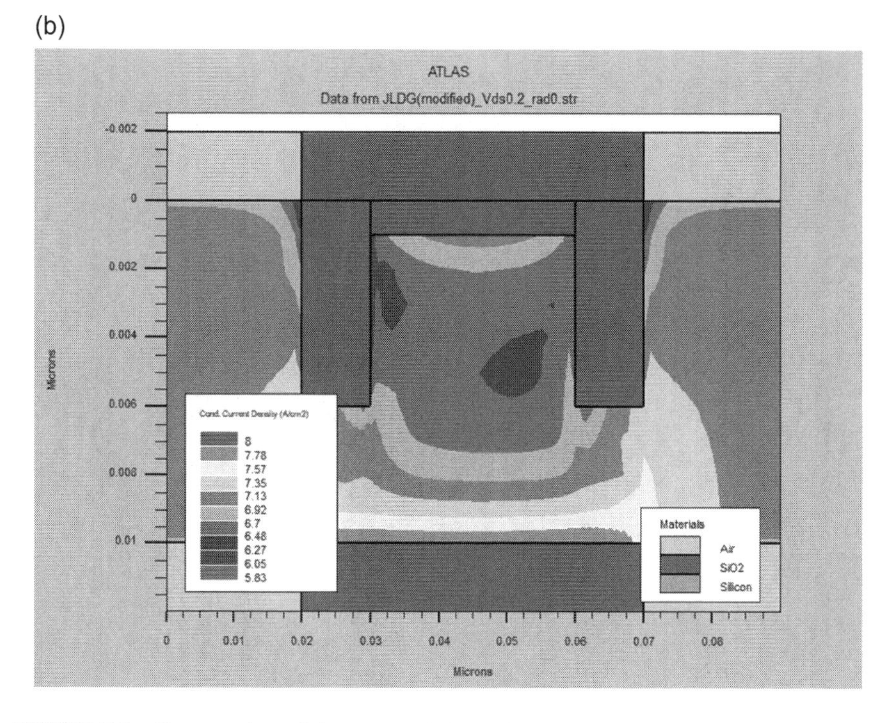

FIGURE 5.2 Contour plots of the proposed channel engineered junctionless DG-MOSFET: (a) electron concentration and (b) conduction current density.

5.2.1 Impact of Irradiation Induced Interface Trap Charges (Uniform) on the Drain Current Characteristics at V_{DS} = 0.2 V

Figure 5.3a–c depict the drain current profile of the proposed MOSFET for different values of interface trap charge densities at V_{DS} = 0.2 V. The drain current characteristics of the proposed device demonstrate that a small variation in N_f (negative to positive charges) causes a significant change in the drain current. Besides, with the change in interface trap charges [13], the off-state current changes largely, whereas the on-current remains almost constant. However, the deformation in the case of the MOSFET surface potential is less for positive trap charges than for the negative ones, which results in more amount of shift in V_{th} for negative fixed trap charges. Therefore, the rise in fixed interface trap charges (negative or positive) changes the flat-band voltage, and thus the threshold voltage shifts. From Table 5.3, it can be seen that the shift in threshold voltage is significant enough with a small variation in N_f and it is nearly 124 mV when the fixed charges change from $1 \times 10^{12}\text{cm}^2$ to $3 \times 10^{12}\text{cm}^2$ and 130 mV for $-1 \times 10^{12}\text{cm}^2$ to $-3 \times 10^{12}\text{cm}^2$ (negative trap charges). Thus, the radiation sensitivity of the proposed MOSFET is quite high. As it can be seen, $S = \Delta V_{th}/D$, where S is the sensitivity parameter, ΔV_{th} is the shift in the threshold voltage of the proposed device and D represents the absorbed dose of radiation. Thus, the sensitivity is directly proportional to the threshold voltage shift, which results in higher sensitivity if the shift in threshold voltage increases keeping D constant.

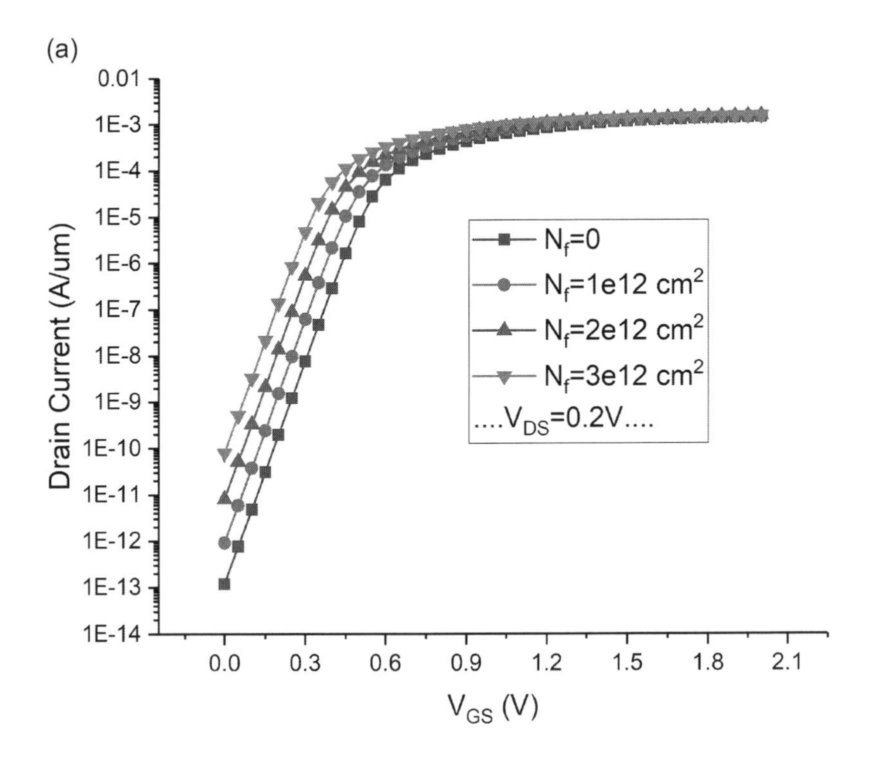

(Continued)

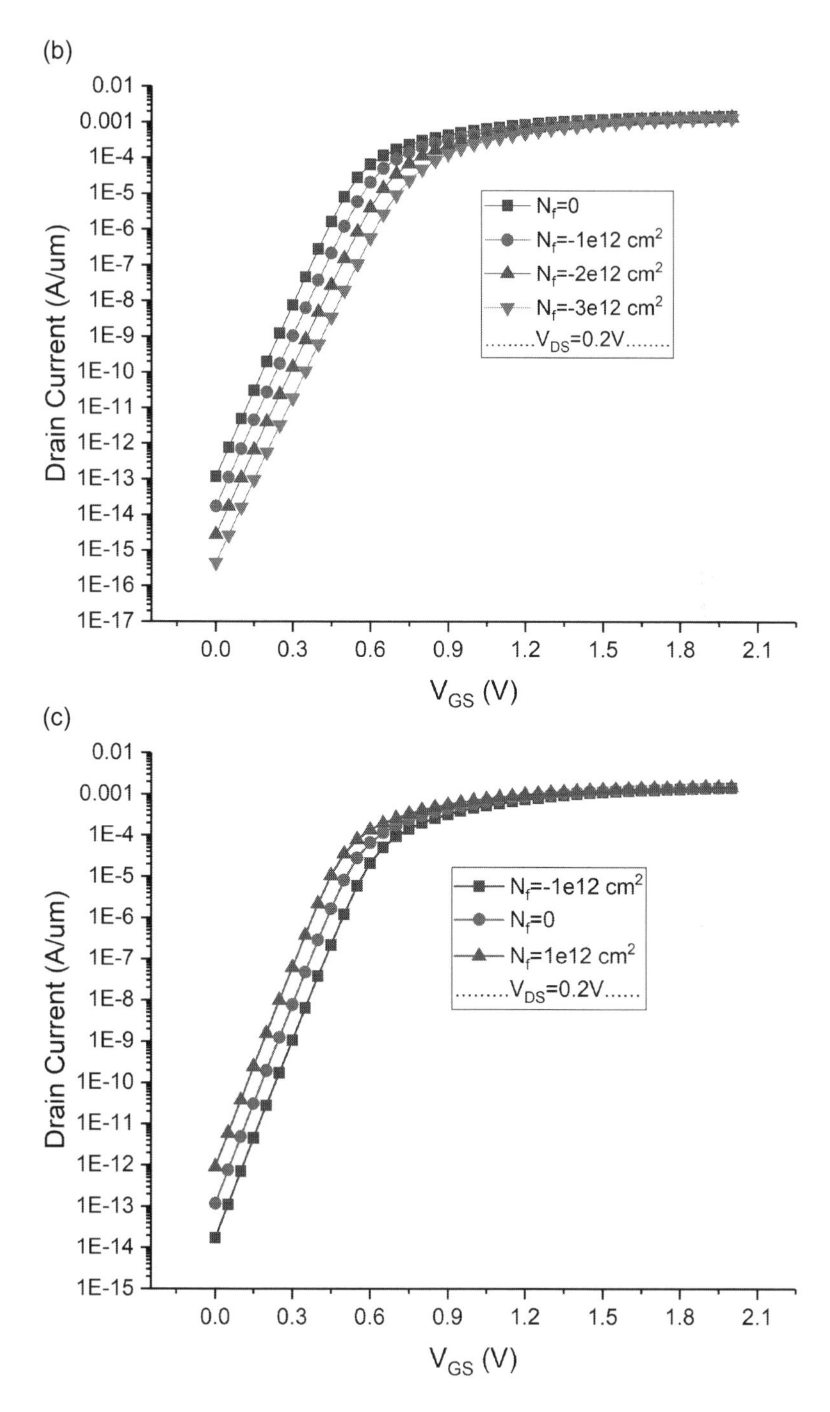

FIGURE 5.3 (CONTINUED) (a, left) (b-c, above) Drain current profile of the proposed channel engineered junctionless DG-MOSFET against gate voltage for different induced interface trap charges.

TABLE 5.3
Threshold Voltage Chart of the Proposed Channel Engineered Junctionless DG MOSFET w.r.t Uniform Charge Distribution by Considering Trap Charges at $V_{DS} = 0.2\,V$

N_f (cm²)	V_{th} (mV)
-3×10^{12}	625
-2×10^{12}	558
-1×10^{12}	496
0	436
1×10^{12}	378
2×10^{12}	317
3×10^{12}	254

5.2.1.1 Impact of Gate Oxide Thickness on the Threshold Voltage Characteristics at $V_{DS} = 0.2\,V$

Figure 5.4 depicts the threshold voltage profile w.r.t the variation of gate oxide thickness for different fixed interface trap charges. The above plot clearly demonstrates that the shift in threshold voltage increases with increasing gate oxide thickness [6].

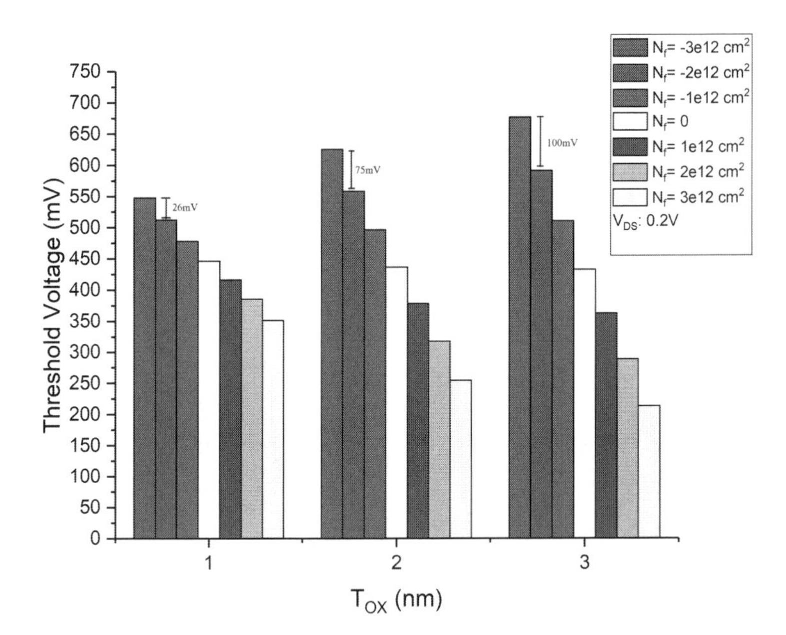

FIGURE 5.4 Threshold voltage profile of the proposed channel engineered junctionless DG-MOSFET against gate oxide thickness variation (T_{ox}) for different induced interface trap charges.

So, the proposed device shows higher sensitivity for thicker gate oxide. In addition to this, the threshold voltage value decreases with increasing positive interface trap charges and it also decreases with increasing gate oxide thickness. Similarly, the threshold voltage rises with negative interface trap charges as shown in Figure 5.4.

5.2.2 Impact of Gate Oxide Dielectrics on the Threshold Voltage Characteristics at $V_{DS} = 0.2\,V$

It is quite evident from Figure 5.5 that the threshold voltage of the proposed device increases significantly with increasing gate oxide dielectric constants or incorporation of high-k dielectrics [13]. It is well known that the downsizing of gate oxide thickness shows detrimental effects such as direct tunnelling of charge carriers, which results in the degradation of device performance. However, incorporating high-k dielectrics suppresses the gate tunnelling current and improves the electrical behaviour of the proposed device such as threshold voltage as shown in Figure 5.5. Therefore, it can be established that the proposed device shows lower gate oxide leakage in the case of high-k dielectrics. In addition to this, the threshold voltage shift is more, which is 41 mV for $K = 3.97$ (SiO_2) to $K = 9.8$ (Al_2O_3) in the case of positive interface trap charges here. However, the threshold voltage slightly decreases for negative charges. Therefore, the high-k gate dielectrics of the proposed MOSFET also provide improved sensitivity for radiation FETs.

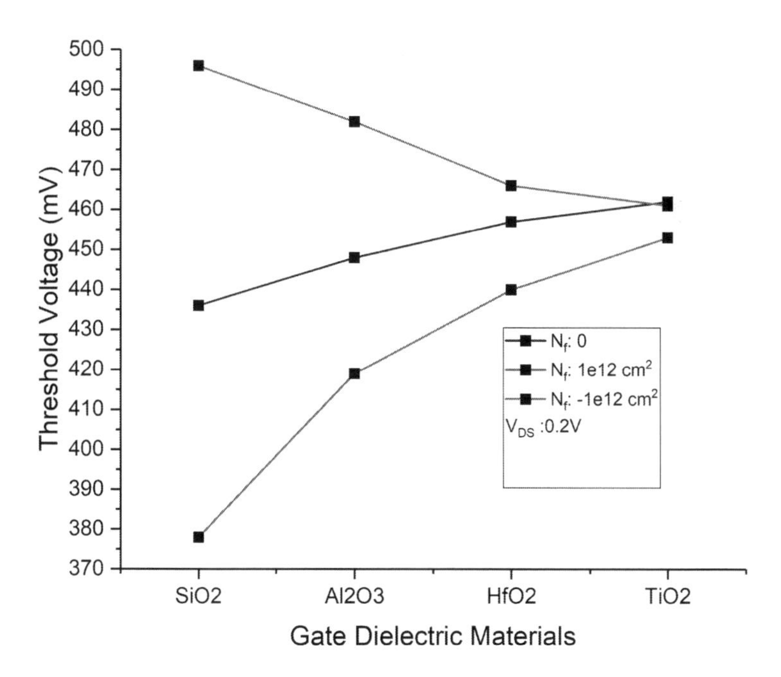

FIGURE 5.5 Threshold voltage characteristics of the proposed channel engineered junctionless DG-MOSFET against gate oxide dielectric variation for different induced interface trap charges.

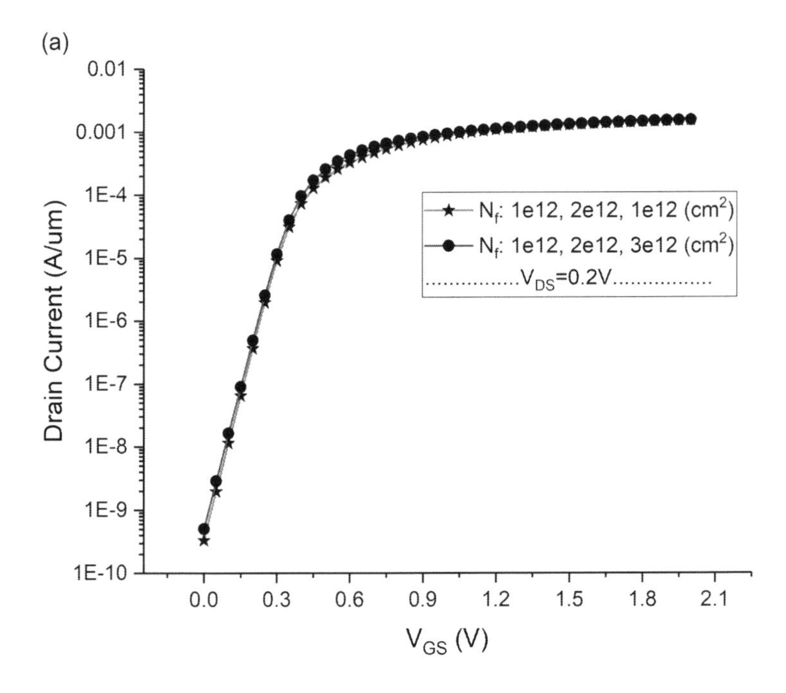

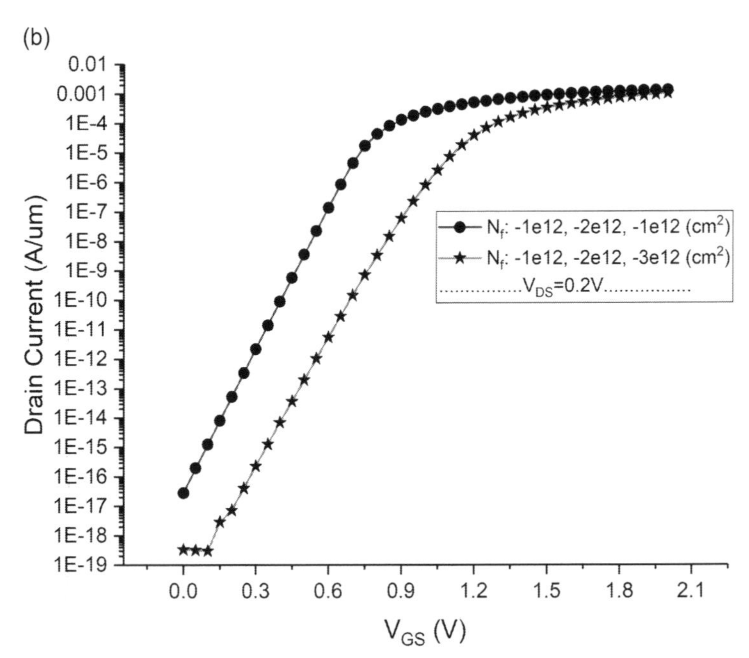

FIGURE 5.6 (a and b) Drain current characteristics of the proposed channel engineered junctionless DG-MOSFET against gate voltage variation for different non-uniform induced interface trap charges.

5.2.3 Impact of Non-Uniform Interface Trap Charge Density on the Drain Current Characteristics at $V_{DS} = 0.2\,\text{V}$

Figure 5.6a and b depict the drain current characteristics for different non-uniform induced interface trap charges and they show significant changes for negative trap charges (Figure 5.6b).The Si–SiO_2 interface (50 nm) is divided into three regions (15, 20 and 15 nm) to incorporate the non-uniform interface trap charges, and the charge distribution has been mentioned in Figure 5.6a and b. Besides, the surface potential deformation is more in the case of negative trap charges, which results in a significant amount of shift in the threshold voltage [14] as depicted by the drain current profile in Figure 5.6b. However, the threshold voltage shift in the case of positive trap charges is negligible as the surface potential deformation is less.

5.3 DC PERFORMANCE OF A PROPOSED MOSFET-BASED CMOS INVERTER

The inverter module is one of the primary building blocks of every possible digital circuit such as logic gates, combinational and sequential circuits, microprocessors etc. and it executes any kind of complementary logic operation (Figure 5.7). Therefore, a CMOS inverter has been realized with the proposed MOSFET, and the performance of the circuit is analysed in terms of power dissipation, propagation delay, noise margin and PDP [15]. Figure 5.8a and b shows the inverter characteristics such as Voltage

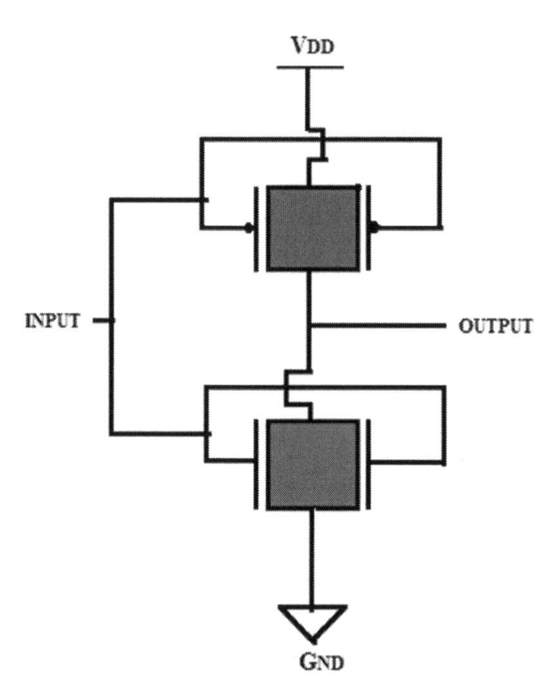

FIGURE 5.7 Schematic representation of the proposed MOSFET-based inverter module.

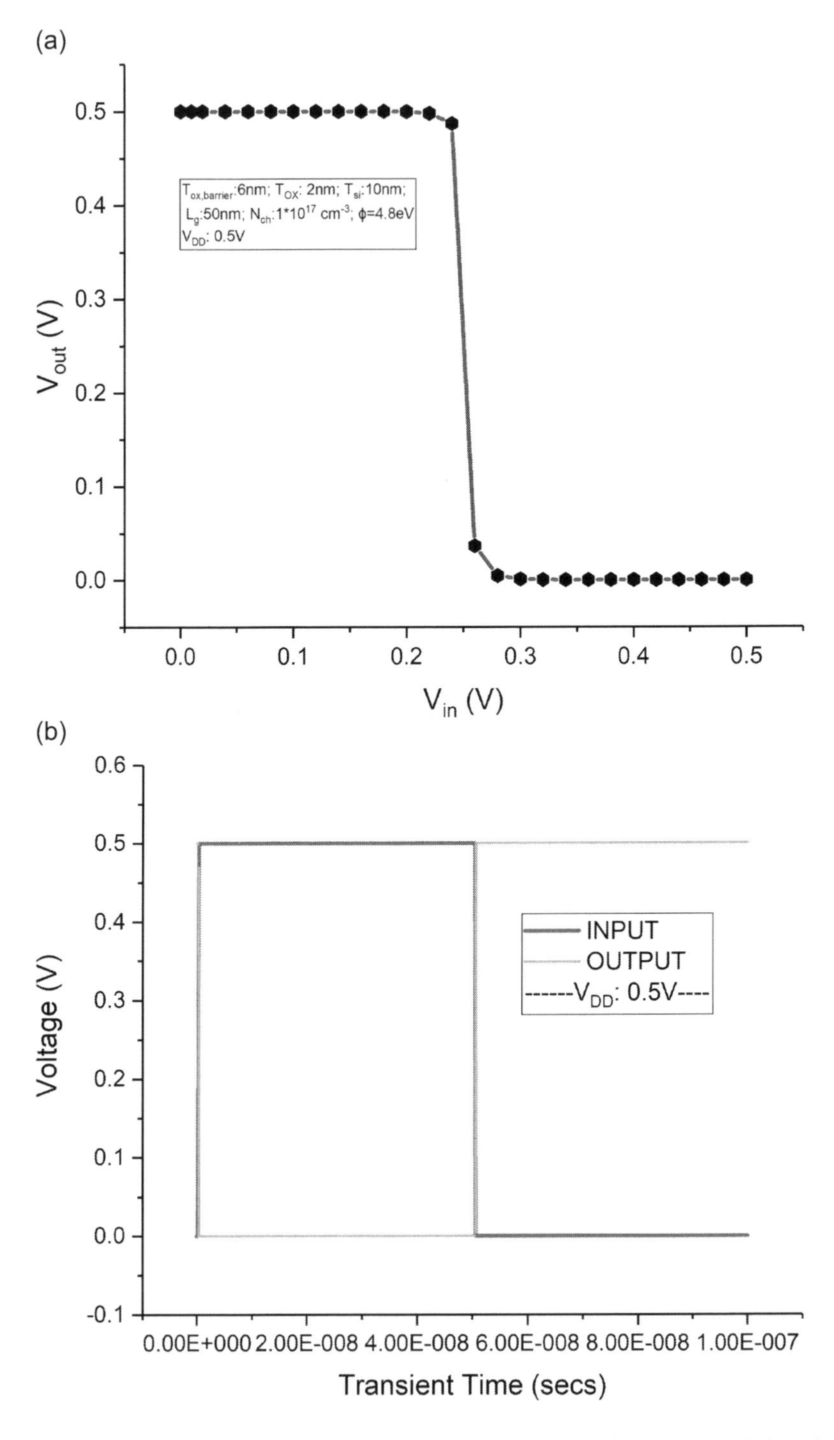

FIGURE 5.8 (a and b) Voltage transfer characteristics and transient characteristics of the proposed MOSFET-based inverter module at $V_{DD} = 0.5\,V$.

Transfer Characteristics (VTC) and the transient plot at $f = 10\,\text{MHz}$ and $V_{DD} = 0.5\,\text{V}$. In addition to this, inverter circuit analyses have been performed with the help of the mixed-mode method in an ATLAS simulator, where n-type and p-type devices are connected in the form of a CMOS architecture. Besides, VTC and transient analyses were performed to calculate the power dissipation, propagation delay [15] etc. for a better understanding of the proposed device behaviour in the case of analysing an inverter.

5.3.1 Impact of Supply Voltage on the Power and Delay Characteristics of an Inverter for Fixed Device Parameters

Figure 5.9 depicts the inverter figures of merit (FoMs; power and delay) with the variation of supply voltage [16] from 0.25 to 1.00 V, where the device parameters are fixed. The power dissipation increases with increasing supply voltage, whereas the delay decreases. It is well established that power dissipation is one of the pivotal parameters for today's high density [16] integrated circuits. It is quite well known that the average power dissipation is the summation of dynamic power dissipation and leakage or stand-by power dissipation. However, the leakage power dissipation should be minimised to achieve ultra-low average power dissipation [17]. Besides, the gate tunnelling current also has a major impact on average power dissipation and should be minimised. Additionally, it can be said that the inverter exhibits ultra-low power dissipation and propagation delay. At $V_{DD} = 0.5\,\text{V}$, the power and delay show optimum values of 50.2 pW and 166.65 pS, which result in an energy consumption

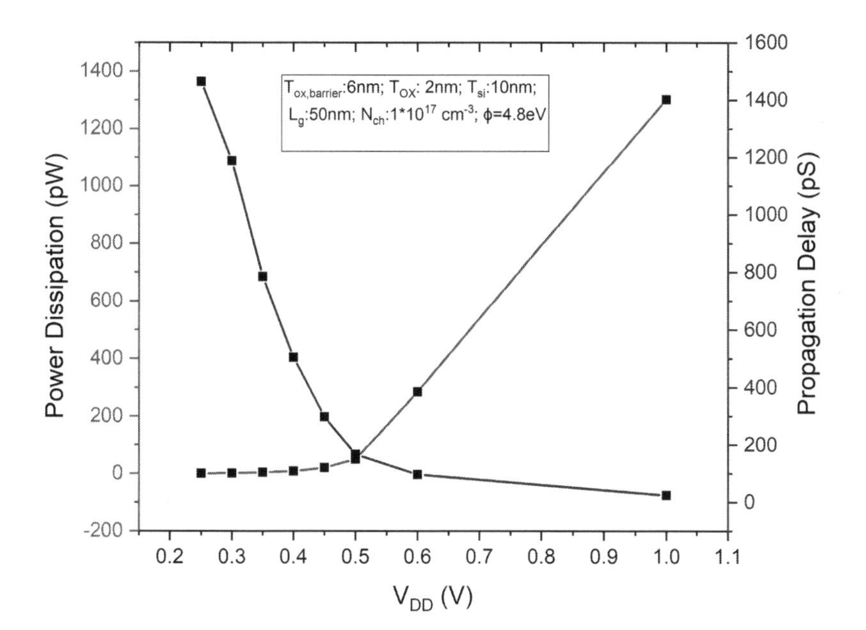

FIGURE 5.9 Power and delay characteristics of the proposed MOSFET-based inverter w.r.t supply voltage (V_{DD}) variation.

TABLE 5.4
Proposed MOSFET-Based Inverter Performance (Power, Delay, NM and PDP) Chart at V_{DD} = 1.0, 0.5 and 0.25 V

V_{DD} (V)	Power (pW)	Delay (pS)	PDP (attojoules)	NM_H, NM_L
1.0	1302	24.22	0.0316	0.45 V, 0.42 V
0.5	50.2	166.65	0.00836	0.26 V, 0.23 V
0.25	0.380	1464	0.00056	0.13 V, 0.12 V

(PDP) of 0.00836 attojoules as shown in Table 5.4. However, the threshold voltage of the proposed device is 436 mV at $N_f = 0$. Therefore, the subthreshold operation (0.25–0.4 mV) of the inverter also shows promising results in terms of power, delay and PDP [17] as shown in Figure 5.9 and Table 5.4. Besides, the optimisation of the device parameters (T_{si}, T_{ox}, $T_{ox,\text{barrier}}$ and N_{ch}) has been performed by analysing the inverter FoMs prior to the fabrication procedure, which is indeed one of the most important performance evaluation methods similar to supply voltage optimisation.

In addition to this, the range of noise or unwanted signal that an electronic module (circuit) can tolerate is denoted as the noise margin [18] or NM. Moreover, the inverter module does not operate properly beyond this level, and thus reliability issues arise. So, the prime focus is to make the noise margins as large as possible. However, large noise margin results in high amounts of voltage excursions which result in a significantly large amount of delay and power dissipation. Therefore, it is desirable to maintain a trade-off between noise margins (NM_H and NM_L), propagation delay and average power dissipation. However, the proposed device shows impressive noise margins at different supply voltages as shown in Table 5.4.

5.3.2 Impact of Silicon Thickness, Gate Oxide Thickness, Oxide Barrier Thickness and Channel Doping Concentration on the Power and Delay Characteristics of an Inverter for V_{DD} = 0.5 V

Figure 5.10a–d depict the power and delay profiles against device parameters such as silicon or body thickness (7–20 nm), gate oxide thickness (0.5–3 nm), oxide barrier thickness (3–8 nm) and channel doping concentration (10^{15}–10^{19} cm^{-3}) at a constant supply voltage of 0.5 V. Here, optimising the device process parameters is one of the prime objectives to achieve the best performance of the proposed work. Figure 5.10a shows that the average power dissipation increases with an increase in the value of silicon thickness as the threshold voltage decreases, resulting in more amount of leakage power dissipation. However, the propagation delay decreases as the driving current (I_{ON}) is more in the case of a thicker silicon body. From Figure 5.10a, it can be noted that the optimum power dissipation (280 pW) and propagation delay (146 ps) can be achieved at a silicon thickness value of 15 nm. In addition to this, Figure 5.10b depicts the change in power and delay against the variation in gate oxide thickness. The optimum values of power and delay have been achieved at a gate oxide thickness of 1.5 nm, but for design feasibility, 2 nm gate oxide thickness is considered,

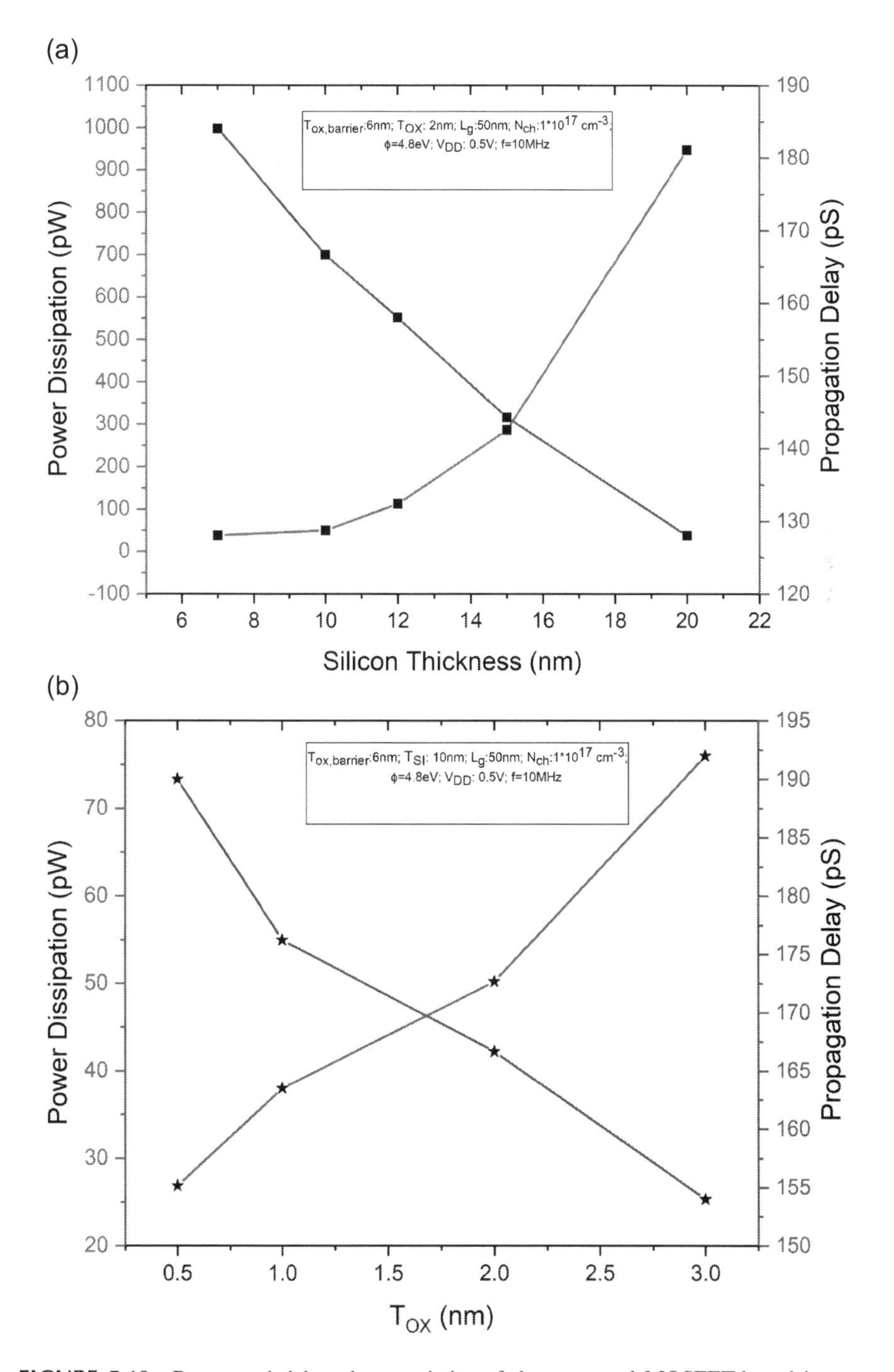

FIGURE 5.10 Power and delay characteristics of the proposed MOSFET-based inverter w.r.t (a) silicon thickness, (b) gate oxide thickness, (c) oxide barrier thickness and (d) channel doping concentration variation at $V_{DD} = 0.5\,\text{V}$.

(*Continued*)

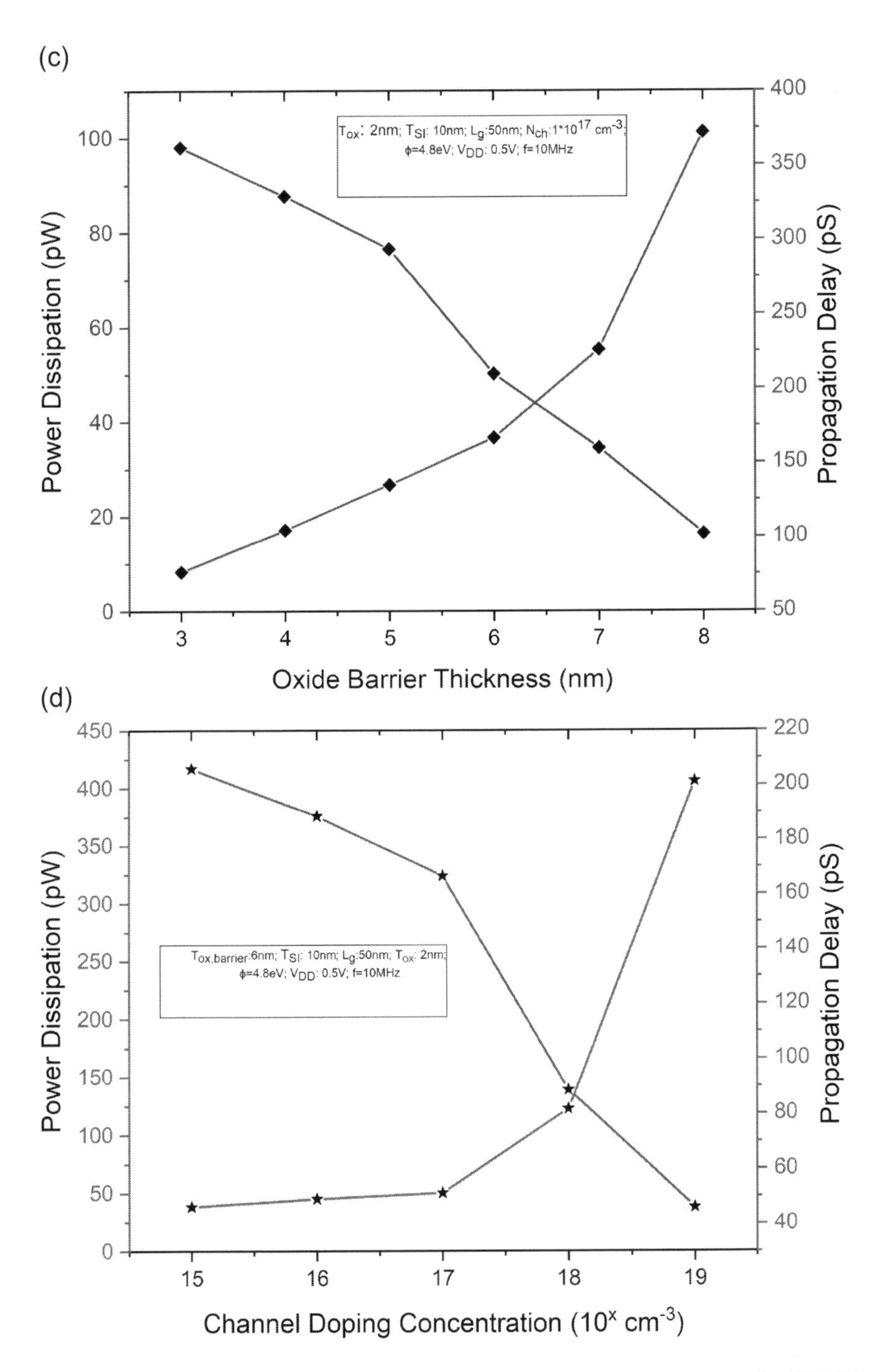

FIGURE 5.10 (CONTINUED) Power and delay characteristics of the proposed MOSFET-based inverter w.r.t (a) silicon thickness, (b) gate oxide thickness, (c) oxide barrier thickness and (d) channel doping concentration variation at $V_{DD} = 0.5\,V$.

which delivers an average power dissipation of 50.2 pW and a propagation delay of 166.65 pS at a supply voltage of 0.5 V. Figure 5.10c depicts that at an oxide barrier thickness of 6 nm delivers an average power dissipation and propagation delay of 50.2 pW and 166.65 pS, which are basically the optimum values shown. Besides, introducing an oxide region within the channel helps in suppressing the surface leakage and provides better short-channel performance. Here, Figure 5.10d demonstrates the inverter DC characteristics (power and delay) [19] against the variation of channel doping concentration. However, a channel doping concentration of 10^{17} cm^{-3} has been considered initially, which gives a power of 50.2 pW and a delay of 166.65 pS, which give a PDP value of 0.00836 attojoules. However, increasing the channel doping concentration results in more energy consumption (PDP), thus degrading the circuit performance. Therefore, the optimisation of device parameters [20] can be extensively performed in terms of analysing an inverter circuit. Moreover, to achieve an optimum average power dissipation and propagation delay of the inverter circuit at $V_{DD} = 0.5$ V, the process parameters should be optimised ($T_{ox} = 2$ nm, $T_{si} = 10$ nm, $T_{ox,\mathrm{barrier}} = 6$ nm and $N_{ch} = 10^{17}$ cm^{-3}) on the basis of PDP analysis [20] as shown in Table 5.4 before the fabrication is performed.

5.4 CONCLUSION

In this chapter, the performance of the proposed MOSFET has been analysed briefly and both the radiation sensing and low-power circuit applications have been depicted. In addition to this, the proposed device shows impressive short-channel performance such as DIBL, SS, I_{ON}/I_{OFF} etc. as shown in Table 5.2. Here, the proposed device shows promising results in terms of radiation sensitivity through V_{th} shift by incorporating uniform charge distribution in the form of trap charges. Besides, the proposed device performance has been investigated in terms of threshold voltage and drain current against the variation of gate dielectrics, gate oxide thickness and non-uniform charge distribution. Also, the MOSFET device shows impressive results in the case of CMOS inverter characteristics, providing ultra-low power, minimum delay, good noise immunity and ultra-low energy consumption. Adding to this, the subthreshold performance is also shown in Table 5.4, which depicts promising characteristics at a low supply voltage of 0.25 V. Therefore, the proposed device can be used as an alternative in nuclear industries (radiation sensor) as well as for low-power biomedical applications.

REFERENCES

1. Lee C.W., Afzalian A., Akhavan N.D., Yan R., Ferain I., & Colinge J.P. (2009). Junctionless multigate field-effect transistor. *Applied Physics Letters*, 94, 053511.
2. Colinge J.-P., Lee C.-W., Ferain I., Akhavan N.D., Yan R., Razavi P., Yu R., Nazarov A.N., & Doria R.T. (2010). Reduced electric field in junctionless transistors. *Applied Physics Letters*, 96, 073510.
3. Rechem D., & Latreche S. (2008). The effect of short channel on nanoscale SOI MOSFETs. *African Physical Review*, 2(38), 80–81.
4. Saha P., Banerjee P., Dash D.K., & Sarkar S.K. (2018). Exploring the short-channel characteristics of asymmetric junctionless double-gate silicon-on-nothing MOSFET. *Journal of Materials Engineering and Performance*, 27(6), 2708–2712.

5. Yeh M.S., Wu Y.C., Wu M.H., Chung M.H., Jhan Y.R., & Hung M.F. (2015). Characterizing the electrical properties of a novel junctionless poly-Si ultrathinbody field-effect transistor using a trench structure. *IEEE Electron Device Letters*, 36(2), 150–152.
6. Avashesh D., Singh A., Narang R., Saxena M., & Gupta M. (2018). Modeling and simulation of junctionless double gate radiation sensitive FET (RADFET) dosimeter. *IEEE Transactions on Nanotechnology*, 17(1), 49–55.
7. Gao H.W., Wang Y.H., & Chiang T.-K. (2018). A new device-parameter-oriented DC power model for symmetric operation of junction-less double-gate MOSFET working on low-power CMOS subthreshold logic gates. *IEEE Transactions on Nanotechnology*, 17(3), 424–431.
8. SILVACO International. (2000). ATLAS, 2-D Device Simulation Software. https://silvaco.com/tcad/
9. Arora N.D., Hauser J.R., & Roulston D.J. (1982). Electron and hole mobilities in silicon as a function of concentration and temperature. *IEEE Transaction on Electron Devices*, 29 (2), 292–295.
10. Fossum J.G., & Lee D.S. (1982). A physical model for the dependence of carrier lifetime on doping density in nondegenerate silicon. *Solid State Electronics*, 25(8), 741–747.
11. Saha S. (2013). MOSFET test structures for two-dimensional device simulation. *Solid State Electronics*, 38(1), 69–73.
12. Saha P., Banerjee P., Dash D. K., & Sarkar S.K. (2018). Modeling Short Channel Behavior of Proposed Work Function Engineered High-k Gate Stack DG MOSFET with Vertical Gaussian Doping. *Electron Devices Kolkata Conference (EDKCON), IEEE*, 32–36, doi: 10.1109/EDKCON.2018.8770421.
13. Saha P., Banerjee P., Dash D. K., & Sarkar S.K. (2020). Interface trap charge induced threshold voltage modeling of WFE high-K SOI MOSFET. *Silicon*, 2020. doi: 10.1007/s12633-020-00386-5.
14. Woo J.-H., Choi J.-M., & Choi Y.-K. (2013). Analytical threshold voltage model of junctionless double-gate MOSFETs with localized charges. *IEEE Transaction on Electron Devices*, 60(9), 2951–2955.
15. Sen D., Sengupta S. J., Roy S., Ray S., & Sarkar S. K. (2019). Impact Analysis of Dual Material Double Gate Oxide-Stack Junction-Less MOSFET in RFID Memory Cell Realisation. *Devices for Integrated Circuit (DevIC), IEEE*, 398–403, doi: 10.1109/DEVIC.2019.8783330.
16. Srivastava N.A., Priya A., & Mishra R.A. (2019). Design and analysis of nano-scaled SOI MOSFET-based ring oscillator circuit for high density ICs. *Applied Physics A*, 125(8), doi: 10.1007/s00339-019-2828-x.
17. Sen D., Banik. B., & Roy S. (2018). Power and Delay Analysis of Junction-Less Double Gate CMOS Inverter in Near and Sub-Threshold Regime. *Electron Devices Kolkata Conference (EDKCON), IEEE*, 367–372, doi: 10.1109/EDKCON.2018.8770468.
18. Massimo A. (2010). Closed-Form Analysis of DC Noise Immunity in Subthreshold CMOS Logic Circuits. *International Symposium on Circuits and Systems, IEEE*, 1468–1471, doi: 10.1109/ISCAS.2010.5537340.
19. Panda S.R. (2017). Device and circuit performance of Si-based accumulation mode CGAA CMOS inverter. *Materials Science in Semiconductor Processing*, 66, 87–91.
20. Sen D., Sengupta S. J., Roy S., Chanda M., & Sarkar S.K. (2019). Analytical modeling of D.C parameters of double gate junctionless MOSFET in near & subthreshold regime for RF circuit application. *Nanoscience & Nanotechnology Asia 09*, doi: 10.2174/221068120966619073017003l.

6 Technological Challenges and Solutions to Advanced MOSFETs

S. Bhattacherjee
JISCE

CONTENTS

6.1 INTRODUCTION

6.1.1 AN OVERVIEW OF MOSFET SCALING AND ADVANCED DEVICES

The field effect transistor was introduced about 20 years earlier to the invention of the bipolar junction transistor. The basic concept of MOSFETs was proposed by J. Lilienfeld in 1925 [1]. He presented a prototype consisting of a semiconductor with two metal contacts on each side and an aluminum plate on the top, which allows the flow of current between the metal plates. Another structure closely resembling a MOS transistor was proposed by O. Heil in 1935. But the lack of proper technological development of appropriate semiconductor materials and quality control challenges delayed the expansion of MOS devices into a variety of commercial uses until 1967. MOS devices, therefore, could not be realized until the innovation of the silicon planar technology during the year 1960 [2]. Fabrication of a Self-Aligned Gate Ion Implanted MOSFET (SAGFET) in a planar process, conceived by Robert Bower in 1965, paved the way for the fabrication of the first integrated circuits (ICs) [3]. Due to the low cost of IC manufacturing, the electronics equipment market became an integral part of the global economy. Research in small-signal MOSFETs continued and new products were introduced. It has progressively captured a higher portion of the global electronics market since 1970 [4]. MOSFETs were used in power electronics in the early 1980s due to their excellent proficiency in carrying current, high off-state and low on-state voltage drop. The performance of the conventional Si complementary MOS has upgraded by 17% per year through proper gate length scaling in association with technology developments over the last 30 years to make more and higher speed compact devices, which consume low power [5]. However, the conventional device dimension scaling could not continue forever. Novel materials and device architectures have become inevitable to boost the device performance.

Silicon-on-Insulator (SOI) MOSFETs started attracting attention due to their sharp subthreshold slope and low body effect in the sub-100 nm regime. SOI microprocessors show 22% improvement in speed over bulk MOSFETs. These MOSFETs also offer outstanding radiation stiffness and controlled second order effects for submicron VLSI applications [6].

Multi-gate transistors such as double-gate (DG) [7], triple-gate [8] and quadruple gate [9] architectures have become popular in the semiconductor industry as well as in the research community. In the sub-100 nm regime of scaling, more non-silicon elements were also introduced to replace Si technology by the development of source/drain, channel region, gate dielectrics, gate materials and isolation techniques of MOSFETs [10,11].

Novel materials provide alternative solutions as they show excellent transport properties. Non-Si materials are mainly Ge, III–V semiconductors, graphene, carbon nanotubes, semiconductor nanowires and high-k dielectrics, which are extensively used in research work [12,13]. At a 90 nm node, SiGe S/D was used to generate uniaxial strain. Oxide thickness less than 1.0–1.5 nm allows huge amounts of tunneling currents at 1 V operating voltage to accommodate standby power necessities in most cases. This shortcoming has been mitigated by the use of a high-k metal gate at a 45 nm node. Beyond 32 nm, high mobility materials along with an improved interconnect technology and lithographic techniques are the foremost criteria of CMOS technology. Beyond the 22 nm technology node, in addition to practical challenges, fundamental restrictions will limit the maximum attainable performance by CMOS. Therefore, performance booster will be required to achieve improved device performance.

6.1.2 Why Si

Since the first experimental demonstration in 1960 [14], Si-based MOSFETs have become the prime driver of the semiconductor industry, which caused substantial improvement in density, switching speed and cost effectiveness. Although Si is hardly found in pure form, in most of the cases it exists in various forms of silicon dioxide or silicates, which are distributed throughout the planetoids and planets as dust or sand. Si is considered as the second most abundant element in Earth's crust, building up 27.7% of the crust by mass [15]. But Si can be easily as well as economically extracted in pure crystalline form and thus it is popular because of its low cost requirements.

The element Si is extensively explored in related industry because of its retention of semiconductor properties at elevated temperature and also due to the unique advantages of its native oxide which is produced in a furnace and acts as a good semiconductor/dielectric interface. The silicon–native oxide interface is atomically abrupt and electrically perfect with very low defect densities. There is a large volume expansion that takes place during thermal oxidation on a Si wafer, leading to a stressed oxide interface. A Si–SiO_2 system has room for these stresses and provides an electrical interface with trapped and fixed charge densities. Any other semiconductor–insulator system cannot provide this level of interface perfection and thus Si–SiO_2 systems opened up new opportunities for widespread application in high-density process integration in the semiconductor industry. Integration in Si technology had its successful journey yielding chronologically LSI, VLSI, ULSI and GSI, which have application in the mobile communication system also. Si CMOS technology serves both the high-performance and low-power IC applications.

Si-based devices have dominated the semiconductor industry during the last two decades. Si-based technologies captured 98% of the IC market in 1995 [16]. Historically, performance improvements in the Si industry were achieved by scaling down the device dimensions, specifically the gate length and oxide thickness. However, the tangible improvements due to scaling are on their last legs, as physical and economic limits have been reached. Innovations of novel solutions will promote the growth of CMOS microelectronics revolution in future.

6.1.3 Non-Si Materials: Are They Compatible in the Nanoscale Regime?

Conventional silicon dioxide for gate oxide scaling has been replaced with high permittivity (high-k) materials. The issue of carrier mobility enhancement is another aspect to ponder over, which is also relevant for improving the device. Recently, Ge has been attracting considerable attention as a channel material as it displays greater mobility and enhanced drive current compared to silicon in the nanometer regime [17]. Ge offers high intrinsic carrier mobilities and high DOS in the conduction band, which make it a promising candidate in Ge CMOS technology for next-generation integrated microelectronics. Ge offers a smaller effective mass for electrons as well as for both heavy and light holes compared to silicon. A reduced effective mass in Ge can contribute to higher carrier mobility and drive currents in Ge compared to Si.

Again, a III–V semiconductor channel device has been a dream for the semiconductor industry for many years. These high speed material-based transistors are the core of many analog/ RF integrated systems. Graphene has been drawing attention since the mid-2000s because of its characteristics such as high electron and hole mobilities. It has the potential to become the prime material for coming years regarding nanodevices. Moreover, it is also a zero band gap material generating electrons and holes with both positive and negative electric fields. These novel material-based advanced MOSFETs and their performance in the circuit may further be investigated with respect to economic viability by a cost optimization approach. However, a higher dielectric constant makes the III–V channel device inferior regarding the electrostatic integrity especially in the sub-100 nm regime, and it can be minimized by introducing innovative architectures such as DG MOSFETs and gate-all-around (GAA) MOSFETs.

A further increase in carrier mobility can be achieved by introducing process-induced and/or global-induced strain which increases the device performance commensurate to scaling. A strained SiGe layer grown onto Si completely changes the material and electronic properties. Heterostructure MOSFETs built with Si and Ge offer high channel mobility while retaining low leakage current, which make them appropriate for scaling into the sub-15 nm regimes. This chapter is a humble endeavor to explore these novel materials and architectures in modern electronic devices.

6.1.4 A General Discussion on MOSFETs

MOSFETs have been the driving force for the last four to five decades. Although the architecture and working principle of MOSFETs have remained the same, the size of devices has been progressively reduced, due to the increase of the number of transistors per chip keeping parity with Moore's law [18].

MOSFETs are made up of a semiconductor substrate, on which a dielectric layer is deposited and a metallic gate is placed above it. The input resistance of MOSFETs is extremely high due to the isolation of the gate. Single crystalline silicon is used as the channel material, whereas for the insulator, a thermally oxidized layer of silicon is grown. Heavily doped drain and source regions are made with a dopant type different from that of the substrate. The structure of a MOSFET is displayed in Figure 6.1. There are two basic forms of MOSFETs.

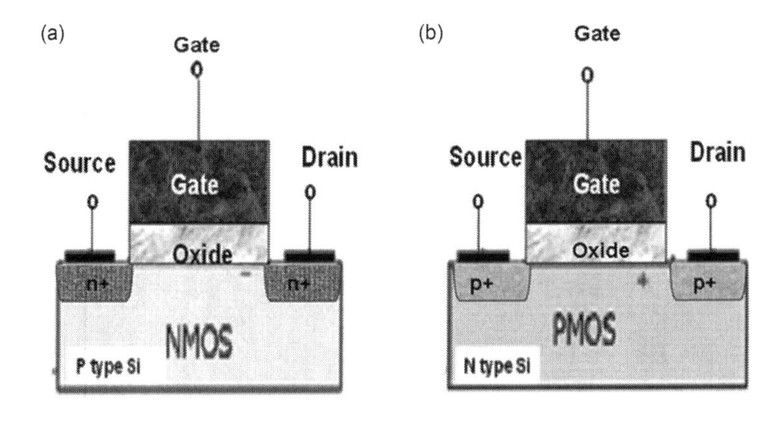

FIGURE 6.1 Schematic diagram of (a) n- and (b) p-channel MOSFETs.

Depletion type: This kind of MOSFET is equivalent to a closed switch, and a suitable gate-to-source voltage is required to switch the device ON.

Enhancement type: This type of MOSFET behaves like an open switch, and a suitable gate-to-source voltage is required to switch OFF the device.

Again, depending on the substrate material, MOSFETs can be classified as p channel and n channel types. For an n-channel device, when a small amount of positive gate bias is given, an electric field is established which generates positive charges that accumulate near the semiconductor–dielectric interface at the p region. As the amount of positive voltage increases, the surface becomes inverted containing an ultra-thin layer of electrons. With the application of suitable drain bias, these electrons are swept out towards the drain end and the current conduction is continued between the n^+ drain and source as demonstrated in Figure 6.2a and b.

6.1.4.1 Discussion on Some Electrical Parameters Related to MOSFETs

6.1.4.1.1 Threshold Voltage

This is one of the important parameters of MOS devices, defined as the minimum gate bias which makes a channel of mobile charges between the insulating layer and the substrate. Various definitions of threshold voltage exist focusing on different aspects of extracting it. Under strong inversion conditions, it is the minimum gate voltage which makes the surface potential two times that in the bulk [19]. For a heavily doped bulk substrate, the inversion charges are positioned next to the surface and the electrostatic integrity is controlled by the surface potential, whereas for an undoped substrate, inversion charges are found everywhere due to the penetration of the gate electric field. Therefore, in the case of undoped channel devices, threshold

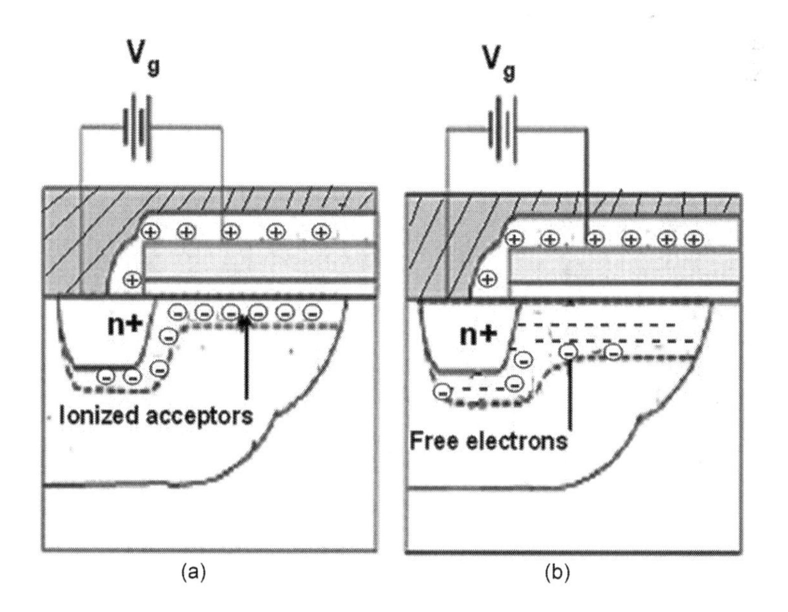

FIGURE 6.2 Development of (a) depletion and (b) inversion layers in an n-channel MOSFET.

voltage is different, and it is the gate voltage for which the sheet density of inversion charges is the same as the critical threshold charge density, which detects the on-state of the device adequately.

6.1.4.1.2 Subthreshold Slope

The subthreshold or below-threshold region is significant for low-power applications. In this region, the surface potential is nearly the same across the channel and minority carriers play the leading role in the conduction of current. As the diffusion current caused by minority carriers is related exponentially to gate bias, the transfer characteristics will therefore be a straight line on a semi-logarithmic scale below the threshold voltage as shown in Figure 6.3a, and the slope of this line is called

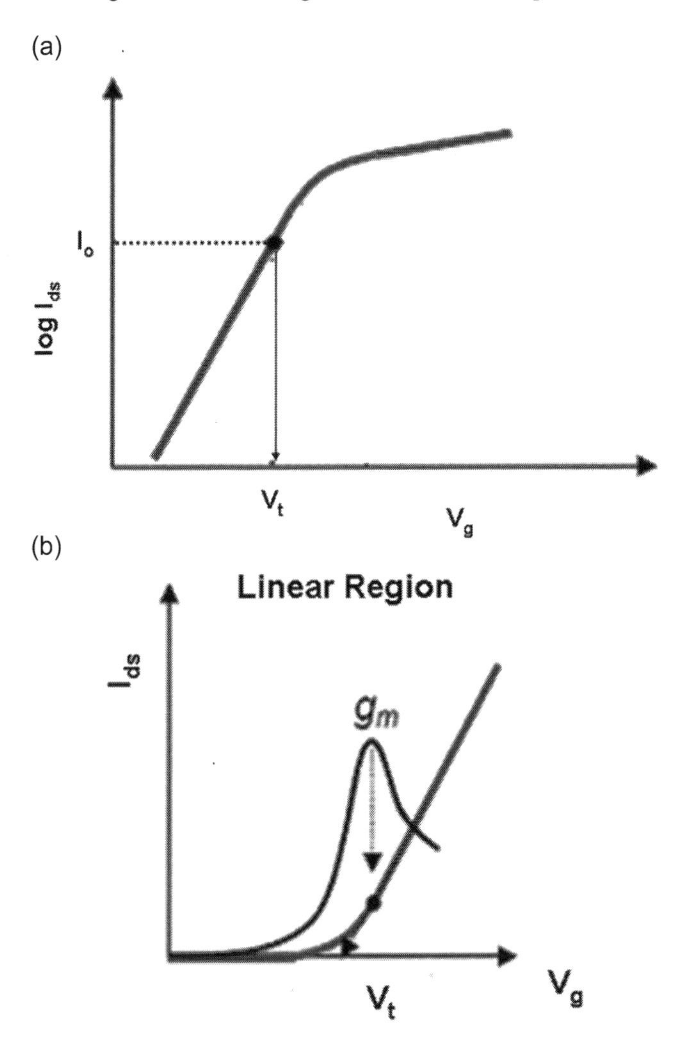

FIGURE 6.3 I–V characteristics of MOSFETs (a) in the log scale and (b) in the linear scale.

subthreshold slope. The reciprocal of this slope is called subthreshold swing (S) [20], which is basically the shift of gate voltage (V_g) required to induce one order of magnitude of drain current (I_{ds}). Mathematically,

$$S \equiv \log_{10}\left\{\frac{dV_g}{d(\ln I_{ds})}\right\} \equiv \left\{\frac{dV_g}{d(\beta\psi_s)}\right\}$$

6.1.4.1.3 Transconductance

This is essentially the gain in a MOSFET. It represents the change in drain current (I_{ds}) for the change in the gate bias (V_g) for a constant drain-to-source voltage (V_{ds}). Mathematically, transconductance (g_m) can be defined as $g_m = \left[\frac{\partial I_{ds}}{\partial V_g}\right]_{V_{ds}}$.

The MOSFET transconductance curve shown in Figure 6.3b is often used to study the effects of scattering caused by surface roughness. It is also used to observe the impacts of interface states and series resistance on the performance of a device. The peak of the curve is frequently utilized to monitor the impact of hot carriers on device reliability.

6.1.4.1.4 Output Conductance

Output conductance (g_d) is a useful figure of merit of MOSFETs and its accurate measurement is also necessary for analog circuit design. It qualifies the drain current variation with the variation of drain–source voltage for a constant gate–source voltage. Mathematically, output conductance can be defined as $g_d = \left[\frac{\partial I_{ds}}{\partial V_{ds}}\right]_{V_g}$.

The inverse of output conductance is called output resistance and is related to the intrinsic voltage gain of the transistor. The output conductance degrades with reducing channel length due to lowering of potential barrier at high drain voltages. A large value of interface trap charges can also deteriorate the output conductance.

6.1.4.2 Short-Channel Effects (SCEs)

The short-channel effects (SCEs) are significant issues in device design and they vary due to process tolerance. Long-channel MOSFETs are defined as devices whose width and length are sufficiently long so that side effects can be ignored. In these devices, the channel length must be sufficiently larger than total depletion widths in source and drain sides. However, for short-channel devices, the channel length is nearly the same as the total depletion widths associated with the drain and source sides [20], and one needs to consider the edge effects. The vital issues to be addressed for device design in the short-channel regime are the following:

- Channel length modulation,
- Threshold voltage roll-off,
- Drain-induced barrier lowering,
- Mobility degradation.

6.1.4.2.1 Channel Length Modulation

It is the shortening of inverted channel length with an increase of drain bias due to the expansion of the non-inverted region towards the source as shown in Figure 6.4a. The channel length reduction further reduces the channel resistance, resulting in an increase of drain current especially in the saturation region of MOSFET operation where the drain bias is kept high. The effect is more obvious for a smaller geometry where the source-to-drain separation is small and the doping concentration of the substrate is kept low. The most severe effect of channel length modulation is merging of depletion layers around the drain and source regions, which is called punch through. This is the most undesirable condition where the effective channel length almost becomes zero causing a rapid increase of drain current with drain voltage.

6.1.4.2.2 Threshold Voltage Roll-Off

Channel length modulation reduces depletion width in the channel and it is caused by charge distribution among the source, drain and gate [20] as schematically shown in Figure 6.4b. Due to the reduction of depletion charges, a small threshold voltage is sufficient to invert the channel, which is termed threshold voltage roll-off. The shared charge becomes a significant factor for low dimensions, and this results in a threshold voltage roll-off with lowering of gate length.

6.1.4.2.3 Drain-Induced Barrier Lowering (DIBL)

In a MOSFET, this effect occurs due to the lack of proper scaling of the device. A potential barrier exists between the source and the channel in the weak inversion region of MOSFET operation. As the barrier height reflects on the drift and diffusion components of current, under off conditions, it prevents the flow of electrons towards the drain. The potential is initially controlled by the gate bias [20], but at high drain bias, the barrier height is decreased, leading to a large drain current as shown in Figure 6.4c.

6.1.4.2.4 Mobility Degradation

In a semiconductor, carriers usually drift in response to an applied electric field, and their velocity is linearly dependent on it. But the relationship no longer remains correct at a high electric field (~10^5 V/cm). The carrier velocity stops increasing at high electric fields and gets saturated [20]. In the channel, a parallel and a vertical component of the electric field are present. The latter causes the bouncing of carriers into the interface causing mobility degradation. Beyond 10^5 V/cm, collisions with the interface diminish the mobility further. Since short channels induce a greater vertical field, mobility degrades more in short-channel devices.

6.1.5 Innovative Architectures as Performance Boosters

The concept of device scaling has always given rise to higher device density and better functionality. As per the suggestion of the industry roadmap of CMOS technology, we are reaching some technological barriers and physical limitations due to continuous scaling, and the need for alternative device structures is foreseen.

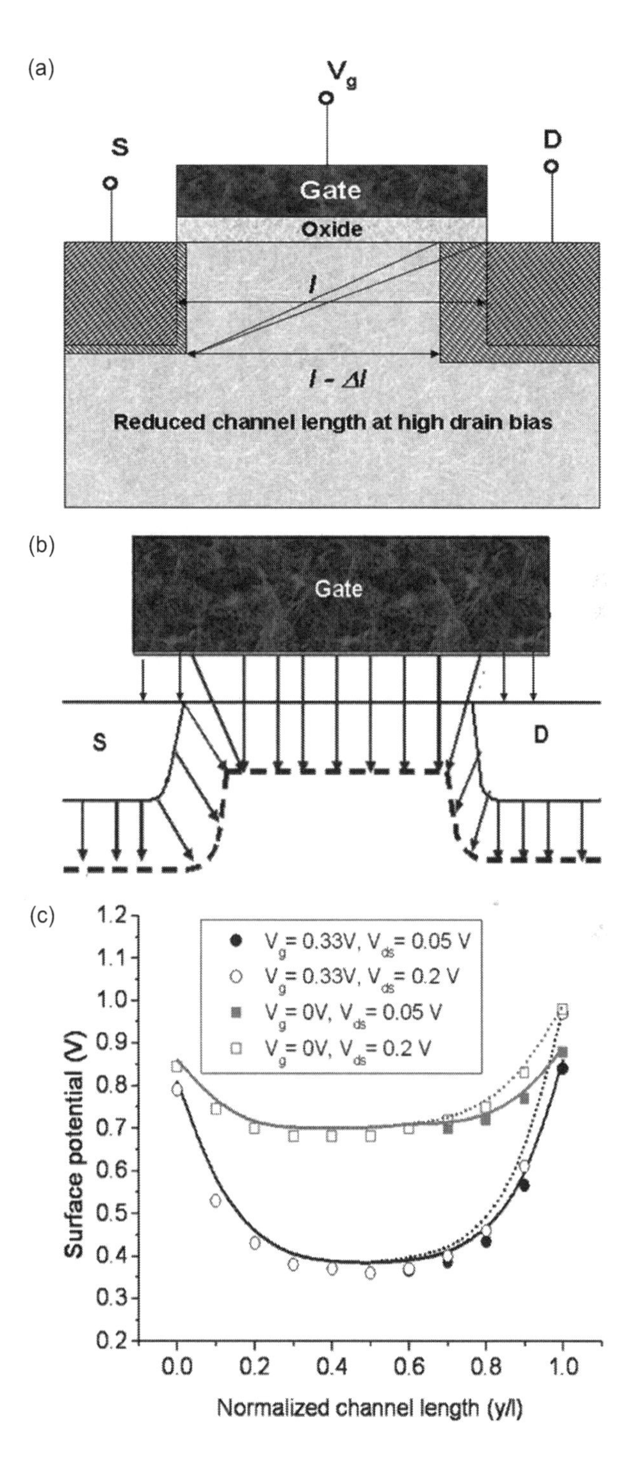

FIGURE 6.4 (a) Channel length modulation. (b) Charge sharing among source, drain and gate. (c) Reduction of potential barrier at high drain bias [21].

6.1.5.1 SOI MOSFETs

A SOI MOSFET was first proposed by Mueller and Robinson in 1964 [22]. It contains a very fine layer of silicon with ~10 nm thickness, which is isolated from the substrate by a relatively thick (~100 nm) layer of silicon oxide deposited on the Si substrate as shown in Figure 6.5a. This separation of a dielectric–semiconductor layer results in a reduction of the parasitic and junction capacitance and improves the device performance. Recent experimental studies have also paid attention to SOI devices because

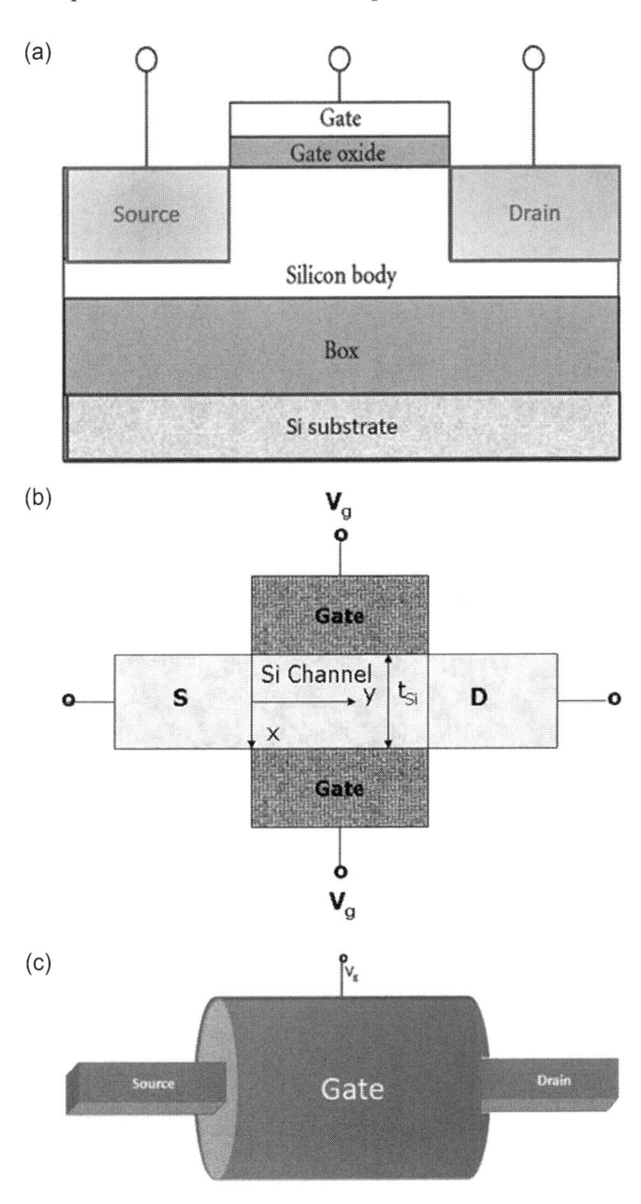

FIGURE 6.5 Schematic diagram of (a) SOI, (b) DG and (c) GAA MOSFETs.

of their low-power applications due to improved isolation, low parasitic capacitance, low leakage current, reduced subthreshold slope and superior scalability compared to bulk silicon CMOS devices. As a consequence of thickness variation of the Si layer, SOI MOSFETs can be classified as fully depleted (FD) or partially depleted (PD). PD transistors are constructed on a comparatively thick (larger than the depletion width) silicon substrate, whereas for a FD SOI, the depletion region spreads throughout the entire substrate. Deficiency of kink effects, large gains, high speed, low power consumption and the utmost level of soft-error immunity make the FD devices more popular than PD devices [6].

6.1.5.2 Double-Gate (DG) MOSFETs

A DG transistor, proposed in the 1980s, comprises a steering channel, enclosed by two gates on opposite sides. A DG MOSFET was first demonstrated in 1987 [23]. A second channel is introduced in a DG MOSFET by adding a lower gate at the lower surface. The two gates are electrically coupled so that both of them are utilized for channel modulation. As every part of the channel is adjacent to the gate electrode, the channel potential is better controlled. Again, two gates can terminate the drain field lines very effectively so that drain potential can be optimized successfully, which further reduces the SCEs [21]. Additionally, a reinforced electrostatic pairing between the gate electrode and the conduction channel is generated in a DG MOSFET architecture, which allows an additional gate length scaling (approximately double) with respect to the single-gate architecture. Other interesting features of these devices are near ideal (~60 mV/dec) subthreshold slope, excellent SCE immunity, volume inversion and low parasites.

6.1.5.3 Gate-All-Around (GAA) MOSFETs

For GAA devices, the entire active region is bounded by insulator couples with the gate electrode [24]. Since each and every part of the channel is surrounded by the gate, the channel potential is effectively controlled with this architecture. Moreover, instead of just in one limited surface, the conduction takes place throughout the entire volume of the device resulting in inversion of total channel mobility leading to superior current drive. DG and GAA MOSFETs are the most appropriate device architectures for conquering DIBL, degradation of subthreshold swing and velocity saturation.

6.1.5.4 Dual Material Gate (DMG) MOSFETs

In 1999, Long et al. [25] suggested a new architecture where two different materials are employed for the gate as depicted in Figure 6.6a. Unlike asymmetric structures, this structure is developed with two gates M_1 and M_2 with dissimilar work functions. The SCEs can be suppressed by a sudden alteration of surface potential along the channel as shown in Figure 6.6b. The solid and the dotted lines represent the surface potential of a DMG and a DG MOSFET, respectively. The step function like nature of the surface potential of the channel region defends the source side region adjacent to the first gate (M_1) by absorbing the additional drain-to-source voltage by M_2. Therefore, the region adjacent to M_1 is separated from the variation of drain potential in the opposite side. Apart from offering better control over channel conduction,

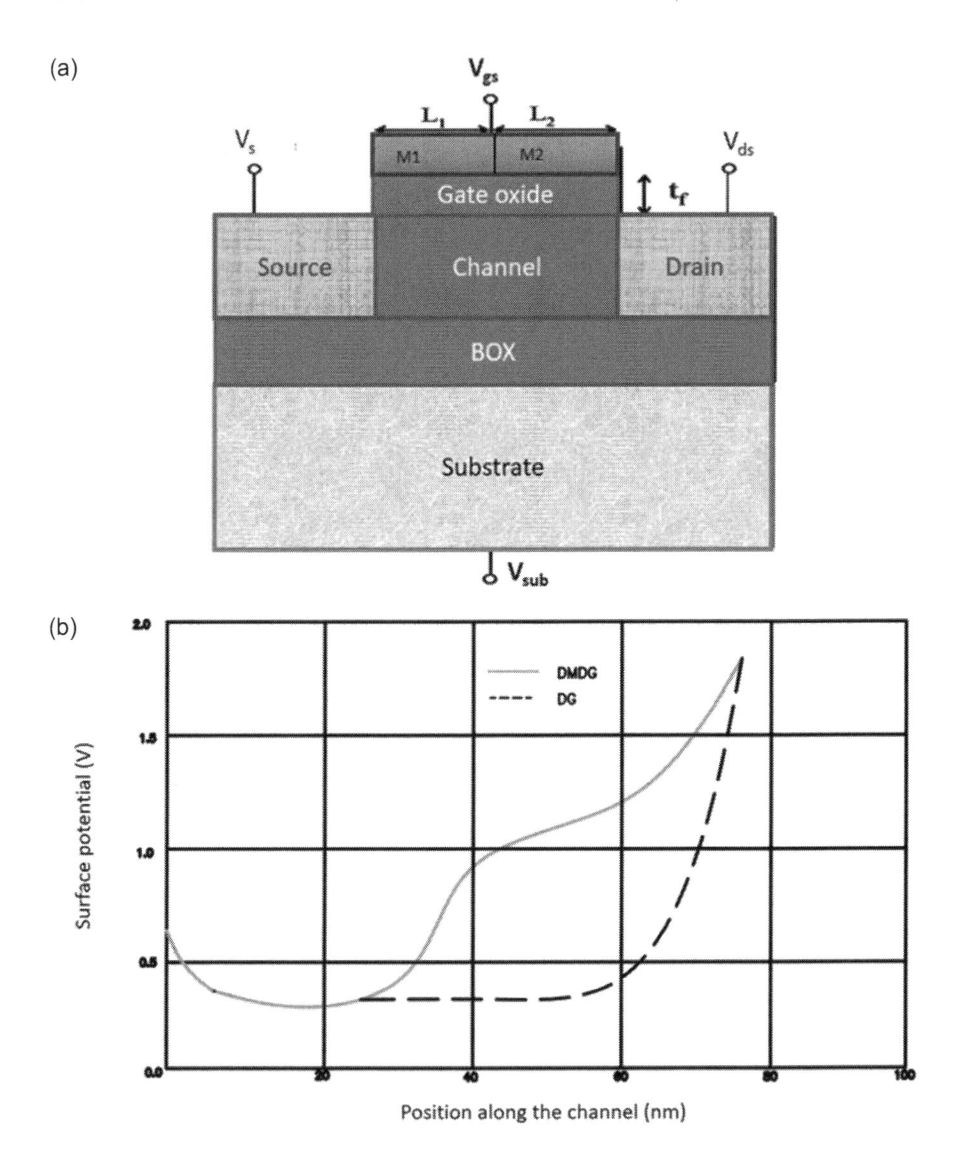

FIGURE 6.6 (a) Schematic diagram and (b) surface potential of DMG MOSFETs (solid line).

DMG MOSFETs also contribute an enlarged gate electric field and remarkable transport efficiency.

6.1.5.5 Quantum Well/Quantum Wire MOSFETs

Apart from the above-mentioned practices to boost up the device performance, quantum mechanical tunneling is often used to improve the device speed. 1-D quantum confinement and 2-D quantum confinement are often used to increase the number of carriers in the channel. The quantum confinement plays a significant role when the thickness of the quantum well is comparable to the de

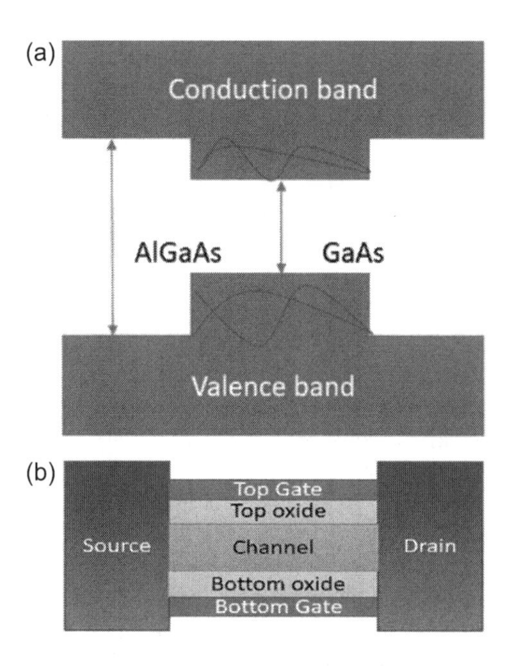

FIGURE 6.7 (a) Quantum well and (b) quantum wire MOSFETs.

Broglie wavelength of the carriers [19]. Quantum wells are formed when a low band gap material, such as gallium arsenide (GaAs), is inserted between two layers of a high band gap material such as aluminum gallium arsenide (AlGaAs). Since the motion of carriers is restricted in 2-D, higher density of carriers is observed in this type of structure. Figure 6.7a shows the formation of quantum wells. The 1-D confinement models are further extended, by 2-D confinement, and quantum wire MOSFET structures [8] are developed in which carrier motion is restricted along a wire-like passage as shown in Figure 6.7b. There are many applications of these types of architectures in high-speed optoelectronic devices.

6.1.6 Novel Material-Based Advanced MOSFETs – A Simulation Study

A III–V semiconductor channel device has been a long-term expectation of researchers. The fruitful application of III–V semiconductors can introduce new applications and topographies such as integrating logic, optoelectronic and communication stages on the same Si wafer. There are a few members of the III–V semiconductor family such as InGaAs, AlGaAs, GaAs, InAsSb and InAs with exceptional electron transport properties [26] such as high electron and hole mobilities with respect to silicon. These materials are the backbone of many high-frequency, high-speed electronic devices and circuits. Actually, there is a vast application of III–V ICs in different domains such as wireless LNA, smart phones, satellite communications, radio astronomy, defense systems and cellular base stations, and a developed industry is required for manufacturing III–V ICs. So there is a huge scope of work in this rising era.

6.1.6.1 Practical Challenges

The foremost challenge in the case of a III–V semiconductor-based MOSFET is development of high-k thin films on III–V substrates for application in future transistors with minimum channel lengths of 22 nm and below. Due to the deficiency of a good native oxide interface, III–V semiconductors often suffer from different limitations such as Coulomb scattering caused by interactions of bulk oxide charges and fixed interface charges, scattering due to surface roughness and remote phonons. Therefore, a broad research area is focused on high-k III–V interface states. Additionally, another problem is also relevant to III–V semiconductor devices, that is, dielectric charge trapping which generates reliability problems like noise [27]. Furthermore, most of the III–V semiconductors have a higher dielectric constant than Si as shown in Table 6.1, which contributes to reduced electrostatic integrity in downscaling. Among all known semiconductors, $InAs_xSb_{1-x}$ has one of the highest electron mobilities as depicted in Table 6.1. A simulation study of an InAsSb channel MOSFET is presented here [28]. The simulation structure is depicted in Figure 6.8a.

For top-gate, a 10-nm-thick ZrO_2 and for bottom gate a 50 nm SiO_2 gate dielectric are used for simulation. Ni is used for the formation of ohmic source and drain contacts. A 2-D numerical device simulator ATLAS is applied for simulation [29]. The device comprises an n-type ultrathin $InAs_{0.7}Sb_{0.3}$ layer of 7 nm thickness. The dielectric constant and band gap are computed as 17.7 and 0.174 eV, respectively, for the $InAs_{0.7}Sb_{0.3}$ channel. The doping concentration is taken as 1×10^{17} cm^{-3}. Figure 6.8b–d show the different parameters extracted by ATLAS. Figure 6.9a shows the variation of drain current with gate-to-source voltage for InAsSb and Si channels in linear and log scales considering various interface trap charge densities for InAsSb channel devices.

Figure 6.9a shows that the InAsSb channel MOSFET offers higher drain current than the Si channel device. But in spite of obtaining a higher value of ON current, a large amount of OFF current relative to the Si device is also observed for the InAsSb channel which makes it inferior for low-power applications. Figure 6.9b depicts the transconductance and output conductance of InAsSb channels for an extensive range of gate biases. Again we can observe a higher transconductance, but at the same time, the output conductance is also found large for InAsSb channel MOSFETs which

TABLE 6.1
Fundamental Electronic Parameters of Si and InAsSb

Parameters	Si	InAsSb
Band gap, E_g (eV)	1.12	0.174
Electron affinity, χ (eV)	4.05	4.9
Hole mobility, μ_h (cm^2/V/s)	450	1,250
Electron mobility, μ_e (cm^2/V/s)	1,500	8,000
Dielectric constant, k	11.9	17.7

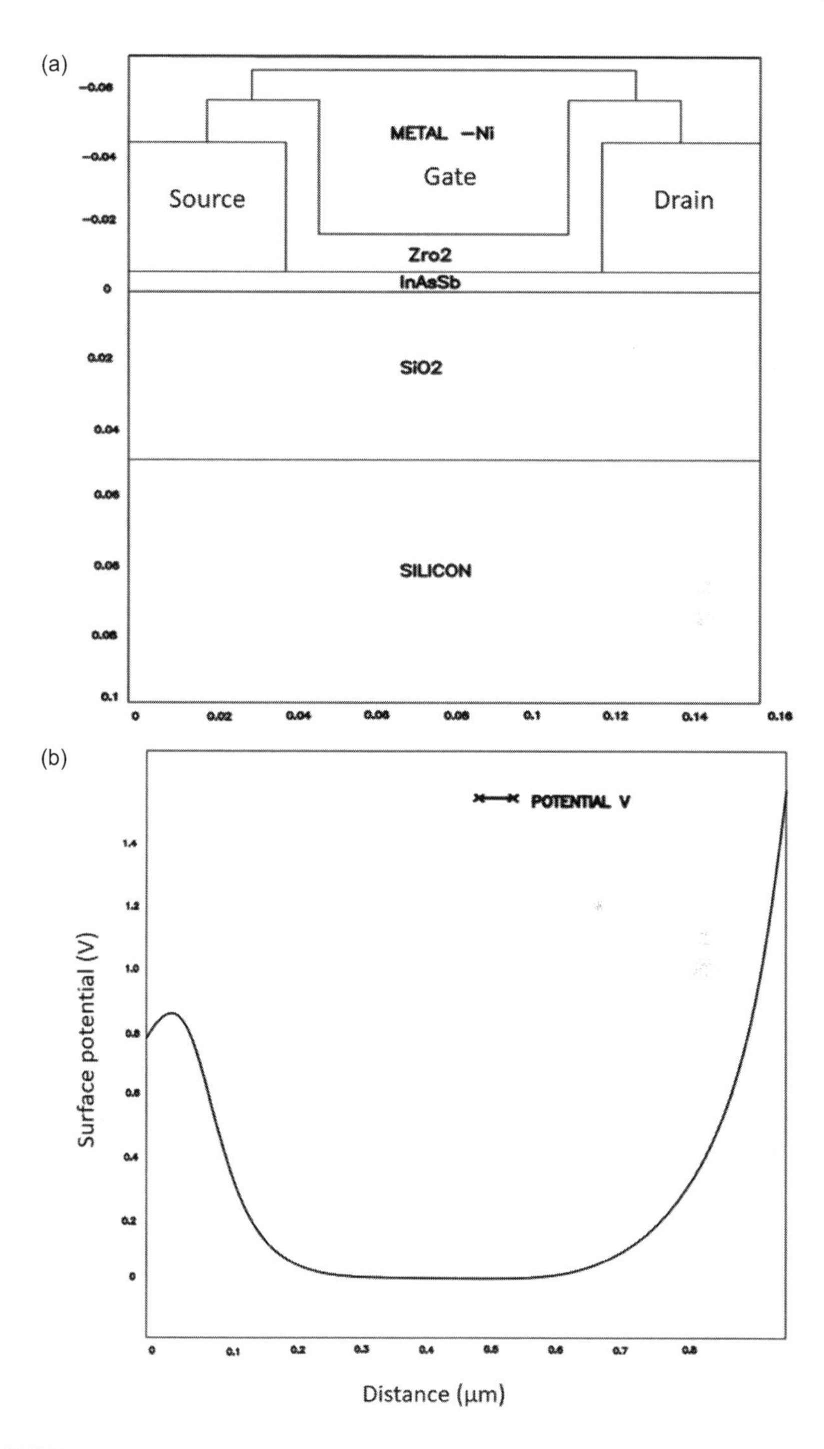

FIGURE 6.8 (a) Structure and (b) surface potential of an InAsSb MOSFET.

(Continued)

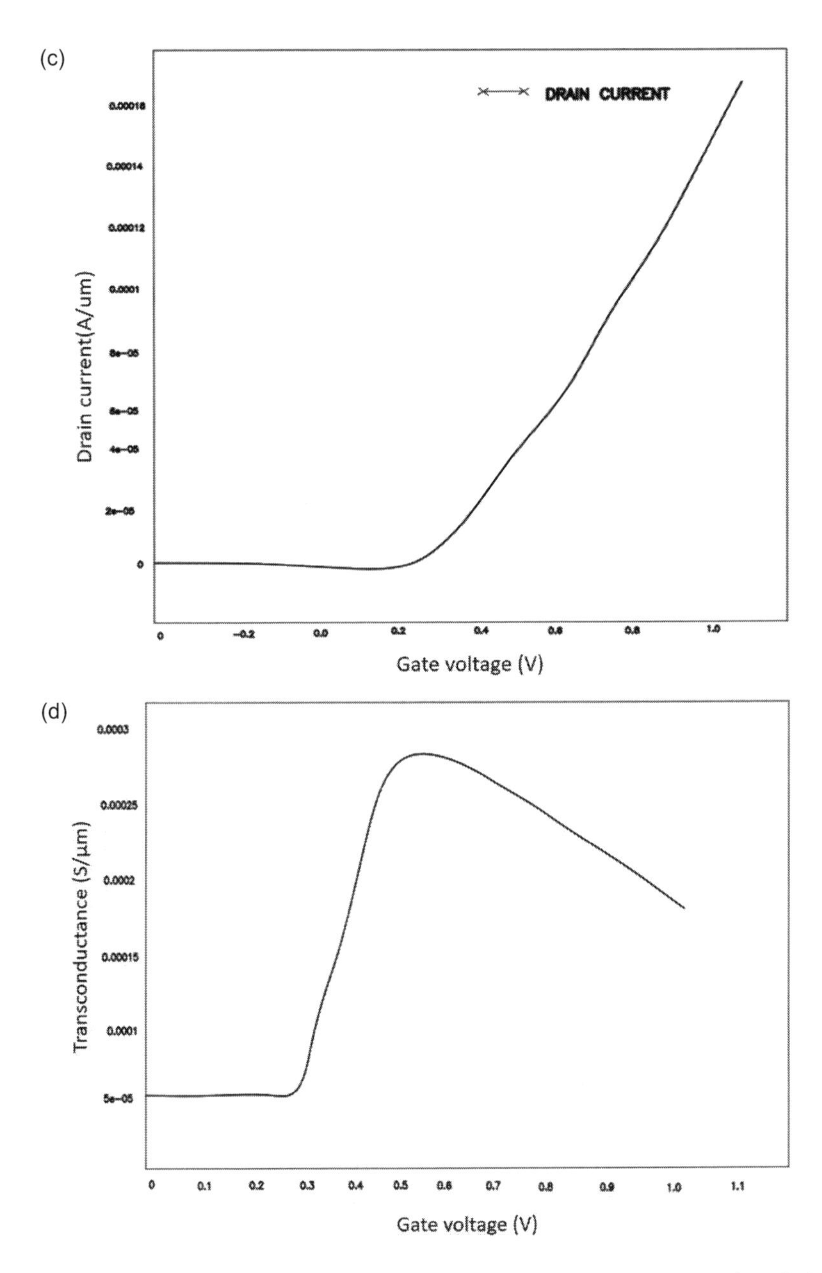

FIGURE 6.8 (CONTINUED) (c) Drain current and (d) transconductance of an InAsSb MOSFET.

indicates high DIBL. Thus the InAsSb channel devices yield better quality in terms of ON current, transconductance etc. But it is lagging behind the Si channel device regarding the SCEs. Needless to mention, continuous research activities are required to stretch beyond these limitations so that implementation of these high mobility novel materials becomes a regular practice in future.

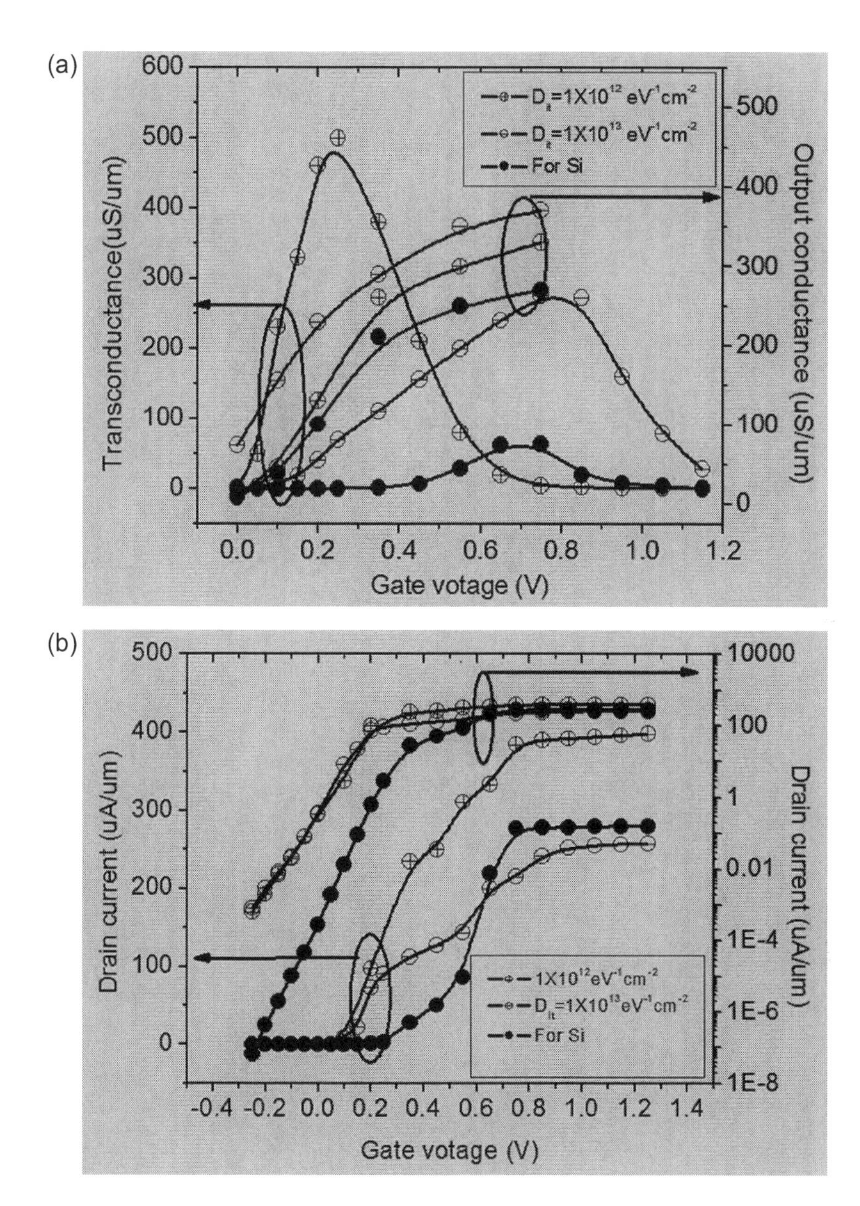

FIGURE 6.9 Variation of (a) drain current, (b) transconductance and output conductance with gate bias.

REFERENCES

1. An outline of the history of the transistor. http://www.pbs.org/transistor/background1/events/miraclemo.html.
2. Integrated Circuit Invention history and story behind it https://www.circuitstoday.com/integrated-circuit-invention-history-and-story-behind-the-scenes.
3. Integrated Circuit https://history-computer.com/ModernComputer/Basis/IC.html.

4. Indian Semiconductor Fabless Startup Ecosystem https://www.iesavisionsummit.org/2018/Documents/files/Final%20executive%20summary%20-%20indian%20semiconductor%20fabless%20startup%20ecosystem.pdf.
5. Park J. T. and Colinge J. P. (2002). Multiple-gate SOI MOSFETs: Device design guidelines. *IEEE Transactions on Electron Devices*, 49(12), 2222–2229.
6. Ernst T., Cristoloveanu S., Ghibaudo G., Ouisse T., Horiguchi S., Ono Y., Takahashi Y. and Murase K. (2003). Ultimately thin double-gate SOI MOSFETs. *IEEE Transactions on Electron Devices*, 50(3), 330–338.
7. Djeffal F., Dibi Z., Hafiane M. L. and Arar D. (2007). Design and simulation of a nano electronic DG MOSFET current source using artificial neural networks. *Materials Science and Engineering C*, 27(5), 1111–1116.
8. Colinge J. P., Baie X., Bayot V. and Grivei E. (1996). A silicon-on-insulator quantum wire. *Solid-State Electronics*, 39(1), 49–51.
9. Hisamoto D., Kaga T., Kawamoto Y. and Takeda E. (1989). A fully depleted lean-channel transistor (DELTA)-a novel vertical ultra thin SOI MOSFET. International Technical Digest on Electron Devices Meeting, 833–836.
10. Hawkins G. A. (1985). Lateral profiling of interface states along the sidewalls of channel-stop isolation. *Solid State Electronics*, 28(9), 945–956.
11. Murarka S. P., Fraser D. B., Sinha A. K. and Levinstein H. J. (1980). Refractory silicides of titanium and tantalum for low-resistivity gates and interconnects. *IEEE Journal of Solid State Circuit*, 15(4), 474–482.
12. Kim D., Krishnamohan T. and Saraswat K. C. (2008). Performance evaluation of III–V double gate n-MOSFETs. Device Research Conference, IEEE, 67–68.
13. Svintsov D. A., Vyurkov V. V., Lukichev V. F., Orlikovsky A. A., Burenkov A., and Oechsner R. (2012). Tunnel field effect transistors with graphene channels. International Conference “SILICON”, 279–284.
14. Kahng D. and Atalla M. M. (1960). Silicon-silicon dioxide field induced surface devices. IRE-AIEEE Solid State Device Research.
15. http://en.wikipedia.org/wiki/Silicon
16. Chapbell S. A. (1996). *The Science and Engineering of Microelectronic Fabrication.* Oxford University Press.
17. Rahman A., Ghosh A. and Lundstrom M. (2003). Assessment of Ge n-MOSFETs by quantum simulation. IEDM Technical Digest, 471–474.
18. Moore G. E. (1998). Cramming more components onto integrated circuits. *Proceedings of IEEE*, 86(1), 82–84.
19. Taur Y. (2001). Analytical solutions of charge and capacitance in symmetric and asymmetric double – Gate MOSFETs. *IEEE Transactions on Electron Devices*, 48(12), 2861–2869.
20. Taur Y. and Ning T. H. (1998). *Fundamentals of Modern VLSI Devices.* Cambridge University Press.
21. Bhattacherjee S. and Biswas A. (2008). Modeling of threshold voltage and subthreshold slope of nanoscale DG MOSFETs. *Semiconductor Science and Technology*, 23(1), 258594.
22. Mueller C. W. and Robinson P. H. (1964). Grown-film silicon transistors on sapphire. *Proceedings of IEEE*, 52(12), 1487–1490.
23. Balestra F., Cristoloveanu S., Benachir M., Brini J. and Elewa T. (1987). Double-gate silicon-on-insulator transistor with volume inversion: A new device with greatly enhanced performance. *IEEE Electron Device Letters*, 8(9), 410–412.
24. Colinge J. P., Gao M. H., Romano-Rodriguez A., Maes H. and Claeys C. (1990). Silicon-on-insulator gate-all-around device. IEDM Technical Digest, 595–598.
25. Long W., Ou H., Kuo J.-M. and Chin K. K. (1999). Dual material gate (DMG) field effect transistor. *IEEE Transactions on Electron Devices*, 46(5), 865–870.

26. Bennett B. R., Ancona M. G., Boos J. B., Shanabrook B. V. (2007). Mobility enhancement in in strained p-InGaSbp- InGaSb quantum wells. *Applied Physics Letters*, 91(4), 042104-3.
27. Bhattacherjee S. and Biswas A. (2019). Development of noise model for InAsSb MOSFETs and their applications in low noise amplifiers. *Microsystem Technologies*, 25(5), 1555–1562.
28. Bhattacherjee S. and Biswas A. (2014). Analog circuit performance of high ultrathin-body InAsSb-on-insulator MOSFETs. IEEE Students' Technology Symposium, 396–401.
29. SILVACO International. ATLAS user's manual. A 2-D Device Simulator Software Package.

7 Energy Storage Device Fundamentals and Technology

Himanshu Priyadarshi, Ashish Shrivastava, and Kulwant Singh
Manipal University Jaipur

CONTENTS

7.1 INTRODUCTION

Futuristic planning is an important aspect for any society that wants to prosper progressively. Energy security is very crucial for any civilization. Substantial reserves of energy are necessary to keep the wheel of civilization pacing, otherwise it might come to a screeching halt. The economies of many societies are abjectly dependent on fossil fuels, and this dependence has caused colossal loss of lives. The increased awareness about the dangers of fossil fuels has forced the intellectual community to think in non-conventional ways, as sticking too much to the convention of fossil fuels has not done any good either to the planet or its people.

TABLE 7.1
Predictability of Solar Photovoltaic Energy

Factors	Specifications	Predictability
Geographic location	Latitude, meridional, and diurnal changes, elevation with respect to the mean sea level – affects the vertical air column, and hence the attenuation	Deterministic
Collector plate disposition	Collector tilt angle, orientation with respect to horizontal	Deterministic
Time of day	Hour angle – represents temporal variation	Deterministic
Time of year	Declination – model seasonal effect on insolation	Deterministic
Atmospheric conditions	Ozone layer status, dust, humidity, and cloud cover	Stochastic

Mother Nature provides abundant energy resources which are much more than that can be ever utilized or even that is being currently harnessed. Hence, it stands to reason that a strong support system be developed for the harvesting of natural energy resources such as solar photovoltaic energy, wind energy, and so on. An important challenge in the harvesting of energy resources such as wind and solar is to support these sources of energy with predictability. The availability pattern of these resources is highly stochastic, i.e., it varies with climatic conditions. A typical case in point is detailed in Table 7.1.

The last item in the table shown above is the atmospheric conditions, which include several important sub-factors, and hence solar and wind energy get very much affected by these atmospheric conditions. Sometimes there is surplus generation and sometimes there may be deficit; the elements in nature balance each other in terms of cause and effect. This provides a clue as to how the variegatedness of the energy procurement quanta can be managed by making use of energy storage devices.

Storage of electrical energy harnessed from renewable sources of energy will provide consistency of availability of electrical power. *Energy storage is a necessity, and not an option.* Right from the discovery of electrical power, batteries have been at the forefront of energy dispensation. At the inception of batteries, their utilization was limited to being used only once, and this limitation in terms of lifetime and non-reusability arrested the growth of this sector. The progression of time unveiled the "reusability" capability of batteries and their corollary devices such as supercapacitors and pseudo-capacitors. This energy storage sector has attracted a lot of investment from advanced nations because they could see the absolute requirement of energy storage devices as the essential support system for the renewable energy sector. Various energy storage technologies are being utilized all over the world, with electrochemical energy storage devices being particularly useful due to their commendable energy storage and power transaction capabilities.

7.2 GENERIC ENERGY STORAGE DEVICE CONCEPTS

The foresighted investors who can envision the worth of the energy storage market are pumping money for the manufacturing of energy storage devices which are being engineered with variegated nomenclatures such as batteries, supercapacitors, ultracapacitors, pseudo-capacitors, fuel cells, etc. The expected outcome of this section is not to present a differentiational discussion of these various devices meant for energy storage. As we go through this section, we intend to present a unified outlook for electrochemical energy storage. The rationale behind prioritizing a unified approach is to develop sound fundamentals, which can then be utilized in the design of devices with niche applications and incumbent features. This synergistic view will also protect the reader from the misconceptions due to the jargon of devices that are being proposed in multitude from the research community. Simplistic understanding of a generic electrochemical energy storage device with regard to the way it exchanges energy with external loading circuits can be initiated with the crude analogy of gravitational potential energy storage and exchange with skateboard, illustrated as follows. An energy storage device is a bidirectional energy exchange device, and the same device must be capable of accumulating as well as dispensing electrical energy as per the situation. Hence, *repetitiveness in reusability with longevity* is an important consideration for the design of electrochemical energy storage devices. The skateboard may go upward as well as downward in the concavity of the gravitational potential well many times if the mechanical design of the skateboard remains unaffected from its locomotion with negligible wear and tear. Similarly, *if the structure and material composition of an electrochemical energy storage device remain preserved in the course of multiple energy exchanges, the energy storage device will have a long lifetime and consistency in performance.*

The *repetitiveness in reusability with longevity* feature is a very crucial engineering consideration because the generic energy storage design technology must be adopted at the global level in this era of increased competitiveness and ambitious business expansion mentality of the investors in technology. Hence, it becomes imperative for us to understand how the energy storage design considerations play an important role in meeting this criterion. To this end, we must understand the most fundamental repetitive unit of the energy storage cell design. A cell is the most fundamental unit of an energy storage system.

In general, the cell acts as a voltage source, of course except solar cells, which are a current source. Strictly speaking, solar cells cannot be termed an energy storage device, as current flow is a kinetic parameter, whereas energy storage devices require a potential parameter. The potential parameter in the electrical domain is the voltage or electromotive force, and for an energy storage cell, the following parameters are very important.

7.2.1 Cell Voltage

It is the potential difference available across the electrical current outlet terminals of a cell. It depends on the active materials used inside the energy storage cell device.

Depending upon the electrochemical material constitution of the energy storage cell, the cell voltage varies.

If we contemplate the chronological evolution of energy storage cells, it gives us the clue of contemporary dominance of lithium ion cells in the market of energy storage. The Daniel cell opened the avenues for utilization of the electrochemical domain for energy storage, and the discovery of lead acid batteries further consolidated the prospects of electrochemical energy storage by increasing the cell voltage and the amount of energy that can be stored. The discovery of nickel metal hydride cells proved to be a milestone in the market of energy storage as it opened up the gateway to the technological landscape for intercalated electrode structures. The phenomenon of intercalation has proved to be very instrumental in the growth of lithium-based energy storage.

In an electrochemical energy storage device, the terminal potential difference depends on the electrochemistry. The difference in the electrode potentials of the positive and negative electrode determines the cell voltage to a considerable extent.

Generally, the elements placed at the left hand side of the periodic table are good candidates for being the active material in positive electrodes, as such elements act as good reducing agents. The elements placed near the right hand side of the periodic table are more suitable for being the active material in negative electrodes, as such materials are good oxidizing agents. However, this tuning between the left and right column elements of the periodic table cannot be done arbitrarily, as the electrolyte must be able to withstand the potential difference across it. An increase in the potential difference may be done by connecting energy storage cells in series; however, stable holistic electrochemistry and the safety cannot be compromised.

7.2.2 Energy Capacity of a Cell

The energy capacity of an energy storage cell is defined as the amount of energy the unit is able to deliver at the nominal cell voltage defined in the foregoing section. Generally, an energy storage device's capacity is commercially proclaimed in terms of ampere-hours. The term ampere-hours gives an index of its charge bearing capacity, and charge multiplied by the potential difference gives the energy capacity.

7.2.3 Cell Charging Rate or C-Rate

The time rate at which the energy capacity of an energy storage cell is being built-up or being discharged is called the cell charging rate or C-rate.

7.2.4 Energy Density and Specific Energy

The energy stored in an energy storage device per unit volume is termed energy density; whereas the energy stored per unit mass is known as specific energy. Thus for a given volume, the device with greater energy density can store more energy; whereas for a fixed energy rating, the device with greater energy density will be more compact. Similarly for a given mass, the device with more specific energy will store more energy; for a fixed energy capacity, the device with higher specific energy will be lighter.

7.2.5 Power Density and Specific Power

The cost of an energy storage device has an energy capacity component and a power component. While the premium on the energy storage is dependent on watt-hour, the premium on the power it is capable of transacting is dependent on the number of watts. In this light, the size and weight of an energy storage device have to be understood with respect to power transaction. The ability to discharge/accept power to a load/from a source per unit volume is known as the power density of an energy storage device, and when this ability is specified per unit mass, it is called the specific power.

7.3 FUNCTIONAL COMPONENTS OF AN ELECTROCHEMICAL ENERGY STORAGE DEVICE

The functional components of electrochemical energy storage devices have been depicted in Figure 7.1. The electrodes are chosen depending on the potential difference criteria.

The purpose of a negative electrode is to give electrons to the load during discharge, whereas the purpose of a positive electrode is to accept the electrons from the source during recharge.

The electrolytes are meant for ionic transport during recharge and discharge to maintain the electrodynamic equilibrium. Moreover, an electrolyte should be able to withstand the potential difference without dissociation. The details regarding the different functionalities with Figure 7.1 as the focal point have been tabulated in Table 7.2.

7.4 CORRELATION WITH POWER GRID

Generally, a storage system reserves energy by accumulating economically available electrical energy and provides electricity when necessary, attracting premium in terms of monetary aspects or for essential services. Energy storage is a malleable resource for a power grid that can be relied upon for delivering a range of services.

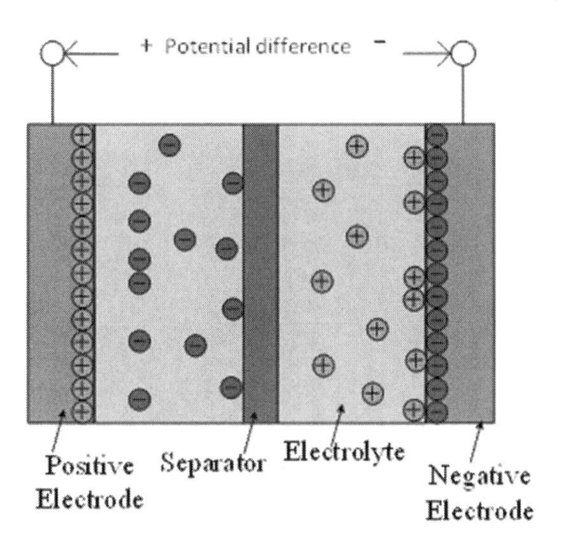

FIGURE 7.1 Charge storage model of a non-faradaic energy storage cell.

TABLE 7.2
Understanding the Essential Functional Components of an Energy Storage Cell

Component's Name	Illustration of the Components	Supplementary Description
Negative electrode	Graphene nano-structures, silicon forests	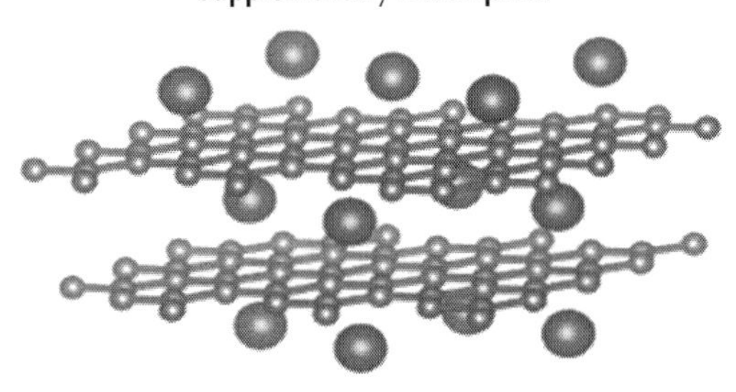 *Graphene layers with intercalated lithium ions [7]. Reproduced with permission of the International Union of Crystallography.* (http://journals.iucr.org/)
Positive electrode	Intercalation structures such as olivine electrodes, spinel electrodes, etc.	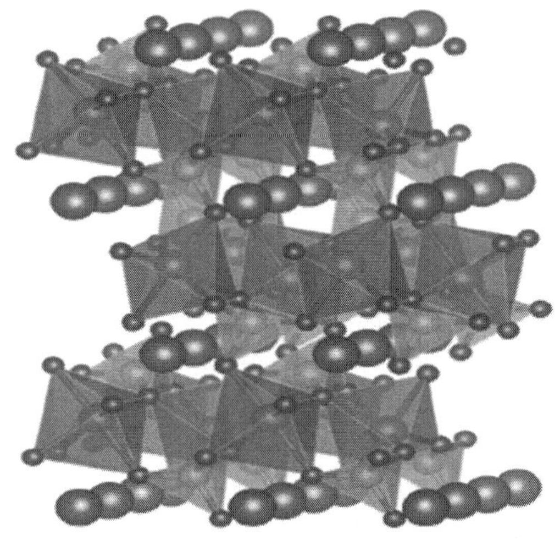 *Low cost, low toxicity lithium ferrous olivine phosphate structure with lithium ions shown as big spheres [7]. Reproduced with permission of the International Union of Crystallography.* (http://journals.iucr.org/)
Electrolyte solvents	Solvents have a common feature of slightly negatively charged oxygen atoms due to the double bonding, which causes polarization-induced dissolution	*Examples of electrolyte solvents: ethylene carbonate, propylene carbonate, dimethyl carbonate, ethyl methyl carbonate, and diethyl carbonate. Solvents do not participate in the cell electrochemistry* However, the wettability of the electrolyte with the electrode plays a very crucial role in enhancing the efficacy of the electrodes for enhancing the mass transfer mobility in the electrolyte This is achieved by building proper interfacial contacts for the electrodes and choosing appropriate electrolytes [8]

(*Continued*)

TABLE 7.2 (*Continued*)
Understanding the Essential Functional Components of an Energy Storage Cell

Component's Name	Illustration of the Components	Supplementary Description
Electrolyte salts	The electrolyte is generally known by its salt name	*Which could be either of the following – hexafluorophosphate, tetrafluoroborate, perchlorate etc.*
Separators	Separation is achieved by dint of nanoscopic structures	It is a permeable nano-membrane with cavities large enough to allow ions to pass through unbridled, but small enough that the bipolar electrode particles are electronically insulated
Current collector	Typically like the aluminum foil used in food packaging, but much reduced thickness	The high potential region uses aluminum foil; whereas the low potential region uses copper foil

7.4.1 Energy Asset Arbitrage

Energy is an invaluable asset, as the strength of the economy of a nation depends upon its energy security and self-dependence. However, the pricing of energy varies a lot across the different contours of its harvesting. Electrical energy being the entropically most favored form of energy facilitates energy exchange across different modes of energy. The different types of energies harnessed from different avenues differ in their prices depending on various factors related to their demand and supply. A conventional power grid has little provision for amply rewarding electrical energy subscribers who might be generating excess energy by renewable energy harvesting. Energy storage devices allow the proprietor or person interested in energy harvesting to gain maximum benefits from energy, by allowing them to store and wait till they get satisfactory premium for their energy harnessing efforts. This is done by utilizing arbitrage concepts from the finance sector, which help a business entity to make profits by capitalizing on the difference in price of a commodity across different sectors.

7.4.2 Demand Side Management

The load consumption pattern at consumer premises varies depending on the requirements of the utilization of the connected capacity. The grid authorities have laid clear-cut guidelines for power consumption pertaining to time-segmentation of the day. During peak hours of load consumption, the current drawn from the grid is very high. Hence it is advisable that non-vital power consumption may be avoided during peak hours. During off-peak hours, when the vital industrial activities are dormant, the essential domestic load utilization may be carried out, and such a practice is very much encouraged and incentivised. However, such practices of demand side management can be efficiently implemented only when the supply and utilization are decoupled through energy storage devices.

7.4.3 Assurance of Reliability

It balances rapid transients in demand and supply, particularly where high-inertia alternators (e.g., steam turbines) have become obsolescent. The temporal decoupling of energy supply and utilization allows us to maintain the electricity utilization if the electricity supply is disrupted by contingencies. The success of distributed energy resources in delivering the promise of reliability depends on energy storage devices.

7.4.4 Infrastructure's Utility Optimization and Infrastructure's Deployment Deferment

Millennial entrepreneurs strongly believe that successful business start-ups operate on the asset-lite model. This means that entrepreneurial endeavors in the field of power sector must be done with minimal infrastructure requirements, focusing more on creating a problem-solving product and platform. Energy storage devices and systems are pivotal in the optimum utilization of the existing infrastructure and delaying the need for infrastructure deployment unless it is absolutely essential. This is true even in the utilization at consumer premises, apart from being witnessed in the generation, transmission, and distribution.

7.4.5 Predictability Enhancement of Non-conventional Resources

The renewable resources of energy which steer the incorporation of distributed generation in the existing power grids have given considerable advantages except for being dependent on geographical conditions. The deployment of energy storage systems has facilitated the accumulation of abundance of energy from renewable sources for reserving and dispatching it when there is a deficit of insolation or windage power.

7.5 CORRELATION WITH E-MOBILITY

Energy storage systems are the foundation to build the e-mobility infrastructure on the technological landscape of transportation. The internal combustion engines of petroleum powered vehicles are being supported/replaced by polyphase induction motors for hybrid electric vehicles/purely electric vehicles. Induction motors need to be operated on AC supply, which is available in the electrical circuitry of electric vehicles after being tapped from the DC supply.

The image shown in Figure 7.2 hints at the immense prospects of utilizing energy storage devices for storing the surplus onsite insolation for powering the commuting needs of the employees in an organization. Studies reveal that such a green infrastructure promotes the credibility of the organization apart from energy savings.

FIGURE 7.2 Top view of the photovoltaic energy charging infrastructure for vehicles.

The emerging technologies, such as vehicle-to-grid energy exchange (V2GEX), vehicle-to-vehicle energy exchange (V2VEX), and wireless inductive power transfer from frictional energy to vehicle for locomotion, depend on energy storage devices and systems for proper implementation.

Apart from batteries, ultracapacitors have come to the fore because of their features compensating for the limitations of batteries. Ultracapacitors or supercapacitors have features such as swift power exchange capability, longer lifetimes, etc. However, their energy density and specific power is not commensurate to those of lithium ion batteries, and that is why the match of supercapacitors and batteries is very important in the globalization of e-mobility infrastructure for decarbonizing the global economy.

7.6 TECHNOLOGICAL CHALLENGES

The landscape of energy storage technology has a lot of challenges to be grappled with. The purpose of discussing these challenges is not to discourage the reader, rather to ignite the spark of innovation for contribution in this immensely impactful and rewarding domain of energy storage, as every challenge opens the avenues for immense opportunities [1–6]. Contributions to meet these challenges will be acclaimed as noble efforts towards the mission of purifying the global economy from the threat of obnoxious carbon emissions.

In this section, we have presented the parameters of performance characterization, so that we know what to look for in an energy storage system and identify the challenges associated with it (Table 7.3).

TABLE 7.3
Exploring the Technological Challenges and Opportunities against Their Performance Indices

Performance Indices	Associated Metrics and Their Significance	Challenges
Size and weight	Specific energy, energy density, specific power, and power density influence the footprint, volume, and weight of the energy storage device	Optimization of specific energy, energy density, specific power, and power density for compact and reliable design
Power and energy ratings	The power rating of an energy storage device (SI unit is watt) is the rate at which electrical power transaction can take place for the device; the energy capacity (SI unit is watt-hour), indicates the time span for which the storage device can supply energy at the rated power	Commissioning and operation of energy storage systems involve cost from both. Energy storage systems must be frugally designed in terms of economy as well as ecology
Ramp rate	The time necessary for dispatching and/or replenishing the energy inside the storage device. It varies depending on the energy storage mechanism. Even within the same family of energy storage devices, the charging/discharging time varies. It may vary from seconds to hours depending on the stakes involved in the application	Swifter charging times are always necessary as time is one of the most precious commodities. Ultracapacitors have been developed to have fast charging, and herein lies a great opportunity for dovetailing this research opportunity with better convenience in the electrification of the transportation sector
Scalability	Capability extension of an energy storage cell is done by serial and parallel connection of multiple units of the energy storage cell	There are many issues pertaining to the capacity enhancement when multiple energy storage units are connected, such as the following: voltage equalization, current sharing equalization, protection issues, and so on. This requires research at the circuit design level, as well as powerful algorithms for measurement, instrumentation, and control
Cycling capability	Technologies may differ in their capability to be repetitively charged and discharged including the time to replenish and the depth of discharge allowed during discharge. Long life of an energy storage cell is very much required for the customer, as well as for ecological perspectives	Ultracapacitors have longer lives than batteries, hence the hybridization of the electro-chemistries of different devices with different approaches is being pursued for achieving this end

(*Continued*)

TABLE 7.3 (*Continued*)
Exploring the Technological Challenges and Opportunities against Their Performance Indices

Performance Indices	Associated Metrics and Their Significance	Challenges
Roundtrip efficiency	The ratio of (useful) energy that can be retrieved from storage compared to the amount of energy that was put into storage. Roundtrip efficiency is about 85% for Li-ion batteries	Advent of symmetrical electrode structures has improved this aspect.
Storage losses	All storage systems do need some energy for their subsistence. Storage losses may include the following: self-discharge, roundtrip efficiency losses, and consumption in auxiliary systems, both during normal operation and during standby	Energy storage device functions can be effectively managed through energy management systems
Capacity loss over time	Energy storage technologies which are based on redox reactions, particularly batteries, get derated in energy capacity with the progression of time	Improvisation in the techniques of electrochemical energy storage through materials science research

REFERENCES

1. https://www.sandia.gov/ess-ssl/global-energy-storage-database/
2. IEEE Std 1679-2020, IEEE Recommended Practice for the Characterization and Evaluation of Energy Storage Technologies in Stationary Applications; https://standards.ieee.org/standard/1679-2010.html
3. IEEE Std 2030.2-2015, IEEE Guide for the Interoperability of Energy Storage Systems Integrated with the Electric Power Infrastructure; https://standards.ieee.org/standard/2030_2-2015.html
4. IEEE Std 1547-2018, IEEE Standard for Interconnection and Interoperability of Distributed Energy Resources with Associated Electric Power Systems Interfaces, https://standards.ieee.org/standard/1547-2018.html
5. DOE OE Global Energy Storage Database, Federal and state Energy Storage Policies; https://www.sandia.gov/ess-ssl/globalenergy-storage-database-home/
6. Rabl, V. et al. (2020) *Energy Storage Primer.* IEEE Power & Energy Society; https://www.ieee-pes.org/.
7. Momma, K. and Izumi, F. (2011) VESTA 3 for three-dimensional visualization of crystal, volumetric and morphology data. *Journal of Applied Crystallography*, 44, 1272–1276.
8. Wu, Z.-S., Ren, W., Xu, L., Li, F., Cheng, H.-M. (2011) Doped graphene sheets as anode materials with superhigh rate and large capacity for lithium ion batteries. *ACS Nano*, 5, 5463–5471.

8 Energy Storage Devices

M. Karthigai Pandian, K. Saravanakumar, J. Dhanaselvam, and T. Chinnadurai
Sri Krishna College of Technology

CONTENTS

8.1 INTRODUCTION TO ENERGY STORAGE DEVICES

The process of harvesting energy in an effective manner and storing it is very much essential for this society. Renewable energy resources such as photovoltaic cells, biogas, hydroelectric and tidal power have recently been proposed to overcome the drawbacks of fossil fuels such as pollution and global warming. Battery banks and fuel cells are employed in modern-day electric vehicles (HEVs and EVs) where they are expected to provide an energy supply of a few kilowatts, with the driving time lasting from a few minutes to a few hours [1]. Similarly, portable electronic devices such as pocket radio, mobile phones and laptops are bound to have an energy capacity in terms of ampere-hours (Ah). These are some of the practical examples that present batteries as the main source for ESDs, but there are various methods and other forms of devices that can be employed in practical applications. The basic classification of ESDs is shown in Figure 8.1. Modern researchers have been actively

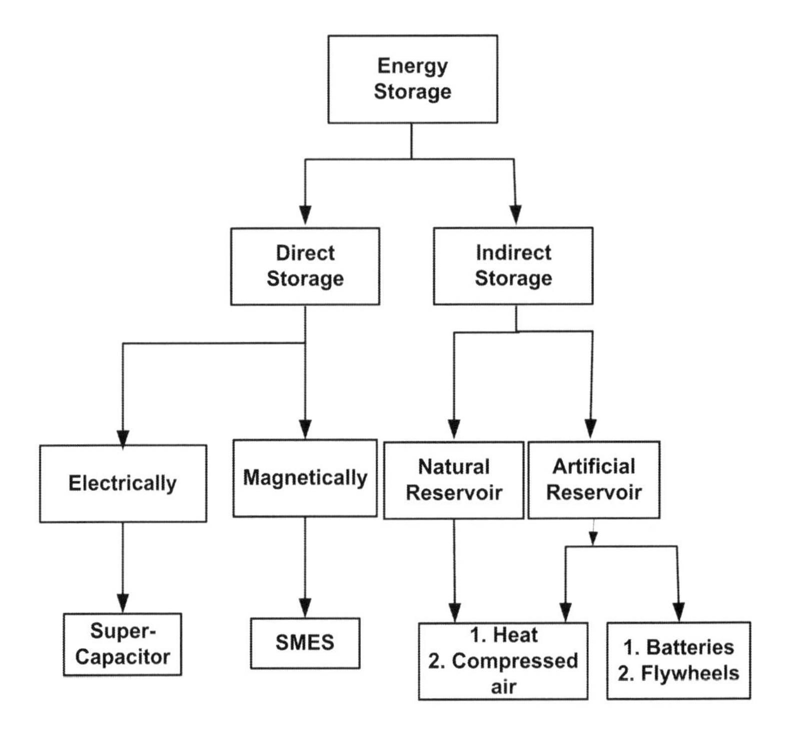

FIGURE 8.1 Energy Storage Devices – Basic Classification.

investigating various opportunities such as molecular physics, flexible nanomaterials and fiber-shaped devices that could actively replace the existing storage devices.

8.2 ENERGY STORAGE DEVICES IN EXISTENCE

To obtain the best performance of electrical and electronic devices, normally the batteries are combined with supercapacitor (SC) packs and employed. Basically, for a huge amount of energy to be stored in capacitors, the basic requirement is that it should have a very huge capacitance or the voltage applied across its terminals should be very huge. In contrast, SCs are bound to have very low voltage abilities [2]. They can be further classified into double-layer capacitors and pseudocapacitors based on the mode of storage as shown in Figure 8.2. A combination of both these modes leads us to hybrid devices.

Flywheels can be touted as the examples of kinetic ESDs [3]. A simple diagram of flywheel's structure is shown in Figure 8.3. Based on the inertia of the rotating mass and the rotor's speed, the energy is stored in the rotating mass. The basic classification of flywheels depends on their speed. They are generally classified as low-speed and high-speed devices. They find their applications in wind turbines, locomotive propulsion systems and in the direction control systems of satellites.

In fuel cells, an electrochemical process is used to convert the fuel's chemical energy into electrical energy directly [4]. The schematic diagram of a basic fuel cell is shown in Figure 8.4. Many types of fuel cells are available with varying energy

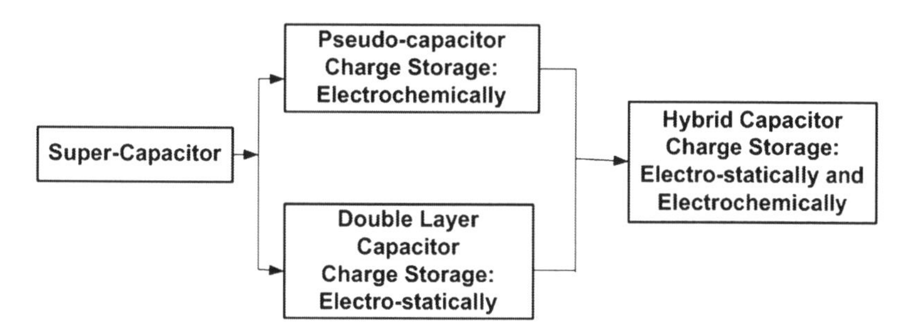

FIGURE 8.2 Classification of SuperCapactitors.

FIGURE 8.3 A Flywheel.

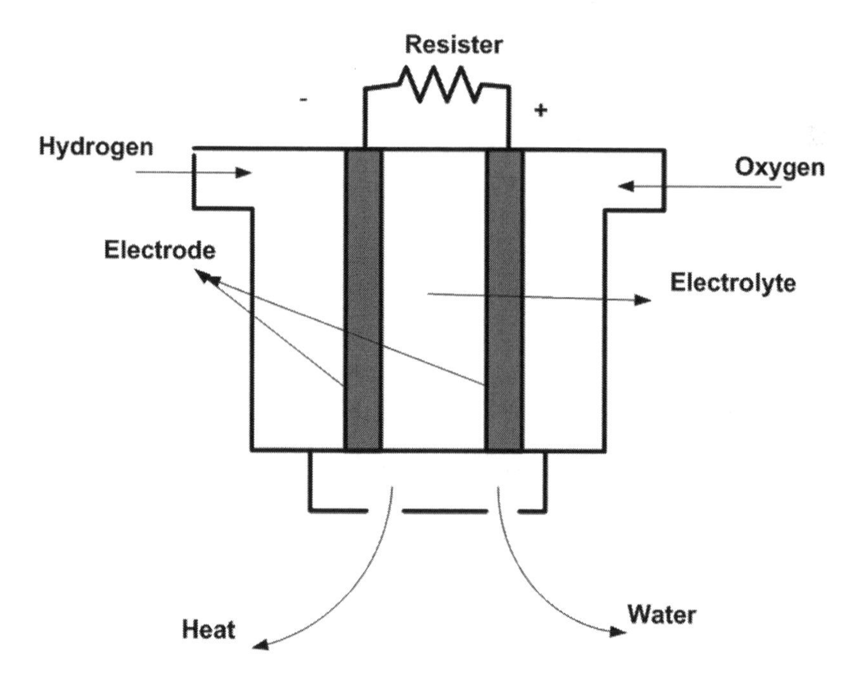

FIGURE 8.4 A Fuel Cell.

storage capabilities. Some of them are Alkaline Fuel Cells (AFCs), Solid Oxide Fuel Cells (SOFCs) and Phosphoric Acid Fuel Cells (PAFCs). Due to their ability to work at very low temperatures and efficient starting characteristics, fuel cells are majorly used in electric vehicle technologies. In compressed air energy storage systems, energy is stored in the form of compressed air for future use. Another well-known technology employed in energy storage is superconductive magnetic energy storage (SMES). It employs a superconducting coil and the direct current (DC) flowing through it is used to create a huge DC magnetic field and this is further used to store the energy [5].

Numerous research studies have been carried out to enhance the characteristics and efficiency of ESDs. Some of the new technologies under consideration are Molecular Solar Thermal Energy Devices (MOSTs), flexible storage devices using nanomaterials and fiber-shaped ESDs.

8.3 MOLECULAR SOLAR THERMAL ENERGY STORAGE DEVICES

Energy storage is the process of capturing energy from one source and the same can be utilized for future purpose. Storing solar energy in the form of heat and chemical energy is generally called molecular solar thermal storage, where chemical bonds are used to store the solar energy [6]. In this process, a catalyst is used to recycle an isomer and convert it into heat where ionized chemical compounds are transformed from chemical isomerization to metastable isomers. There are various methods available for thermal energy storage where the molten-salt technique is the simplest one, but it experiences thermal losses due to insulation problems [7]. Another method is the solar-driven conversion type in which dicyclopentadiene is converted into cyclopentadiene, which is thermodynamically the most favorable at increased temperature.

8.3.1 Small Molecule Organic Solar Cells

Cost reduction is the main factor that makes silicon solar cells most suitable for electricity generation. But comparatively, 50% of silicon is wasted during ingot preparation. This initiates the search for new solar harvesting materials. The main advantages of carbon technology are its transparency and lightweight properties [8]. The three factors that determine the materials used in energy devices are high efficiency, long life and low cost. But organic photovoltaic cells are found to have certain limitations in these criteria. So, to meet the criteria, an alternative option called Organic Photo Voltaic (OPV)-based oligomers is suggested, which can be normally implanted using vacuum sublimation [9].

Carbon atoms are the fundamental units of small molecule organic solar cells (OSCs) which have 100 complex atoms compared to atomic materials and have a weight of less than 1,000 amu. Based on van der Waals force, a solid crystal organic molecular complex is formed, and this is represented by the highest occupied molecular orbital and the lowest occupied molecular orbital, and the corresponding energy gap exists between these two orbitals [10]. Sandwiching two electrodes between two organic materials forms an OSC. In general, these devices are fabricated in the form of flat heterojunctions where the conversion of photons into electrons happens [11].

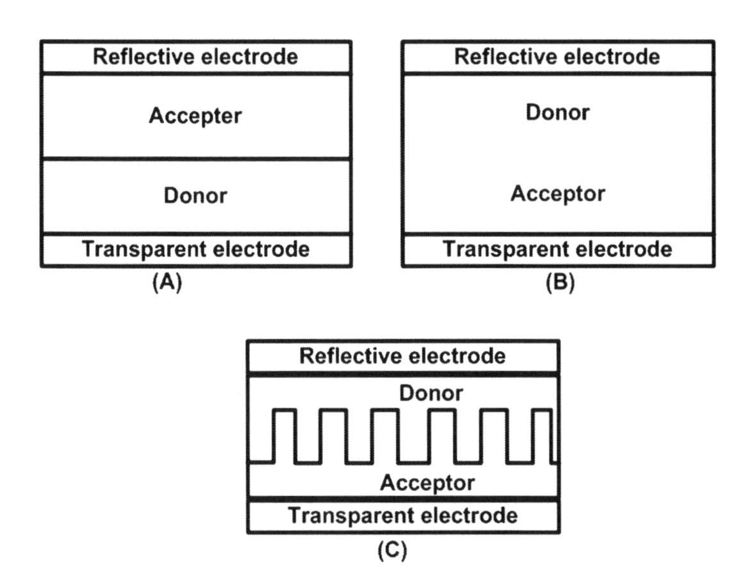

FIGURE 8.5 Different Configurations of OSC.

The various geometrics of an OSC are shown in Figure 8.5. From the configurations, one can understand that the donor electrode is the light absorbing material, whereas the second electrode acts as an acceptor. By using bulk heterojunctions (BHS), efficiency can further be improved, but the thermodynamics complexity level will be increased leading to energetic and structural disorders. This is considered as a major drawback of BHJ OSCs. These problems can be resolved by increasing the interface area and using an enclosed communication channel [12].

In general, the efficiency level of a small molecule OSC is 13.2% when processed under vacuum. These molecules have certain merits in all aspects such as molecular structure, purity level, molecular weight and also morphological control. Merocyanines (MCs), phthalocyanines (Pcs), borondipyrromethenes (BODIPYs), diindenoperylene (DIP) and oligothiophenes are a few materials used for small molecule solar cells [13]. The molecular structures of various materials used in the manufacture of OSCs are depicted in Figure 8.6. MCs are high polarization and high absorption materials that possess high thermal stability and a good absorption level when phenyl-C61-butyric acid methyl ester (PCBM) is used as the active material instead of the C_{60} acceptor. In organic electronics, phthalocyanines (Pcs) are another important class of materials which play a vital role in terms of stability and synthetic versatility [14].

Other OSC materials include BODIPYs, which have a better absorption coefficient and photostability. These derivatives have a reasonable near infrared range that is capable of harvesting more photons even at small energy levels of the spectrum. A very simple molecular structure and enhanced mobility of charge carriers make DIP and its derivatives the most suitable candidates for organic electronic applications like OLED. At the same time, a low extinction coefficient is a major drawback of these devices, and this results in reduced photocurrents of OSCs [15].

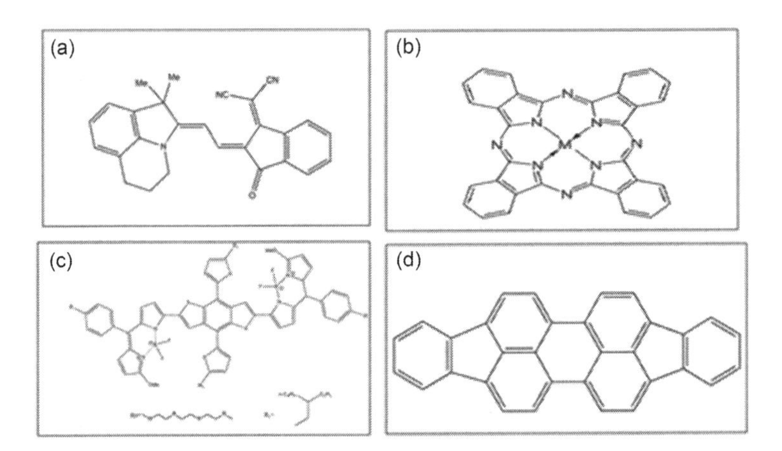

FIGURE 8.6 Molecular structure of (a) Merocyanines derivative HB194 (b) MPCs (c) donor molecule BDTT-BODIPY and (d) donor molecule DIP.

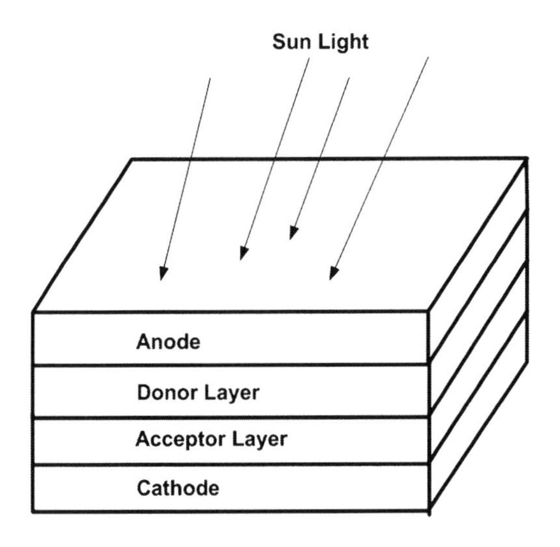

FIGURE 8.7 Constructional Structure of PSC.

8.3.2 Polymer Solar Cells

Polymer solar cells (PSCs) have a very simple construction in which an active layer is sandwiched between a cathode and an anode. The active layer is designed with an acceptor and donor combination that performs functions such as formation of an electric field internally in the device and generation of charge carriers. In the transportation of charge carriers, the following layers called the hole transport or extracting layer (HTL or HEL) and an electron transport or extracting layer (ETL or EEL) help the active layer during charge transformation [16]. The construction structure of a simple PSC is shown in Figure 8.7.

PSCs are an alternative option for OSCs and are solution processable. They contain polymers and fullerene. For both single and multiple cells, the efficiency is about 12% which helps in more production. In previous decades, solar cells were made using bilayers of polymer and buckminsterfullerene. Performance efficiencies of these devices were pretty low owing to the fact that generation of charges can occur only at the interfaces. This problem is resolved, and the efficiency is increased by using soluble fullerene derivatives (PC61BM) [17]. The fundamental principle of PSCs is that the absorption takes place through a larger transparent electrode active material, and the photons are reflected by the electrode surface when the source light falls on it [18]. Opto-electric features are fundamental properties that are mainly used in active layers for better performance.

Conjugated polymers are generally used as donor derivative materials because of their rich characteristics such as robustness, flexibility and less weight. Polythiophene (PT) derivatives are the most preferable polymers among which poly-3-hexylthiophene (P3HT) is the most used polymer in the design of PSCs [19]. The molecular structure of a poly-3-hexylthiophene is shown in Figure 8.8. The next prominent donor material is poly-p-phenylenevinylenes (PPVs) with very high solubility. These are highly processable polymers and can be made more suitable for photovoltaic applications. A PSC using a PC61BM active layer has a Power Conversion Efficiency (PCE) value range of 1.1%–1.3%. Other than this, important materials used as PSC-donor materials are D–A conjugated polymers and D–A copolymers [20]. The molecular structure of a PPV is shown in Figure 8.9.

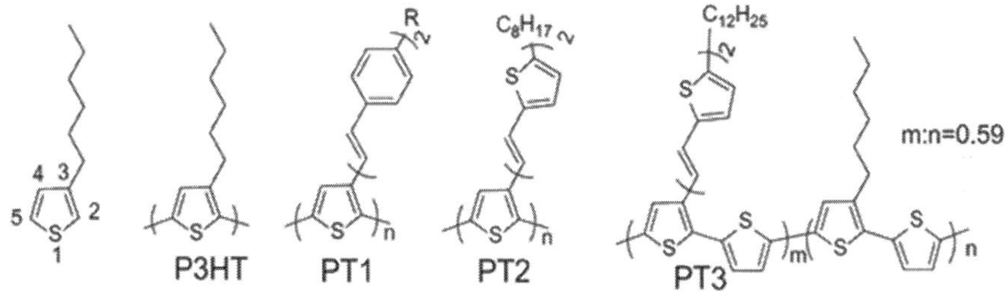

FIGURE 8.8 Molecular structure of poly 3-hexylthiophene.

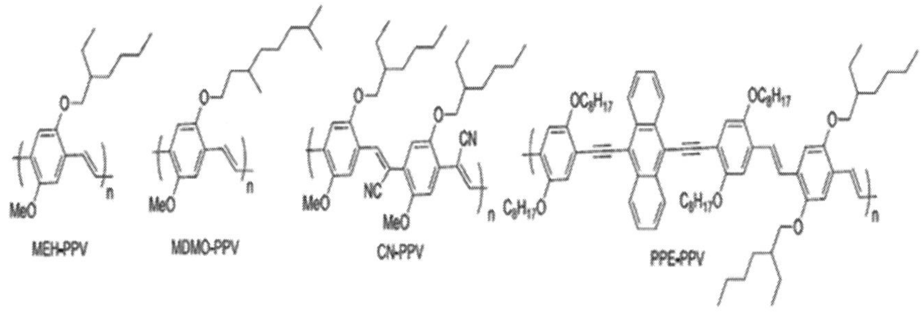

FIGURE 8.9 Molecular structure of poly-p-phenylenevinylene.

There are many acceptor materials used in designing PSCs. Among these, fullerenes play a vital role because of their electron mobility features and efficient charge transfer mechanisms. Under ideal conditions, they have good solubility in many solvents, and these are the most suitable donor materials to increase the open circuit voltage of a device. The main drawbacks of fullerene-based derivatives are energy loss, which is about 0.6 eV, and their lower absorption power [21]. To overcome this, non-fullerene derivatives are also preferred for PSCs. Perylene diimide (PDI) is a non-fullerene derivative which is mostly preferred as an acceptor material because it can be made as small-sized crystals during the process. Other significant materials that are closely associated with PSCs are polymer/polymer blend materials [22].

8.4 FLEXIBLE ENERGY STORAGE DEVICES USING NANOMATERIALS

Flexible electronics is the most recent upcoming technology in the electronics world. These flexible products include mobile phones, displays, surgical tools, measuring sensors, implantable sensors, automobile sensors, agricultural devices, environmental monitoring sensors and strain gauges [23]. In real-time operation, a flexible device has to satisfy major demands such as performance, low cost, flexible electrodes, more reliability, safety, high stability and longer life time. However, the structural design can also hamper the device stability, performance and its durability. The recent advancements in battery technology have been provoked by new technologies such as portable electronic devices with a compact size and low cost [24]. Reducing the battery weight and size, increasing the capacity and performance and obtaining more flexibility are possible when nanotechnology can be incorporated with battery technology [25]. By combining these two methods, the battery efficiency is improved drastically by exploiting various materials at the nanoscale that include nanoparticles, nanotubes and nanowires [26]. Nanomaterial research has changed the way things operate in the energy sector, particularly in the harvested energy storage, flexible electronics and wearable electronics.

8.4.1 Materials for Flexible Electrodes

Li-ion batteries are viewed as the new pitch of experimentation for flexible ESDs [27]. Li-ion batteries do not change their crystalline structure while charging and discharging. If a change occurs in the crystalline structure of the material, the life time and number of cycles will reduce to a great extent. So, in maintaining the electrode plate properties in a stable manner, the nanomaterials play a crucial role [28]. Nanomaterials such as nanotubes are easily implanted with Li-ions to achieve improved charge capacity of batteries, lack of reactivity with materials, long lifecycle and higher electron mobility. Also, they provide better flexibility under pressure.

8.4.2 Carbon Nanotubes

Carbon nanotubes (CNTs) are used in batteries due to their high elasticity, less weight and low density [29]. CNTs naturally exhibit good structural stability under loading conditions without structural deformation [30]. The nature of CNTs is

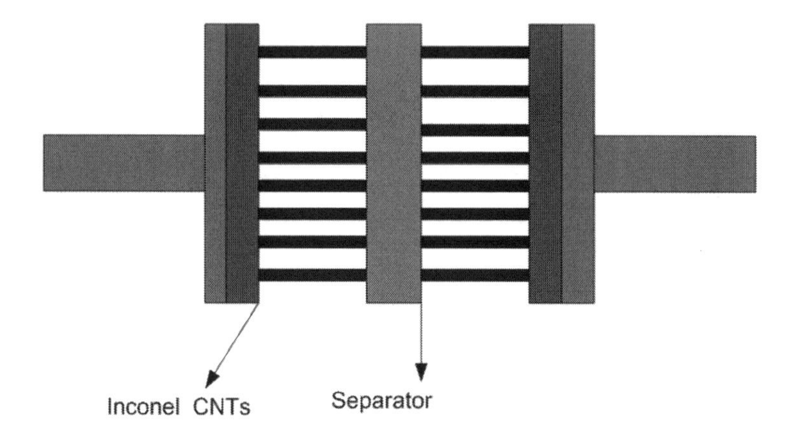

FIGURE 8.10 Inconel CNTs implantation on electrode.

super-compressibility under 15% strain, and once the strain is removed, the structure will retain its original position. Hence, CNTs can turn out to be the ideal materials for flexible battery applications that would benefit from their characteristics of light weight, structural stability and a combination of electrical conductivity, chemical and thermal stability [31]. The deposition of CNTs into a current collector is very difficult because the development of CNTs is a high temperature process and it requires less catalytic interaction. The substrate needs to be grown under well aligned conditions at high temperature. There are many successful CNT implementations reported with different materials such as Inconel and stainless-steel substrates [32]. A well-grown CNT anchored into a substrate is expected to provide excellent electrical conductivity. Figure 8.10 shows the Inconel implementation on an electrode. A large temperature range between 500°C and 820°C should be available for the growth of an Inconel substrate. The maximum growth rate is attained at 770°C at a rate of 2.8 mm/min [33].

8.4.3 Nanocomposites

CNTs are integrated into nano-polymer composites that are bound to have extreme mechanical properties for flexible storage devices. These composites are employed as various smart flexible devices in a wide range of electrical and electronic products [34]. In recent years, CNT–polymer (CNT-P) composites have been tested in supercapacitors and batteries to improve their mechanical stability [35]. A CNT-P consists of an ionic liquid at room temperature, cellulose and CNT, and it fits the characteristics of a spacer, an electrode and an electrolyte, and this provides better flexibility to the batteries [36,37]. The flexible battery devices are assembled as shown in Figure 8.11. A thin film of Li deposited by thermal evaporation is used as an electrode. A Li-based flexible battery is manufactured using a basic building block method. This battery consists of a Room Temperature Ionic Liquid (RTIL), nano-composite film and Li-metal layer. The cellulose layer acts as a spacer between the electrodes. So far, we have discussed the non-conducting polymers mixed with

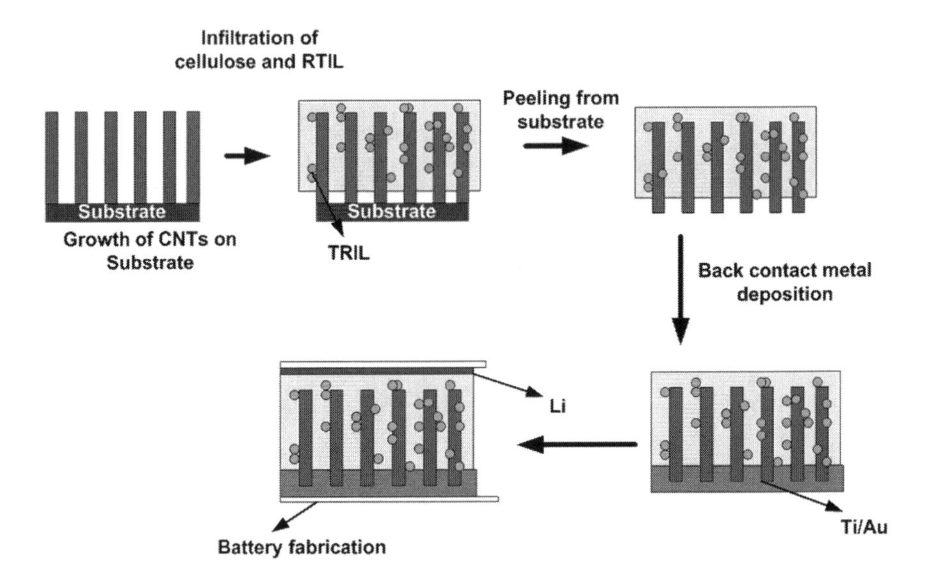

FIGURE 8.11 Flow diagram for the fabrication of flexible battery devices.

CNTs. An improvisation in this technology has been proposed where conducting polymers are integrated with CNTs to improve conductivity [38].

8.4.4 Hybrid Nanostructures

Hybrid nanostructures can be effectively used in flexible battery manufacturing. In this process, nanotubes developed using multi-segmented nanowires are implanted in electrodes, and this helps in reducing the internal resistance, increasing the conductivity power density. In the manufacture of nanotubes, different methods using nanowires and nanoparticles are developed. The most commonly used method for developing a single dimension crystalline is the template method. This method is supposed to have more flexibility in terms of different varieties of nanomaterial arrangements. But growing of CNTs in a metal layer is very challenging, but this unique method of fabricating multi-segmented nanowire/nanotubes gives a pathway to develop CNTs on metal sheets, which can be further used for manufacturing flexible batteries [39]. Two and three segmented hybrid structures of CNTs are also made using the template method. This is achieved with the help of chemical vapor deposition and electro chemical deposition. A metal–CNT hybrid gives good conductivity as the metal sheet gets connected with all CNTs. This enables reduction of resistance and internal heat; hence the devices are able to store high energy [40]. As shown in Figure 8.12, porous anodized alumina (AAO) is prepared using a modified two-step anodizing process. Then, Au nanowires are deposited by the electrodeposition method on the AAO sheet. The remaining portion of the AAO template is deposited with CNTs by pyrolysis of acetylene at 650°C at a rate of 35 ml/min [41].

The intermetallic alloys are also explored for manufacturing flexible batteries, as they are expected to have large charge capacity but their number of cycles is lower.

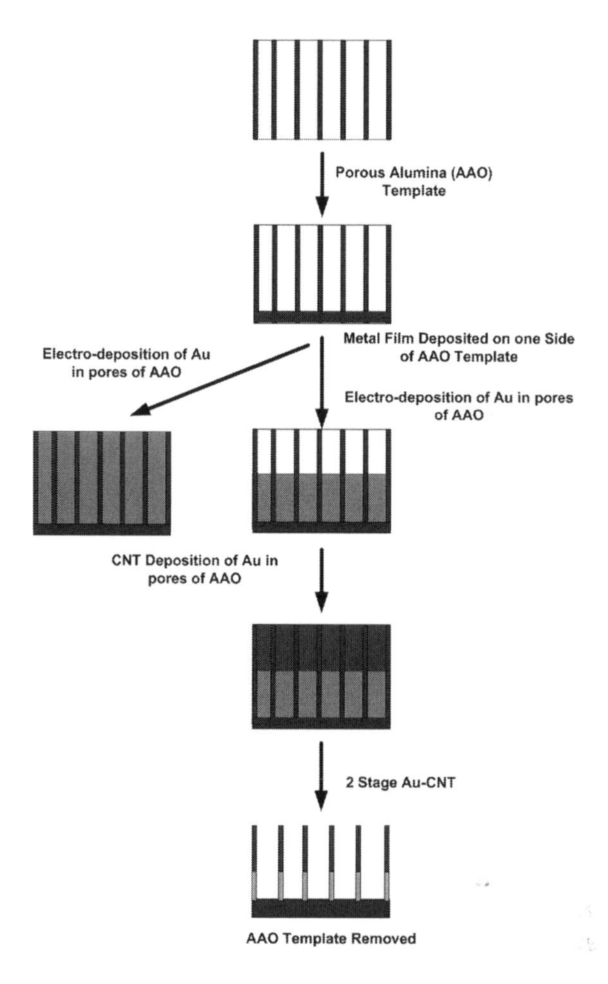

FIGURE 8.12 CNTs deposited in the pores of an AAO template.

There are different suitable elements available that can form an alloy with Li, B, As, Sb, Sn, Pb and Mg. Likewise, different inactive materials such as Fe, Ni and Co are used in battery manufacturing. Therefore, still lots of elements are required to be explored to make more flexible batteries along with large charge capacity.

8.5 FIBER-SHAPED ENERGY STORAGE DEVICES

The rapid increase in energy requirements has pushed us to search for renewable energy resources. The main disadvantage of silicon solar cells is that they are vulnerable to temperature fluctuations. They literally become inactive in the areas with low temperature ranges. Fabrication of silicon solar panels is also very complex and too costly. To overcome these drawbacks, dye-sensitized solar cells (DSCs) are introduced that are very easy to fabricate at low cost and they can produce good energy even in low light intensities.

8.5.1 Fiber-Shaped Solar Cells

A fiber-shaped DSC provides a very good flexible energy system for wearable devices due to its easy fabrication and operation, light weight and low cost. In this fiber-shaped thin-film type solar cell structure, the working electrode is made of either stainless steel or platinum and is coated with a layer of dye-absorbed TiO_2 particles, and the counter electrode is formed using a conductive polymer that is transparent in nature. Even though the conductivity of DSCs is poor compared to that of silicon solar cells, recent advancements in enhancing the performance of these devices have proven that it is a cost-effective technique [42]. A novel variation in this technology that has recently been proposed is polymer-based solar cell devices [43].

8.5.2 Fiber Electrodes

The main components that form a major part of fiber-shaped devices are fiber electrodes [44]. They should have very good conductivity, should be electrochemically active, have considerably good mechanical strength and should be low cost. In the present fiber-shaped devices, fiber electrodes are made up of metal, carbon and polymers. Metal wire fiber electrodes are found to have good conductivity, but they are expensive, very heavy and stiff. It will cause major problems during installation. Polymers can be a good replacement for metal wires in terms of flexibility, but polymers are intrinsically nonconductive in nature, which makes them not suitable to be used as electrodes. CNTs and graphene are bound to have very good conductivity, mechanical strength and electrochemical features, which make them the appropriate candidates to be exercised as electrodes in fiber-shaped devices [26].

8.5.3 Fiber-Shaped Dye-Sensitized Solar Cells

A conventional DSC has a planar structure and the structure of a fiber-shaped DSC depends on the fiber electrode used. Diagrammatic representations of a conventional DSC and a fiber-shaped DSC are shown in Figure 8.13. A fiber-shaped DSC can also be called a solar cell in the fiber form [42]. A couple of metal electrodes are twisted together in the fabrication of these devices. A stainless wire coated with dye-sensitized TiO_2 acts as the working electrode and a platinum wire as the counter electrode. An alternative option for the expensive platinum (Pt) electrode is to use a metal wire on which platinum particles are deposited electrochemically. Considering the high cost of Pt as a counter electrode, fibrous carbon materials are introduced to replace the Pt wire. A CNT fiber is another counter electrode used in fiber-shaped DSCs and has very good conductivity and mechanical strength. For the better performance of fiber-shaped DSCs, fabrication of counter electrodes and fiber working electrodes should be done very carefully. The fiber working electrodes normally used are carbon nanotube fibers, carbon nanotube fiber coated with TiO_2 nanoparticles and Ti wires grown with aligned TiO_2 nanotube arrays. The working electrodes used are carbon nanotube fibers, carbon-nanostructured fibers and carbon/platinum composite fibers. These devices are basically classified into coaxial fiber-shaped DSCs and twisted fiber-shaped DSCs [45].

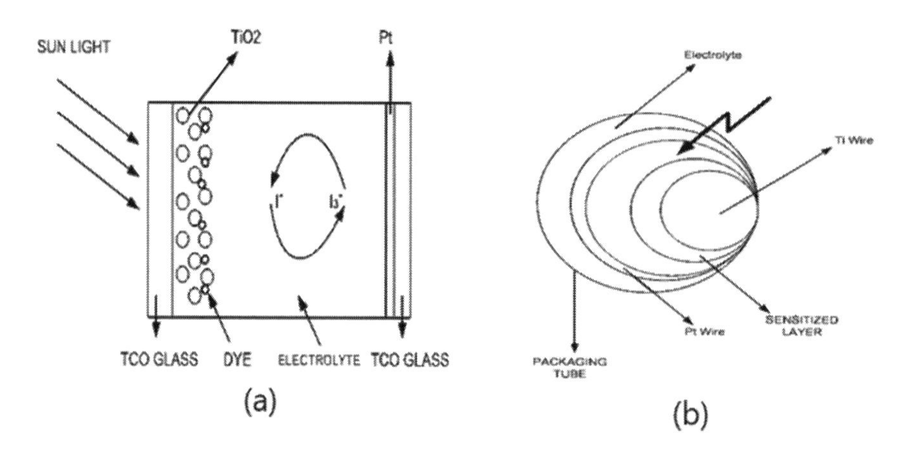

FIGURE 8.13 Structures of (a) Conventional DSC and (b) Fiber-shaped DSC.

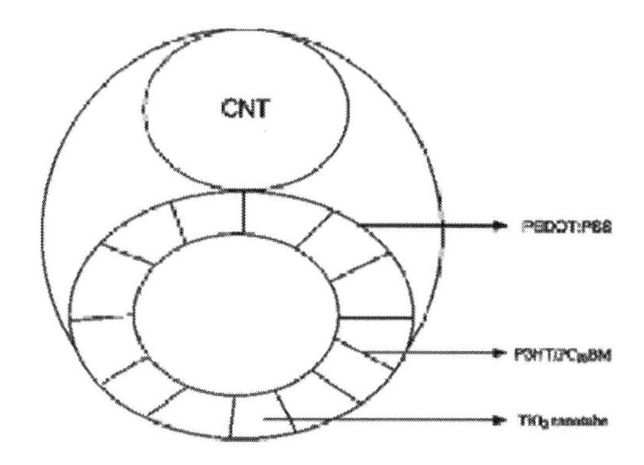

FIGURE 8.14 Fiber Shaped Polymer Solar Cell.

8.5.4 Fiber-Shaped Polymer Solar Cells

Fiber-shaped PSCs shown in Figure 8.14 are light weight and easy to fabricate. Hence these fiber-shaped PSCs are used in electronic fabrics. Like DSCs, they are also classified as both twisted and coaxial structures [46]. A TiO_2 nanotube-modified titanium wire coated with a P3HT:PC70BM blend followed by an immersion in PEDOT:PSS solution acts as the primary electrode. Finally, the primary electrode is wrapped around with a CNT fiber material. P3HT molecules absorb incident photons and generate excitons that are segregated when bonding with PC70BM and TiO_2.

8.5.5 Fiber-Shaped Supercapacitors

SCs and lithium-ion batteries are low-cost and environmentally friendly energy conversion/storage devices. Gel electrolytes are used in these devices to override the safety issues of liquid electrolytes [47]. Fiber-shaped SCs are constructed on suitable fiber electrodes. Among various fiber electrodes, CNT fibers have many unique

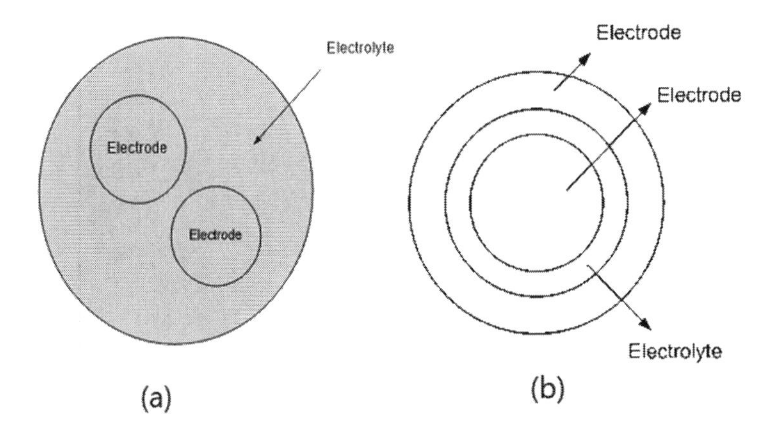

FIGURE 8.15 Fiber Shaped Supercapacitors (a) Twisted Pair and (b) Coaxial.

advantages such as light weight, excellent mechanical properties and high thermal and electrical conductivity. Polyvinyl alcohol or inorganic acid gel electrolyte is normally used in SCs. The two types of fiber-shaped supercapacitors are shown in Figure 8.15. Twisted fiber-shaped devices and coaxial fiber-shaped devices are the two major types of fiber-shaped supercapacitors. A carbon nanotube fiber/Ti wire twisted pair fiber-shaped SC is found to achieve a specific capacitance of 64.9 mF/cm or 1.65 mF/cm^2 at a scan rate of 10 mV/s. Nevertheless, a mismatch in the capacitances of the carbon nanotube fiber and titanium wire is expected to affect its performance. Carbon nanotube fiber/conducting polymer composite fiber twisted fiber exhibits high performance due to the deposition of polyaniline on the surface of CNT fibers. This fiber-shaped SC produces a specific capacitance of 274 F/g or 263 mF/cm. Coaxial fiber-shaped SCs exhibit lots of advantages compared to a twisted pair structure. In a twisted pair structure, electrodes are kept in parallel but the coaxial structure of SCs comprises an inner electrode, outer electrode and electrolyte. Graphene, CNT films or CNT sheets can be used as the outer electrode while the inner electrode is made up of materials such as nanowire-modified plastic wires, CNTs, graphene or metal wires with porous carbon [48]. By increasing the quality of the fiber-shaped SCs, they are employed in electronic devices. Stretchable SCs are one special type used in practical applications where elongation problems may occur. A chromatic SC is another special function device which can respond to the variations in charge states by changing colors reversibly, as polyaniline switches its oxidation states to the completely reduced state (yellow), the completely oxidized state (blue) and the mediate state (green).

8.5.6 Fiber-Shaped Integrated Devices

An integrated device tries to combine the functions of solar energy conversion and storage of electrical energy. It is generally classified into two types known as all-in-one devices and assembled devices. An all-in-one device employs an electrode which is capable of absorbing solar light and also has the ability to store the generated charge.

Energy conversion and storage in assembled devices are carried out using two different elements that are connected together using an electrode. A solar cell is used to harvest the solar energy while a capacitor is used to store the charges in these devices. Modern day integrated devices are built with electrochemical SCs as their basic components.

A fiber-shaped integrated device was first successfully realized by Wong et al. in 2011. This device is found to be a combination of a DSC, an electrochemical capacitor and a nanogenerator. Various research studies have been carried out since then to improve the structure of this device and its performance. An electrode is shared between the solar cell and the capacitor in this device, very much like a normal three-electrode integrated device. Hence the material for this electrode is chosen in such a way that it is compatible with both the solar cell and the electrochemical capacitor. Some of the materials used for this electrode include CNTs, zinc oxide and titanium oxide. These fiber-shaped devices are also implemented in two different forms, either as a coaxial structure or a twisted structure. Another important characteristic of a fiber-shaped device is that it is stretchable and hence can be used successfully in wearable devices. Practical difficulties faced in the implementation of these devices are the electrolyte stability and scale-up fabrication issues. The overall energy conversion efficiency achieved by these integrated devices is found to be 11.2% and research studies are being carried out to improve the performance of the devices.

8.6 SUMMARY

As we start to move away from conventional energy sources such as fossil fuels, we need to understand that these resources provided two functions in one. They acted as both energy sources and also as energy storing devices. In the modern world that employs renewable energy sources, these two functions are to be attended separately. The basic idea behind modern storage technologies is to keep on harvesting the energy sources at a constant rate, and not based on their demand. This chapter intends to study the advancements in the field of modern ESDs and to analyze their performance characteristics. These advanced devices include molecular solar thermal energy storage, nanomaterial-based flexible devices and fiber-shaped devices. Structure, device materials, characteristics and performance of each of these devices have been discussed in detail. We can conclude that these devices are going to rule the market of ESDs in near future.

REFERENCES

1. Vazquez, S, Lukic, SM, Galvan, E, Franquelo, LG, & Carrasco, JM. 2010. Energy storage systems for transport and grid applications. *IEEE Transactions on Industrial Electronics*, 57(12), 3881–3895.
2. Libicj, J, Maca, J, Vondrak, J, Cech, O, & Sedlarikova, M. 2018. Supercapacitors: properties and applications. *Journal of Energy Storage*, 17, 224–227.
3. Amiryar, ME, & Pullen, KR. 2017. A review of flywheel energy storage system technologies and their applications. *Applied Sciences*, 7(3), 286–307.
4. Wang, S, & Jiang, SP. 2017. Prospects of fuel cell technologies. *National Science Review*, 4(2), 163–166.

5. Buckles, W, & Hassenzabl, WV. 2000. Superconducting magnetic energy storage. *IEEE Power Engineering Review*, 20(5), 16–20.
6. Himmelberger, S, Vandewal, K, Fei, Z, Heeney, M, & Salleo, A. 2014. Role of molecular weight distribution on charge transport in semiconducting polymers. *Macromolecules*, 47(20), 7151–7157.
7. Peumans, P, & Forrest, SR. 2004. Separation of geminate charge-pairs at donor–acceptor interfaces in disordered solids. *Chemical Physics Letters*, 398(1–3), 27–31.
8. Braun, CL. 1984. Electric field assisted dissociation of charge transfer states as a mechanism of photocarrier production. *The Journal of Chemical Physics*, 80(9), 4157–4161.
9. Kronenberg, NM, Steinmann, V, Bürckstümmer, H, Hwang, J, Hertel, D, Würthner, F, & Meerholz, K. 2010. Direct comparison of highly efficient solution-and vacuum-processed organic solar cells based on merocyanine dyes. *Advanced Materials*, 22(37), 4193–4197.
10. Hinderhofer, A, & Schreiber, F. 2012. Organic–organic heterostructures: concepts and applications. *ChemPhysChem*, 13(3), 628–643.
11. Kronenberg, NM. 2010. Organic bulk heterojunction solar cells based on merocyanine colorants (Doctoral dissertation, University of Cologne).
12. Schünemann, C, Elschner, C, Levin, A, Levichkova, M, Leo, K, & Riede, M. 2011. Zinc phthalocyanine—influence of substrate temperature, film thickness, and kind of substrate on the morphology. *Thin Solid Films*, 519(11), 3939–3945.
13. Tress, W, Leo, K, & Riede, M. 2013. Dominating recombination mechanisms in organic solar cells based on ZnPc and C60. *Applied Physics Letters*, 102(16), 163901.
14. Morita, S, Zakhidov, AA, & Yoshino, K. 1992. Doping effect of buckminsterfullerene in conducting polymer: change of absorption spectrum and quenching of luminescene. *Solid State Communications*, 82(4), 249–252.
15. Dennler, G, Scharber, MC, & Brabec, CJ. 2009. Polymer-fullerene bulk-heterojunction solar cells. *Advanced Materials*, 21(13), 1323–1338.
16. Kim, Y, Cook, S, Tuladhar, SM, Choulis, SA, Nelson, J, Durrant, JR, & Ree, M. 2011. A strong regioregularity effect in self-organizing conjugated polymer films and high-efficiency polythiophene: fullerene solar cells. In: Materials for Sustainable Energy: A Collection of Peer-Reviewed Research and Review Articles from Nature Publishing Group (pp. 63–69).
17. Li, W, Hendriks, KH, Furlan, A, Wienk, MM, & Janssen, RA. 2015. High quantum efficiencies in polymer solar cells at energy losses below 0.6 eV. *Journal of the American Chemical Society*, 137(6), 2231–2234.
18. Scharber, MC. 2016. On the efficiency limit of conjugated polymer: fullerene-based bulk heterojunction solar cells. *Advanced Materials*, 28(10), 1994–2001.
19. Ala'a, FE, Sun, JP, Hill, IG, & Welch, GC. 2014. Recent advances of non-fullerene, small molecular acceptors for solution processed bulk heterojunction solar cells. *Journal of Materials Chemistry A*, 2(5), 1201–1213.
20. Hou, J, Chen, TL, Zhang, S, Huo, L, Sista, S, & Yang, Y. 2009. An easy and effective method to modulate molecular energy level of poly (3-alkylthiophene) for high-VOC polymer solar cells. *Macromolecules*, 42(23), 9217–9219.
21. Wang, E, Wang, L, Lan, L, Luo, C, Zhuang, W, Peng, J, & Cao, Y. 2008. High-performance polymer heterojunction solar cells of a polysilafluorene derivative. *Applied Physics Letters*, 92(3), 23.
22. He, Y, & Li, Y. 2011. Fullerene derivative acceptors for high performance polymer solar cells. *Physical Chemistry Chemical Physics*, 13(6), 1970–1983.
23. Li, H, Zhang, X, Zhao, Z, Hu, Z, Liu, X, & Yu, G. 2020. Flexible sodium-ion based energy storage devices: recent progress and challenges. *Energy Storage Materials*, 26, 83–104.

24. Wang, X, & Shi, G. 2015. Flexible graphene devices related to energy conversion and storage. *Energy & Environmental Science*, 8, 790–823.
25. Bruce, PG, Scrosati, B, & Tarascon, JM. 2008. Nanomaterials for rechargeable lithium batteries. *Angewandte Chemie International Edition*, 47(16), 2930–2946.
26. Iijima, S. 1991. Helical microtubules of graphitic carbon. *Nature*, 354(6348), 56–58.
27. Gao, B, Kleinhammes, A, Tang, XP, Bower, C, Fleming, L, Wu, Y, & Zhou, O. 1999. Electrochemical intercalation of single-walled carbon nanotubes with lithium. *Chemical Physics Letters*, 307(3–4), 153–157.
28. Lee, KT, & Cho, J. 2011. Roles of nano size in lithium reactive nanomaterials for lithium ion batteries. *Nano Today*, 6(1), 28–41.
29. Liu, XM, dong Huang, Z, woon Oh, S, Zhang, B, Ma, PC, Yuen, MM, & Kim, JK. 2012. Carbon nanotube (CNT)-based composites as electrode material for rechargeable Li-ion batteries: a review. *Composites Science and Technology*, 72(2), 121–144.
30. Baughman, RH, Zakhidov, AA, & De Heer, WA. 2002. Carbon nanotubes – the route toward applications. *Science*, 297(5582), 787–792.
31. Lourie, O, Cox, DM, & Wagner, HD. 1998. Buckling and collapse of embedded carbon nanotubes. *Physical Review Letters*, 81(8), 1638.
32. Pushparaj, VL, Shaijumon, MM, Kumar, A, Murugesan, S, Ci, L, Vajtai, R, Linhardt, RJ, Nalamasu, O, & Ajayan, PM. 2007. Flexible energy storage devices based on nanocomposite paper. *Proceedings of the National Academy of Sciences*, 104(34), 13574–13577.
33. Talapatra, S, Kar, S, Pal, SK, Vajtai, R, Ci, L, Victor, P, Shaijumon, MM, Kaur, S, Nalamasu, O, & Ajayan, PM. 2006. Direct growth of aligned carbon nanotubes on bulk metals. *Nature Nanotechnology*, 1(2), 112–116.
34. Niu, C, Sichel, EK, Hoch, R, Moy, D, & Tennent, H. 1997. High power electrochemical capacitors based on carbon nanotube electrodes. *Applied Physics Letters*, 70(11), 1480–1482.
35. Chen, J, Minett, AI, Liu, Y, Lynam, C, Sherrell, P, Wang, C, & Wallace, GG. 2008. Direct growth of flexible carbon nanotube electrodes. *Advanced Materials*, 20(3), 566–570.
36. Frackowiak, E, Metenier, K, Bertagna, V, & Beguin, F. 2000. Supercapacitor electrodes from multiwalled carbon nanotubes. *Applied Physics Letters*, 77(15), 2421–2423.
37. Frackowiak, E. 2007. Carbon materials for supercapacitor application. *Physical Chemistry Chemical Physics*, 9(15), 1774–1785.
38. Song, HK, & Palmore, GTR. 2006. Redox-active polypyrrole: toward polymer-based batteries. *Advanced Materials*, 18(13), 1764–1768.
39. Ou, FS, Shaijumon, MM, Ci, L, Benicewicz, D, Vajtai, R, & Ajayan, PM. 2006. Multisegmented one-dimensional hybrid structures of carbon nanotubes and metal nanowires. *Applied Physics Letters*, 89(24), 243122.
40. Masuda, H, & Fukuda, K. 1995. Ordered metal nanohole arrays made by a two-step replication of honeycomb structures of anodic alumina. *Science*, 268(5216), 1466–1468.
41. Meng, G, Jung, YJ, Cao, A, Vajtai, R, & Ajayan, PM. 2005. Controlled fabrication of hierarchically branched nanopores, nanotubes, and nanowires. *Proceedings of the National Academy of Sciences*, 102(20), 7074–7078.
42. Baps, B, Eber-Koyuncu, M, Koyuncu, M. 2001. Ceramic based solar cells in fiber form. *Key Engineering Materials*, 206, 937–940.
43. Liu, J, Namboothiry, MAG, & Carroll, DL. 2007. Optical geometries for fiber-based organic photovoltaics. *Applied Physics Letters*, 90(13), 133515.
44. Thomas, S, Deepak, TG, Anjusree, GS, Arun, TA, Nair, SV, & Nair, AS. 2014. A review on counter electrode materials in dye-sensitized solar cells. *Journal of Materials Chemistry A*, 2(13), 4474–4490.
45. Sun, H, Li, H, You, X, Yang, Z, Deng, J, Qiu, L, & Peng, H. 2014. Quasi-solid-state, coaxial, fiber-shaped dye-sensitized solar cells. *Journal of Materials Chemistry A*, 2(2), 345–349.

46. Kim, JY, Lee, K, Coates, NE, Moses, D, Nguyen, TQ, Dante, M, & Heeger, AJ. 2007. Efficient tandem polymer solar cells fabricated by all-solution processing. *Science*, 317(5835), 222–225.
47. Jost, K, Dion, G, & Gogotsi, Y. 2014. Textile energy storage in perspective. *Journal of Materials Chemistry A*, 2(28), 10776–10787.
48. Bae, J, Park, YJ, Lee, M, Cha, SN, Choi, YJ, Lee, CS, Kim, JM, & Wang ZL. 2011. Single-fiberbased hybridization of energy converters and storage units using graphene as electrodes. *Advanced Materials*, 23(30), 3446–3449.

9 A Heuristic Approach for Modelling and Control of an Automatic Voltage Regulator (AVR)

Rishabh Singhal
Roorkee Institute of Technology

Abhimanyu Kumar and Souvik Ganguli
Thapar Institute of Engineering and Technology

CONTENTS

9.1 INTRODUCTION

An automatic voltage regulator (AVR) is an intelligent device which ensures that the voltage and reactive power variations in an exciter occur in a desired style. In the current power industry, an AVR feels the output voltage and starts a curative mechanism of adjusting the exciter, and thus it is a controller system. It is, therefore, a complicated system that requires several electrical components. In order to investigate such a method, rigorous mathematics is necessary during modelling, making control algorithms more complicated. At this point, control engineers/practitioners aim to reduce the complexity by employing the principle of reduced-order modelling. Model Order Reduction (MOR) is an approximation scheme which preserves the inherent features in the course of modelling, whereas the unnecessary or the less important characteristics are being denied [1].

In an AVR system, the use of conventional tuning techniques is inadequate under certain operating conditions due to non-linear load characteristics. Since last two decades, researchers have preferred an optimization technique-based self-tuning strategy. Proportional–integral–derivative (PID) parameters are tuned using techniques such as the particle swarm optimization (PSO) technique [2], Sugeno fuzzy logic (SFL) employing crazy PSO [3], chaotic ant swarm algorithm [4], artificial bee colony (ABC) algorithm [5], many optimizing liaisons (MOL) algorithm [6], local unimodal sampling (LUS) optimization [7], etc.

Due to non-linearity and variable operating conditions, there exists parameter uncertainty to some extent, but there is no effective strategy to remedy this problem. For a better outcome, a grey PID controller was designed in Ref. [8] where the parameters of the GPID were optimized using the imperialist competitive algorithm (ICA). As the primary purpose has always been to enhance AVR performance, so this again brings us back to the fundamental problem of employing an optimization technique that leads to the best tuning. Gizi et al. [9] proposed to employ a combination of SFL approach, genetic algorithm and radial basis function neural network (RBF-NN) for optimal design purposes. In Ref. [10], a teaching–learning based (TLBO) optimization algorithm was suggested for AVR tuning.

Based on better transient and steady-state outcome, the literature also suggests the use of fractional-order PID (FOPID) in AVR systems [11]. Since a FOPID controller has more parameters than PID, its tuning process is different and comparatively more complicated. Odili et al. [12] suggested the use of an African buffalo optimizer (ABO) for PID parameter tuning to have an effective AVR, after comparison with the performance of Genetic Algorithm (GA), PSO, Bacterial Foraging Optimization (BFO), ant colony optimization (ACO), etc. With the development of several new techniques, the past three years record the application of the grasshopper optimization algorithm (GOA) [13], stochastic fractal search (SFS) [14], non-dominated shorting GA-II (NSGA-II) [15], symbiotic organisms search (SOS) algorithm [16], hybrid of SOS and simulated annealing (hSOS-SA) technique [17], sine-cosine algorithm (SCA) [18], cuckoo search [19], etc. for developing efficient AVRs.

The comparison of performances of the designed PIDs in the literature reveals scope for further development. Although the control aspect of an AVR is more emphasized in the literature, its modelling and development of its reduced-order system for further analysis are less reported in the literature. Hence this work is carried out in the present chapter. Further, a new treatment is given to its controller synthesis.

Although Ref. [1] narrates a comprehensive study of the model reduction techniques on an AVR system with a handful of classical and heuristic methods, the controller synthesis part is missing in this paper. Our chapter provides a competitive reduced-order model through a mixed approach combining the stability equation method (SEM) [20] with the grey wolf optimizer (GWO) [21], along with a controller design for the second-order model thus obtained. The controller parameters are obtained by applying the approximate model matching (AMM) [22] technique via GWO. The proposed method is compared with a sufficient number of metaheuristic algorithms [23–29] for both model reduction and controller design.

The rest of the chapter is structured into the following sections. In Section 9.2, a brief overview of AVR systems is given to familiarize the readers with the topic.

In Section 9.3, the suggested technique is deliberated. Section 9.4 reports the simulated outcomes and their interpretations. In Section 9.5, the concluding remarks are provided indicating some directions for future research.

9.2 A BRIEF OVERVIEW OF THE AUTOMATIC VOLTAGE REGULATOR

Synchronous generators often encounter disturbances (such as changes in load) which make the synchronous generators oscillate around the equilibrium. Consequently, the terminal voltage cannot remain constant any longer. Thus, AVRs are used for the regulation of the output voltage of synchronous generators as well as for enhancing their transient stability, because if it is not dealt properly then it degrades the power quality and may even turn out to be harmful for power system stability [30,31]. There are broadly three categories of AVRs:

1. A DC machine in the form of a generator coupled to the field of the synchronous generator.
2. An alternator combined with the field of a synchronous generator with the help of a rectifier.
3. Rectifier units connected to the field of a synchronous generator and supplied from the terminal voltage.

For appropriate AVR tuning and modelling, we shall look at some general features of exciters such as the electrical properties of the field coil. The field voltage E_{fd} (the output of exciters) is proportional to the exciter angular velocity ω_{ex} and total flux linkage λ in the field coil. The modelling of the saturation term in the current–flux relationship in coils can be done using an exponential function (or a quadratic relationship) expressed by an extra term S_e.

Now, to comprehend the basics of exciter tuning, let us observe the first-order AVR model represented by

$$G(s) = \frac{K_A}{1 + sT_A} \tag{9.1}$$

for the discussion of the tuning process. Tuning is extremely necessary to ensure that the device meets the desired performance specifications. For the purpose of AVR tuning, the values corresponding to compensation for loading and rectifier voltage drop are assumed to be one per unit and thus removed from the representation. The methods of AVR tuning are discussed below.

9.2.1 OPTIMIZING THE K_A AND THE PHASE-MARGIN

The immediate problem is to select the gain K_A which provides a small steady-state error and fast rise-time. Figure 9.1 shows the block diagram of the system, where $d(s)$ represents the disturbance term.

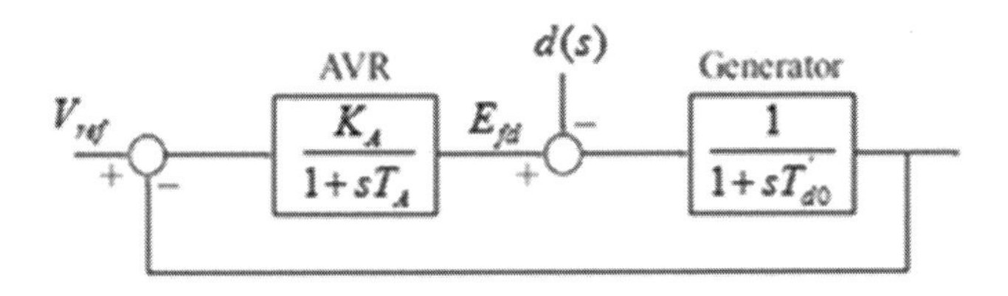

FIGURE 9.1 AVR tuning – steady-state error.

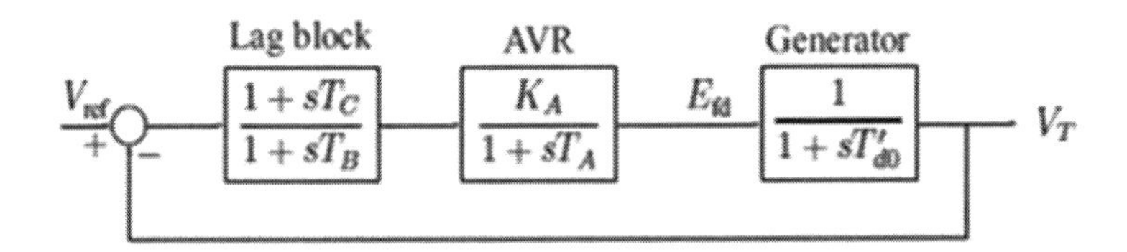

FIGURE 9.2 AVR tuning – lag compensator.

From theoretical analysis, a higher K_A value is expected to yield the desired results; however, this also implies mean oscillations. Employing optimization techniques on an appropriate objective function, we can determine the value of K_A that helps to meet both requirements of exciter gain and damping. Still an AVR with constant gain is not suitable in most cases. Therefore, a lag block can be used to augment the controller.

9.2.2 Use of the Lag Compensator Block

A lag compensator block, also known as the transient gain reduction block, is employed to reduce gain (at higher frequency) without any reasonable reduction in the DC gain. Figure 9.2 shows the block diagram with a lag block.

Here, the time constants T_B and T_C are also a design issue such that $T_B > T_C$. Suppose the open loop gain is around 20 dB (without the lag block) at 10 rad/s, then T_B and T_C are chosen in such a way that at higher frequencies the gain is reduced by 20 dB, which is by placing the zero and pole one decade apart. Furthermore, the net phase due should be approximately zero near the cross-over frequency. It is clear from theoretical considerations that a lag compensator acts as a damper without alteration in the steady-state error.

9.2.3 Rate-Feedback Method

In practice, several AVRs currently employ the rate feedback method, as shown in Figure 9.3.

Now, appropriate values of T_F and K_F need to be evaluated/chosen. The reason for having an inner loop is to get high DC gain while reducing high frequency gain. The crossover frequency in this case can simply be taken as the geometric mean of the two end-points of the flat section of the inner loop gain. This is known to produce a stable closed loop system.

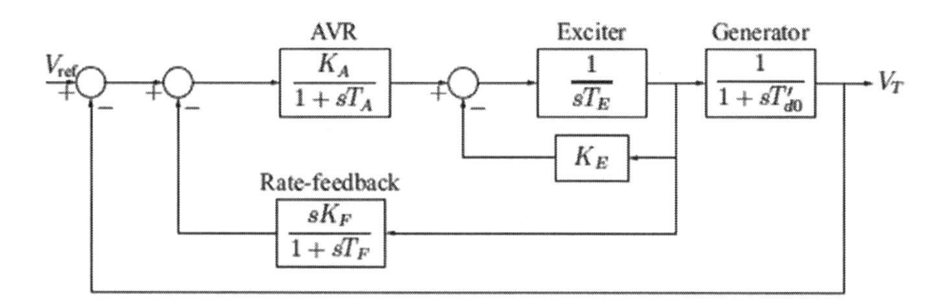

FIGURE 9.3 AVR tuning – rate-feedback compensator.

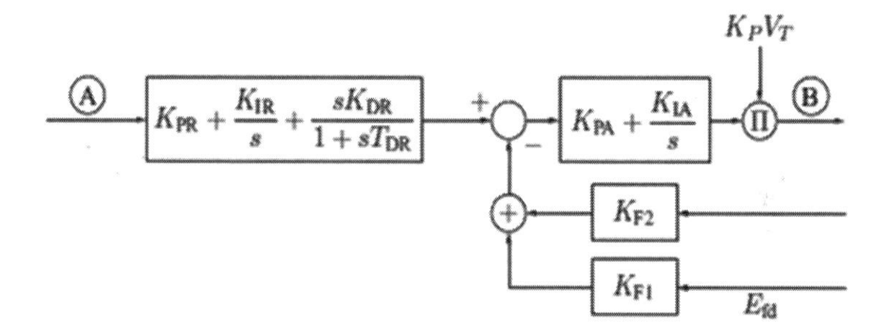

FIGURE 9.4 Segment of the IEEE AC7B compensator with a PID controller.

9.2.4 Use of PID Controller

The most preferable method is to use a PID controller because of its simple design and great performance [32]. For a step input, the steady-state error of a PID controller is zero, making it more favourable for use in AVRs. Figure 9.4 shows the use of a PID controller in an AVR with the example of an IEEE AC7B compensator.

For designing AVRs, methods such as optimal control, robust control and adaptive control are available [33,34]. Due to the ease of implementation and robustness to disturbances, researchers mostly opt for the use of PID controllers. The discussion above covers the information about the general methods for an efficient AVR system; further adjustments may be needed while implementation. Recent research has paid attention to parameter tuning to meet the optimal criterion for AVR design using a PID controller.

9.3 PROPOSED METHODOLOGY

The work carried out in this chapter deals with two important aspects. In the first part, MOR of the AVR model is performed using a mixed method. The second part uses a solo heuristic approach. In the model reduction phase, the denominator polynomial is determined by the SEM prescribed by Chen et al. [20]. The purpose of choosing this approach for obtaining the coefficients of the denominator is to provide stability to the reduced model. The coefficients of the numerator polynomial are found by employing a popular heuristic technique, namely the GWO [21]. The GWO developed by Seyedali performs through the guidance of the hierarchy in leadership and prey catching ability of the grey wolves. Interested readers may refer to the details of the GWO technique in Ref. [21]. To determine the coefficients of numerator, the sum of square error (SSE) is minimized using the GWO method. Unlike the step response matching, pseudo random binary sequence (PRBS) is used as an input signal to obtain the unknowns in the numerator. Our method also maintains the DC gain matching and also retains the non-minimum phase feature of the original test system taken up for the study. Additionally, characteristics of the parent system in both time and frequency domains are also preserved. Moreover, the dominant pole location has also been retained. The mixed method developed for the model reduction part is thus denoted as GWO-SEM in the subsequent simulation results section. A host of new algorithms such as the ant lion optimization (ALO) [23], dragonfly algorithm (DA) [24], moth flame optimization (MFO) [25], multi-verse optimizer (MVO) [26], GOA [27], SCA [28] and salp swarm algorithm (SSA) [29] are taken up for the purpose of comparison with the suggested technique. Unlike our method, the methods used for comparison are solely heuristic approaches for reducing the high order AVR test model.

The AVR model is now reduced to a second-order system by applying a hybrid approach constituting of a heuristic technique mixed with a standard classical approach. One of the purposes of the MOR technique is to develop a feasible low-order PID controller for which the controller synthesis is carried out in this chapter. Conventionally the PID parameters are obtained using either auto-tuning or some standard age-old PID tuning rules [35]. The Truxal method [36] is also an established procedure for controller synthesis on the basis of exact model matching (EMM) [37] in which the coefficients of the test system are calculated to meet the desired performance, and then the controller parameters are computed such that the complete closed-loop system closely resembles the reference model. The major disadvantage of this EMM technique is that the controller's hardware implementation is not guaranteed. To overcome this limitation of the EMM method, an AMM [22] may serve as a substitute, also applicable to the design scheme. Thus, the AMM technique is utilized to obtain the controller coefficients using the GWO method. On a similar note, the GWO technique is compared with a handful of other algorithms to test the efficacy of our technique.

9.4 SIMULATION RESULTS

A popular higher-order test system for the AVR model [1] is considered below

$$G(s) = \frac{5.994s^2 + 825.2}{0.573s^5 + 7.176s^4 + 51.06s^3 + 451.1s^2 + 876.6s + 260.8} \tag{9.2}$$

Applying the SEM, a second-degree polynomial for the denominator is evaluated as follows:

$$D_r(s) = s^2 + 1.9431s + 0.5781 \tag{9.3}$$

The numerator polynomial is obtained with the help of the widely cited GWO technique. The number of search agents considered is 20 while the greatest number of iterations is taken as 500 for the test system whose unknown coefficients are to be determined. Other algorithm parameters of the GWO method are assumed as per literature to calculate the numerator polynomial. Other methods such as DA, ALO, MFO, MVO, GOA, SCA and SSA are used for comparison. A similar assumption has been carried out for the choice of algorithm parameters used for the purpose of comparison. The numerator polynomial is thus obtained as

$$N_r(s) = -9.8047 \times 10^{-6} s + 1.8292 \tag{9.4}$$

The reduced-order model applying the mixed technique, namely GWO-SEM, is given by

$$G_r(s) = \frac{-9.8047 \times 10^{-6} s + 1.8292}{s^2 + 1.9431s + 0.5781} \tag{9.5}$$

Thus, the DC gain of the reduced model is represented by $\frac{1.8292}{0.5781} = 3.1641$ equalling that of the original test system as given by $\frac{825.2}{260.8} = 3.1641$ calculated at $s = 0$ for both the test systems. The non-minimum phase feature of the original model is usually secured in the reduced-order system, indicated by the negative sign in the expression for the numerator polynomial. The convergence characteristics for the GWO technique are obtained and are represented in Figure 9.5.

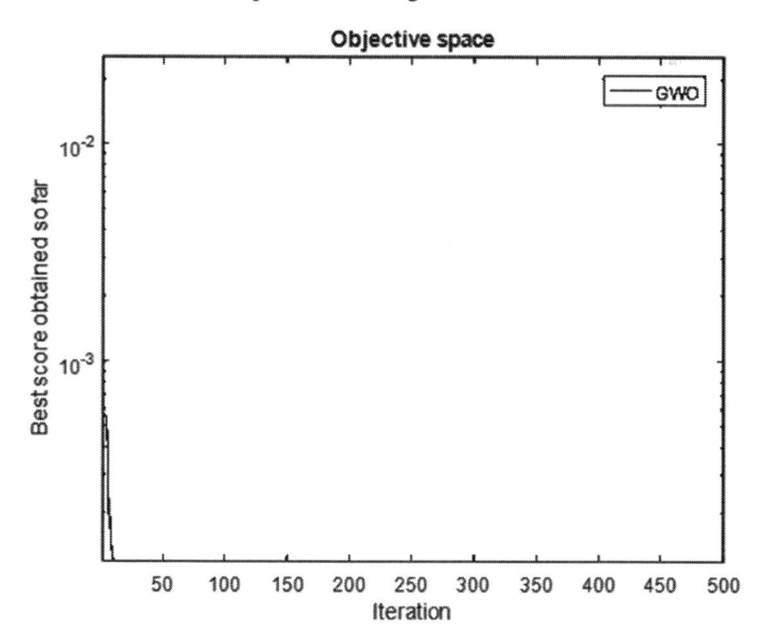

FIGURE 9.5 Convergence plot of the GWO technique to obtain the numerator polynomial.

TABLE 9.1
Reduced-Order Models in the Continuous-Time Domain and Their Fitness Functions

Methods	Reduced-Order Models	Fitness Function
Proposed (GWO-SEM)	$\frac{-9.8047\times10^{-6}s+1.8292}{s^2+1.9431s+0.5781}$	**0.0001**
ALO	$\frac{8.09s+9.74}{s^2+10.45s+3.08}$	0.3938
DA	$\frac{0.019s+15}{s^2+6.76s+4.74}$	0.0081
MFO	$\frac{-2.52\times10^{-5}s+15}{s^2+15s+4.74}$	0.0044
MVO	$\frac{0.22s+5.89}{s^2+14.67s+1.86}$	0.0013
GOA	$\frac{4.38s+6.31}{s^2+14.94s+1.99}$	0.0872
SCA	$\frac{1.54s+8.93}{s^2+15s+2.82}$	0.0165
SSA	$\frac{0.55s+7.49}{s^2+2.29s+2.37}$	0.0109

The bold value represents the best result

The convergence took place within the first 50 iterations to very close to the zero-value indicating faster convergence as well as accuracy of results. The other reduced systems are also obtained by applying constrained heuristic approaches and are shown in Table 9.1. The minimum error values are reported in this table.

A higher-order test system has poles on the $j\omega$ axis, so it is expected that the reduced model will be the non-minimum phase in nature. The second-order model determined by our method thus has right-half plane zeros. Only the MFO method used for comparison has the non-minimum phase. The rest of the algorithms violate this constraint. Moreover, the minimum fitness function is also produced by the proposed method. The result of the MVO method in terms of the error value is only close by. So, it is proved that the GWO-SEM method yields the best fitted reduced-order model with a minimum error value. However, only a single test run with a sufficient number of function evaluations has been considered for this investigation. Statistical analysis of the test results could also be carried out. In that case, the model with the minimum fitness function value could be reported in the table. Even some non-parametric tests such as Wilcoxon test and Kruskal Wallis test could be conducted for significance validation of the results. The specifications viz. rise time, settling time, overshoot and undershoot are calculated in Table 9.2 corresponding to the models developed. The comparison is carried out not only with some standard heuristic methods but also with the original test model considered in this study.

TABLE 9.2
Measures for Some Important Time Domain Specifications

Test System	Algorithms	Rise Time (s)	Settling Time (s)	Overshoot (%)	Undershoot (%)
Original		6.2420	11.6452	0	0
Reduced	GWO-SEM	**6.2655**	**11.3895**	**0**	**0**
	ALO	6.6884	12.0451	**0**	**0**
	DA	2.8089	5.1046	**0**	**0**
	MFO	6.8032	12.1812	**0**	**0**
	MVO	17.1446	30.5591	**0**	**0**
	GOA	16.1587	28.3760	**0**	**0**
	SCA	11.5276	20.4177	**0**	**0**
	SSA	1.4608	3.6966	3.0947	**0**

The bold values represent the closest values to the original system.

TABLE 9.3
Gain and Phase Margin Calculations

Test System	Algorithms	Gain Margin	Phase Margin
Original		9.2472	78.9266
Reduced	GWO-SEM	1.9818e+05	80.9933
	ALO	Inf	144.0809
	DA	Inf	89.4908
	MFO	5.9622e+05	104.6959
	MVO	Inf	107.7419
	GOA	Inf	122.3961
	SCA	Inf	111.7911
	SSA	Inf	63.5549

In view of settling time and rise time given in Table 9.2, it is evident that our method performs better than all the other heuristic methods. The results produced by the GWO-SEM method are the closest to the dimensions of the original test system taken up for the study. Only ALO and MFO methods produce nearly close results. However, all the methods have matching values of overshoot and undershoot percentages except that produced by the SSA technique. In Table 9.3, the frequency domain parameters of the original and reduced models are enumerated. Normally, the gain margin value is reported in decibels, whereas the phase margin is in degrees as given in Table 9.3.

From Table 9.3 results, it is found that our method and the other heuristic methods used for comparison have either a very high value or infinite gain margin. This justifies the fact that the reduced models produced are all inherently stable and are quite different from the gain margin of the original test system considered in this chapter. The phase margin of the proposed approach is in close proximity to

TABLE 9.4
Pole and Zero Locations of Higher- and Reduced-Order Models

Test System	Algorithms	Poles	Zeros
Original		−10.0007 + 0.0000*i*	0.0000 +11.7333*i*
		−0.0802 + 7.9227*i*	0.0000 − 11.7333*i*
		−0.0802 − 7.9227*i*	
		−2.0000 + 0.0000*i*	
		−0.3625 + 0.0000*i*	
Reduced	GWO-SEM	−1.5764	1.8656e+05
		−0.3667	
	ALO	−10.1485	−1.2032
		−0.3032	
	DA	−5.9696	−769.3963
		−0.7941	
	MFO	−14.6770	5.9622e+05
		−0.3230	
	MVO	−14.5405	−26.6894
		−0.1281	
	GOA	−14.8081	−1.4403
		−0.1347	
	SCA	−14.8094	−5.7861
		−0.1906	
	SSA	−1.1428 + 1.0309*i*	−13.5209
		−1.1428 − 1.0309*i*	

that of the original model. The results produced by DA and SSA methods are only near the actual value. Thus, it can be concluded that the suggested technique shows close resemblance with the original system in terms of frequency domain measures. Further, the pole–zero locations of the original test system and the proposed model are reported in Table 9.4. The pole–zero locations implicate two salient points: whether the systems have right-half s-plane zeros and poles. Right-half poles indicate the unstable nature of the system while right-half plane zeros imply a non-minimum phase. These two properties are verified through the outcomes reflected in Table 9.4. In addition to this, the closest pole and zero locations in comparison to the original model are also being identified.

From the pole locations of Table 9.4, it is clear that our model reaches the closest to the dominant pole location. But the results produced by ALO, MFO, MVO, GOA and SCA techniques are not far behind. Regarding the zero location of the test model, only the proposed technique and MFO method follow the non-minimum phase characteristics. In contrast, the other methods have a minimum phase nature. Some of the performance indices that are widely popular are assessed and the outcomes can be visualized in Table 9.5. The minimum value in each column of this table is considered to be the best value and, therefore, represented with the help of bold-faced letters.

TABLE 9.5
Calculation of Some Popular Error Indices

Methods	IAE	ITAE	ISE	ITSE	H_∞ Norm
Proposed method	**1.0357e−04**	**8.9350e−06**	**1.1872e−07**	**1.1368e−08**	**0.0016**
ALO	0.0066	3.9866e−04	3.9379e−04	2.3490e−05	0.0938
DA	8.6201e−04	7.3852e−05	8.1072e−06	7.7136e−07	0.0127
MFO	6.5253e−04	5.4662e−05	4.4581e−06	4.1160e−07	0.0089
MVO	3.7845e−04	2.7160e−05	1.2686e−06	9.7206e−08	0.0045
GOA	0.0031	1.8188e−04	8.7236e−05	4.9372e−06	0.0458
SCA	0.0013	8.0994e−05	1.5661e−05	9.6294e−07	0.0183
SSA	0.0011	8.2659e−05	1.0933e−05	8.9672e−07	0.0138

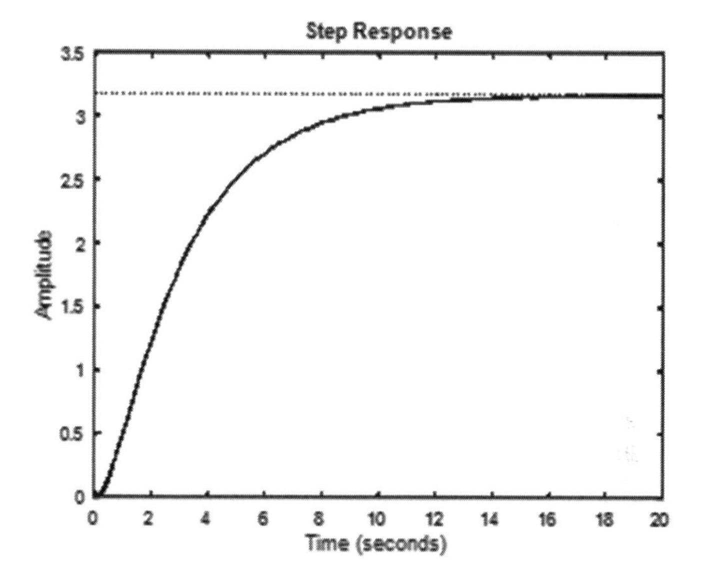

FIGURE 9.6 Step response study of parent and reduced models.

Table 9.5 gives clear indication of the fact that our technique surpasses other techniques used for comparison with respect to all the error indices considered for the investigation. Only the results of DA, MVO and MFO techniques are close to that of our method for the IAE and ISE indices. In addition to DA, MFO and MVO, the methods SCA and SSA are also close to the GWO-SEM method in terms of ITAE. MVO is somewhat close to the suggested approach in terms of ITSE, while DA, MFO, SCA and SSA techniques slightly fall behind. In terms of the H_∞ norm, the performances of MVO and MFO are only in close proximity to that of the proposed mixed approach. Furthermore, the step and Bode responses of the parent and reduced models applying the GWO-SEM technique are also plotted in Figures 9.6 and 9.7, respectively.

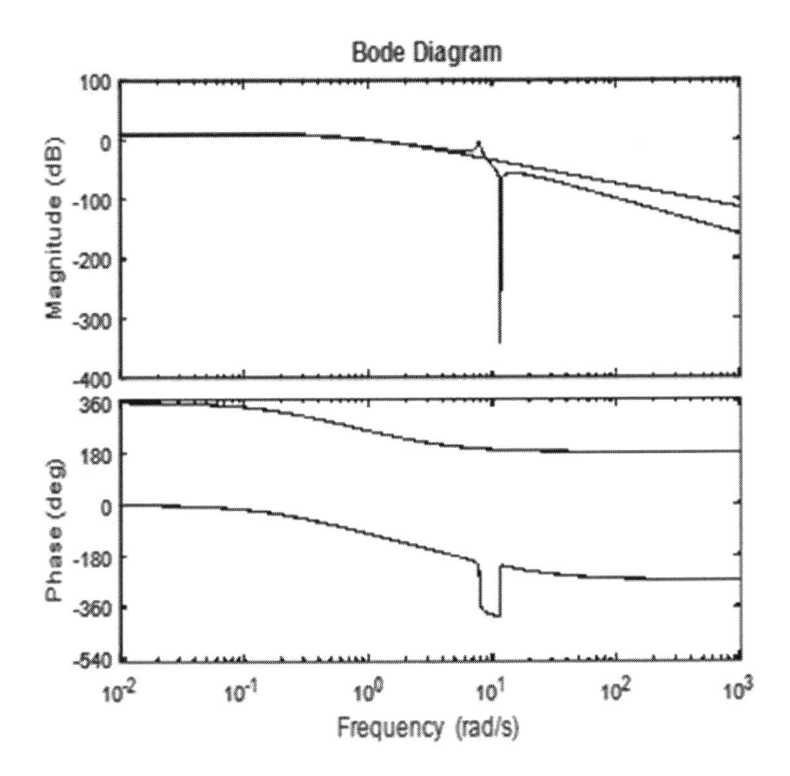

FIGURE 9.7 Frequency response study of original and reduced models.

The step response offered by the proposed GWO-SEM technique shows a very close match with respect to the higher-order system under test. The frequency response of the lower-order model, however, deviates slightly from that of the original model necessitating the use of a controller. For the design of a controller using AMM, a model is selected as per Ref. [38]. A PID controller is usually denoted by

$$G_c(s) = K_P + \frac{K_I}{s} + K_D s \tag{9.6}$$

The response of the cascaded plant and the controller is compared with the reference model's response to extract the unknown viz. K_P, K_I and K_D of the controller, considered as decision variables of the objective to minimize the integral of square error (ISE). The tuning parameters of the controller obtained by applying the GWO technique are provided in Table 9.6. Some popular metaheuristic algorithms are used for comparison purpose. The fitness function value is also provided to check the efficiency of the suggested GWO approach.

From the table it is evident that the GWO technique obtains the controller parameters with the least fitness function value and hence is considered to be the best amongst the algorithms used. Thus, the GWO-SEM method has been able to reduce effectively the AVR model, and finally a reference model AMM matching evaluates successfully its controller parameters with the least ISE value.

TABLE 9.6
Comparative Performance Analysis of Controller Tuning Parameters and Fitness Function

Algorithms	K_P	K_I	K_D	ISE
Proposed	0.0618	0.0059	2.3493	**3.7616e−09**
ALO	4.1553	3.5721	0.0011	0.0052
DA	1.5760	3.6891	0.2598	3.4444e−06
MFO	3.3906	2.4831e−09	0.2939	2.2672e−08
MVO	5.0000	5.0000	0.0087	7.4998e−05
GOA	1.2264	5.0000	0.0005	6.1514e−06
SCA	3.2704	5.0000	0.0004	1.8923e−06
SSA	3.8879	4.4822	0.0048	1.9556e−05

The bold value represents the minimum or best result.

9.5 CONCLUSIONS

This chapter effectively demonstrates the MOR and designing of an AVR controller. The methodology adopted in the order reduction part is a mixed technique comprising a classical method namely the stability equation approach to obtain the coefficients of the denominator polynomial, whereas the numerator polynomial is determined by the GWO. In the controller synthesis part, the controller parameters are obtained using the GWO technique by applying AMM. A handful of meta-heuristic algorithms are considered for comparison in both model reduction and in controller design. The proposed technique outperforms the algorithms used for comparison. The proposed methodology can further be utilized for modelling and control of different complicated power system models. Controllers available nowadays are digital in nature. Hence a discrete-time modelling and control is expected to be communicated in future. Some typical variants of the GWO technique can be tried out to yield even better modelling and control in the discrete-time domain. Hybrid computation algorithms will of course pose new challenges to parameter estimation and control.

REFERENCES

1. Biradar, S., Saxena, S., & Hote, Y. V. (2015). Simplified model identification of automatic voltage regulator using model-order reduction. In *2015 International Conference on Power and Advanced Control Engineering (ICPACE)*, 423–428.
2. Gaing, Z. L. (2004). A particle swarm optimization approach for optimum design of PID controller in AVR system. *IEEE Transactions on Energy Conversion*, *19*(2), 384–391.
3. Mukherjee, V., & Ghoshal, S. P. (2007). Intelligent particle swarm optimized fuzzy PID controller for AVR system. *Electric Power Systems Research*, *77*(12), 1689–1698.
4. Zhu, H., Li, L., Zhao, Y., Guo, Y., & Yang, Y. (2009). CAS algorithm-based optimum design of PID controller in AVR system. *Chaos, Solitons & Fractals*, *42*(2), 792–800.

5. Gozde, H., & Taplamacioglu, M. C. (2011). Comparative performance analysis of artificial bee colony algorithm for automatic voltage regulator (AVR) system. *Journal of the Franklin Institute*, *348*(8), 1927–1946.
6. Panda, S., Sahu, B. K., & Mohanty, P. K. (2012). Design and performance analysis of PID controller for an automatic voltage regulator system using simplified particle swarm optimization. *Journal of the Franklin Institute*, *349*(8), 2609–2625.
7. Mohanty, P. K., Sahu, B. K., & Panda, S. (2014). Tuning and assessment of proportional–integral–derivative controller for an automatic voltage regulator system employing local unimodal sampling algorithm. *Electric Power Components and Systems*, *42*(9), 959–969.
8. Tang, Y., Zhao, L., Han, Z., Bi, X., & Guan, X. (2016). Optimal gray PID controller design for automatic voltage regulator system via imperialist competitive algorithm. *International Journal of Machine Learning and Cybernetics*, *7*(2), 229–240.
9. Al Gizi, A. J., Mustafa, M. W., Al-geelani, N. A., & Alsaedi, M. A. (2015). Sugeno fuzzy PID tuning, by genetic-neutral for AVR in electrical power generation. *Applied Soft Computing*, *28*, 226–236.
10. Chatterjee, S., & Mukherjee, V. (2016). PID controller for automatic voltage regulator using teaching–learning based optimization technique. *International Journal of Electrical Power & Energy Systems*, *77*, 418–429.
11. Li, X., Wang, Y., Li, N., Han, M., Tang, Y., & Liu, F. (2017). Optimal fractional order PID controller design for automatic voltage regulator system based on reference model using particle swarm optimization. *International Journal of Machine Learning and Cybernetics*, *8*(5), 1595–1605.
12. Odili, J. B., Kahar, M. N. M., & Noraziah, A. (2017). Parameters-tuning of PID controller for automatic voltage regulators using the African buffalo optimization. *PloS One*, *12*(4), e0175901.
13. Hekimoğlu, B., & Ekinci, S. (2018). Grasshopper optimization algorithm for automatic voltage regulator system. In *2018 5th International Conference on Electrical and Electronic Engineering (ICEEE)*, 152–156.
14. Çelik, E. (2018). Incorporation of stochastic fractal search algorithm into efficient design of PID controller for an automatic voltage regulator system. *Neural Computing and Applications*, *30*(6), 1991–2002.
15. Yegireddy, N. K., Panda, S., Papinaidu, T., & Yadav, K. (2018). Multi-objective non dominated sorting genetic algorithm-II optimized PID controller for automatic voltage regulator systems. *Journal of Intelligent & Fuzzy Systems*, *35*(5), 4971–4975.
16. Çelik, E., & Durgut, R. (2018). Performance enhancement of automatic voltage regulator by modified cost function and symbiotic organisms search algorithm. *Engineering Science and Technology, an International Journal*, *21*(5), 1104–1111.
17. Çelik, E., & Öztürk, N. (2018). A hybrid symbiotic organisms search and simulated annealing technique applied to efficient design of PID controller for automatic voltage regulator. *Soft Computing*, *22*(23), 8011–8024.
18. Hekimoğlu, B. (2019). Sine-cosine algorithm-based optimization for automatic voltage regulator system. *Transactions of the Institute of Measurement and Control*, *41*(6), 1761–1771.
19. Sikander, A., & Thakur, P. (2019). A new control design strategy for automatic voltage regulator in power system. *ISA Transactions*, *100*, 235–243.
20. Chen, T. C., Chang, C. Y., & Han, K. W. (1979). Reduction of transfer functions by the stability-equation method. *Journal of the Franklin Institute*, *308*(4), 389–404.
21. Mirjalili, S., Mirjalili, S. M., & Lewis, A. (2014). Grey wolf optimizer. *Advances in Engineering Software*, *69*, 46–61.
22. Nijmeijer, H., & Savaresi, S. M. (1998). On approximate model-reference control of SISO discrete-time nonlinear systems. *Automatica*, *34*(10), 1261–1266.

23. Mirjalili, S. (2015). The ant lion optimizer. *Advances in Engineering Software*, *83*, 80–98.
24. Mirjalili, S. (2016). Dragonfly algorithm: a new meta-heuristic optimization technique for solving single-objective, discrete, and multi-objective problems. *Neural Computing and Applications*, *27*(4), 1053–1073.
25. Mirjalili, S. (2015). Moth-flame optimization algorithm: a novel nature-inspired heuristic paradigm. *Knowledge-Based Systems*, *89*, 228–249.
26. Mirjalili, S., Mirjalili, S. M., & Hatamlou, A. (2016). Multi-verse optimizer: a nature-inspired algorithm for global optimization. *Neural Computing and Applications*, *27*(2), 495–513.
27. Mirjalili, S. Z., Mirjalili, S., Saremi, S., Faris, H., & Aljarah, I. (2018). Grasshopper optimization algorithm for multi-objective optimization problems. *Applied Intelligence*, *48*(4), 805–820.
28. Mirjalili, S. (2016). SCA: a sine cosine algorithm for solving optimization problems. *Knowledge-Based Systems*, *96*, 120–133.
29. Mirjalili, S., Gandomi, A. H., Mirjalili, S. Z., Saremi, S., Faris, H., & Mirjalili, S. M. (2017). Salp swarm algorithm: a bio-inspired optimizer for engineering design problems. *Advances in Engineering Software*, *114*, 163–191.
30. Demello, F. P., & Concordia, C. (1969). Concepts of synchronous machine stability as affected by excitation control. *IEEE Transactions on Power Apparatus and Systems*, *88*(4), 316–329.
31. Khezri, R., & Bevrani, H. (2015). Voltage performance enhancement of DFIG-based wind farms integrated in large-scale power systems: coordinated AVR and PSS. *International Journal of Electrical Power & Energy Systems*, *73*, 400–410.
32. Ang, K. H., Chong, G., & Li, Y. (2005). PID control system analysis, design, and technology. *IEEE Transactions on Control Systems Technology*, *13*(4), 559–576.
33. Aguila-Camacho, N., & Duarte-Mermoud, M. A. (2013). Fractional adaptive control for an automatic voltage regulator. *ISA Transactions*, *52*(6), 807–815.
34. Shayeghi, H., Younesi, A., & Hashemi, Y. (2015). Optimal design of a robust discrete parallel FP+ FI+ FD controller for the automatic voltage regulator system. *International Journal of Electrical Power & Energy Systems*, *67*, 66–75.
35. O'Dwyer, A. (2009). *Handbook of PI and PID Controller Tuning Rules*. London: Imperial College Press.
36. Lago, V. G., & Truxal, G. J. (1954). The design of sampled-data feedback systems. *Transactions of the American Institute of Electrical Engineers, Part II: Applications and Industry*, *73*(5), 247–253.
37. Wolovich, W. A. (1972). The use of state feedback for exact model matching. *SIAM Journal on Control*, *10*(3), 512–523.
38. Rana, J. S., Prasad, R., & Agarwal, R. P. (2016). Designing of a controller by using model order reduction techniques. *International Journal of Engineering Innovations and Research*, *5*(3), 220.

10 Reduced-Order Modelling and Control of a Single-Machine Infinite Bus System with the Grey Wolf Optimizer (GWO)

Rishabh Singhal
Roorkee Institute of Technology

Saumyadip Hazra, Sauhardh Sethi, and Souvik Ganguli
Thapar Institute of Engineering and Technology

CONTENTS

10.1 INTRODUCTION

Analysing the stability issues of power systems is one of the key aspects of research. But most power system models are of higher order. Hence in order to evaluate the reliability of a power system model, the system can be reduced to either a two-machine equivalent device or a single machine connected to an infinite bus. The total electrical power network connected to the machine under study can be modelled using the analogous method of Thevenin. The benefit of a single machine infinite bus (SMIB) power system is that it allows the tuning of controllers on one machine without the influence of other electrical devices on power systems. The effect is distributed among the different machines in an interconnected power system [1].

Modelling of a SMIB leads to a higher-order system, hence suitable model reduction schemes to address this problem are already available in the literature. Some of the notable studies are highlighted as follows to create awareness amongst readers. Arredondo et al. [2] applied the genetic algorithm and the Levenberg–Marquardt algorithm to reduce a power system model having 68 nodes and 16 generators connected through 86 transmission lines retaining closely the electromechanical modes related to the generators of the system considered for the investigation through the minimisation of the root mean square error. Đukić and Sarić [3] reviewed some of the model reduction techniques applicable to large scale linear as well as non-linear dynamic power system models. Methods such as singular perturbation, modal analysis, singular value decomposition and moment matching were thoroughly discussed from the view point of linear model reduction problems, whereas techniques such as the proper orthogonal decomposition, trajectory piecewise linear method and balancing-based methods were covered in the context of non-linear reduction study. A New England 10-generator, 39-bus test system was considered to validate the above-mentioned methods. Ghosh and Senroy [4] compared balanced truncation with the Krylov subspace method to reduce a large-scale power system model. The prime motive behind this study was to obtain the dynamic pattern like coherency between generators. The achievement of the methods was measured under varying the damping conditions of the system. Ghosh and Senroy [5] further applied the balanced truncation technique to solve the order reduction of a test system representing the northern grid of India considering changing conditions of the system with preserving the input–output relations in between the areas. Coherency of the generators was very well captured in their equivalent model. Ramirez et al. [6] applied the balanced realization technique to determine the reduced systems from dynamic system equivalents of electrical power systems. Four test systems, including one transmission line network and three models involving wind power plants, were considered to prove the efficacy of the proposed method. Cheng and Scherpen [7] developed an order reduction methodology according to clustering separately the generators and the loads based on their behavioural mismatches using H_2-norm. A power system network with distributed controllers considering an IEEE 30-bus system was taken up for the study. Osipov and Sun [8] presented a novel dynamic technique for model reduction of power system networks to establish not only fast but also accurate time-domain simulation. The method considered a linear reduction scheme for quicker simulations while a non-linear reduction technique was applied to obtain better accuracy of the simulation. The work was validated with the help of a system from Northeast Power Coordinating Council comprising 148 buses and 48 machines.

From the literature it is found that mostly the model reduction approaches based on classical techniques have been applied to reduce different power system models. Hardly any heuristic techniques were employed to simulate the reduced systems. The SMIB system is one such test system whose order reduction and control scheme can be devised with the help of the grey wolf optimizer (GWO) [9]. Several new metaheuristic techniques [10–16] developed during the period 2015–2018 are used for the purpose of comparison. Further, the literature discussed above only considered the model reduction issues of various power system models without highlighting their controller design. This chapter also takes up the problem of controller synthesis of the reduced models using the GWO technique. The approximate model matching (AMM) [17] framework was considered to yield the unknown controller parameters. A similar comparison follows for the controller design problem as well. Some statistical measures and non-parametric tests were conducted to validate the test outcomes.

The rest of the chapter is structured as mentioned in the following lines. Section 10.2 narrates the modelling of a SMIB system to expose the readers to the problem. In Section 10.3, a thorough literature survey is conducted on the modelling and control of SMIB systems carried out by several researchers across the globe, while Section 10.4 presents the proposed methodology of work. Section 10.5 enumerates the results of the reduction part as well as the controller design. Section 10.6 draws the salient inferences. It also outlines some future scope of the study as well.

10.2 A BRIEF OVERVIEW OF SMIB SYSTEMS

SMIB is a bus system in which the transient nature of a synchronous generator along with the nature and effects of the faults can be studied in a transmission line. It is well known that these machines supply power to a network which is much greater than the rating of the machine, and this is what happens in the real-world power systems where a huge constructed network experiences a little impact due to any kind of change in the status of the machine. This concept introduces the term 'infinite' in its name (Figure 10.1).

In order to mathematically model the SMIB, Park's equations are used in which the axes of the machine are described with the help of d-axis and q-axis. The original equations are then changed into the new form with the help of these variables considering the effect of damper winding of the machine. The equations are described

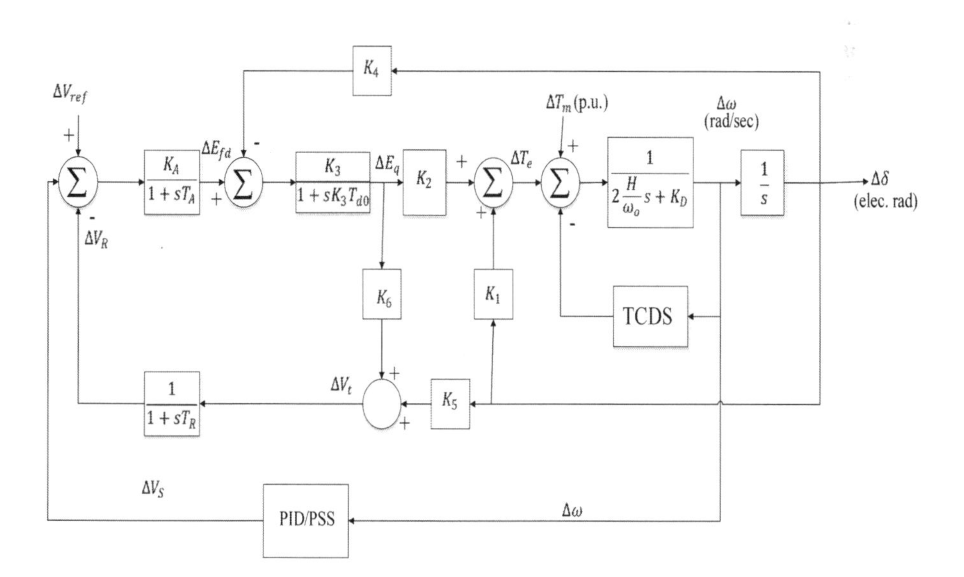

FIGURE 10.1 Block diagram of a SMIB power system.

by drawing the relationship between their flux linkage ψ, current, voltage and resistance. For any SMIB system, the voltage equation is written as:

$$\overline{V_t} = jx_t\overline{I_t} + \overline{V_b} \tag{10.1}$$

which is then further written in terms of d and q variables or can be separated on their basis. The electrical useful torque produced is equal to the power (P_t) that is delivered by the machine to the busbar (infinite). P_t is given by:

$$P_t = v_d i_d + v_q i_q \tag{10.2}$$

For the linearization of the mathematical model, a small change in each variable is considered with respect to the values at the steady-state operating conditions and is divided with the total voltage of the busbar. For the study of the oscillations that occur in the system, generally the Heffron-Phillips model is used for simplification. As per this model, the effect of damper winding of the synchronous machine is removed from the damper coefficient, the resistances that are offered by the d and q axes are neglected and in the presence of a very small signal, the change in the speed of the rotor of the machine is considered very small. Therefore, the Heffron-Phillips model can be considered as the steady-state model of the system. The Heffron-Phillips model can only be applied after the network has been linearized. Modal analysis generally includes the study of a system in the frequency domain. In modal analysis, the system given is converted first into a state-space model from where the Eigen solutions are calculated. Once the values of Eigen solutions are obtained, they are used to decompose the network into several parallel branches representing an open loop model of the system. The analysis is further performed for determining the stability of the system by calculating the locations of poles. A system is said to be stable if it does not contain any pole on the right-hand side. Usually, the Heffron-Phillips model is combined with modal analysis to determine the stability and calculate the oscillations. When oscillations are considered for a power system, they are associated with the generator, which is associated with a rotor, and are called electromechanical oscillations. With their help, the effect of damping is calculated.

The Heffron-Phillips model also plays an important role in determining the other important features of a SMIB system, one of which is the analysis of damping torque. The basic principle behind it is that all the oscillations which are electromagnetic in nature and are occurring in the generator decide the power oscillations' damping in SMIB dynamically. Besides, the low-frequency oscillations, which may be caused by the distortions occurring in the system or some other kind of small transient operation, can be ruled out with the help of a Power System Stabilizer (PSS). This technology has been used for a long period and is considered as one of the most general solutions to this problem. A PSS may also employ an additional component to increase the damping even further, which is known as the Automatic Voltage Regulator (AVR). Sometimes in the recent technology, a PSS is used with dual inputs, which are the speed of the machine and the power given out as output. The main advantage of this method is that it becomes more robust, and hence the damping can be done using a large gain factor. When only the effect of AVR is considered, then the output equation of the SMIB system after linearization is given as follows:

$$\Delta\delta(t) = ae^{-\frac{D}{2m}t}\cos(\omega_{NF}t) + b \tag{10.3}$$

In this equation, a and b are constants and ω_{NF} is the natural frequency of the system in radians. This equation represents the behaviour of a generator when active power is supplied by it during transient conditions and when the system is undergoing some kind of small distortion. In this equation, mainly one factor is to be kept in mind, which is the ratio of $D/2m$. When the magnitude of this factor is very small or even negative, then the power system would have a poor damping capacity and the magnitude of the amplitude of the oscillation would keep on increasing. At this point, it becomes important to note that the oscillating natural frequency is very close to the oscillation of the power system. When a PSS is connected with an AVR, it provides an additional electromagnetic torque, which is responsible for further dampening of the oscillations. For the design of actual analysis of the damping torque, first the effect of the PSS is neglected. When it is neglected, the torque developed due to the presence of the AVR only, that is, the electromagnetic torque, is complex in the frequency domain, and its solution indicates that it affects or it damps the real part of the power oscillation. When the effect of the PSS is considered and the AVR is neglected, the PSS only works in the steady state. When it is disturbed from the steady state (whether upward or downward from the operating point), then it exerts opposite electromagnetic torque in the opposite direction to compensate for it. The best thing about the PSS is that it can be applied to any kind of system for suppressing the low-frequency oscillations. Hence, any type of model which is linearized first can be applied even to a multi-machine bus [18,19].

10.2.1 Design of Power System Stabilizers (PSSs)

The PSSs that were designed using the conventional method were based on the linear control theory. For doing this, a system used to be linearized based on a calculated operating point, and at that point, an optimisation technique was used for the calculation of the optimal parameters. But the problem faced due to the use of this method is that any power system model is usually a non-linear model and the operating points vary over a wide range. So, to conclude any particular point as the operating point would be wrong. Hence a system should be designed which, according to the conditions, calculates the operating conditions and also calculates the values of the optimal parameters. In order to overcome these limitations, a self-tuning method, a lead compensator-based stabilizer and a gradient method have been implemented, but they are very complex to design and require a lot of processing time. Also, these types of methods work under completely constant conditions. They are not robust and when there is any change in the physical conditions then they mostly either fail to work or do not give satisfactory results. Even if some changes are made in the system, such as the system is changed configurationally, then also they fail to give proper results, and this directly affects their convergence towards the result. These conditions lead to the use of an adaptive controller which takes into account the gain scheduling of the system and all the non-linear components of the power system. These controllers adjust themselves when subjected to any changes and are able to provide damping over a wide range [20].

10.3 RELATED STUDIES ON SMIB

Some relevant studies pertaining to the modelling and control of SMIB are discussed in this section for the interested readers. Yang [21] studied the H_∞ optimisation method for power system stabilizer design. The weighting function selection and the application to multi-machine power systems were developed. A new method was presented for the selection of weighted functions and an extended PSS was presented. Wang and Swift [22] established a unified model of Static Var compensator (SVC), Controllable series compensator (CSC) and Phase shifter flexible AC transmission systems (PS FACTS) and studied their oscillation damping capability using their damping torque contribution in power systems. The results were obtained for damping control and their robustness were compared for all the FACTS devices. Ford et al. [23] examined the transient stability of a SMIB system using non-linear dynamic programming. The results explained the stability of many approaches and presented a novel solution to the transient stability problem. Ghfarokhi et al. [24] studied the dynamic behaviour and transient stability of SMIB using eigenvalue analysis and discussed the variations in PSS parameters using the system dynamic performance response with their simulation. Sambariya and Prasad [25] applied the Routh approximation method to reduce a SMIB model containing a PSS. Sambariya and Prasad [26] developed methods using the stability equation-based reduction technique for obtaining a simplified model of a system and analysed them using the Routh Stability Array Method in time and frequency domains. The results were encouraging, showing the preservation of stability and reduction of second and third order parameters of SMIB.

Sambariya and Prasad [27] described a differentiation method based stable reduced model of a SMIB system with a PSS and analysed it in time and frequency domains. The result proved the preservation of stability and reduction of the second and third order parameters of SMIB-PSS. Wan and Zhao [28] proposed an extended backstepping method for studying the stability of SMIB with Superconducting magnetic energy storage (SMES) using the generator excitation control and SMES control. A class K function was also introduced for improving transient stability, and simulation results proved its effectiveness. Sambariya and Prasad [29] presented the evaluation of membership functions on SMIB-PSS and analysed it using CPSS, FPSS and without PSS. The simulation obtained for different plants indicated that FPSS gave better results than CPSS. Sambariya and Prasad [30] described and analysed the use of four different Model order reduction (MOR) methods in time and frequency domains with a PSS. The results were encouraging and proved the preservation of stability, and parameters of SMIB-PSS whose second and third order reduced models were obtained. Wan et al. [31] framed a new control synthesis for the robust stabilisation problem of SMIB-SVC. A class K function was introduced in the result which improved the transient stability and the simulations depicted the same. Perev [32] presented a problem of MOR for a SMIB system using a Legendre polynomial approximation based balanced residualization method. The numerical performance of the method proved good approximation properties.

Wang et al. [33] studied the non-linear dynamic characteristics of a SMIB using the swing equation. The results obtained showed a detailed explanation of non-linear dynamic response of the SMIB power system placed in a periodic load disturbance. Sambariya and Arvind [34] reduced a large order system using the firefly algorithm. The numerator was reduced and optimized by integral square minimisation and the denominator by the stability equation. The results obtained satisfied the Routh equation and stability criteria in terms of minimum error. Sambariya and Arvind [35] reduced a large order system using a self-adaptive bat algorithm. The numerator was reduced and optimized by integral square minimisation and the denominator by the stability equation. The results obtained satisfied the Routh equation and stability criteria in terms of minimum error. Milla and Duarte-Mermoud [36] described the improvement of oscillations of a SMIB using a predictive optimized adaptive PSS. The results proved that the POA-PSS was far better than the conventional PSS. Yilmaz and Savacı [37] investigated the stability of a SMIB using the alpha-stable Levy type load fluctuations over the parameter space of mechanical power and damping parameter. Alrifai et al. [38] used the discrete time sliding model control technique to control a chaotic power system and eliminate the chaotic oscillations. Two techniques were used, first was the exponential reaching law and the second was the double power reaching law. The results obtained showed that both the techniques gave good results and the second one proved better.

Farhad et al. [39] presented the design of a Proportional-integral-derivative (PID) PSS using the firefly algorithm. The calculations were carried out with eigenvector analysis and the results were compared with those of a bat algorithm optimized conventional PSS. The results proved that FA-PID-PSS gave better results in the system stabilisation. Bux et al. [40] aimed to study the damping effect of a voltage source converter (VSC) based stabilizer and simulated it and investigated its impact on electromechanical oscillation modes (EOMs). Simulation and investigation revealed that the VSC successfully stabilized EOMs. Roy et al. [41] presented a new form of non-linear control scheme and combined feedback linearization and an adaptive control scheme to design an adaptive partial feedback linearizing controller. It was tested and verified successfully on a SMIB. Kim [42] proposed the design of a non-linear algorithm for controlling a SMIB system, including all of its non-linear components to control the power angle and regulate the output voltage of the generator. The simulated results proved the correctness of the algorithm. Salik et al. [43] investigated a novel approach Invasive weed optimization (IWO) for system stability enhancement by fast damping with optimal parameter setting by a PSS and compared it with a GA-PSS. The IWO-PSS showed better damping characteristics than GA-PSS.

10.4 METHODOLOGY

This chapter considers the modelling and control issues of a SMIB system. Thus, the work is split into two sections. The first section deals with the reduced-order modelling of a SMIB system considered from the literature. The SMIB presents a higher-order system when modelled. However, to design its controller, a low order model is required due to implementation issues of the controller. The GWO is taken up to reduce the parent SMIB model. Both the numerator and denominator polynomial coefficients are considered as unknowns and are determined with the help of the GWO applying constraints to retain dc gain, minimum phase and stability of the higher-order model. The pseudo random binary sequence (PRBS) was used as the input for both the original and reduced-order models containing unknown numerator terms. The difference between the responses is considered as the error function. Sum of square error (SSE) was minimized to obtain the numerator polynomial. It is worth mentioning here that the GWO was developed to emulate the leadership hierarchy and the hunting activity of grey wolves found in the northern part of America. The detailed algorithm along with its pseudo code can be found in Ref. [9] for interested researchers. Once the reduced system is developed, the second section of this work involves the controller design where the three-term controller coefficients are determined using the GWO method. The concept of the AMM technique is followed for the controller synthesis. As many as seven heuristic techniques such as Ant lion optimization (ALO), Dragonfly algorithm (DA), Moth flame optimization (MFO), Multiverse optimizer (MVO), Grasshopper optimization algorithm (GOA), Sine cosine algorithm (SCA) and Salp swarm algorithm (SSA) are taken up for comparison to justify the effectiveness of the proposed methodology.

Initially, the reduced order systems are developed with the said methods and the error functions are evaluated. Further, these algorithms are run at least 30 times to provide a statistical measure of the fitness function. Once again, the representations of the minimum, maximum, average and standard deviation are tabulated. Since multiple data sets are compared, the Kruskal Wallis test [44] is carried out to test the validity of the outcome. In addition to this, the p-values are calculated by applying the rank-sum test of Wilcoxon [45], which is once again a non-parametric test. Usually, the p-values are considered to be meaningful if they are less than 0.05 for 95% confidence interval. If the p-values are found to be greater than 0.05, then the outcomes of the experiments are insignificant. Since the proposed method is compared with manifold metaheuristic techniques, the Holm-Bonferroni correction [46] is incorporated in the Wilcoxon test to get modified p-values. Moreover, quite a sufficient number as well as relevant time-domain and frequency-domain parameters are assessed for the reduced models in comparison to the original system. As many as five benchmark error performance indices are evaluated to show comparison with some of the recently developed metaheuristic approaches. Thereupon, the controller parameters are estimated with the help of the GWO in the AMM framework. In the controller synthesis problem, the responses of the plant–controller cascade and the chosen reference model are compared, setting to minimize the integral of square error (ISE) in order to determine the controller gains. In addition, the minimum fitness value is also reported.

10.5 EXPERIMENTAL RESULTS AND DISCUSSION

A SMIB model, considered from the literature [47], is represented as

$$G_{\text{SMIB}}(s) = \frac{33s^7 + 1{,}086s^6 + 13{,}285s^5 + 82{,}402s^4 + 27{,}8376s^3 + 511{,}812s^2 + 482{,}964s + 194{,}480}{s^8 + 33s^7 + 437s^6 + 3{,}017s^5 + 11{,}870s^4 + 27{,}470s^3 + 37{,}429s^2 + 28{,}880s + 9{,}600} \quad (10.4)$$

Since an eighth-order model is presented, it is quite obvious that the system needs to be reduced to a relatively lower order in order to develop a suitable controller. Applying the GWO technique, the reduced-order model is obtained as follows

$$G_{r\text{SMIB}}(s) = \frac{34.2199s + 4.5813}{s^2 + 1.76324s + 0.01329} \quad (10.5)$$

In the above model, dc gain, minimum-phase feature and stability are preserved. Several methods such as ALO, DA, MFO, MVO, GOA, SCA and SSA are used for comparison whose models are given in Table 10.1. Further, the least fitness function value, SSE in this case, is also reported in this table.

TABLE 10.1
Reduced Model Representations and Their Error Function

Method	Reduced-Order Models	SSE
Proposed	$\frac{34.2199s+4.5813}{s^2+1.76324s+0.01329}$	**0.19142**
ALO	$\frac{0.395379s+7.83716}{s^2+48.1192s+0.386861}$	3.4751
DA	$\frac{3.21776s+5.43847}{s^2+50s+0.26845}$	3.3577
MFO	$\frac{50s+19.7491}{s^2+46.2446s+0.974877}$	2.6270
MVO	$\frac{49.8878s+2.24418}{s^2+4.83956s+0.109919}$	2.7145
GOA	$\frac{8.02426s+0.909401}{s^2+27.0173s+0.0448965}$	2.3921
SCA	$\frac{41.3002s+0.25706}{s^2+2.39749s+0.0130317}$	1.3534
SSA	$\frac{16.396s+48.6917}{s^2+19.0759s+2.40354}$	3.0516

The bold value represents the minimum error value.

It is seen from Table 10.1 that the reduced system obtained by the proposed approach produces the least SSE. Thus, it can be inferred that the GWO method generated the best reduced model amongst the methods compared. Further, statistical measures of the error function are also studied, and the results of which are shown in Table 10.2. The four most popular indices are considered for the study, namely the lowest, highest, mean and the standard deviation. The first two basically denote the span of the error function while the standard deviation accounts for the stability of the algorithm. The lower the value of the standard deviation, the more stable is the algorithm. The best results i.e. the least value obtained in each column of the table are marked with the aid of bold letters.

From Table 10.2, it is apparent that the GWO outperforms all the reported algorithms in terms of best, worst, average and standard deviation values and hence indicated by bold letters in the table. Only the results of SCA are close to those of the GWO method in terms of the minimum fitness value. The standard deviation of the ALO algorithm is nearly close to that of the proposed technique, indicating that the algorithm is stable enough. Usually, the standard deviation value is less than the average value as found in the table, validating the theoretical concept as well. Since multiple algorithms are used for comparison with our method, the Kruskal Wallis test is carried out as a measure of non-parametric statistical inference for the validity of the results obtained. The test diagram is shown in Figure 10.2.

The Kruskal Wallis test diagram shown in Figure 10.2 clearly indicates that out of the seven metaheuristic algorithms used for comparison, the suggested GWO method proved significant compared to the six algorithms. In only one algorithm, namely Group 7, there seems to be some closeness of the data set. To check further, a Wilcoxon test based on rank-sum was conducted on the data samples and the p-values are reported in Table 10.3. A value less than 0.05 will be considered to be meaningful while a value than 0.05 will be taken up as insignificant. This limit is considered for 95% confidence interval, a popular one amongst the different confidence intervals.

All the p-values in Table 10.3 are less than 0.05. Thus, the results obtained by the proposed technique are significant with respect to all other algorithms. The same p-values in the table are merely accidental. A value lower or higher than that may not

TABLE 10.2
Statistical Analysis of the Fitness Function, SSE

Methods	Lowest	Highest	Mean	Std. Deviation
Proposed	**0.19142**	**0.2635**	**0.2133**	**0.0083**
ALO	3.4751	3.9430	3.7604	0.2034
DA	3.3577	4.5830	3.9883	0.5096
MFO	2.6270	3.6789	3.2688	0.4876
MVO	2.7145	4.2911	3.4013	0.5802
GOA	2.3921	4.4221	3.9015	0.8536
SCA	1.3534	3.5973	2.3523	0.8322
SSA	3.0516	3.9563	3.6230	0.3685

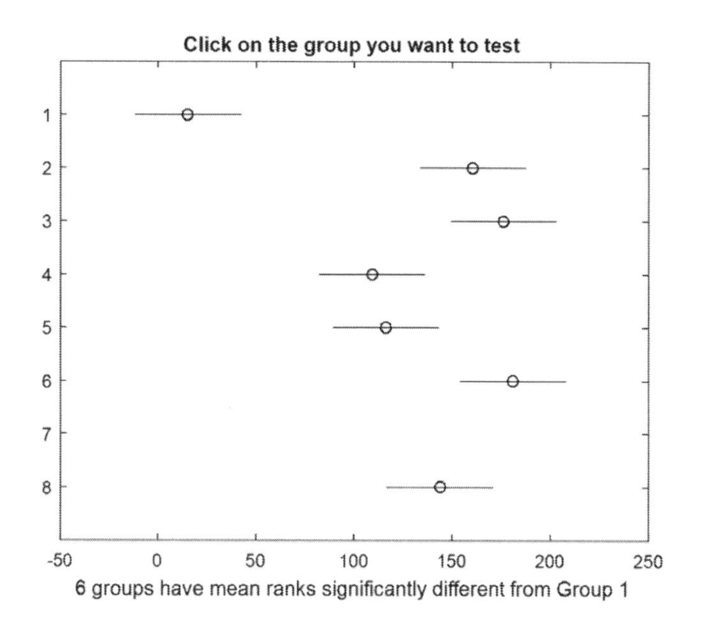

FIGURE 10.2 Kruskal Wallis test for checking the significance of mean ranks.

TABLE 10.3
Calculations Showing p-Values Using the Non-parametric Wilcoxon Rank-Sum Test

Algorithm	ALO	DA	MFO	MVO
Proposed	1.0496e−12	1.0496e−12	1.0496e−12	1.0496e−12
	GOA	SCA	SSA	–
Proposed	1.0496e−12	1.0496e−12	1.0496e−12	–

TABLE 10.4
Modified Wilcoxon Test Results with Holm-Bonferroni Corrections

Algorithm	p-Values after Holm-Bonferroni Corrections	h-Values
Proposed	$10^{-11} \times \begin{bmatrix} 0.7347 & 0.7347 & 0.6298 & 0.5248 & 0.4198 & 0.3149 & 0.2099 \end{bmatrix}$	$\begin{bmatrix} 1 & 1 & 1 & 1 & 1 & 1 & 1 \end{bmatrix}$

be achievable from the given data set. The p-values are further modified applying Holm-Bonferroni corrections and quoted in Table 10.4. The h-values representing whether the test of hypothesis is true or false are marked in the table. 1's recorded against each h-value correspond to statistically significant results, whereas 0's denote insignificant results.

It is observed from the p-values in the table that they are quite less than 0.05 and hence significant. This is also indicated by the h-values where the 1's represent meaningful outcomes, and 0's on the other hand indicate that they are insignificant. Moreover, the important time and frequency domain parameters of the reduced systems are provided in Table 10.5 to make a fair comparison with the original system.

From the results shown in Table 10.5, it is clearly visible that the GWO method produces the closest match to the original higher-order model in terms of both time-domain and frequency-domain parameters. The other methods with which the comparison is carried out are only close in terms of overshoot, undershoot, gain and phase margins. The rise time and settling time of the ALO, DA, MFO, MVO, GOA and SCA are really huge, thereby suggesting sluggish response. Only the rise time and the settling time of the SSA method are slightly better. The phase margins of the MFO, GOA and SSA show a wide deviation from that of the original model. The step responses of the original system and the reduced test system are shown in Figure 10.3 to further validate the results shown in Table 10.5 in terms of time-domain specifications.

It is quite clear from the time response curves of the original and the reduced systems in Figure 10.3 that the reduced-order model produced by the suggested technique closely matches the parent model. The frequency response is further plotted in Figure 10.4 to check the closeness of the proposed model in terms of the magnitude and phase plot of the Bode diagram.

The Bode diagrams of the parent and reduced system models show a very close resemblance as observed in Figure 10.4. Some of the well-known errors widely popular in the literature of control are calculated for each of the reduced order test systems and their outcomes are enumerated in Table 10.6. The best rather than the least values reported for each of these errors are indicated with the help of bold letters for the proper understanding of the readers.

TABLE 10.5

Quantitative Time- and Frequency-Domain Measures and Their Comparison with the Original System

Test System	Methods	Rise Time (s)	Settling Time (s)	Overshoot (%)	Undershoot (%)	Gain Margin (dB)	Phase Margin (°)
Original		1.0692	1.5686	0.7852	0	Inf	90.8510
Reduced	Proposed	1.1967	5.1427	2.2227	0	Inf	92.7294
	ALO	273.2268	486.4874	0	0	Inf	93.1089
	DA	409.1763	728.0001	0	0	Inf	96.3845
	MFO	104.1833	182.9161	0	0	Inf	156.4374
	MVO	70.0453	140.6177	0	0	Inf	95.5164
	GOA	1.3223e+03	2.3453e+03	0	0	Inf	109.8870
	SCA	40.9063	336.3791	0	0	Inf	93.3197
	SSA	17.3216	30.5515	0	0	Inf	135.1152

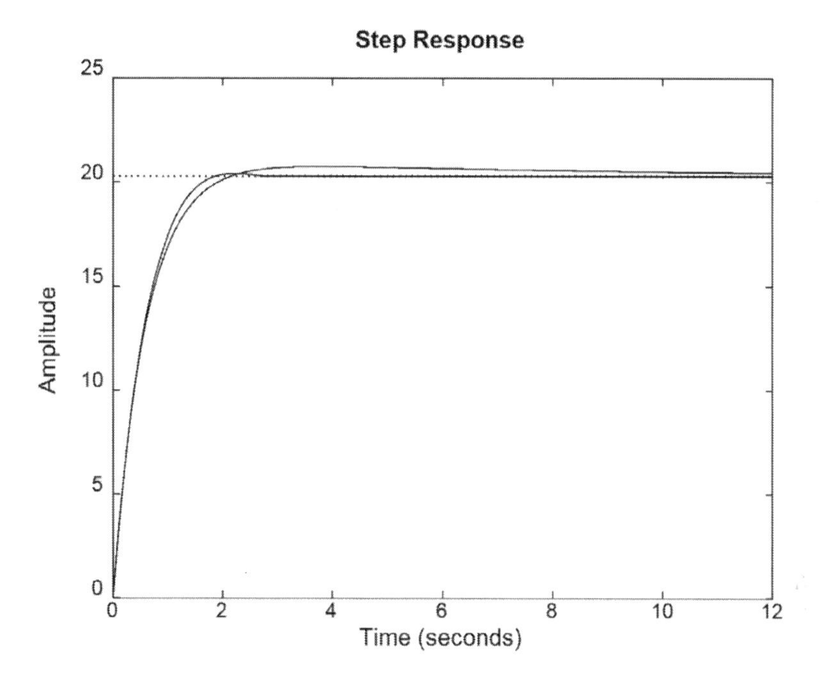

FIGURE 10.3 Step response matching of parent and reduced models using the GWO technique.

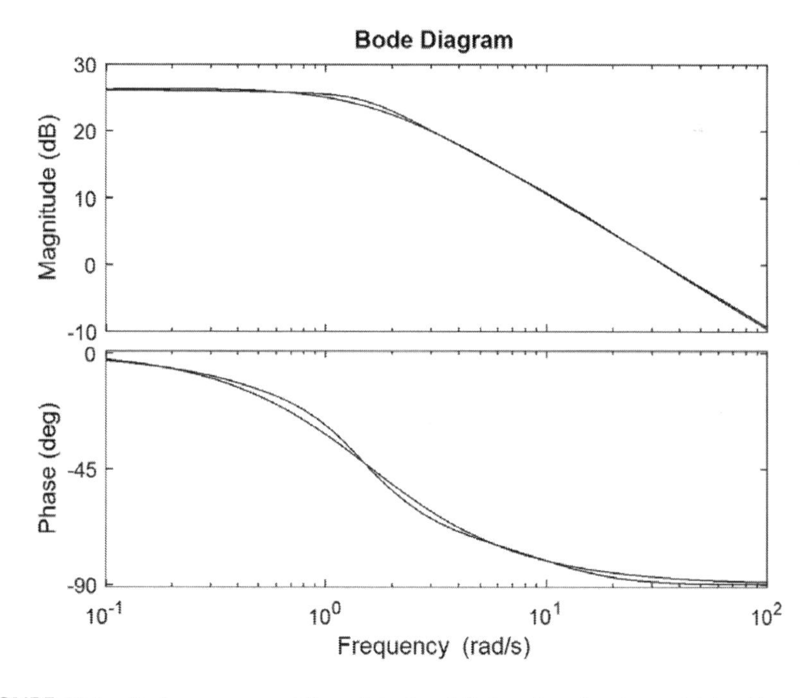

FIGURE 10.4 Bode responses of the original and their reduced systems obtained by applying the proposed method.

TABLE 10.6
Representation of Popular Error Indices Using Different Methods

Methods	IAE	ITAE	ISE	ITSE	H_{inf} Norm
Proposed	**0.0677**	**0.0503**	**0.0046**	**0.0037**	**0.1136**
ALO	2.2142	1.2810	4.5083	2.4893	3.3427
DA	2.2095	1.2789	4.4813	2.4750	3.3155
MFO	2.0110	1.1807	3.7716	2.1086	2.9623
MVO	1.0732	0.8342	1.1443	0.9601	1.6727
GOA	2.1770	1.2630	4.3310	2.3989	3.2233
SCA	0.3876	0.3214	0.1766	0.1672	0.8452
SSA	1.8383	1.0611	3.1013	1.7002	2.7011

From Table 10.6, it is observed that the proposed GWO technique surpasses all other algorithms in terms of the error indices considered. The least value in each column is thus indicated with the help of bold letters. A reference model is then selected as per [48] to constitute the controller design of the reduced SMIB model. Generally, the response of the plant and controller with unknown parameters, connected in cascade, is compared to the response of the reference model. The objective is set in such a way that the plant and controller combination follows the response of the reference model approximately. The ISE is minimized to determine the controller parameters whose input–output relationship is defined by

$$G_c(s) = K_P + \frac{K_I}{s} + K_D s \tag{10.6}$$

The controller gains are thus determined and the results are shown in Table 10.7 along with the fitness function values. Similar to the model order reduction problem, in the controller design problem, the proposed methodology is compared with

TABLE 10.7
Different Controller Gains and Their Fitness Values

Methods	K_P	K_I	K_D	J_{min}
Proposed	0.12589	0.048128	3.0368e−06	**6.6008e−09**
ALO	2.8592	1.0965	1.1993e−06	0.00015018
DA	5.0000	2.4788	3.3968e−06	0.00020168
MFO	0.095166	3.2931	5.0366e−06	1.741e−07
MVO	0.05926	0.014016	0.00063065	0.00017156
GOA	2.9233	5.0000	0.21605	0.38052
SCA	0.10152	0.075265	4.6343e−05	4.5961e−07
SSA	4.2953	0.78715	0.0010865	0.016981

some of the widely cited heuristic techniques. The least fitness values reported in this table give an indication of the best controller parameters obtained, marked in bold.

In Table 10.7, the GWO method reports the least ISE value and hence the best choice for the controller amongst all other compared techniques. MFO and SCA methods only provide close match. The methods such as ALO, DA and MVO also yield satisfactory results. The results of GOA and SSA are far away from the best reported results. The one producing the best fitness value is also expected to closely match the response of the reference model. Thus, the GWO technique successfully addresses the modelling and control of SMIB test systems. Multiple test runs could also be performed for the controller synthesis part. In that case, parametric and non-parametric tests could be conducted to get better assessments. Moreover, new optimisation techniques such as the whale optimisation algorithm (WOA), Harris hawks optimisation (HHO), equilibrium optimizer (EO), marine predator algorithm (MPA) etc. may be applied to get still better performance of modelling and control aspects of SMIB systems. Obvious variants such as chaotic form, opposition-based methods and new hybrid combinations of these algorithms could be employed for the reduced-order modelling and controller synthesis problem.

10.6 CONCLUSIONS

This chapter effectively and efficiently utilized a heuristic approach-based technique to perform model reduction and determine the controller parameters. The GWO, a widely popular metaheuristic technique, is employed to carry out the order reduction and controller synthesis of a SMIB test system. Constrained optimisation is adopted to satisfy dc gain, minimum phase and stability requirements of the model reduction problem. A handful of new approaches are used for comparison. The step and Bode responses, time- and frequency-domain measures and non-parametric tests proved the superiority of the proposed approach over the algorithms considered for this work. Thus, the GWO mostly gives satisfactory outcomes compared to all other algorithms. Moreover, the controller parameters viz. K_P, K_I and K_D are determined by applying the method of AMM. Here also, the GWO technique wins over all the other algorithms in terms of the least error function. While SSE is used as the error function for the model reduction problem, ISE is taken up as the objective for the controller design problem. A mix of two approaches such as the stability equation method, time moment matching etc. may be combined with the GWO to develop new reduction models for suitable controller design. Further, many new algorithms such as HHO, EO, political optimizer (PO), MPA etc. and their variants developed recently may further be applied to device new reduced models. Additional constraints to meet time- and frequency-domain specifications may also be imposed. More complicated problems involving right-half plane zeroes of the SMIB model can also be handled. Discrete-time modelling and the relevant control aspects may be taken up further to provide a unified modelling approach.

REFERENCES

1. Wang, H., & Du, W. (2016). A single-machine infinite-bus power system installed with a power system stabilizer. In Wang, H., & Du, W. (Eds.), *Analysis and Damping Control of Power System Low-Frequency Oscillations* (pp. 17–79), Springer, Boston, MA.
2. Arredondo, J. M. R., & Valle, R. G. (2004). An optimal power system model order reduction technique. *International Journal of Electrical Power & Energy Systems*, *26*(7), 493–500.
3. Đukić, S. D., & Sarić, A. T. (2012). Dynamic model reduction: an overview of available techniques with application to power systems. *Serbian Journal of Electrical Engineering*, *9*(2), 131–169.
4. Ghosh, S., & Senroy, N. (2012). A comparative study of two model order reduction approaches for application in power systems. In *2012 IEEE Power and Energy Society General Meeting*, 1–8.
5. Ghosh, S., & Senroy, N. (2013). Balanced truncation approach to power system model order reduction. *Electric Power Components and Systems*, *41*(8), 747–764.
6. Ramirez, A., Mehrizi-Sani, A., Hussein, D., Matar, M., Abdel-Rahman, M., Chavez, J. J., … & Kamalasadan, S. (2015). Application of balanced realizations for model-order reduction of dynamic power system equivalents. *IEEE Transactions on Power Delivery*, *31*(5), 2304–2312.
7. Cheng, X., & Scherpen, J. M. (2018). Clustering approach to model order reduction of power networks with distributed controllers. *Advances in Computational Mathematics*, *44*(6), 1917–1939.
8. Osipov, D., & Sun, K. (2018). Adaptive nonlinear model reduction for fast power system simulation. *IEEE Transactions on Power Systems*, *33*(6), 6746–6754.
9. Mirjalili, S., Mirjalili, S. M., & Lewis, A. (2014). Grey wolf optimizer. *Advances in Engineering Software*, *69*, 46–61.
10. Mirjalili, S. (2015). The ant lion optimizer. *Advances in Engineering Software*, *83*, 80–98.
11. Mirjalili, S. (2016). Dragonfly algorithm: a new meta-heuristic optimization technique for solving single-objective, discrete, and multi-objective problems. *Neural Computing and Applications*, *27*(4), 1053–1073.
12. Mirjalili, S. (2015). Moth-flame optimization algorithm: a novel nature-inspired heuristic paradigm. *Knowledge-Based Systems*, *89*, 228–249.
13. Mirjalili, S., Mirjalili, S. M., & Hatamlou, A. (2016). Multi-verse optimizer: a nature-inspired algorithm for global optimization. *Neural Computing and Applications*, *27*(2), 495–513.
14. Mirjalili, S. Z., Mirjalili, S., Saremi, S., Faris, H., & Aljarah, I. (2018). Grasshopper optimization algorithm for multi-objective optimization problems. *Applied Intelligence*, *48*(4), 805–820.
15. Mirjalili, S. (2016). SCA: a sine cosine algorithm for solving optimization problems. *Knowledge-Based Systems*, *96*, 120–133.
16. Mirjalili, S., Gandomi, A. H., Mirjalili, S. Z., Saremi, S., Faris, H., & Mirjalili, S. M. (2017). Salp swarm algorithm: a bio-inspired optimizer for engineering design problems. *Advances in Engineering Software*, *114*, 163–191.
17. Nijmeijer, H., & Savaresi, S. M. (1998). On approximate model-reference control of SISO discrete-time nonlinear systems. *Automatica*, *34*(10), 1261–1266.
18. Shah, N. N., Joshi, S. R., & Jadav, S. S. (2018). An application of 100 MW DFIG based wind model for damping oscillation in SMIB. In *2018 IEEE Texas Power and Energy Conference (TPEC)* (pp. 1–6). IEEE.
19. Chen, L., Min, Y., Chen, Y. P., & Hu, W. (2013). Evaluation of generator damping using oscillation energy dissipation and the connection with modal analysis. *IEEE Transactions on Power Systems*, *29*(3), 1393–1402.

20. Sambariya, D. K., & Prasad, R. (2013). Design of PSS for SMIB system using robust fast output sampling feedback technique. In *7th international conference on intelligent systems and control (ISCO)*, 166–171.
21. Yang, T. C. (1997). Applying H∞ optimisation method to power system stabiliser design. Part 1: single-machine infinite-bus systems. *International Journal of Electrical Power & Energy Systems, 19*(1), 29–35.
22. Wang, H. F., & Swift, F. J. (1997). A unified model for the analysis of FACTS devices in damping power system oscillations. I. Single-machine infinite-bus power systems. *IEEE Transactions on Power Delivery, 12*(2), 941–946.
23. Ford, J. J., Ledwich, G., & Dong, Z. Y. (2006). Nonlinear control of single-machine-infinite-bus transient stability. In *2006 IEEE Power Engineering Society General Meeting* (p. 8). IEEE.
24. Ghfarokhi, G. S., Arezoomand, M., & Mahmoodian, H. (2007). Analysis and simulation of the single-machine infinite-bus with power system stabilizer and parameters variation effects. In *2007 International Conference on Intelligent and Advanced Systems* (pp. 167–171). IEEE.
25. Sambariya, D. K., & Prasad, R. (2012). Routh approximation based stable reduced model of single machine infinite bus system with power system stabilizer. In *DRDO-CSIR Sponsered: IX Control Instrumentation System Conference (CISCON-2012)* (pp. 85–93).
26. Sambariya, D. K., & Prasad, R. (2012). Routh stability array method based reduced model of single machine infinite bus with power system stabilizer. In *International Conference on Emerging Trends in Electrical, Communication and Information Technologies (ICECIT-2012)* (pp. 27–34).
27. Sambariya, D. K., & Prasad, R. (2012). Differentiation method based stable reduced model of single machine infinite bus system with power system stabilizer. *International Journal of Applied Engineering Research, 7*(11), 2116–2120.
28. Wan, Y., & Zhao, J. (2012). Extended backstepping method for single-machine infinite-bus power systems with SMES. *IEEE Transactions on Control Systems Technology, 21*(3), 915–923.
29. Sambariya, D. K., & Prasad, R. (2013). Robust power system stabilizer design for single machine infinite bus system with different membership functions for fuzzy logic controller. In *2013 7th International Conference on Intelligent Systems and Control (ISCO)* (pp. 13–19). IEEE.
30. Sambariya, D. K., & Prasad, R. (2013). Stable reduced model of a single machine infinite bus power system with power system stabilizer. In *2013 International Conference on Advances in Technology and Engineering (ICATE)* (pp. 1–10). IEEE.
31. Wan, Y., Zhao, J., & Dimirovski, G. M. (2014). Robust adaptive control for a single-machine infinite-bus power system with an SVC. *Control Engineering Practice, 30*, 132–139.
32. Perev, K. (2015). Model order reduction for a single machine infinite bus power system. *Electrotechnica & Electronica*, 1–2, 42–49.
33. Wang, X., Chen, Y., Han, G., & Song, C. (2015). Nonlinear dynamic analysis of a single-machine infinite-bus power system. *Applied Mathematical Modelling, 39*(10–11), 2951–2961.
34. Sambariya, D. K., & Arvind, G. (2016). Reduced order modelling of SMIB power system using stability equation method and firefly algorithm. In *2016 IEEE 6th International Conference on Power Systems (ICPS)* (pp. 1–6). IEEE.
35. Sambariya, D. K., & Arvind, G. (2016). Reduced order model of single machine infinite bus power system using stability equation method and self-adaptive bat algorithm. *Universal Journal of Control and Automation, 4*(1), 1–7.

36. Milla, F., & Duarte-Mermoud, M. A. (2016). Predictive optimized adaptive PSS in a single machine infinite bus. *ISA Transactions*, *63*, 315–327.
37. Yılmaz, S., & Savacı, F. A. (2018). Basin stability of single machine infinite bus power systems with Levy type load fluctuations. In *2017 10th International Conference on Electrical and Electronics Engineering (ELECO)* (pp. 125–129). IEEE.
38. Zribi, M., Alrifai, M. T., & Smaoui, N. (2018). Control of chaos in a single machine infinite bus power system using the discrete sliding mode control technique. *Discrete Dynamics in Nature and Society, 2018.*
39. Farhad, Z., Ibrahim, E. K. E., Tezcan, S. S., & Safi, S. J. (2018). A robust PID power system stabilizer design of single machine infinite bus system using firefly algorithm. *Gazi University Journal of Science*, *31*(1), 155–172.
40. Bux, R., Xiao, C., Hussain, A., & Wang, H. (2019). Study of single machine infinite bus system with VSC based stabilizer. In *Proceedings of the 2019. The 2nd International Conference on Robotics, Control and Automation Engineering* (pp. 159–163).
41. Roy, T. K., Mahmud, M. A., Shen, W. X., & Oo, A. M. (2019). An adaptive partial feedback linearizing control scheme: an application to a single machine infinite bus system. *IEEE Transactions on Circuits and Systems II: Express Briefs*. 1–5, doi: 10.1109/TCSII.2019.2962098.
42. Kim, S. K. (2019). Proportional-type non-linear excitation controller with power angle reference estimator for single-machine infinite-bus power system. *IET Generation, Transmission & Distribution*, *13*(18), 4029–4036.
43. Salik, M., Rout, P. K., & Mohanty, M. N. (2020). Inter-area and intra-area oscillation damping of power system stabilizer design using modified invasive weed optimization. In Mohanty, M. N., & Das, S. (Eds.), *Advances in Intelligent Computing and Communication* (pp. 347–359). Springer, Singapore.
44. McKight, P. E., & Najab, J. (2010). Kruskal-wallis test. In Weiner, I. B., & Craighead, W. E. (Eds.), *The Corsini Encyclopedia of Psychology* (pp. 1–1). Wiley.
45. Rosner, B., Glynn, R. J., & Ting Lee, M. L. (2003). Incorporation of clustering effects for the Wilcoxon rank sum test: a large-sample approach. *Biometrics*, *59*(4), 1089–1098.
46. Abdi, H. (2010). Holm's sequential Bonferroni procedure. *Encyclopedia of Research Design*, *1*(8), 1–8.
47. Balaga, R. (2012). Model order reduction of electrical power system (Masters dissertation).
48. Rana, J. S., Prasad, R., & Agarwal, R. P. (2016). Designing of a controller by using model order reduction techniques. *International Journal of Engineering Innovations and Research*, *5*(3), 220–223.

11 Internet of Things (IoT) with Energy Sector-Challenges and Development

Arun Kumar
Panipat Institute of Engineering and Technology

Sharad Sharma
Maharishi Markandeshwar (Deemed to be University)

CONTENTS

11.1 INTRODUCTION

The Internet of Things, or IoT, alludes to billions of physical sensor devices around the globe that are currently associated with the web, all gathering and sharing information. On account of the appearance of super-modest PC chips and the omnipresence of remote systems, it is conceivable to turn anything, from something as little as a pill to something as large as a plane, into a piece of IoT. Interfacing up all these various items and adding sensors to them provide a degree of advanced insight to gadgets that would in any case lack intelligence, empowering them to convey continuous information without including a person. With the coming of innovation, there is an immense spread in the utilization of Internet of Things (IoT) in different sections, for example, vitality, urban communities, coordinations, homes, enterprises, wellbeing, and agribusiness. As per Gartner, there will be 26 billion gadgets utilizing this innovation by 2020. IoT is an innovation that includes interfacing physical things to the internet in this manner, empowering them to impart and move information over the internet as shown in Figure 11.1. The presentation of IoT has made the matrix more brilliant, dependable, effective, and strong. Mechanical upsets can be separated into four stages. In the primary transformation, new wellsprings of vitality were found to run the machines. The mass extraction of coal and the innovation of steam power plants were noteworthy improvement arrangements in this stage [1].

The subsequent upset known as large-scale manufacturing and power age was a time of quick improvement in the industry, recognized by the enormous scope of iron and steel creation. During this stage, some large-scale processing plants with their sequential construction systems were built up, which shaped new organizations [2]. The third unrest presented PCs and the original correspondence advancements, e.g., the communication framework, which empowered mechanization in graceful chains [3]. A wide assortment of present-day innovations, for example, correspondence frameworks (e.g., 5G), wise robots, and the Internet of Things (IoT), are expected to enable the fourth industrial revolution known as 4.0 [4–6]. IoT interconnects various gadgets, individuals, information, and procedures, by permitting them to speak with one another consistently. Thus, IoT can assist with improving various procedures to be progressively quantifiable by gathering and handling a lot of information [7]. IoT can upgrade personal satisfaction in various zones including clinical administrations, savvy urban communities, the development industry, agribusiness, and the vitality area [8]. This is empowered by giving expanded mechanized dynamics continuously and encouraging instruments for enhancing such choices.

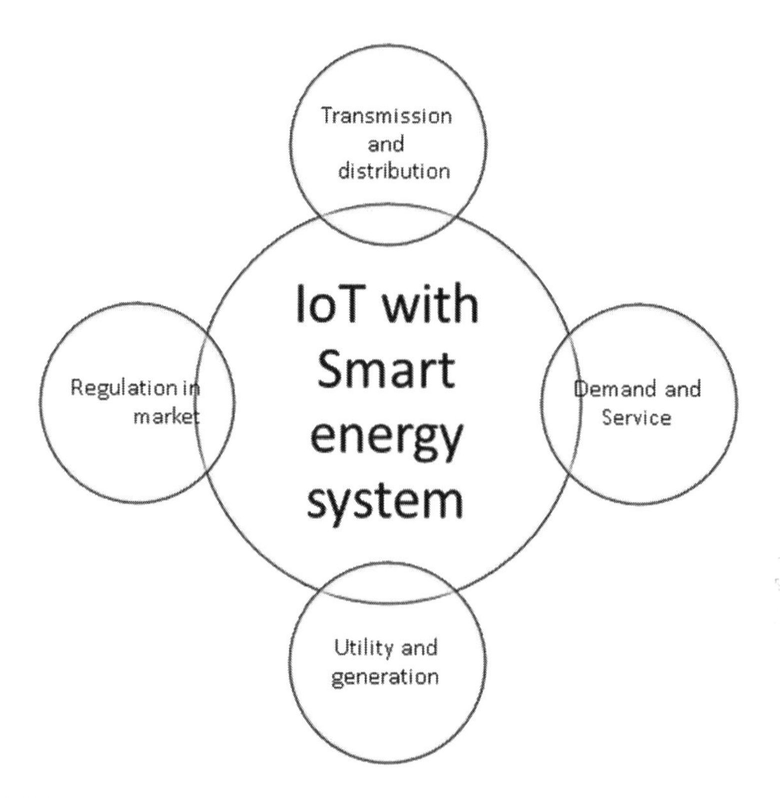

FIGURE 11.1 IoT with the energy system.

11.2 ARCHITECTURE OF IoT

In the new arrangement for IoT, an alteration on the traditional thought of the web is basic. In the customary adjustment, the web is a system that gives the terminals to end-customers, while inside IoT it gives the interconnection of savvy questions inside an unavoidable handling condition [9]. The web establishment will expect a basic activity as the overall stage to engage the correspondence capacity of physical articles. The curiosity will be engaged by introducing devices into objects, making them smart while being joined in the general physical establishment. The IoT should be fit for interfacing billions or trillions of heterogeneous contraptions through the web, so there is a fundamental prerequisite for versatile layered building [10,11].

The IoT region encases a wide extent of standardized or unstandardized headways, programming stages, and different applications. Along these lines, single reference designing cannot be used as an arrangement for all possible strong executions. Despite the possibility that a reference model can be considered for IoT, a couple of reference structures will surely concur [12–15]. Here, we portray structure as a framework in which things, people, and cloud organizations are joined to support application tasks. In this manner, the reference model for the IoT can schematically be portrayed as in Figure 11.2.

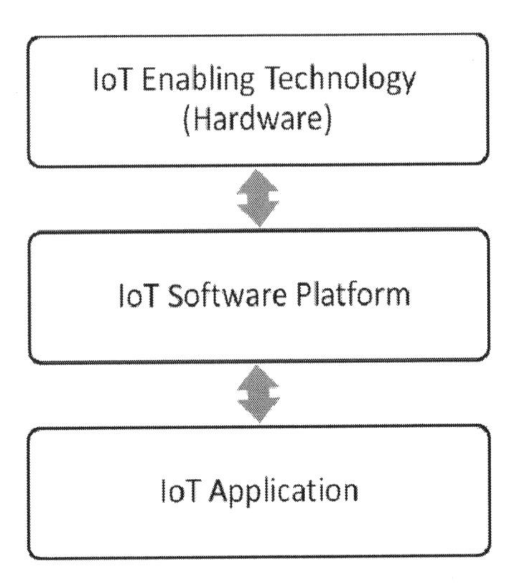

FIGURE 11.2 IoT architecture.

11.3 INTERNET OF THINGS (IoT) ENERGY MARKET

The vitality advertises is expected to develop at a huge rate during the estimated time frame 2019–2025. Internet of Things (IoT) has been seeing expanding usage in various end-client ventures around the globe. The vitality division is one such territory in which the execution of IoT is viewed as one of the ways, breaking open doors for the expansion of the market. IoT is being utilized in creating inventive frameworks that are expected to benefit vitality organizations and clients. The key advantage of IoT is the decrease in the utilization of vitality combined with the expansion of the operational proficiency of a plant. Key IT companies are offering numerous solutions ranging from asset management to energy analysis which is expected to create opportunities for the growth of the market shortly [16]. Organizations are centered around creating advanced IoT energy solutions to remain successful in the market. Topographical development, merger, procurement, finding another market, or advance in their center of competency to overall industry is a key procedure embraced by significant market players [17]. Oil fields in remote areas have segregated server farms, which are critical to getting offloaded into a brought-together vault, say a cloud, for better administration and preparation of data across flexible chains. IoT accommodates this uniform conveyance of information progressively for steady upgrades in vitality usage and proficiency, disposing of any chance of wastage. Organizations are additionally disposed to incorporate savvy matrix frameworks for restricting vitality utilization. Brilliant lattices permit administrators to present the timings and the measure of vitality gracefully in offices even with a cell phone.

11.4 IoT ENABLING TECHNOLOGY

A schematic diagram of IoT enabling technology is shown in Figure 11.3.

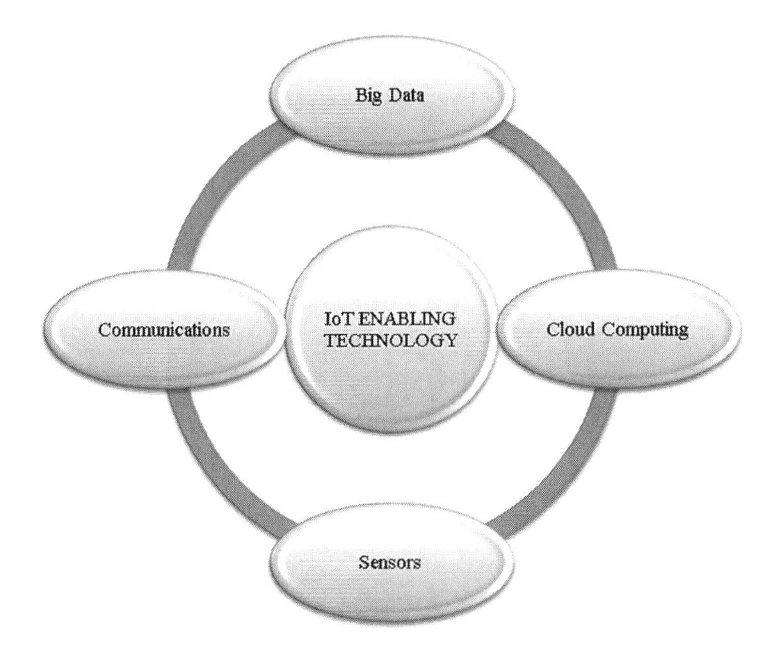

FIGURE 11.3 IoT enabling technology.

11.4.1 Big Data

As more things (or "smart objects") are related to the IoT, more data are accumulated from them to play out an assessment to choose examples and affiliations that lead to bits of information. For example, an oil well all around outfitted with 20–30 sensors can deliver 5 mega byte data concentrates every 15–20 seconds, a jetliner with 6,000 sensors produces 2.5 terabytes of data consistently [18], and the more than 46 million insightful utility meters present in the U.S. produce more than 1 billion data concentrates each day [19]. Thus, the expression "big data" alludes to these huge informational collections that should be gathered, stored, questioned, broken down, and for the most part overseen to convey the guarantee of the IoT [20].

11.4.2 Cloud Computing

Cloud computing provides "the virtual infrastructure for utility computing integrating applications, monitoring devices, storage devices, analytics tools, visualization platforms, and client delivery... [to] enable businesses and users to access [IoT-enabled] applications on-demand anytime, anyplace, and anywhere" [21,22].

11.4.3 Sensors

Today, with the huge development in the field of sensors, there is a much improvement in the field of technologies such as microelectromechanical systems (MEMS). "Tiny sensor provides a complete embedded system for all processing [smart objects]" [23].

11.4.4 Communications

The conventional communication technologies are RFID, NFC, Wi-Fi, and NuelNET, as well as satellite communications and versatile systems utilizing GSM, GPRS, 3G, LTE, or WiMAX [24]. Wired technologies, that are used as fixed broadband products, include Ethernet, HomePlug, HomePNA, HomeGrid/G.hn, and LonWorks, as well as customary phone lines [25,26].

11.5 IoT IN RENEWABLE ENERGY INDUSTRY

The following describes how the IoT innovation can drive transformation worldwide in the sustainable power source industry [27]:

- With the quickening power age limits of sustainable power sources, force must be represented and stored proficiently. This stored vitality can be utilized during a time of emergency. To more readily deal with this developing force limit, a brilliant framework of the board enabled with IoT will help.
- IoT can assist with getting information from remote ranches, solar-based homesteads, or hydro stations continuously. This advantage of IoT will likewise help laborer wellbeing, as they can screen gear, for example, huge wind turbines to resolve structural problems.
- IoT will help deal with the conveyance of vitality dependent on constant information, rather than authentic information used at present.
- Furthermore, the predictive investigation will caution administrators in advance if a part needs to be fixed, requires prompt consideration for examination.
- The adaptive examination will permit frameworks to consequently adjust vitality loads and abatement weight on the hardware and forestall overheating.
- Forecasts identified with the yield of the framework will permit suppliers to offer information to affiliates continuously and assist in selling fuel on the open market.
- The predictive investigation will likewise assist suppliers with driving yield and convey increasingly reasonable support of the market.
- Availability of ongoing information will support straightforwardness in the sustainable power source industry.
- IoT will chip away at a full-scale level and small-scale level. It can help purchasers with housetop photovoltaic (PV) establishments to deal with their framework better.

The reception of smart meters is expanding quickly. In any case, private, smart meter projects can be mind-boggling, as huge number of meters can be tested. With IoT, the information from the savvy meter readings can enable a supplier to assemble the information and offer customers all the more engaging rates and important suggestions to spare vitality.

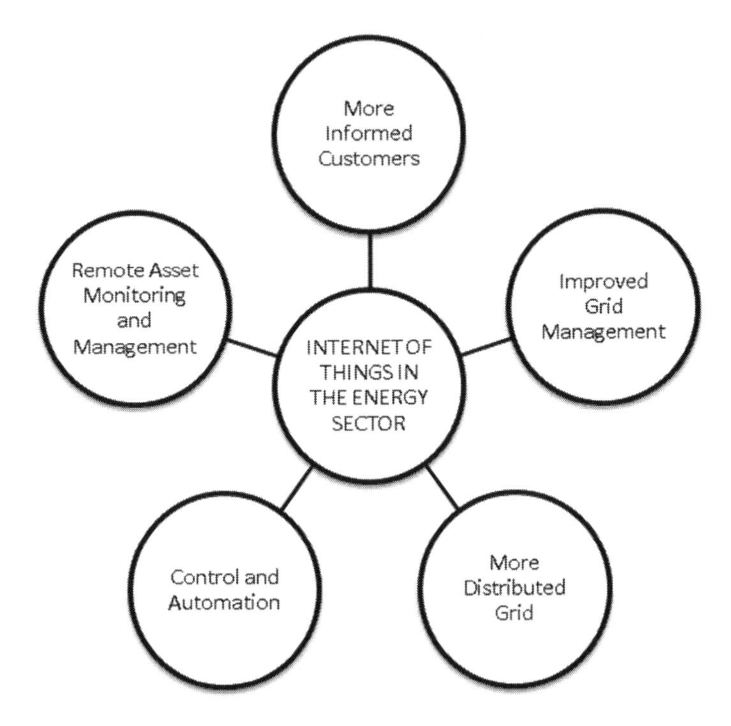

FIGURE 11.4 Internet of things in the energy sector.

11.6 INTERNET OF THINGS IN THE ENERGY SECTOR

The energy sector is undergoing a massive transformation. Along with solar, wind, storage, and other technologies, the IoT is helping to drive this transformation [28]. It is revolutionizing nearly every part of the industry from generation to transmission to distribution and changing how energy companies and customers interact. Here is how the IoT is transforming the energy industry as shown in Figure 11.4.

11.6.1 Remote Asset Monitoring and Management

Affixing IoT sensors to generation, transmission, and distribution equipment can enable energy companies to monitor it remotely. These sensors measure parameters such as vibration, temperature, and wear to optimize maintenance schedules. This preventative maintenance approach can significantly improve reliability by keeping equipment in optimal state and providing the opportunity to make repairs before it fails. Digital twin technology, which involves creating an advanced digital model of an existing piece of equipment, could help with this as well. IoT sensors attached to the physical unit collect data about its performance, which they feed to the digital twin. In addition to supporting preventative maintenance programs, this technology enables virtual troubleshooting and support from remote locations. IoT sensors can also help improve safety. Affixing internet-connected sensors to pipelines can help detect leaks that, if left unaddressed, may result in fires or explosions.

11.6.2 Control and Automation

IoT and energy solution helps to automate the management of wind farms, optimize preservation, and minimize the charge prominently. The same is applicable for solar fields, geothermal plants, and traditional oil and gas deposits.

11.6.3 A More Distributed Grid

The energy grid is becoming more dispersed due to the rise of residential solar energy and other technologies. The residential solar capability has grown quickly in recent years and could increase by more than three times to 41 GW by 2025, as per a report on analysis from Credit Suisse. Homeowners and businesses can now generate their electricity by placing solar panels on their rooftops or even by building small wind turbines on their properties. This increasingly distributed power system represents a major change for energy companies. In addition to managing a few large generators, they must also now manage a growing number of small generation resources located across the grid. This presents a challenge to grid operators, but smart grid technology powered by the IoT is helping to enable this distributed energy transformation. A smart grid uses IoT technology to detect changes in electricity supply and demand. It can react to these changes autonomously or provide operators with the information they need to more precisely manage demand.

11.6.4 More Informed Customers

In addition to providing more information to utilities, IoT technology can help customers to be more informed about their energy usage. Internet-connected smart meters collect usage data and send it to both utilities and customers remotely. Thanks to smart meter technology, many energy companies now send their customers detailed reports about their energy usage. Customers can also install smart devices in their homes or commercial buildings that measure the power consumed by each appliance and device. They can use this information to identify waste and especially power-hungry appliances to save on their energy bills. Other IoT devices, such as thermostats, can automatically optimize their operation to reduce energy use. Residential customers could potentially benefit the most from these technologies, as the U.S. residential sector represents 37% of energy usage. The commercial and industrial sectors, which use 35% and 27%, respectively, could benefit substantially as well.

11.6.5 Improved Grid Management

IoT technology can enable the integration of more distributed resources into the grid, but it can also improve grid management in other ways. Placing sensors at substations and along with distribution, lines provide real-time power consumption data that energy companies can use to make decisions about voltage control, load switching, network configuration, and more. Some of these decisions can be automated.

Sensors located on the grid can alert operators to outages, allowing them to turn off power to damaged lines to prevent electrocution, wildfires, and other hazards. Smart switches can isolate problem areas automatically and reroute power to get the lights back on sooner.

Power usage data can also serve as the basis for load forecasting. It can help in managing congestion along transmission and distribution lines and help ensure that all of the connected generation plants meet requirements related to frequency and voltage control. This power consumption data can also help companies decide where to build new infrastructure and make infrastructure upgrades.

The IoT is transforming nearly every sector of our economy, including the one that powers — the energy sector. Over the coming years, the energy industry is going to get smarter, more efficient, more distributed, and more reliable, thanks in part to the IoT.

11.7 CONSUMER-ORIENTED IoT ENERGY DEVICES AND CASES

11.7.1 Smart Meters

This smart energy device is directly linked to the power conveyance station, enabling two-way correspondence. As a result, they can send critical activity data to utility organizations progressively. This causes utility offices to quickly address any concerns, including blackouts, and limit the framework vacation. Smart meters can likewise perceive and naturally separate the harmed segment of a line without upsetting the exhibition of the remainder of the system [29].

11.7.2 Developing Intelligent Energy-Efficient Buildings with IoT Technology

The Green Building term isn't something new and has been around for some time. To keep up vitality utilization expenses and ozone-depleting substance discharges to a base seem to be the greatest test. Building Energy Management Systems or BEMS assume a crucial job in the structures' operational efficiencies. The IoT is changing the vitality of the executives' landscape in the smaller place of business and even home conditions. An IoT stage in the BEMS condition contains passages, sensors, and remote correspondences. To give better information to the examination motor that thus gives better experiences and activities to clients with decrease in cost from this point of view of innovation. IoT-empowered smart structure frameworks are protected, adaptable, and good. IoT innovation empowers the office administrator by bringing together the building tasks and joining all the information related to structure [30].

11.7.3 Grid Balancing and Contribution

IoT can give the ongoing data required to viably deal with the blockage on transmission and appropriation lines. With IoT, the lattice can guarantee that the associated age stations have met the necessities from recurrence to voltage control to forestall shakiness. One of the noteworthy future patterns in power age is the commitment of normal homes to the vitality network.

11.8 FOUR PHASE OF POWER GENERATION

The smart grid can be segregated into four stages as shown in Figure 11.5 and the job of IoT in all the four components has been described below.

11.8.1 Power Generation

In recent years, nations over the globe have understood the prompt need of setting up sustainable power source frameworks, for example, solar based and wind vitality frameworks. If there should be an occurrence of solar powered vitality, a lot of regions accepting legitimate daylight during the time is picked and PV cells are introduced subsequently, making a solar-powered homestead. As of late, India has disclosed the world's biggest solar-powered ranch: Pavagada Solar Park with a limit of 2 GW. Likewise, the legislature of India has been advancing the establishment of PV boards on housetops of structures to satisfy expanding vitality needs regularly. Essentially, an enormous number of wind turbines are introduced in territories where wind speed is generally high. Be that as it may, solar power and wind power are discontinuous and rely profoundly upon climate conditions and area. Along these lines, it causes difficulties for unwavering quality and consistency of intensity flexibly. Here, the IoT-based framework utilizes sensors, for example, temperature, moistness, and wind speed to gather climate data continuously and store this information.

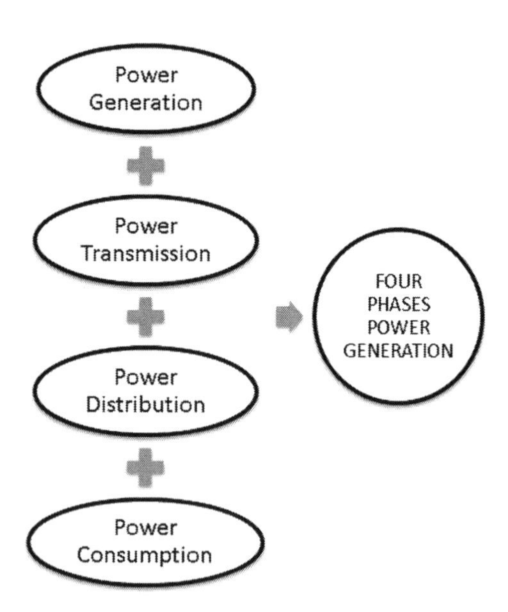

FIGURE 11.5 Four phases of power generation.

The information helps in anticipating climate change and subsequently helps in determining vitality accessibility in the future.

11.8.2 Power Transmission

Customarily, the checking of the transmission framework was performed physically. Occasional visits were made to check the gear status. In contrast to age and dispersion frameworks, the transmission framework is broadly spread over an immense zone. This postures extraordinary test for manual watching as towers and lines are situated in remote regions. IoT-based web-based checking framework for transmission towers gives an increasingly powerful and dependable method. The pinnacle faces harm because of numerous elements, for example, tornado, tempest, tremor, unlawful development, and robbery which may prompt crumbling of the pinnacle. Tilt, vibration, climate sensors, and cameras introduced on towers can provide the precise status of tower information to the authorities [31].

11.8.3 Power Distribution

The specialists can convey power all the more productively in this IoT savvy network time. Smart vitality meters send utilization information to the cloud. Subsequently, the specialists have total data concerning the heap utilization in every area and at every moment. They can revamp their booking plans depending on load necessity and make the dispersion productive and in an improved manner. This will improve the unwavering quality and cost viability [32].

11.8.4 Power Consumption

The job of IoT in making the power consumption savvy is discussed. Passive infrared sensors are utilized to screen movement in a room and the lights are turned off subsequently if there is no movement for a predetermined timeframe. The light switch is connected to the web and the end user can remotely screen the status of lights in each room and can turn on/off [33].

11.9 CONCLUSION

Many energy companies are developing solar energy, which is among the most energy-efficient and cost-effective origins of renewable electricity in the market. The main advantage of using IoT in solar energy is that you can observe exactly what is happening with all your assets from one central control panel. By attaching your devices to a cloud network, you can recognize where the problem has arisen and assign a technician to fix the issue before it will damage your entire system. The main aim of power generation is the elimination of fossil fuels, but, meanwhile, stations generating power are capable of cutting down emissions by integrating energy generated through renewable means such as wind and solar with the standard coal or gas stations. We can accomplish constant remote observation and assemble data through sensors. The dataset can be dissected by utilizing AI calculations to streamline planning, consideration, and foreseeing load designs, and expanding productivity. The popularity of IoT with the energy system is increasing day by day in India and the implementation of IoT in the smart grid will benefit the utilities.

REFERENCES

1. Stearns PN. (2011). "Reconceptualizing the Industrial Revolution." *Journal of Interdisciplinary History*, vol. 42, pp. 442–443.
2. Mokyr J. (2020). "The Second Industrial Revolution, 1870–1914." In Castronovo V. (Ed.) *Storia dell'Economia Mondiale*. Rome: Laterza Publishing, pp. 219–245.
3. Jensen M. (1993). "The Modern Industrial Revolution, Exit, and the Failure of Internal Control Systems." *The Journal of Finance*, vol. 48, pp. 831–880.
4. Kagermann H, Helbig J, Hellinger A, and Wahlster W. (2013). Recommendations for Implementing the Strategic Initiative Industrie 4.0: Securing the Future of German Manufacturing Industry; Final Report of the Industrie 4.0 Working Group; Forschungsunion: Frankfurt/Main, Germany.
5. Witchalls C and Chambers J. (2013). The Internet of Things Business Index: A Quiet Revolution Gathers Pace; The Economist Intelligence Unit: London, pp. 58–66.
6. Datta SK and Bonnet C. (2018). MEC and IoT Based Automatic Agent Reconfiguration in Industry 4.0. In: Proceedings of the 2018 IEEE International Conference on Advanced Networks and Telecommunications Systems (ANTS), Indore, India, pp. 1–5.
7. Shrouf F, Ordieres J, and Miragliotta G. (2014). Smart Factories in Industry 4.0: A Review of the Concept and of Energy Management Approached in Production Based on the Internet of Things Paradigm. In: Proceedings of the 2014 IEEE International Conference on Industrial Engineering and Engineering Management (IEEM), Selangor Darul Ehsan, Malaysia, pp. 697–701.
8. Bandyopadhyay D and Sen J. (2011). "Internet of Things: Applications and Challenges in Technology and Standardization." *Wireless Personal Communications*, vol. 58, pp. 49–69.
9. Miorandi D, Sicari S, De Pellegrini F, and Chlamtac I. (2012). "Internet of Things: Vision, Applications and Research Challenges." *Ad Hoc Network*, vol. 10, no. 7, pp. 1497–1516.
10. Santucci G. (2009). From Internet of Data to Internet of Things. In: Paper for the International Conference on Future Trends of the Internet 2009 Jan 28 (Vol. 28).
11. Digital Agenda for Europe: IoT Architecture. Available online: https://ec.europa.eu/digital-single-market/en
12. Sharma S and Kumar A. (2019). "Enhanced Energy-Efficient Heterogeneous Routing Protocols in WSNs for IoT Application." *IJEAT*, vol. 9, no. 1. ISSN: 2249-8958.
13. Kumar K, Gupta S and Rana A. (2018). "Wireless Sensor Networks: A Review on 'Challenges and Opportunities for the Future World-LTE'." *AJCS*, vol. 1, no. 2. ISSN: 2456-6616.
14. Sharma S, Kumar A, Krishna R and Dhawan S. (2019). Review on Artificial Intelligence with the Internet of Things - Problems, Challenges and Opportunities. In: 2nd International Conference on Power Energy, Environment and Intelligent Control (PEEIC), Greater Noida, India, pp. 383–387.
15. Rana A and Salau A. (2019). *Recent Trends in IoT, Its Requisition with IoT Built Engineering: A Review*. Singapore: Springer. ISBN: 978-981-13-2553-3.
16. Barreto C, Neema H and Koutsoukos X. (2020). "Attacking Electricity Markets Through IoT Devices." *Computer*, vol. 53, no. 5, pp. 55–62. doi: 10.1109/MC.2020.2973951.

17. Bopape LP, Nleya B and Khumalo P. (2020). A Privacy and Security Preservation Framework for D2D Communication Based Smart Grid Services. In: 2020 Conference on Information Communications Technology and Society (ICTAS), Durban, South Africa, pp. 1–6. doi: 10.1109/ICTAS47918.2020.233995.
18. Marr B. (2015). That's Data Science: Airbus Puts 10,000 Sensors in Every Single Wing!, Data Science Central.
19. Tweed K. (2013). Smart Meters Deliver 1 Billion Data Points Daily, Greentech Media.
20. Kaur K, Garg S and Kaddoum G. (2020). "A Big Data-Enabled Consolidated Framework for Energy Efficient Software Defined Data Centers in IoT Setups." *IEEE Transactions on Industrial Informatics*, vol. 16, no. 4, pp. 2687–2697. doi: 10.1109/TII.2019.2939573.
21. Natl. Inst. of Standards & Tech. (U.S. Dept. of Commerce). (2011). The NIST Definition of Cloud Computing, Special Publ. 800-145.
22. Canellos D. (2013). How the "Internet of Things" Will Feed Cloud Computing's Next Evolution, Cloud Security Alliance Blog.
23. Combaneyre F. (2015). Understanding Data Streams in IoT, SAS White Paper.
24. See generally, Wikibooks, I Dream of IoT (last visited Apr. 12, 2016) at Chap. 6.
25. Malibari NA. (2020). A Survey on Blockchain-Based Applications in Education. In: 2020 7th International Conference on Computing for Sustainable Global Development (INDIACom), New Delhi, India, pp. 266–270.
26. Tripathi S and De S. (2020). "Channel-Adaptive Transmission Protocols for Smart Grid IoT Communication." *IEEE Internet of Things Journal*. doi: 10.1109/JIOT.2020.2992124.
27. Jiang J. (2020). "A Novel Sensor Placement Strategy for an IoT-Based Power Grid Monitoring System." *IEEE Internet of Things Journal*. doi: 10.1109/JIOT.2020.2991610.
28. Mishra R, Pandey A and Savariya J. (2020). Application of Internet of Things: Last Meter Smart Grid and Smart Energy Efficient System. In: 2020 First International Conference on Power, Control and Computing Technologies (ICPC2T), Raipur, India, pp. 32–37. doi: 10.1109/ICPC2T48082.2020.9071503.
29. Ray A and Goswami S. (2020). IoT and Cloud Computing Based Smart Water Metering System. In: 2020 International Conference on Power Electronics and IoT Applications in Renewable Energy and its Control (PARC), Mathura, Uttar Pradesh, India, pp. 308–313, doi: 10.1109/PARC49193.2020.236616.
30. Metallidou C, Psannis KE and Egyptiadou E. (2020). "Energy Efficiency in Smart Buildings: IoT Approaches." *IEEE Access*, vol. 8, pp. 63679–63699. doi: 10.1109/ACCESS.2020.2984461.
31. Kale A. (2019). Collaboration of Automotive, Connected Solutions and Energy Technologies for Sustainable Public Transportation for Indian Cities. In: 2019 IEEE Transportation Electrification Conference (ITEC-India), Bengaluru, India, pp. 1–6. doi: 10.1109/ITEC-India48457.2019.ITECINDIA2019-85.
32. Starke M. (2020). Real-Time MPC for Residential Building Water Heater Systems to Support the Electric Grid. In: 2020 IEEE Power and Energy Society Innovative Smart Grid Technologies Conference (ISGT), Washington, DC, USA, 2020, pp. 1–5. doi: 10.1109/ISGT45199.2020.9087716.
33. Samani E. (2020). Anomaly Detection in IoT-Based PIR Occupancy Sensors to Improve Building Energy Efficiency. In: 2020 IEEE Power and Energy Society Innovative Smart Grid Technologies Conference (ISGT), Washington, DC, USA, pp. 1–5. doi: 10.1109/ISGT45199.2020.9087681.

12 Automatic and Efficient IoT-Based Electric Vehicles and Their Battery Management System

A Short Survey and Future Directions

Parag Nijhawan, Manish Kumar Singla, and Souvik Ganguli
Thapar Institute of Engineering and Technology

CONTENTS

12.1 INTRODUCTION

The first fully powered electric vehicle (EV) was made in the 1830s, which was the first small-scale electric car developed by Porsche but it got popular in the 21st century. As that time there were no battery charging stations so the automobile industry research shifted towards the petrol- and diesel-driven vehicles. With the advancement in the technology, the research and development have moved in the direction of EV Tesla Roadster, thus creating a huge impact in the market. There is a growing public awareness of the environmental pollution caused by crude oil engines, and a shift to EVs is due to the limited traditional means. Slowly, several companies, including General Motors, Tata, Honda, Tesla, and Toyota, have begun mass production of EV and hybrid vehicles to address the gasoline-related issues.

Although EV approval has been delayed, sales have increased significantly by 39% over the past 2 years [1]. Over the past few years, the global electric automotive industry has seen significant changes, many different countries have adopted the complete electric route, but the Indian market has not changed significantly. One of the main reasons behind this is the lack of EV charging infrastructure in India. However, as the Indian government innovates to build the ideal infrastructure for EVs by 2030 and completely changes the engine's power, the future is expected to change. Sales of classic, proprietary gasoline and diesel cars are down globally. Customer choice has changed in recent years while picking a vehicle, and they have been shifting towards the EV. The decision to use environment-friendly drive machines is definitely a good choice; however, how costly it is to drive sustainable vehicle becomes another question. Consumers have a variety of options to choose from zero-emission EVs, hybrid vehicles, and plug-in hybrid vehicles (PHEVs). Each option has its pros and cons.

Energy is the determinant of a nation's economy, infrastructure, transport, and quality of life. A main problem which the world is facing these days is the gap between energy consumption and accessibility. Currently, fossil fuel is the only means to generate energy globally, which is a conventional source of energy. In order to meet rapidly growing energy needs of the global population, it must upgrade to the alternative sustainable energy sources that do not have an adverse environmental impact [2,3]. Till now, the usage of conventional source of energy is still very high and expected to account for 75% usage of the energy production by 2050 [4,5]. Generally, current energy solutions have manifold disadvantages. Experts predict that the worst global warming and its effects cannot be achieved by various measures. Over the past two decades, cars have become more economical and hybrid cars have become more common. Electricity is one of the fastest growing alternative energy sources in the car. Traditional energy sources, such as coal and oil, are the primary sources of conventional energy. In an EV, the energy source is the fully charged battery. Battery electric vehicles (BEVs) are very effective at supplying power to the pallets from the mains power supply and can be energized during operation using brakes. One of the biggest drawbacks of BEVs is that their range is usually limited due to the size and battery costs required for the engine power and the energy needs. Refueling a battery system can take several hours as compared to minutes with a conventional engine (CV).

Hydrogen is able to generate chemical energy carrier, which has the capacity to generate electricity, exceeding the highest energy density of batteries. An Internal combustion (IC) engine converts the chemical energy stored in the fuel supplied to the engine to rotational mechanical energy [6]. Power generated by the rotation is used to drive the vehicle or is concentrated via a generator and converted to electricity. Fuel cell (FC) works much like Internal combustion engine (ICE), which converts the generated power directly to FC electricity with chemical energy, and does not affect the environment [7–12]. ICE and fuel cells serve as a continuous source of energy provided that fuel is supplied continuously, unlike batteries that deplete when providing power to electrical components [13,14]. IC engine uses less than 20% of

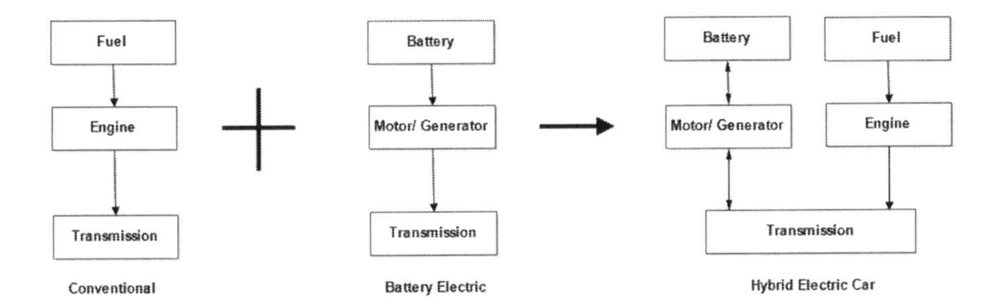

FIGURE 12.1 Schematic of hybrid electric vehicle.

fuel and fuel cell uses almost 60% of fuel – that is why the efficiency of fuel cell is better as compared to that of the IC engines. Therefore, hydrogen fuel cells are expected to overcome the deficiencies of BEVs and convert hydrogen into future transportation fuels. Figure 12.1 represents the schematic of hybrid electric vehicle (HEV). The combination of BEV and CV is known as hybrid electric vehicle (HEV).

In a very smart way, the IoT can be considered as a global dynamic network communication. This will allow connecting IoT devices to share the information and create new topologies that can advance people's routine [15–17]. The IoT concept was first evaluated in 1999 by Kevin Ashton, the founder of the MIT Automatic Identification Center [18–20]. The IoT was officially announced by the International Telecommunications Union (ITU) in 2005 [21]. IoT can be defined in different ways by many researchers and organizations. However, the most commonly used definition is provided by the ITU in 2012. IoT is thus defined as the "Global Information Community that enables advanced services by integrating infrastructure (physical and virtual) based on existing and emerging operational information and communication technologies" [22]. In addition, Ref. [23] provide general definitions that clearly explain IoT. It is acknowledged that the "The IoT allow people and thing to stay in touch anytime, anywhere, on any road/network, and on any service." Though, most or all researchers agree that the IoT creates a better world for all [24–26].

Thus, the remainder of this chapter is discussed as follows. Section 12.2 lays the foundation with a discussion on the development of EVs and their battery management system. Section 12.3 elaborates the fuel cell system an alternative for lithium-ion batteries. Section 12.4 deliberates on the production of hydrogen in a fuel cell system. Section 12.5 explores the application of IoT in EVs. Section 12.6 concludes the outline of some future indications.

12.2 ELECTRIC VEHICLES (EVs)

EVs have really become a buzz word in the recent years. Scientists and engineers had already made many advances in the technology of EVs way back in the 1830s. The first small-scale model EV was invented by the countries like the Hungary and the United States. Around 1832, the initial functioning of the EV was developed by the British Scientist Robert Anderson [27]. In the end of the 19th century, the first EV was introduced by the inventor William Morrison in the United States. According to the personal vehicles, three technologies were emerged, namely, steam, petrol/gasoline, and the electrically powered. Due to the long start times – taking 45 minutes, especially in the colder months – and the refilling of the water in the tank, the steam engine was not much used in personal vehicles.

On the other hand, the electric motor did not encounter any of the above collisions. In the early 1900s, EVs gained significant advantages along with ease of use. The next advantage over gasoline engines is that it is perfect for pollution-free driving of the engines and short-distance urban transport. These facts have very big impact on market, and EVs have entered a new road. Based on US off-road vehicles, the market coverage rate is 28% [28]. Inventors of this era noticed this. In 1891, the founder of the then successful namesake Ferdinand Porsche launched the P1 car. This special model is the company's first car to use electric power. In 1914, Thomas Edison and Henry Ford joined forces to develop a cheap EV for a large audience [29–31]. Ford Model T was the first mass-produced automobile in 1908. It was considered a quite inexpensive car, and suitable for the big mass, with a price of $650, while electric road cars cost about three times the $1,750. In addition, an electric starter motor was created in 1912, causing the original damage to disappear. Sales of hand crank and gasoline vehicles increased. The biggest developments in the market, especially in the United States, were the decline in gas prices across the country and the rising gasoline network in the 1920s. On the other hand, electricity is not yet available in rural areas at the moment, and eventually around 1935, EVs eventually disappeared from the market after being hit by vehicles equipped with combustion engines [32–35]. Figure 12.2 represents the architecture of an EV. The charger interfaces with the battery management system to ensure that the cell's energy is charged correctly before it meets the high-voltage (HV) specifications. Solutions can be external to or incorporated in a vehicle. These solutions can also differ depending on the charging mode (slow, medium, or fast) and the technology (wired or wireless).

Figure 12.3 represents the block diagram of charging of EV. The EV system contains a protection circuit for any transient surges. It also includes a rectification unit for conversion of AC voltage into DC. It also takes care of power factor and corrects it if required. Switching inverter and DC–DC converter are also part of the circuit. The output of the DC–DC converter is fed back to gate drive circuit which is controlled using Pulse width modulation (PWM) technique. The controlling aspect of the gate drive circuit can be carried out by personal computer connected via RS-232 or through USB. It is possible to charge an EV either at home or at work with the

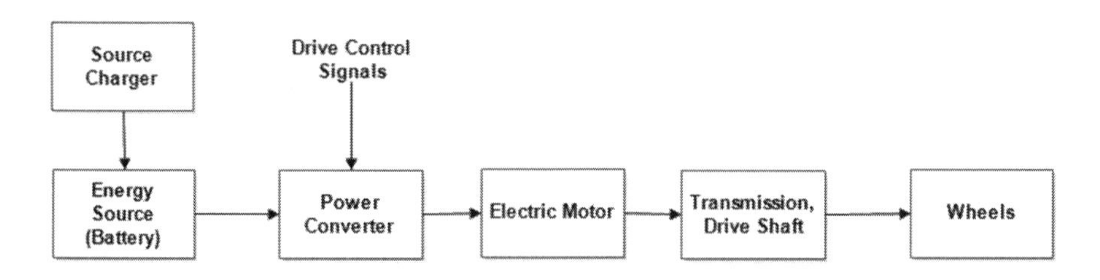

FIGURE 12.2 Architecture of an electric vehicle converter charger.

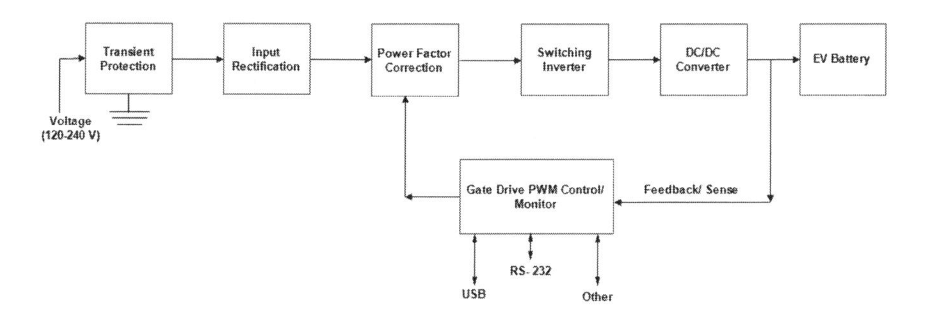

FIGURE 12.3 Block diagram representing charging of electric vehicles [36].

help of a normal electrical plug point supplying an AC voltage of 240 V and 15 A current-carrying capacity. The charge rate will, of course, rely on the EV on-board charger having rating varying between 2.5 and 7 kW. This is the normal rating of the on-board charger.

EVs are of two types, namely, BEVs and PHEVs.

12.2.1 Battery Production

PHEV and BEV use much related batteries, and the most common chemical used in these batteries is lithium ion. Sporamin and petaloid ores are extracted from Salt Lake evaporation ponds, and these extracted ores are used in batteries to remove lithium present in them. Most lithium salts are obtained from water treatment [37]. Battery systems are an important technology that defines the efficiency/performance and range characteristics of EVs. The battery also acts like an inverter by converting chemical energy into electrical energy. In the future, lithium ion is expected to be the main chemical for BEV and PHEV. It offers relatively very high energy and power for a given size or weight, and can significantly reduce costs compared to other battery available. Battery energy density is estimated to increase approximately 300-fold per kilogram between 2007 and 2030 [38]. The few disadvantages are the possibility of overload and limited use of time, and if rechargeable batteries are used, they also take time to recharge, which can be a big hindrance in case of an emergency. Owing to these disadvantages of battery, the world is moving to the fuel cells. Fuel cells have the potential to replace the lithium-based batteries that are responsible for water and air pollution. The cost of producing a recycled lithium-based battery is approximately five times that of producing a freshly extracted lithium-based battery. That is why, all the battery manufacturing units are extracting more and more lithium from the Earth, and in turn, they are unconsciously converting the mother Earth into a "LITHIUM DUMP." Fuel cells have a real potential to qualify as technology from which electricity can be generated with harmless by-products. Other automotive battery concepts include non-electromagnetic alternatives like supercapacitors – which have low energy density and the capacity of charging fast – and another form of batteries such as sodium-nickel chloride ($Na/NiCl_2$) and nickel metal hydride (Ni-MH) [38–40].

12.3 FUEL CELL SYSTEM

Hybrid fuel cell electric vehicle (HFCEV) is evaluated with the basic technology of fuel cell. It mainly consists of a fuel cell layer and various auxiliary devices, also known as balance of plant (BOP). There are various types of fuel cells based on temperature/electrolyte, fuel–oxidizer electrolyte, and state of aggregation of the reactants, as shown in Figure 12.4. The most common type of fuel cell layer used for various HFCEVs is the polymer–electrolyte membrane (PEM) fuel cell. The most common fuel cell from researcher's point of view is PEM fuel cell because it has low operating temperature, high power density, and low starting time as compared to other types of fuel cells. Just like the ICE fuel tank, internal storage tank is also used to store the hydrogen. With the advancement in technologies, nowadays hydrogen is stored as compressed gas.

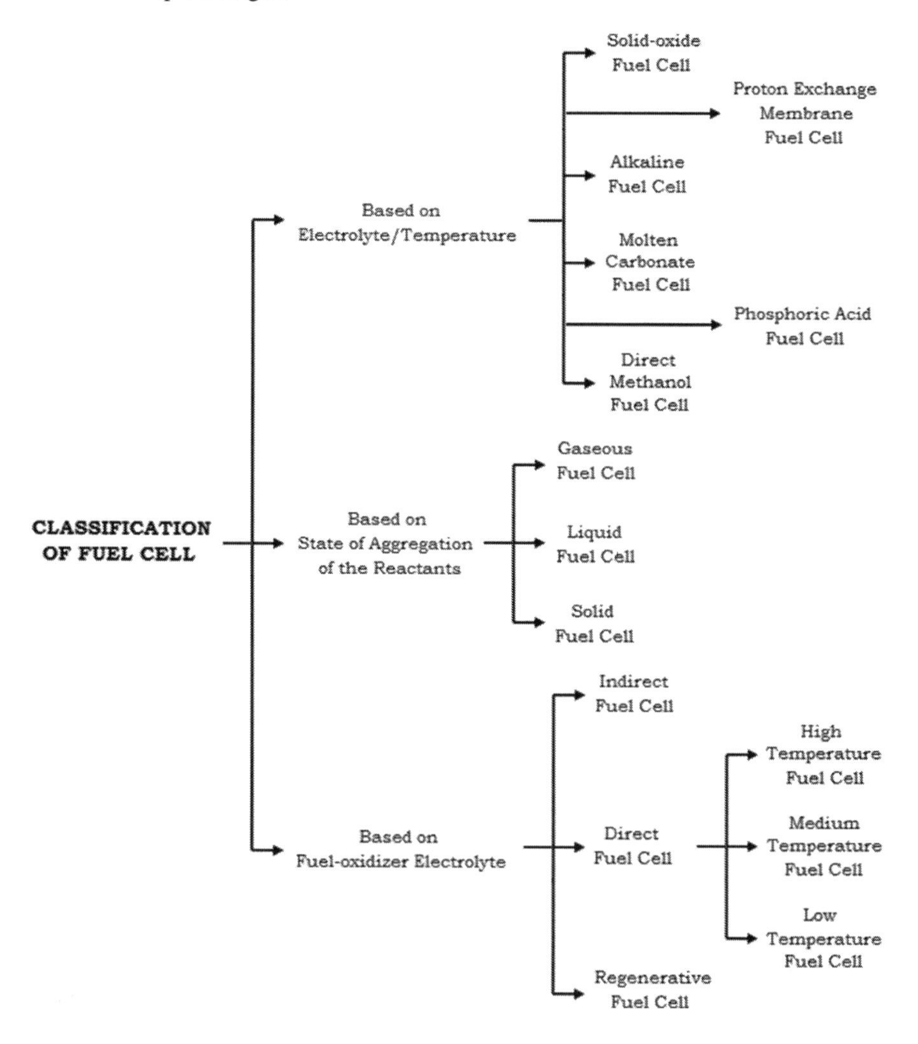

FIGURE 12.4 Classification of fuel cells.

The fuel cell, like the electric battery, is manufactured based on the anode and cathode principle. Hydrogen is supplied from a built-in storage tank and surrounding oxygen for feeding as a cathode fuel. Hydrogen electrons act as external circuits that produce electricity. Energy efficiency of the fuel cells is determined based on the BEV and internal combustion engine batteries, which has improved in recent years.

12.4 HYDROGEN PRODUCTION FOR FUEL CELL TECHNOLOGY

Electrolysis and reforming are capable of generating hydrogen. Currently, hydrogen is primarily generated by small generators through small-scale natural gas reforms. Other production methods include the electrolysis of water and modification of biofuels. In the future, large plants may generate low-cost hydrogen using a variety of methods, such as natural gas and coal gas [41]. Fossil fuel is considered as the lowest cost method of generating hydrogen and is widely used in industry. In the United States, 95% of hydrogen is produced from the natural gas.

The recent developments in field of energy are explored by combining the renewable solar and wind with electrolysis to generate the power. These developments provide a convenient and synergistic way to store intermittent wind and solar energy through hydrogen production or charging EVs. Hydrogen is produced by many other sources, as shown in Figure 12.5.

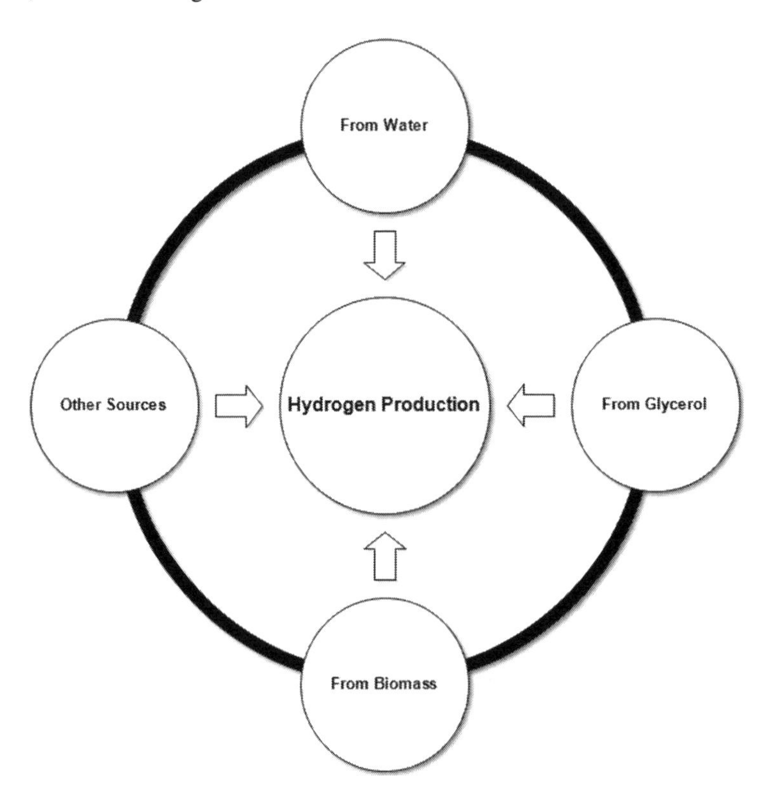

FIGURE 12.5 Various forms of hydrogen production.

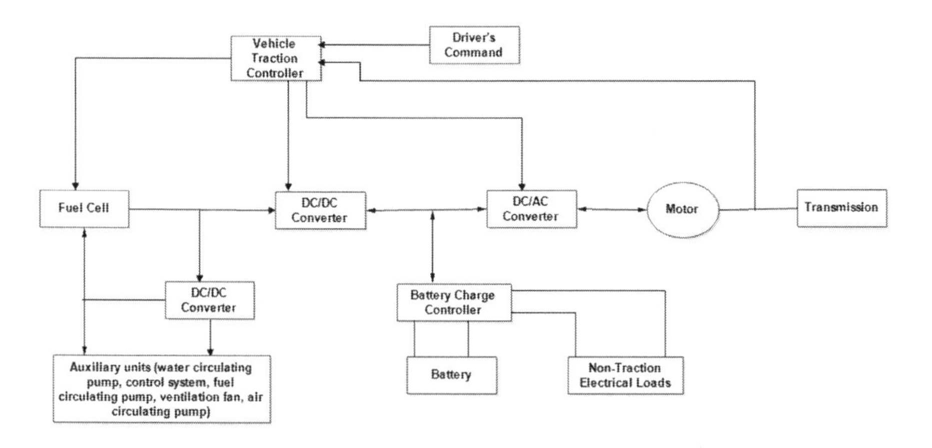

FIGURE 12.6 Block diagram of hybrid fuel cell electric vehicles [45].

12.4.1 Hybrid Fuel Cell Electric Vehicles (HFCEVs)

HFCEV produces the electric energy from hydrogen and air which are powered by fuel cell. The power produced from the fuel cell could be used to directly drive the EV and also can charge the battery, if needed. The latest fuel cell vehicle detains regenerative brake to capture energy, and a battery pack will help accelerate the fuel cell, as shown in Figure 12.6. The size of the battery is a bit bulky, or about the same area when used in general HEV. HFCEVs lead to higher conversion efficiency and have a high growth cost than ICEVs [42,43]. HFCEV's refueling is much faster than the battery charging. In the United States, commercially sold model is Toyota Mirai, which was introduced in 2015, and in Europe, the most sold model is ix35. Tucson is the model of Hyundai and is the first commercially produced hydrogen vehicle, which is manufactured in Japan and which was introduced in 2013 with a 24 kW battery capacity and a 100 kW fuel cell system. Now, the production of fuel cells is much lower than that of batteries. In 2015, approximately 700 FCEVs were produced by the Toyota, and more than 26,000 vehicles in the same year were produced by the manufacturers.

FCV's driving range is over 300 miles and can be loaded in less than 10 minutes at a hydrogen station. This is similar to the conventional fossil fuel vehicles. Hydrogen has great potential as future vehicle fuel. By 2030, it is estimated that fuel cell costs will compete with ICE based on advanced technologies and improved accessibility [44]. One of the biggest obstacles to using large amounts of hydrogen is more efficient storage.

12.5 IoT: A SUPPLEMENT FOR SMART ELECTRIC VEHICLES

Internet of Things (IoT) means the network connection between daily routine items. For devices connected to household items, this is called a temporary wireless connection. It uses electronic identifiers, sensors, and QR codes on the interface to communicate with wireless networks on the interface. IoT technology helps communicate between people and machines, or between machines. The IoT Services Platform should have the capability to deploy, install, troubleshoot, safeguard, manage and monitor IoT devices [46,47]. The IoT has four functions: data integration, two-way communication, processing, and feedback control. ITU formally introduced the "Internet of Things" concept in 2005 [48]. From the very beginning, the concept of smart grids that have been an integral part of IoT has been widely promoted by public applications. Integrated intelligent power systems in IoT increase energy productivity, reduce environmental impact, increase safety, reduce vulnerability to external interference, and improve the coordination of energy delivery.

Today, among the researchers, the widely studied topic is IoT. In the field of Internet, this is considered to be the next era. Since its discovery in the mid-1980s, the Internet has gone through many periods, but has evolved from billions of PCs to billions of computing devices and billions of mobile phones. The IoT enters an era in which all elements of the environment are connected to the Internet and can communicate with each other with minimal human effort, as shown in Figure 12.7 [46,47]. The main cause for this great interest in IoT is the unlimited possibilities it can offer. For example, it can provide a truly smart platform for collaboration between wireless or distributed smart objects [49].

The main evolution which IoT has brought in field of science is the Internet of Energy (IoE). Along with all the renewable sources of energy such as solar, wind, and geothermal, batteries and fuel cell play the most primary role in energy storage. However, using renewable or "free" energy sources, fuel cells are one of the most important sources of IoE. As fuel cell is the part of green energy, it does not pollute the environment and has no direct carbon footprint. Its construction is very simple and operates quietly. However, hydrogen production methods release carbon in atmosphere and have an initial effect. In addition, it also produces carbon dioxide. But unlike most fossil fuels and battery technologies (which will eventually cost too much), they are a source of energy as much as the sun and wind. In IoE facilities, such as smartphones and tablets, portable devices, medical devices, sensors, and smart homes, fuel cells are a viable source of energy. Figure 12.7 summarizes some of the significant applications of IoT with the help of a diagram to encourage the readers to explore more on IoT applications.

IoT has stretched its wings across every sphere of human interventions. EV is no exception in this regard. The charging of an EV can be monitored and controlled in a smarter way using IoT-based devices. Figure 12.8 shows a smart and efficient way of monitoring of EV charging using IoT.

The IoT apps can change our lives today tremendously. Nowadays, IoT device users can assess the performance of the engine, monitor ambient temperature

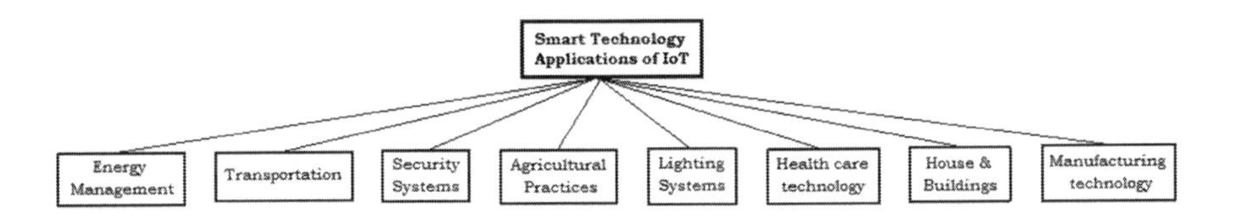

FIGURE 12.7 IoT and its various applications [46].

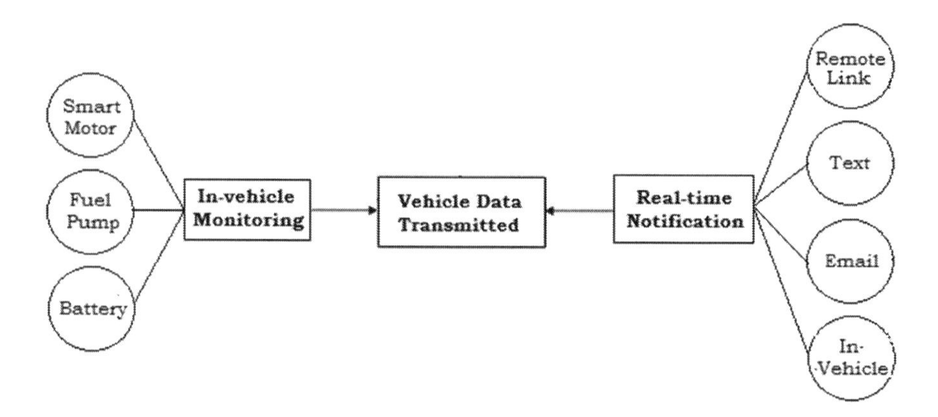

FIGURE 12.8 Monitoring of electric vehicle charging using IoT [26].

within the car, and invigilate the health monitoring of the various indicators with just a few clicks.

The traditional views of the automotive industry are dramatically shifting with the inclusion of IoT. With the advent of IoT, the car maintenance can now be done through predictive algorithms. Wi-Fi control powered by 3G or 4G or 5G in time to come will also be an inherent feature of the car. Networking between cars and cutting-edge fleet management are just a few scenarios of IoT-based applications which will rule the modern automotive era.

The IoT technology in electric cars will definitely involve complex circuits, some dedicated software, intelligent sensing devices, actuation tools, and communication equipment. Seasoned techniques and the speed at which IoT-based applications and automobile sector are emerging can scare people at a first glance. One of the fastest growing markets is the automotive industry providing an IoT-based solution. More than 250 million cars are expected to be linked by 2020, which illustrates the effect of IoT on the automotive industry. Nearly 67% further increase is projected in the number of installed connectivity units in vehicles in the next 2 years. The customers are expected to pay double on vehicle connectivity by the end of the decade. The drivers around the world look forward to their cars to turn into smartphones on wheels. Thus, the IoT-borne technology is at present showing that car networking is the most exciting and the innovative technology.

12.6 CONCLUSIONS

The IoT and more recently IoE have immensely contributed to the development of EVs. The commonly used lithium-ion batteries possess the following demerits. The best performance of batteries is obtained only within a narrow temperature range. Further, these batteries can easily catch fire during charging and discharging conditions. Moreover, they have lower power densities, which lead to increase in their size. In addition, they have lower charging rates, which contribute to larger charging time. Thus, they are gradually getting replaced by fuel cell system with hydrogen being used as a fuel. Fuel cell-powered EVs can thus have greater penetration in the automobile sector. Moreover, IoT has occupied every sector of human life. The automotive industry is not far behind in this competition. Maintenance of cars through prediction algorithms, Wi-Fi connections in a car, car-to-car networking, and other smartphone app-based services will galore automobile industries in years to follow. So IoT-based EVs powered by fuel cell are expected to flood the market in the next decade. Although robust, the automotive and the IoT-enabled smart EVs may at times suffer from the problem of data privacy for which adequate security protocols need to be developed.

REFERENCES

1. Bansal, A., & Agarwal, A. (2018). Comparison of electric and conventional vehicles in indian market: total cost of ownership, consumer preference and best segment for electric vehicle. *International Journal of Science and Research*, *7*(8), 683–695.
2. Granovskii, M., Dincer, I., & Rosen, M. A. (2007). Greenhouse gas emissions reduction by use of wind and solar energies for hydrogen and electricity production: economic factors. *International Journal of Hydrogen Energy*, *32*(8), 927–931.
3. Derbeli, M., Barambones, O., & Sbita, L. (2018). A robust maximum power point tracking control method for a PEM fuel cell power system. *Applied Sciences*, *8*(12), 2449.
4. Hosseini, S. E., Wahid, M. A., & Aghili, N. (2013). The scenario of greenhouse gases reduction in Malaysia. *Renewable and Sustainable Energy Reviews*, *28*, 400–409.
5. Eriksson, E. L. V., & Gray, E. M. (2017). Optimization and integration of hybrid renewable energy hydrogen fuel cell energy systems – a critical review. *Applied Energy*, *202*, 348–364.
6. Reitz, R. D. (2015). Grand challenges in engine and automotive engineering. *Frontiers in Mechanical Engineering*, *1*, 1.
7. Costilla Reyes, A., Erbay, C., Carreon-Bautista, S., Han, A., & Sánchez-Sinencio, E. (2018). A time-interleave-based power management system with maximum power extraction and health protection algorithm for multiple microbial fuel cells for internet of things smart nodes. *Applied Sciences*, *8*(12), 2404.
8. Kerviel, A., Pesyridis, A., Mohammed, A., & Chalet, D. (2018). An evaluation of turbocharging and supercharging options for high-efficiency fuel cell electric vehicles. *Applied Sciences*, *8*(12), 2474.
9. Wang, C., Nehrir, M. H., & Gao, H. (2006). Control of PEM fuel cell distributed generation systems. *IEEE Transactions on Energy Conversion*, *21*(2), 586–595.
10. Somekawa, T., Nakamura, K., Kushi, T., Kume, T., Fujita, K., & Yakabe, H. (2017). Examination of a high-efficiency solid oxide fuel cell system that reuses exhaust gas. *Applied Thermal Engineering*, *114*, 1387–1392.
11. Ayad, M. Y., Becherif, M., & Henni, A. (2011). Vehicle hybridization with fuel cell, supercapacitors and batteries by sliding mode control. *Renewable Energy*, *36*(10), 2627–2634.
12. Chakraborty, U. K. (2019). A new model for constant fuel utilization and constant fuel flow in fuel cells. *Applied Sciences*, *9*(6), 1066.
13. Giorgi, L., & Leccese, F. (2013). Fuel cells: technologies and applications. *The Open Fuel Cells Journal*, *6*, 1–20.
14. Offer, G. J., Howey, D., Contestabile, M., Clague, R., & Brandon, N. P. (2010). Comparative analysis of battery electric, hydrogen fuel cell and hybrid vehicles in a future sustainable road transport system. *Energy Policy*, *38*(1), 24–29.

15. Shanbhag, R., & Shankarmani, R. (2015). Architecture for internet of things to minimize human intervention. In *2015 International Conference on Advances in Computing, Communications and Informatics (ICACCI)*, 2348–2353.
16. Ibrahim, M., Elgamri, A., Babiker, S., & Mohamed, A. (2015). Internet of things based smart environmental monitoring using the Raspberry-Pi computer. In *2015 Fifth International Conference on Digital Information Processing and Communications (ICDIPC)*, 159–164.
17. Perera, C., Zaslavsky, A., Christen, P., & Georgakopoulos, D. (2013). Context aware computing for the internet of things: a survey. *IEEE Communications Surveys & Tutorials*, *16*(1), 414–454.
18. Elkhodr, M., Shahrestani, S., & Cheung, H. (2013). The internet of things: vision and challenges. In *IEEE 2013 Tencon-Spring*, 218–222.
19. Ashton, K. (2009). That 'internet of things' thing. *RFID Journal*, *22*(7), 97–114.
20. Joshi, G. P., & Kim, S. W. (2008). Survey, nomenclature and comparison of reader anti-collision protocols in RFID. *IETE Technical Review*, *25*(5), 234–243.
21. Jing, Q., Vasilakos, A. V., Wan, J., Lu, J., & Qiu, D. (2014). Security of the internet of things: perspectives and challenges. *Wireless Networks*, *20*(8), 2481–2501.
22. Atlam, H. F., Alenezi, A., Alassafi, M. O., & Wills, G. (2018). Blockchain with internet of things: benefits, challenges, and future directions. *International Journal of Intelligent Systems and Applications*, *10*(6), 40–48.
23. Guillemin, P., & Friess, P. (2009). Internet of things strategic research roadmap. *The Cluster of European Research Projects, Tech. Rep.*
24. Li, S., Da Xu, L., & Zhao, S. (2015). The internet of things: a survey. *Information Systems Frontiers*, *17*(2), 243–259.
25. Evans, D. (2011). The internet of things: how the next evolution of the internet is changing everything. *CISCO White Paper*, *1*, 1–11.
26. https://www.intellias.com/how-can-the-automotive-industry-use-internet-of-things-iot-technology/ (accessed on September 2018).
27. Palinski, M. (2017). A comparison of electric vehicles and conventional automobiles: costs and quality perspective. Bachelor Thesis in Business Administration.
28. Anderson, C. D., & Anderson, J. (2010). *Electric and Hybrid Cars: A History*. California: McFarland.
29. Strohl, D. (2010). *Ford, Edison and the Cheap EV That almost Was*. https://www.pinterest.com/pin/117797346486465764/
30. Liang, J. S. (2010). A web-based training framework in automotive electric education. *Computer Applications in Engineering Education*, *18*(4), 619–633.
31. Vidal-Bravo, S., De La Cruz-Soto, J., Arrieta Paternina, M. R., Borunda, M., & Zamora-Mendez, A. (2020). Light electric vehicle powertrain: modeling, simulation, and experimentation for engineering students using PSIM. *Computer Applications in Engineering Education*, *28*(2), 406–419.
32. Wolfram, P., & Lutsey, N. (2016). Electric vehicles: literature review of technology costs and carbon emissions. *The International Council on Clean Transportation: Washington, DC, USA*, 1–23.
33. "2010 Prius Plug-in Hybrid Debuts at Frankfurt Motor Show". Toyota. 2009-09-09. Archived from the original on 2010-02-17. Retrieved 2010-02-03.
34. Bastani, P., Heywood, J. B., & Hope, C. (2012). The effect of uncertainty on US transport-related GHG emissions and fuel consumption out to 2050. *Transportation Research Part A: Policy and Practice*, *46*(3), 517–548.
35. https://www.nxp.com/applications/solutions/automotive/powertrain-vehicledynamics/hybrid-electric-vehicle-hev-converter-and-charger:HEV-CONVERTER-CHARGER (accessed on May 2019).
36. http://www.smcdiodes.com/artilce.asp?m_id=283 (accessed on December 2018).

37. National Research Council. (2013). *Transitions to Alternative Vehicles and Fuels.* Washington, DC: National Academies Press.
38. Cookson, C. (2015). Cambridge chemists make super battery breakthrough. *CNBC.com.*
39. Hacker, F., Harthan, R., Matthes, F., & Zimmer, W. (2009). Environmental impacts and impact on the electricity market of a largescale introduction of electric cars in Europe-critical review of literature. *ETC/ACC Technical Paper, 4*, 56–90.
40. Imanishi, N., & Yamamoto, O. (2014). Rechargeable lithium–air batteries: characteristics and prospects. *Materials Today, 17*(1), 24–30.
41. Edwards, R., Larivé, J., & Beziat, J. C. (2011). Well-to-wheels analysis of future automotive fuels and powertrains in the European context. *JRC, CONCAWE and Renault/EUCAR*, 74.
42. Jiménez, F., López, J. M., Sánchez, J., & Cobos, P. (2011). Simulation and testing of hybrid vehicle function as part of a multidisciplinary training. *Computer Applications in Engineering Education, 19*(3), 604–614.
43. Manoharan, Y., Hosseini, S. E., Butler, B., Alzhahrani, H., Senior, B. T. F., Ashuri, T., & Krohn, J. (2019). Hydrogen fuel cell vehicles; current status and future prospect. *Applied Sciences, 9*(11), 2296.
44. Choi, H., Shin, J., & Woo, J. (2018). Effect of electricity generation mix on battery electric vehicle adoption and its environmental impact. *Energy Policy, 121*, 13–24.
45. Das, H. S., Tan, C. W., & Yatim, A. H. M. (2017). Fuel cell hybrid electric vehicles: a review on power conditioning units and topologies. *Renewable and Sustainable Energy Reviews, 76*, 268–291.
46. Chen, M., Wan, J., & Li, F. (2012). Machine-to-machine communications: architectures, standards and applications. *KSII Transactions on Internet & Information Systems, 6*(2), 480–497.
47. Dengfeng, K. G., & Shan, X. (2010). The internet of things hold up smart grid networking technology. *North China Electric, 2*, 59–63.
48. Fang, X., Misra, S., Xue, G., & Yang, D. (2011). Smart grid—the new and improved power grid: a survey. *IEEE Communications Surveys & Tutorials, 14*(4), 944–980.
49. Roberts, B. P., & Sandberg, C. (2011). The role of energy storage in development of smart grids. *Proceedings of the IEEE, 99*(6), 1139–1144.

13 A Hybrid Approach for Model Order Reduction and Controller Design of Large-Scale Power Systems

Rishabh Singhal
Roorkee Institute of Technology

Yashonidhi Srivastava, Shini Agarwal, Abhimanyu Kumar, and Souvik Ganguli
Thapar Institute of Engineering and Technology

CONTENTS

13.1 INTRODUCTION

Higher-order differential equations transform most physical phenomena into a mathematical model. They are usually preferred to reduce the order of this model while preserving the behaviour of the original system. It helps in complexity reduction of its hardware, which in turn makes designing of controller feasible [1]. Several methods have been developed for reducing certain systems in the domain of both time and frequency [2,3]. Numerous composite techniques have also been proposed [4,5]. Soft computing methods have been applied in the field of model reduction [6–8].

Large systems are exposed to declination in performance due to hindrance caused by load fluctuations, variations in parameters, and other uncertainties. Generation takes place in different areas, and transmission takes place over huge distances. In this entire interconnected system, both frequency and power variations occur due to imbalance in power demand and generation. This mismatch may be treated by kinetic energy extraction, which gradually decreases the frequency. But the gamble of frequency reduction to obtain equilibrium seems huge [9].

In this regard, the field of load frequency control (LFC) aims to provide an effective solution. The principal roles of LFC are the prevention of sudden load disturbances, ensuring zero steady-state error, minimizing unscheduled power exchanges, and ensuring system nonlinearities to lie within the specified tolerance [9].

With the fast headway in electrical power technology, the complete power system has developed into a complicated entity and is hence of higher order. Consequently, its order reduction has become equally important. Some related works are discussed below.

Gallehdari et al. [10] applied particle swarm optimization (PSO) to address the order reduction of a power system model. Outcomes of PSO were compared with those of Hankel norm method as well. Sturk et al. [11] proposed a structured model reduction scheme. The algorithm was tested on a three-machine, nine-bus system. Saxena and Hote [12] adapted Routh approximation method to reduce a single-area model and further proposed an internal model control (IMC)-based approach for smooth LFC operation. Kumar and Nagar [13] developed a new version of balanced truncation method to reduce large-scale power system model preventing the interaction between the study and the external area. Biradar et al. [14] compared around ten model reduction schemes to simplify the automatic voltage regulator model. Sambariya and Arvind [15] proposed a mixed method to reduce single-machine infinite bus system. The coefficients of the denominator polynomial were obtained by the stability equation method, whereas those of the numerator polynomial were determined using firefly algorithm (FA). Semerow et al. [16] stated a modal analysis approach based on the known dominant modes to reduce single- and multi-machine infinite bus systems. Singh et al. [17] applied a balanced realization method to reduce an inherently unstable power system model having several input–output states. Saxena [9] further developed reduced model and its controller for multi-area network. Acle et al. [18] presented a new method to reduce higher-order practical power system stabilizers.

From the literature, it seems that only few works have been reported the use of soft computing for reduced-order modelling (ROM). Moreover, efficacy was not tested using some of the recently developed metaheuristic algorithms. Only step/impulse responses of the original and reduced order systems were considered to estimate the unknown model parameters. Being an unbiased signal, pseudo-random binary sequence (PRBS) has been taken up to obtain the ROM parameters.

Thus, a composite method for order reduction has been employed combining stability equation approach [19] and grey wolf optimizer (GWO) [20] for a two-area system. Stability equation method is used to obtain the coefficients of denominator polynomial, while GWO is used to determine the coefficients of the numerator polynomial. A proportional integral derivative (PID) controller is synthesized using

GWO technique by applying approximate model matching (AMM) framework [21]. Further, the proposed technique is compared with some of the latest heuristic methods used and cited in the literature [22–28].

The remainder of this chapter is structured as follows. Section 13.2 gives an overview of the modelling issues of the single-area and two-area power system networks. Section 13.3 briefs the proposed methodology of work. Section 13.4 presents the relevant results. Section 13.5 concludes this chapter with a discussion on the future scope work.

13.2 AN OVERVIEW OF LOAD FREQUENCY CONTROL SCHEME

On the basis of scale, power systems are classified into single-area, two-area, and so on. Before proceeding to overview on LFCs, consider the following description on single-area and dual-area systems.

13.2.1 LFC – Single-Area Plant

A model providing power to one service region by one generator is treated under a single-area system. Figure 13.1 depicts this system, where $G_g(s)$ refers to governor's transfer function, $G_t(s)$ non-reheated turbine, and $G_p(s)$ load and machine. The term '1/R' in the feedback is to enhance the damping characteristics [9].

For simplicity, these transfer functions can be considered unity order functions with nonzero gain. As LFC essentially presents a disturbance rejection issue, so the purpose is to find $u(s) = -K(s)\Delta f(s)$ such that compensator $K(s)$ controls plant $G(s)$ to reduce the influence on $\Delta f(s) = G(s)u(s) + G_d(s)\Delta P_D(s)$ in the presence of load disturbance $\Delta P_D(s)$ [9].

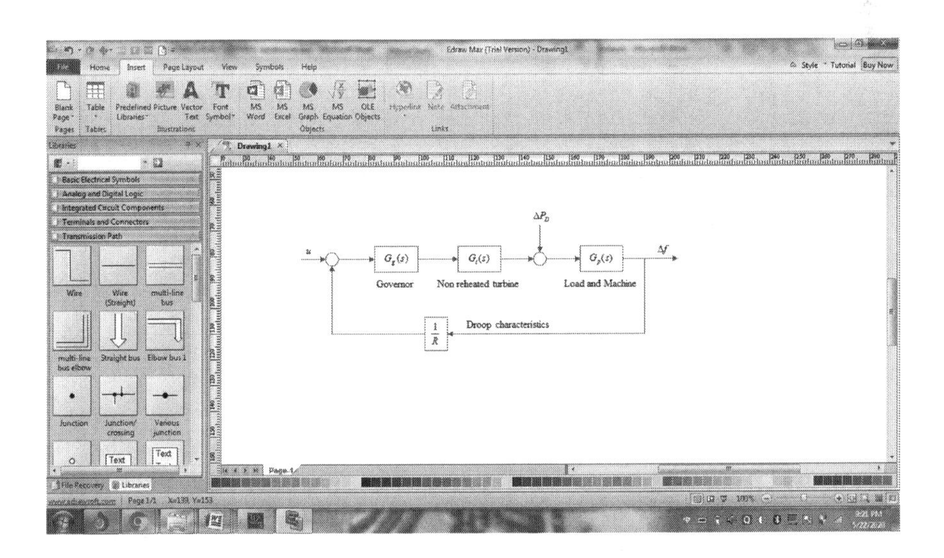

FIGURE 13.1 Block diagram of single-area system [9].

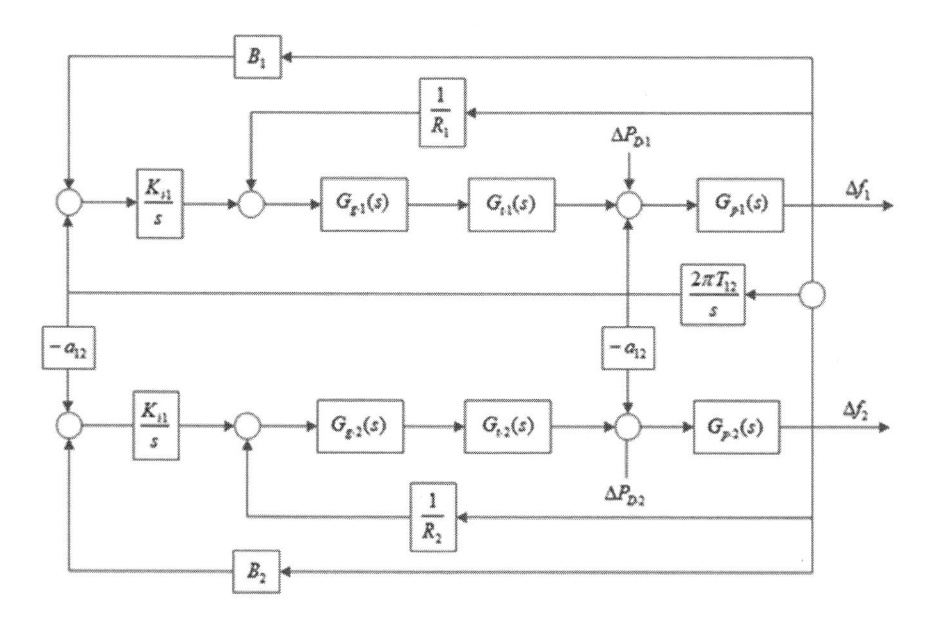

FIGURE 13.2 Block diagram of dual-area system [9].

13.2.2 LFC – Dual-Area Plant

This system is shown in Figure 13.2, where $G_g(s)$ is governor's transfer function, $G_t(s)$ non-reheated turbine, and $G_p(s)$ load and machine. The term '1/R' is again present in the feedback for the same purpose. The quantity $\Delta P_D(s)$ is the load disturbance. Unlike Figure 13.1, these quantities are associated with the subscripts 1 or 2 which refer to the area [9].

Control system in each of the two-area plant monitors the system frequency and tries to restore to normal operation in case it senses deviation. So, in the block diagram, Δf_i refers to the frequency error and B_i is the frequency bias coefficient for the ith area. In dual-area system, LFC model for multi-source can also be marked out. This entire discussion can be easily extended to multi-area system and frequency control.

Approaches known for LFCs can be broadly categorized into (i) modern approaches (like sliding mode scheme, adaptive approach, and fuzzy control), (ii) designing PID controllers, and (iii) controller's parameter adjustment using soft computing technique.

13.3 PROPOSED METHODOLOGY OF WORK

A dual-area power network is controlled in this chapter. This technique involves two segments. Initially, the model taken up for the study is reduced to a second-order system for the suitable controller design. The reduction algorithm is a hybrid approach consisting of a classical technique in fusion with a popular heuristic method. Model order reduction of the original power system model is performed in two parts.

The prior part deals with parameter estimation of the denominator polynomial with a view to stabilize it. This implies that just like the poles of the parent model, the poles of the reduced model must also be present in the left half of the s-plane. To achieve this, the stability equation method developed by Chen et al. [19] is adopted. To satisfy the matching of the dc gain, the coefficient of s^0 is selected. Finally, the other coefficient of the numerator section is determined using GWO [20]. GWO works on two main principles. The first one is the leadership hierarchy of the grey wolves found to be moving in the forests in small groups. The second mechanism involves the prey searching and hunting pattern of the grey wolves. The quest for prey detection, chasing them, and finally pouncing on them to complete the attack involves a search process which is employed to determine the numerator coefficient. The optimization to determine the unknown numerator coefficient satisfies the minimization of sum of square error (SSE) using PRBS matching from the higher and lower models. Quite a sufficient number of algorithms such as ant lion optimization (ALO), grasshopper optimization algorithm (GOA), salp swarm algorithm (SSA), dragonfly algorithm (DA), moth flame optimization (MFO), multi-verse optimizer (MVO), and sine–cosine algorithm (SCA) are used for comparison with the suggested method.

The multi-area system model is thus reduced with the help of a mixed method described above. Next the controller synthesis is performed to control the load frequency model. By controller synthesis, we mean to find out the three parameters, namely, proportional, integral, and derivative gains. Lago and Truxal [29] formulated a means of determining the controller coefficients by applying exact model matching (EMM) [30]. With the help of EMM, the controller does not always ensure a realizable hardware. Hence, another popular technique, namely, the AMM [21], may be considered a suitable alternative to overcome this demerit of EMM approach. Hence, AMM technique is employed to determine the PID controller's parameters. Likewise, to the model order reduction problem, the controller design is realized through comparison with quite a good number of recently developed metaheuristic algorithms.

13.4 RESULTS AND DISCUSSIONS

LFC model for a two-area system is considered [9], which is represented by

$$G_{LFC}(s) = \frac{87.5s + 59.52}{\left(s^4 + 16.12s^3 + 46.24s^2 + 48.65s + 25.3\right)}. \tag{13.1}$$

The stability equation approach reduces the denominator of the above model to a second-order polynomial, which is denoted by

$$D_r(s) = s^2 + 3.0179s + 1.5694. \tag{13.2}$$

Since the dc gain of the parent model is 2.3525, the numerator polynomial in the reduced system is determined by

$$N_r(s) = x(1) \times s + 3.6922, \tag{13.3}$$

where $x(1)$ is the unknown parameter to be obtained by GWO technique. The population size of the GWO algorithm is fixed as 50, while the maximum value for the number of iterations is set as 200 to determine the decision variable $x(1)$. A search bound of ±15 is considered as the limit for this variable. Some of the popular metaheuristic methods such as ALO, DA, GOA, MFO, MVO, SSA, and SCA are used for comparison. Unlike the proposed method, both the numerator and denominator polynomials are found out in the case of the metaheuristic algorithms mentioned above. The expression of the numerator applying GWO technique is given by

$$N_r(s) = -7.7536 \times 10^{-7} s + 3.6922. \tag{13.4}$$

The lower-order model by the mixed method is thus represented as

$$G_r(s) = \frac{-7.7536 \times 10^{-7} s + 3.6922}{s^2 + 3.0179s + 1.5694}. \tag{13.5}$$

It is quite apparent – from the second-order model developed – that the system is having non-minimum phase. This means that the system has a right-half plane zero. However, the poles of the system are found to be stable. For comparison, the reduced models are also obtained using other metaheuristic algorithms as well. Suitable design constraints on maintaining dc gain and preserving the stability of the lower-order models are ensured. A comparison is provided in Table 13.1. Further, the minimum error value is also quoted to show the best method amongst them. The lowest error value is marked in boldface.

TABLE 13.1
Reduced-Order Systems in the Continuous-Time Domain and Their Fitness Function Values

Methods	Continuous-Time Reduced Systems	Fitness Function Value (J)
Proposed	$\frac{-7.7536 \times 10^{-7} s + 3.6922}{s^2 + 3.0179s + 1.5694}$	**6.3518E–05**
ALO	$\frac{0.084606s + 4.06292}{s^2 + 14.2812s + 1.72701}$	0.030663
DA	$\frac{2.265s + 3.4461}{s^2 + 0.23305s + 1.4648}$	0.020236
GOA	$\frac{4.53046s + 6.04476}{s^2 + 12.194s + 2.56942}$	0.014898
MFO	$\frac{-2.12852 \times 10^{-6} s + 15}{s^2 + 15s + 6.37601}$	0.0014735
MVO	$\frac{0.395075s + 4.54042}{s^2 + 14.6664s + 1.9299}$	0.0010015
SSA	$\frac{6.31538s + 6.1376}{s^2 + 11.7349s + 2.60889}$	0.01758
SCA	$\frac{0.11814s + 0.00012168}{s^2 + 0.0024321s + 5.37 \times 10^{-5}}$	0.025603

From the results of Table 13.1, only the MFO and MVO come closer to the proposed method in terms of SSE. The convergence plot of the GWO method meant for determining the numerator coefficient is shown in Figure 13.3. The other methods employed for determining the reduced-order systems are purely heuristic methods and are hence not included in the convergence curve.

Since only one decision variable is to be obtained using GWO, it hardly took ten iterations to converge to the lowest value of the error function. Since heuristic methods are involved to estimate the unknown parameters, repeated test runs are required, suitable statistical measures of the error function are obtained and the outcomes of the error function are shown in Table 13.2. Thirty test runs have been

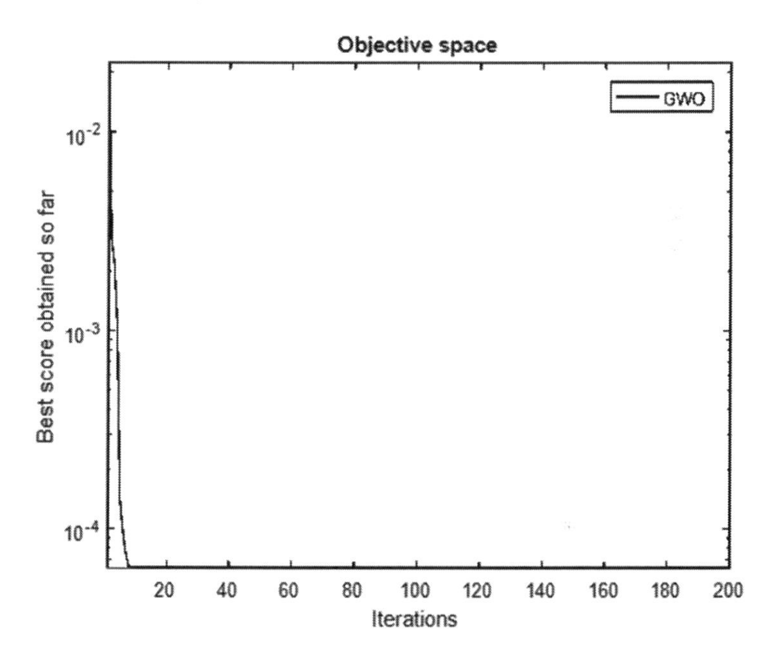

FIGURE 13.3 Convergence curve of GWO technique to determine the numerator coefficient.

TABLE 13.2
Statistical Outcomes of Error Function

Methods	Best	Worst	Average	Std. Deviation
Proposed	**6.3518E−05**	**6.3518E−05**	**6.3518E−05**	**2.7568E−20**
ALO	0.030663	0.63332	0.2465	0.2260
DA	0.020236	0.074737	0.0399	0.0202
GOA	0.014898	0.44315	0.1973	0.1868
MFO	0.0014735	0.0083489	0.0041	0.0029
MVO	0.0010015	0.0069407	0.0029	0.0025
SSA	0.01758	0.27581	0.1506	0.1144
SCA	0.025603	0.3809	0.1295	0.1276

considered to draw meaningful statistical inferences. The best results of each column are specified with the aid of bold letters.

The proposed method outperforms all other algorithms used for comparison. Kruskal–Wallis test [31] is performed to validate the significance. Usually, this test is performed where multiple data sets are involved. The suggested technique is compared with seven other algorithms; hence, the above-mentioned test is aptly justified. The p-value obtained using this test is 3.0322E−37. Usually, a value lower than 0.05 is considered significant for 95% confidence interval, which is a common practice to conduct this test. Moreover, the group mean ranks of this test are also depicted in Figure 13.4, which clearly shows that the proposed technique is different from the five groups. MFO and MVO, marked by groups 4 and 5 in the y-axis of the graph, results do not have a significant difference from the proposed method.

Further, another test, namely, Wilcoxon test [32], is performed on the available data samples to double-check the validity of results. Similar condition is kept for finding the p-value; i.e., p-value of less than 0.05 is taken up the insignificant result. The p-values are enumerated in Table 13.3.

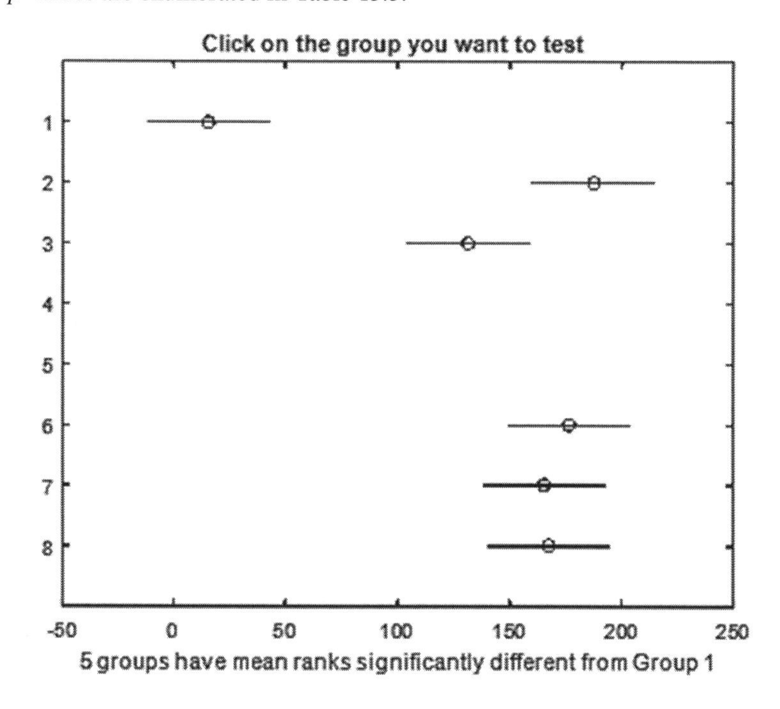

FIGURE 13.4 Kruskal–Wallis results for test of significance.

TABLE 13.3
Wilcoxon Test Outcomes for Significance of Results

Algorithms	ALO	DA	GOA	MFO	MVO	SSA	SCA
Proposed	1.0983E−12	1.0983E−12	9.9054E−13	1.0983E−12	1.0983E−12	1.0983E−12	1.0983E−12

It is evident from Table 13.3 that all p-values are quite less than 0.05. Hence, the results are meaningful. Holm–Bonferroni corrections [33] are added to the Wilcoxon test results to obtain further the corrected p-values, which are shown in Table 13.4.

In Table 13.4, the p-values are less than 0.05, justifying the significance of experiment's outcome. The time and frequency domain parameters of the original, proposed model and models generated from the methods used for comparative study are provided in Tables 13.5 and 13.6, respectively.

From Table 13.5, it is found that our method closely matches the original system parameters in comparison with the metaheuristic algorithms considered. Only no overshoot is observed in the reduced models. The amplitude produced by ALO, GOA, MFO, MVO, and SSA nearly matches that generated by the proposed method.

From the results of Table 13.6, it is observed that the gain margin of the reduced systems is mostly on the higher side, justifying the stability of the models. DA produces the closest phase margin to the parent model. The proposed technique and DA also generated very close phase crossover frequencies in comparison with the original higher-order system. The step and frequency responses of our method are given in Figures 13.5 and 13.6, respectively.

The magnitude part of the Bode diagram shows a close match in Figure 13.6, while the phase part shows quite a significant amount of deviation which has to be coped up with the use of a suitable controller. Moreover, the pole and zero locations of the original system and the reduced model are compared in Table 13.7.

TABLE 13.4
Holm–Bonferroni Correction-Added p-Values

Method	Corrected p-Values
Proposed	$10^{-11} \times \begin{bmatrix} 0.6934 & 0.6590 & 0.6934 & 0.5491 & 0.4393 & 0.3295 & 0.2197 \end{bmatrix}$

TABLE 13.5
Time Domain Specifications of the Test System

Test System	Algorithms	Rise Time (s)	Settling Time (s)	Overshoot (%)	Undershoot (%)	Peak	Peak Time (s)
Original		1.1435	5.5005	21.8343	0	2.8662	2.9013
Reduced	Proposed	3.5078	6.3587	0	0	2.3489	10.1505
System	ALO	18.0132	32.1242	0	0	2.3525	86.4620
	DA	0.6313	33.8485	95.9057	0	4.6088	2.0766
	GOA	9.9274	17.5063	0	0	2.3518	36.5892
	MFO	5.0196	9.0050	0	0	2.3510	16.7232
	MVO	16.5457	29.4428	0	0	2.3526	79.4179
	SSA	9.0257	16.1726	0	0	2.3502	29.3640
	SCA	15.9506	3.3454E+03	538.6099	218.7777	14.4701	227.2152

TABLE 13.6
Frequency Domain Specifications of the Test System

Test System	Algorithms	Gain Margin (dB)	Phase Margin (°)	Gain Crossover Frequency (rad/s)	Phase Crossover Frequency (rad/s)
Original		5.9831	49.6143	5.9577	2.3054
Reduced	Proposed	3.8923E+06	91.1362	3.7909E+03	1.2232
System	ALO	Inf	114.4260	NaN	0.2596
	DA	Inf	68.4772	NaN	3.0129
	GOA	Inf	131.4367	NaN	0.4928
	MFO	7.0431E+06	111.5602	1.0278E+04	0.9299
	MVO	Inf	115.4533	NaN	0.2827
	SSA	Inf	139.2745	NaN	0.5764
	SCA	Inf	90.6819	NaN	0.1186

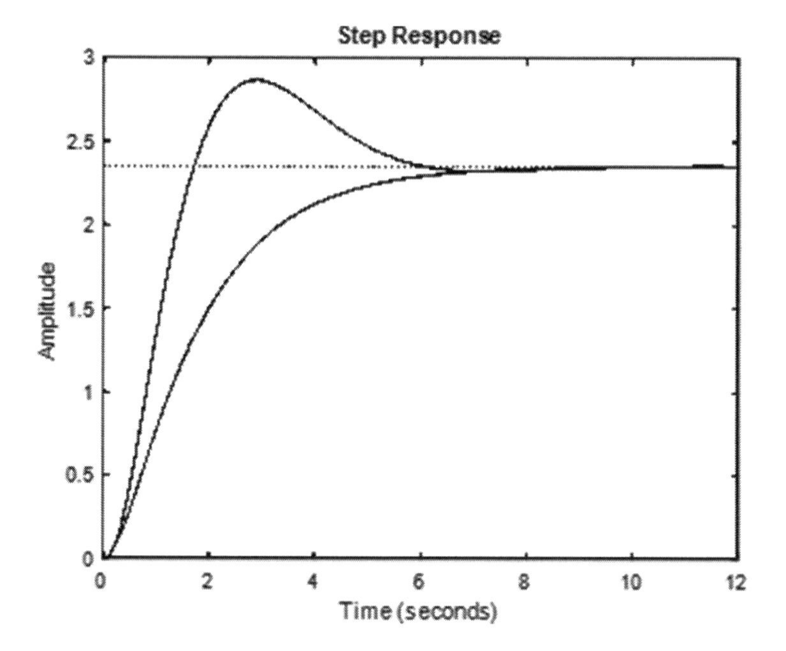

FIGURE 13.5 Step response matching of parent and reduced system.

In Table 13.7, all the poles are present in the left half so the reduced systems produced by the proposed technique as well as the other algorithms are all stable. The proposed system is having right-half plane zero similar to MFO technique. SSA method produces a nearly close zero location. The error indices are calculated in Table 13.8. The best performance is marked in boldface.

In Table 13.8, the suggested technique surpasses all other methods in terms of Integral of absolute error (IAE), integral of square error (ISE), Integral of time

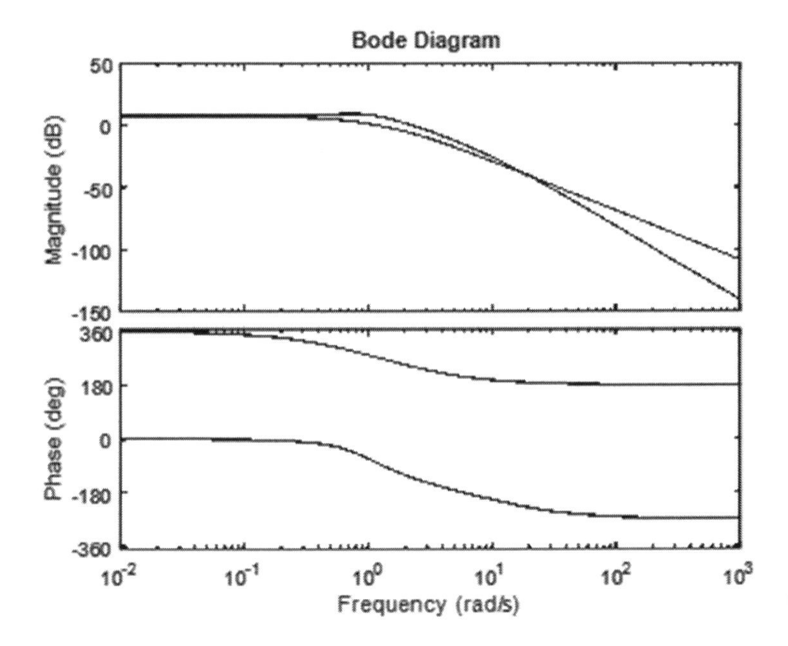

FIGURE 13.6 Frequency response of parent and reduced system.

TABLE 13.7
Pole and Zero Locations of the Parent and Reduced-Order Models

Test System	Methods	Pole Locations	Zero Locations
Original		−12.7900 + 0.0000i −2.0000 + 0.0000i −0.6650 + 0.7395i −0.6650 − 0.7395i	−0.6802
Reduced	Proposed	−2.3501 −0.6678	4.7619E+06
	ALO	−14.1592 −0.1220	−48.0217
	DA	−0.1165 + 1.2047i −0.1165 − 1.2047i	−1.5214
	GOA	−11.9795 −0.2145	−1.3342
	MFO	−14.5622 −0.4378	7.0472E+06
	MVO	−14.5336 −0.1328	−11.4925
	SSA	−11.5082 −0.2267	−0.9719
	SCA	−0.0012 + 0.0072i −0.0012 − 0.0072i	−0.0010

TABLE 13.8
Performance Indices with Different Reduction Methods

Methods	IAE	ISE	ITAE	ITSE	H_{inf} Norm
Proposed	**8.1503E−05**	**6.3517E−08**	**5.8844E−06**	**4.7447E−09**	**9.7749E−04**
ALO	1.0726E−04	1.1028E−07	6.6533E−06	7.0077E−09	0.0018
DA	0.0029	7.4737E−05	1.9899E−04	5.3535E−06	0.0346
GOA	0.0034	1.0510E−04	2.0150E−04	6.0300E−06	0.0498
MFO	5.1147E−04	2.5932E−06	4.1211E−05	2.2681E−07	0.0063
MVO	3.2364E−04	9.6290E−07	1.9103E−05	5.5219E−08	0.0048
SSA	0.0048	2.0680E−04	2.8324E−04	1.1904E−05	0.0697
SCA	1.1183E−04	1.6560E−07	8.7281E−06	1.6245E−08	0.0035

weighted absolute error (ITAE), Integral of time weighted square error (ITSE), and H_{inf} norm. A reference model is selected as per literature [34] to design the controller of reduced systems. A trusted name in the controller topology is the PID controller given by

$$G_{contr}(s) = K_P + \frac{K_I}{s} + K_D s. \tag{13.6}$$

The cascade of the plant and controller with unknown parameters is formed. The response of this cascade is compared to that of the reference model chosen. The error constitutes the difference of the response. ISE is taken as the fitness function for the estimation of controller parameters. PID controller's tuning parameters along with the fitness function values are provided in Table 13.9.

It is clearly seen from the Table 13.9 that GWO method outperforms other metaheuristic techniques in terms of achieving minimum fitness value. The tuned parameters can effectively be utilized to match the reference model approximately.

TABLE 13.9
Controller Tuning Parameters and Their ISE Values

Algorithms	K_P	K_I	K_D	Minimum Fitness Value
Proposed	1.1003	0.058059	1.1681	**4.0436E−09**
ALO	5.0000	2.5493	0.060767	0.0001183
DA	2.1606	0.0029739	0.00068983	3.0525E−05
GOA	5.0000	5.0000	1.5247	6.058
MFO	3.4047	0.11292	0.2939	2.118E−08
MVO	5.0000	5.0000	0.00029006	5.6911E−05
SSA	4.2096	2.3989	0.00097081	0.0024553
SCA	5.0000	5.0000	9.4813E−07	0.00014768

13.5 CONCLUSIONS

This chapter quite efficiently presents both the reduced-order modelling and the controller synthesis of a dual-area power network. This work comprises hybrid approach to perform the order reduction of a load frequency model. The coefficients of the denominator polynomial are determined by stability methodology, whereas GWO technique after dc gain adjustment is employed for numerator polynomial. The controller parameters are tuned by the GWO algorithm by applying the principle of reference model matching. About seven metaheuristic algorithms are used for comparison. The suggested method supersedes them on most occasions. The technique discussed in this chapter can further be employed for order reduction and control of multi-input–multi-output power system models. Normally, the input signals are continuous time in nature, while system components are getting digitized; hence, a unified approach of modelling and control can be taken up in future to address this non-uniformity. New algorithms such as equilibrium optimizer (EO), Harris Hawks optimization (HHO), marine predator algorithm (MPA), and their new variants can be developed and applied to the modelling and control of load frequency model. Even complicated models such as three- and four-area models can be considered to model the reduced system and synthesize their controller parameters.

REFERENCES

1. Abu-Al-Nadi, D. I., Alsmadi, O. M., Abo-Hammour, Z. S., Hawa, M. F., & Rahhal, J. S. (2013). Invasive weed optimization for model order reduction of linear MIMO systems. *Applied Mathematical Modelling*, *37*(6), 4570–4577.
2. Nagar, S. K., & Singh, S. K. (2004). An algorithmic approach for system decomposition and balanced realized model reduction. *Journal of the Franklin Institute*, *341*(7), 615–630.
3. Mukherjee, S., & Mittal, R. C. (2005). Model order reduction using response-matching technique. *Journal of the Franklin Institute*, *342*(5), 503–519.
4. Parmar, G., Mukherjee, S., & Prasad, R. (2007). System reduction using factor division algorithm and Eigen spectrum analysis. *Applied Mathematical Modelling*, *31*(11), 2542–2552.
5. Shamash, Y. (1975). Linear system reduction using Pade approximation to allow retention of dominant modes. *International Journal of Control*, *21*(2), 257–272.
6. Sharma, H., Bansal, J. C., & Arya, K. V. (2012). Fitness based differential evolution. *Memetic Computing*, *4*(4), 303–316.
7. Desai, S. R., & Prasad, R. (2013). A novel order diminution of LTI systems using Big Bang Big Crunch optimization and Routh approximation. *Applied Mathematical Modelling*, *37*(16–17), 8016–8028.
8. Biradar, S., Hote, Y. V., & Saxena, S. (2016). Reduced-order modeling of linear time invariant systems using big bang big crunch optimization and time moment matching method. *Applied Mathematical Modelling*, *40*(15–16), 7225–7244.
9. Saxena, S. (2019). Load frequency control strategy via fractional-order controller and reduced-order modeling. *International Journal of Electrical Power & Energy Systems*, *104*, 603–614.
10. Gallehdari, Z., Karrari, M., & Malik, O. P. (2009). Model order reduction using PSO algorithm and it's application to power systems. In 2009 *International Conference on Electric Power and Energy Conversion Systems, (EPECS)*, 1–5.

11. Sturk, C., Vanfretti, L., Milano, F., & Sandberg, H. (2012). Structured model reduction of power systems. In 2012 *American Control Conference (ACC)*, 2276–2282.
12. Saxena, S., & Hote, Y. V. (2013). Load frequency control in power systems via internal model control scheme and model-order reduction. *IEEE Transactions on Power Systems*, *28*(3), 2749–2757.
13. Kumar, D., & Nagar, S. K. (2014). Order reduction of power system models using square-root balanced approach. In 2014 *Eighteenth National Power Systems Conference (NPSC)*, 1–6.
14. Biradar, S., Saxena, S., & Hote, Y. V. (2015). Simplified model identification of automatic voltage regulator using model-order reduction. In 2015 *International Conference on Power and Advanced Control Engineering (ICPACE)*, 423–428.
15. Sambariya, D. K., & Arvind, G. (2016). Reduced order modelling of SMIB power system using stability equation method and firefly algorithm. In 2016 *IEEE* 6th *International Conference on Power Systems (ICPS)*, 1–6.
16. Semerow, A., Wolf, T., Wellhöfer, S., & Luther, M. (2017). Power system model order reduction based on dominant modes in modal analysis. In 2017 *IEEE Power & Energy Society General Meeting*, 1–5.
17. Singh, A., Yadav, S., Singh, N., & Deveerasetty, K. K. (2018). Model order reduction of power plant system by balanced realization method. In 2018 *International Conference on Computing, Power and Communication Technologies (GUCON)*, 1014–1018.
18. Acle, Y. G. I., Freitas, F. D., Martins, N., & Rommes, J. (2019). Parameter preserving model order reduction of large sparse small-signal electromechanical stability power system models. *IEEE Transactions on Power Systems*, *34*(4), 2814–2824.
19. Chen, T. C., Chang, C. Y., & Han, K. W. (1979). Reduction of transfer functions by the stability-equation method. *Journal of the Franklin Institute*, *308*(4), 389–404.
20. Mirjalili, S., Mirjalili, S. M., & Lewis, A. (2014). Grey wolf optimizer. *Advances in Engineering Software*, *69*, 46–61.
21. Nijmeijer, H., & Savaresi, S. M. (1998). On approximate model-reference control of SISO discrete-time nonlinear systems. *Automatica*, *34*(10), 1261–1266.
22. Mirjalili, S. (2015). The ant lion optimizer. *Advances in Engineering Software*, *83*, 80–98.
23. Mirjalili, S. (2016). Dragonfly algorithm: a new meta-heuristic optimization technique for solving single-objective, discrete, and multi-objective problems. *Neural Computing and Applications*, *27*(4), 1053–1073.
24. Mirjalili, S. (2015). Moth-flame optimization algorithm: a novel nature-inspired heuristic paradigm. *Knowledge-Based Systems*, *89*, 228–249.
25. Mirjalili, S., Mirjalili, S. M., & Hatamlou, A. (2016). Multi-verse optimizer: a nature-inspired algorithm for global optimization. *Neural Computing and Applications*, *27*(2), 495–513.
26. Mirjalili, S. Z., Mirjalili, S., Saremi, S., Faris, H., & Aljarah, I. (2018). Grasshopper optimization algorithm for multi-objective optimization problems. *Applied Intelligence*, *48*(4), 805–820.
27. Mirjalili, S. (2016). SCA: a sine cosine algorithm for solving optimization problems. *Knowledge-Based Systems*, *96*, 120–133.
28. Mirjalili, S., Gandomi, A. H., Mirjalili, S. Z., Saremi, S., Faris, H., & Mirjalili, S. M. (2017). Salp Swarm algorithm: a bio-inspired optimizer for engineering design problems. *Advances in Engineering Software*, *114*, 163–191.
29. Lago, V. G., & Truxal, G. J. (1954). The design of sampled-data feedback systems. *Transactions of the American Institute of Electrical Engineers, Part II: Applications and Industry*, *73*(5), 247–253.
30. Wolovich, W. A. (1972). The use of state feedback for exact model matching. *SIAM Journal on Control*, *10*(3), 512–523.

31. Breslow, N. (1970). A generalized Kruskal-Wallis test for comparing K samples subject to unequal patterns of censorship. *Biometrika*, *57*(3), 579–594.
32. Wilcoxon, F., Katti, S. K., & Wilcox, R. A. (1970). Critical values and probability levels for the Wilcoxon rank sum test and the Wilcoxon signed rank test. *Selected Tables in Mathematical Statistics*, *1*, 171–259.
33. Aickin, M., & Gensler, H. (1996). Adjusting for multiple testing when reporting research results: the Bonferroni vs Holm methods. *American Journal of Public Health*, *86*(5), 726–728.
34. Rana, J. S., Prasad, R., & Agarwal, R. P. (2016). Designing of a controller by using model order reduction techniques. *International Journal of Engineering Innovations and Research*, *5*(3), 220.

14 Day-Ahead Electricity Price Forecasting for Efficient Utility Operation Using Deep Neural Network Approach

K. Arya and K.R.M. Vijaya Chandrakala
Amrita School of Engineering

CONTENTS

14.1 INTRODUCTION

The electric industry operates on a just-in-time basis. This is because unlike solid, liquid, or gaseous forms of energy, there is as of yet no practical and economic way to store electrical energy in the amounts that modern society uses on a second-by-second basis. Consequently, the supply of electric energy must be produced so as to always meet demand. It is the job of electric system operators to keep electric supply balanced with demand, while ensuring there are sufficient system backups in place to keep the grid functionality intact even if generators or grid components have to come off-line or even fail. To accomplish this, the system operator must forecast the demand for electricity 1 day ahead. Day-ahead forecasting of demand is done using models that draw upon historical demand data for that particular time of year. These models will be re-run on the next day, while electricity is being demanded, so that the forecast can adjust the real-time response to the day's actual demand need. Day-ahead forecast in hand will help the system operator to schedule generation to meet anticipated demand and also a certain level of unexpected demand or loss of generation at optimum electricity price. Therefore, system operators are responsible for their own electric grids, as well as for coordinating operations with the operators of adjacent interconnected grids.

14.1.1 BACKGROUND STUDY

Price forecasting of electricity has been done using different varieties of forecasting methods. Traditionally, many methods are implemented for the price forecasting accuracy. Many researchers have discussed about the forecasting and methods of electricity forecasting. One of the authors [1] has put forth a detailed survey regarding the different market power areas and significance of market price forecasting. Different forecasting approaches such as load forecasting and its impact under short term, forecasting of electricity price, demand response, and forecasting of renewable generation are explained [2]. The in–out hidden Markov model and autoregressive integrated moving average (ARIMA) models using wavelet transform have been proposed for the study and prediction of spot prices for electricity. These models provide both good accuracy predictions and dynamic market information [3,4], which paved the way for the researchers to introduce an adaptive neural wavelet network (AWNN) for temporary demand forecasting (TDF) in the energy markets but with errors when compared to statistical methods [5]. Due to the disadvantage of the above models, a hybrid time sequence and AWNN model for day-to-day energy price clearance forecasting is suggested by the researchers [6]. And a new method having novel hybrid solution incorporating wavelet conversion, particle swarm optimization (PSO), and adaptive fuzzy technique towards less time energy prices predicting in a dynamic market is proposed, and two methods of price forecasting approaches are used to perform the day forecast in advance electricity price and demand from the past information [7,8]. An attempt defining the periodic autoregressive neural network (PARNN) as a complex feed forward artificial neural network (ANN) for forecasting electricity prices is made in terms of homologous models [9–11]. From these methods, time sequence-based forecast method is having high precision and neural network model is having accuracy. There were different models developed using ANNs – some methods include exogenous factors – to set day-to-day demand forecast errors lower than the ones obtained with time sequence models [12,13]. Among the deep neural network (DNN) method, a Back Propagation (BP) based neural network enhanced by firefly algorithm is suggested for dynamic-step forward electricity price forecasts [14]. This hybrid method has good forecasting accuracy. A new wavelet transform (WT) hybrid system is proposed, and its performance is validated using an ARIMA predicting electricity prices [15,16]. And the hourly electricity prices are forecasted using DNN. Review of the work done by the above researchers has paved the way towards price forecasting as a major challenge, and deep learning-based methods will be the futuristic models to be adapted. In this work, deep learning-based methods on a realistic model of the system is used for the 24-hour ahead price forecasting to achieve high accuracy.

14.2 METHODS ADOPTED IN PRICE FORECASTING

14.2.1 Introduction

Energy demand forecasting is a more common concept when addressing electrical power systems, with electricity marketing only recently emerging as research into electricity price forecasting. The accuracy of the forecast of future electricity prices will have practical significance for an efficient market operation. Different methods have established over a period of time, which are mainly categorized into forecasting electricity prices as time-series model and simulation-based model. Among this time-series model, forecasting is mostly used for day-ahead forecasting.

14.2.2 Price Forecasting

The price information comes from the generators. Electricity demand strongly depends on peaks during evening time, weekends, and working hours. The supply is distinguished by the types or sources of electricity generation, i.e., coal, nuclear power plants, and renewable energy resources (solar and wind energy). There are various types of forecasting methods: spot price forecasting, 24-hour ahead price forecasting, monthly forecasting, and seasonal forecasting. These models are classified into six groups, as shown in Figure 14.1. There are different categories under these forecasting methods [17]. For our requirements, the forecasting methods are selected [18,19].

14.2.3 Forecasting Algorithms

There are many old price forecasting methods available, such as time of day methods, regression methods, stochastic time-series method, state space methods, expert system methods, and modern methods (fuzzy logic, genetic algorithm, neural network-based methods). So statistical approaches are not reliable for the price prediction. Artificial intelligence and neural network-based methods are more suitable for this forecasting [20]. Some of the methods are explained here.

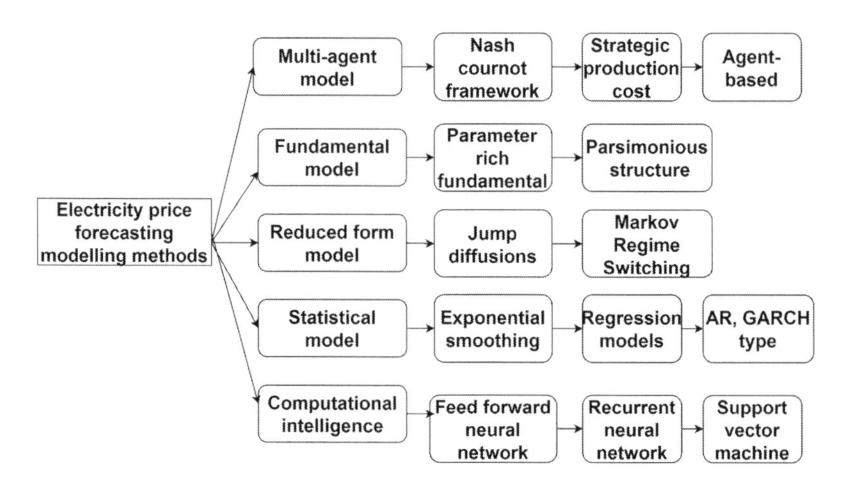

FIGURE 14.1 Electricity price forecasting-based models.

14.2.3.1 Linear Regression-Based Models

14.2.3.1.1 Autoregressive Integrated Moving Average (ARIMA) Model

ARIMA is a statistical price forecasting modelling, which is used for the time sequence prediction. This model has a strong capability to forecast short-time electricity prices. In the ARIMA model, upcoming cost of a variable is a linear combination of previous values and previous errors. In general, this method is expressed as

$$\phi(\beta)p_t = \theta(\beta)\varepsilon_t, \tag{14.1}$$

where p_t is the time t related to price, $\phi(\beta)$ and $\theta(\beta)\varepsilon$ are the backshift operators, and ε_t is the error term. ARIMA method is usually represented using the typical forms of ARIMA (P, D, Q) and (p, d, q) [21]. This model is difficult to understand some conventional model recognition strategies for distinguishing the correct model from the class of possible models.

14.2.3.1.2 GARCH Model

Generalized autoregressive conditional heteroskedasticity (GARCH) is an abbreviation of GARCH model. While the ARIMA model aims at predicting and forecasting the market transition itself, GARCH model aims at predicting demand fluctuations. GARCH model (p, q) is well defined as,

Assume a time sequence y_t,

$$y_t = \beta + \varepsilon_t, \tag{14.2}$$

where β is the offset and $\varepsilon_t = \sigma_t z_t$.

$$\sigma_t = e + \sum_{i=1}^{q} \mu_i \varepsilon_{t-i}^2 + \sum_{i=1}^{p} \phi_i \sigma_{t-i}^2 \tag{14.3}$$

p represents the terms of orders σ^2 and q represents the terms of order ε^2 [22].

14.2.3.2 Based on Nonlinear Heuristics

14.2.3.2.1 Artificial Neural Network (ANN) Model

One application of ANNs is time-series modelling, such as stock price prediction, future demand, and sales promotion. The ANN's architecture can be specified by three variables, namely input neurons, hidden layers, and output neurons. The number of output neurons represents the problem in time-series forecasting for which the prediction to be addressed. So this architecture is well suited for time-series forecasting problems, which is shown in Figure 14.2, with back propagation technique, as shown in Figure 14.3.

Recurrent neural network (RNN) architecture will be explained in upcoming session. ANN has some drawbacks in the instance of forecasting; i.e., a broad sample data size is needed to produce a reliable and consistent forecast performance.

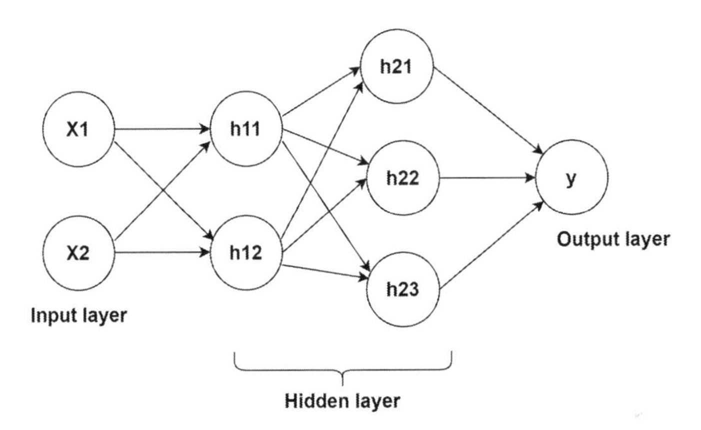

FIGURE 14.2 ANN architecture.

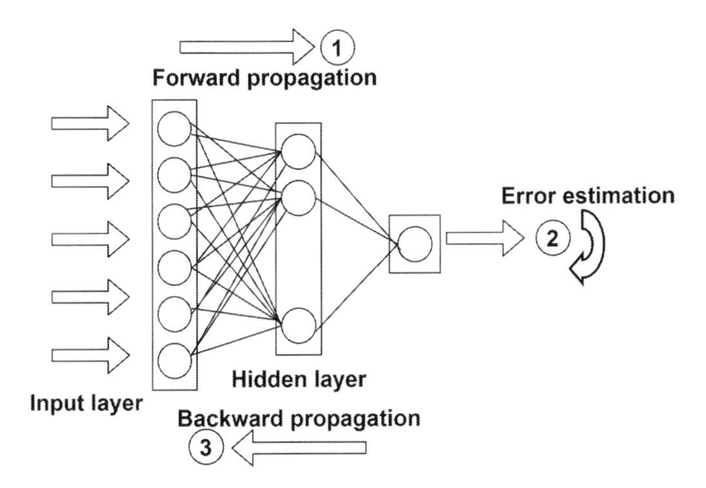

FIGURE 14.3 Back propagation network.

14.2.3.3 Deep Learning-Based Models

Among the current machine learning techniques, the deep learning method can be considered as one of the most promising tools so far, especially in the fields of image and text mining [23].

14.2.3.3.1 Recurrent Neural Network (RNN)

RNN is particularly significant in forecasting of time series. Each neuron in an RNN is capable of keeping preceding input information using an internal memory. Architecture of RNN is shown in Figure 14.4.

In this network, $x_0, x_1, \ldots x_t$ are the stock prices today, and $h_0, h_1 \ldots h_t$ are the hidden states of recurrent network. Circles represent the layers of the recurrent network. Usually, an RNN has three groups of constraints: input to hidden weights (w), hidden to hidden weight (u), and hidden to output weight (v). The property of weight sharing makes our network ideal for inputs with variable dimensions. The hidden networks are recursively stated as.

$$f(x) = vh_t \tag{14.4}$$

$$h_t = \lambda(uh_{t-1} + wx_t) \tag{14.5}$$

$$h_o = \lambda(wx_o). \tag{14.6}$$

One can minimize the cost function to get correct weight. For using back propagation, one has to calculate the gradient of RNN.

14.2.3.3.2 Long Short-Term Memory (LSTM)

RNN with LSTM paved the way as an efficient as well as accessible paradigm for many sequential data-related learning problems [23]. The fundamental concept of LSTM design is a memory cell that can uphold its position over time, and nonlinear gating units control stream of info flowing into the cell and out of the cell. The main awareness behind the LSTM is to control the cell positions using dissimilar gate forms, namely the input gate, the forget gate, and the output gate. Figure 14.5 shows the gate levels of LSTM architecture.

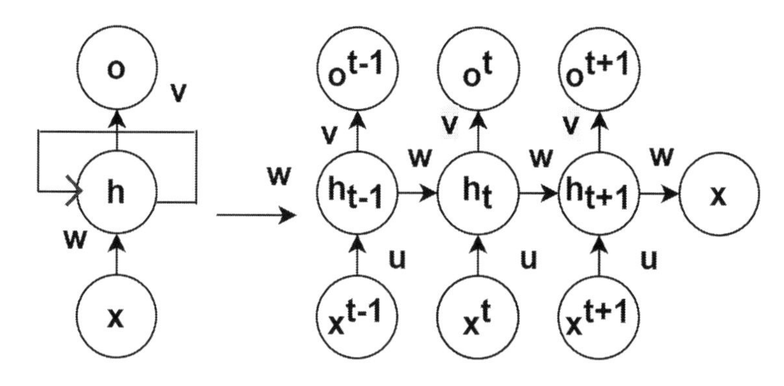

FIGURE 14.4 Recurrent neural network architecture.

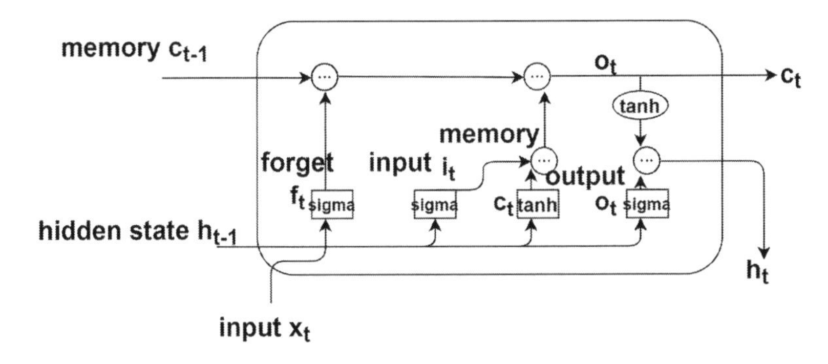

FIGURE 14.5 LSTM architecture.

Every cell's state $(c_t - 1)$ permits over the LSTM cells to produce state related to the subsequent stage c_t. Mathematical equations of three gates are defined as follows:

$$i_t = \sigma\left(w_{pi} p_t + w_{hi} h_{t-1} + \beta_i\right) \tag{14.7}$$

$$f_t = \sigma\left(w_{pf} p_t + w_{hf} h_{t-1} + \beta_f\right) \tag{14.8}$$

$$o_t = \sigma\left(w_{po} p_t + w_{ho} h_{t-1} + \beta_o\right) \tag{14.9}$$

$$c_t = f_t \odot c_t - 1 + i_t \odot \tanh\left(w_{pc} p_t + w_{hc} h_{t-1} + \beta_c\right) \tag{14.10}$$

$$h_t = o_t \odot \tanh(c_t), \tag{14.11}$$

where i_t is the input gate that regulates data flow of inputs p_t and the preceding hidden layer h_{t-1} is permissible to pass the memory cell. f_t is the forget gate, which is regulated by means based on evidence elapsed passing to the cell, o_t is the output gate that governs and estimates evidence which can be transmitted from the present memory cell to a hidden-layer state. c_t is the memory cell state, w is the matrix weight, and β is the bias to the memory cell. $\odot$ Symbol signifies the multiplication of individual parameters. Every gate is viewed as the layer of the neural network. LSTM memory cells may also arrange to form a network organized in several layers [23].

14.2.4 Conclusion

Various price forecasting approaches in the deregulated scenario is analysed in this chapter. Time sequence-based approaches are most widely used for forecasting electricity prices because of their simplicity and ease of execution. Among these models, currently deep learning model is much efficient to provide great accuracy and precision, which is applied towards price forecasting.

14.3 DAY-AHEAD PRICE FORECASTING USING DEEP NEURAL NETWORK

14.3.1 Introduction

Need for the current discussion is to forecast the price of electricity on the day-ahead market service, 24 hours in advance. Deep learning approaches are often used to gain highly precise outcomes in demand forecasting [23]. Many neural network-based approaches are used for the forecasting; among all the methods, RNN has generally used in time sequence forecasting. During this learning process, to get rid of problem related to vanishing and gradient exploding, the LSTM system is invented.

14.3.2 Technique Proposed

The objective of the system is to forecast 24 hours in advance electricity price using the historic price and multivariate dependent features of the electricity price. Nordic pool-based historical data is taken as the multivariate input. Data preprocessing is the important process before forecasting. 2013–2019 (6-year) price data has given the input to the network.

14.3.2.1 Deep Recurrent Neural Network (DRNN)

There are many architectures which are of profound learning, i.e., DNN, deep belief network (DBN), and RNN. In this work, deep recurrent neural network (DRNN) for electricity price forecasting is developed and analysed to predict day-ahead price of the electricity. Simple RNN consists of single-layer neural network, and DRNN has a multilayer architecture. Figure 14.6 shows the multilayer architecture of RNN.

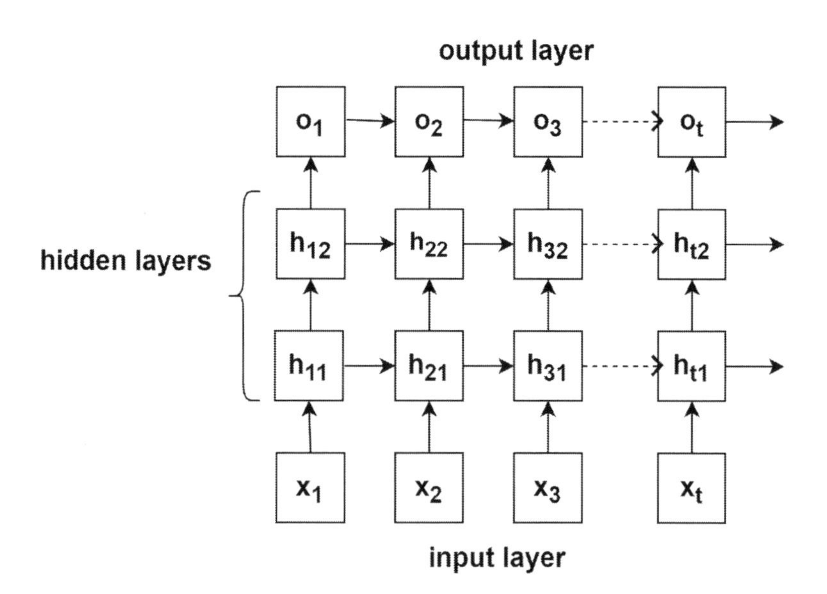

FIGURE 14.6 Deep recurrent neural network (DRNN) architecture.

Input sequence data,

$$x(t) = x_1, x_2, \ldots.x(T) \text{ with time step index } t, \text{ range from 1 to } T \quad (14.12)$$

$$h(t) = f(ux(t) + wh(t-1)) \quad (14.13)$$

$h(t)$ is calculated based on the present input and preceding time step hidden state $h(t)$.

u, v, and w are the weight matrices related to input, hidden, and output layers.

$$a(t) = b + wh(t-1) + ux(t) \quad (14.14)$$

$$h(t) = \tanh a(t) \quad (14.15)$$

$$o(t) = c + vh(t) \quad (14.16)$$

$$y(t) = \text{soft}\max o(t) \quad (14.17)$$

$o(t)$ is the network output, and the softmax operation is applied as post-processing step to obtain a vector $y(t)$ with a consistent output possibility. The loss function for an RRN 'L' is defined for all time steps based on the loss at-time step:

$$L(\tilde{y}, y) = \sum_{t=1}^{T} L(\tilde{y}^{<t>}, y^{<t>}), \quad (14.18)$$

$\tilde{y}$ is the actual value and y is the predicted value.

In DRNN as shown in Figure 14.7, multivariate features with historical price data are given as the input. Six input features are taken, which are the number of days, hours of days, temperature, oil prices, natural gas prices, and energy prices. This model has two hidden layers with 20 numbers of cells. And the output layer gives the result of 24-hour ahead forecasted values. First 80% of the Nordic pool data is taken as training set and rest 20% as test data. DRNN has a sequential input of past 1 week of data and given the output of 24-hour ahead predicted value. Prediction accuracy is calculated using mean absolute percentage error (MAPE)

$$MAPE = \frac{1}{n} \sum_{t=1}^{n} \left| \frac{\tilde{y} - y}{\tilde{y}} \right|. \quad (14.19)$$

14.3.2.2 Long Short-Term Memory (LSTM)

In RNN, the gradient descent algorithm (GDA) is used to enhance the weights between network layers on training [23]. The weights upgrade scheme could avoid further training of the neural network. This is because the gradient value within a range (0, 1) will become small after a long time. This is because of gradient vanishing problem. To overcome this problem, LSTM is proposed for the forecasting of electricity price. Here, simple LSTM (SLSTM) network with single layer and stacked layers of network with deep LSTM (DLSTM) is considered. Both layers of the LSTM network have three gates and memory cell.

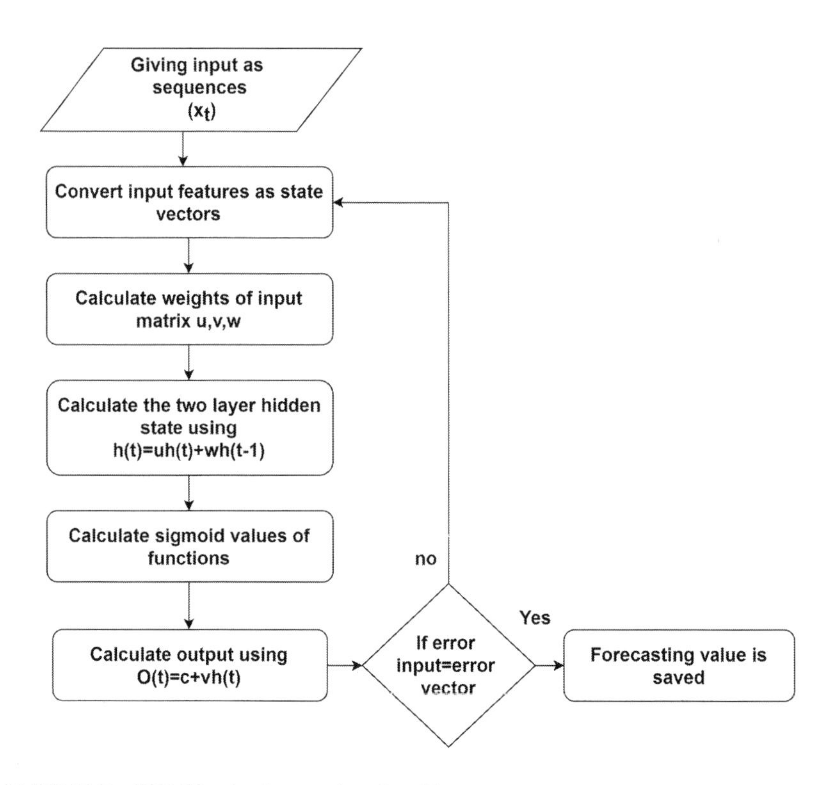

FIGURE 14.7 DRNN price forecasting algorithm.

$$i_t = \sigma\left(w_{pi}p_t + w_{hi}h_{t-1} + \beta_i\right) \tag{14.20}$$

$$f_t = \sigma\left(w_{pf}p_t + w_{hf}h_{t-1} + \beta_f\right) \tag{14.21}$$

$$o_t = \sigma\left(w_{po}p_t + w_{ho}h_{t-1} + \beta_o\right) \tag{14.22}$$

$$c_t = f_t \odot c_t - 1 + i_t \odot \tanh\left(w_{pc}p_t + w_{hc}h_{t-1} + \beta_c\right) \tag{14.23}$$

$$h_t = o_t \odot \tanh(c_t). \tag{14.24}$$

As per the discussion in Chapter 2 and Figure 2.6, the equations are developed related to LSTM gate input equations. The actual price and predicted price values in a day for both the layers of network are represented.

Actual price value of the day d is represented as

$$pr^d = (pr_1^d, pr_2^d, \ldots, pr_t^d, \ldots, pr_n^d). \tag{14.25}$$

Predicted price value of the day d is denoted as

$$pr^{\tilde{d}} = (pr_1^{\tilde{d}}, pr_2^{\tilde{d}}, \ldots, pr_t^{\tilde{d}}, \ldots, pr_n^{\tilde{d}}) \tag{14.26}$$

where $pr_t^{\tilde{d}}$ is the price predicted at time step t, and n is the 24-hour prediction value.

Input and output are used here as the multivariate features such as historical price data, hourly temperature, oil prices, coal prices, natural gas, and uranium gas prices. The input to the LSTM for one day d is as follows:

$$In^d = (In_1^d, In_2^d, ..., In_t^d, ..., In_n^d). \tag{14.27}$$

In the LSTM method, time step 'u' is used to evaluate the price in the following step. A price sequence of 'u' time steps should memorize the statistics in preceding 'u' steps throughout the LSTM training phase, and the trained LSTM method is used to estimate the price rate for subsequent step.

The input of the LSTM has 'u' time step as given by,

$$i_{t(u)}^d = (i_{t-u}^d, i_{t-u+1}^d, ..., i_{t-2}^d, i_{t-1}^d, i_t^d). \tag{14.28}$$

This model will store information in series inside 'u' previous steps. Owing to its advanced accuracy and quicker merging than SGD (stochastic gradient descent), the Adam procedure is chosen as the optimizer. The loss function is stated as follows:

$$\text{Loss}_{\text{fun}} = \sum_{t=1}^{n} (pr_t - \tilde{pr_t})^2. \tag{14.29}$$

Loss function gives the idea about weights and biases, by calculating the actual and predicted price values. LSTM architecture is explained in Figure 14.8.

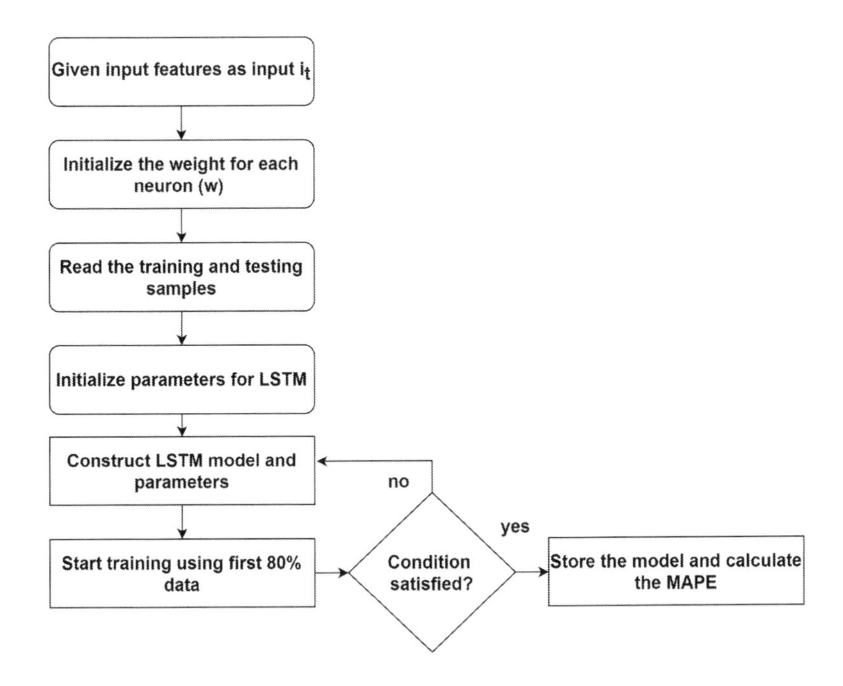

FIGURE 14.8 LSTM algorithm for forecasting.

SLSTM comprises a single hidden layer of 20 cells having six input features that predict the output of 24-hour ahead price. In DLSTM, inputs of six features having previous 1 week of price data with three hidden layers (80, 40, and 30) as the number of cells is considered. The output provides the accurate prediction result of 24-hour ahead price.

14.3.3 Conclusion

Deep learning method has given the idea about time-series forecasting more efficiently. This algorithm has the capacity to predict the future sequence of data using the previous data. The sequence prediction will give accurate results compared to the traditional methods. RNN and LSTM methods are more efficiently used for the time sequence prediction problems.

14.4 RESULTS

14.4.1 Introduction

In the early 1990's, the Nordic countries deregulated their power markets and put together their independent markets into a single Nordic economy. Nord Pool provides day-to-day and intra-day dealing, clearing and payment, data and arbitration, and advisory services. The system price for the Nordic zone is an unconstrained market clearing reference price.

14.4.2 Case Study

For electricity price forecasting, deep learning methods such as DRNN and DLSTM methods are implemented and tested using Nordic pool (2013–2019) year electricity price market data. Input to this network contains historic prices, number of days, temperature, coal prices, oil prices, and natural gas prices. Nordic pool input data information is taken from the reference [24]. A sample of 24 hours' price data is highlighted in Table 14.1.

DRNN, SLSTM, and DLSTM methods have taken these six features as their depending factors of the price forecasting. In DRNN, six input neurons are given as the input on the basis of time sequence. Recurrently, these inputs are measured and extracted using the weights and bias in the networks. To apply these methods, data features are splitted into training, validation, and testing. The training set is

TABLE 14.1
Sample Input Data for 24 Hours in a Day

Methods Used for Forecasting	MAPE (%)
DLSTM	10.13
SLSTM	9.29
DRNN	7.49

again spilt into training and validation set. During the process of training, the data is trained according to the designed model. Training loss and validation loss are calculated according to the number of epochs in each model. During the initial training stage, validation loss and training loss are slightly different due to the slight overfitting of the model. Further proceeding with the number of epochs, this model is properly trained and yields the accurate forecasting results. Every model is trained according to the training and validation set of data.

SLSTM model highlighted in Figure 14.9 is also trained and validated; these two losses are around the same and the model is trained correctly. When the number of epochs increases, the accuracy of the model also increases. So the LSTM model shows good accuracy in forecasting.

As seen from Figure 14.10, the training and validation losses show that training and validation have been done properly, and when the number of epochs increases, accuracy also increases.

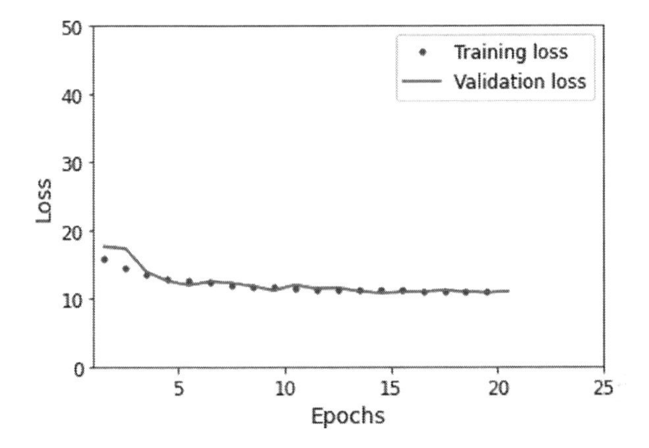

FIGURE 14.9 SLSTM model loss curve.

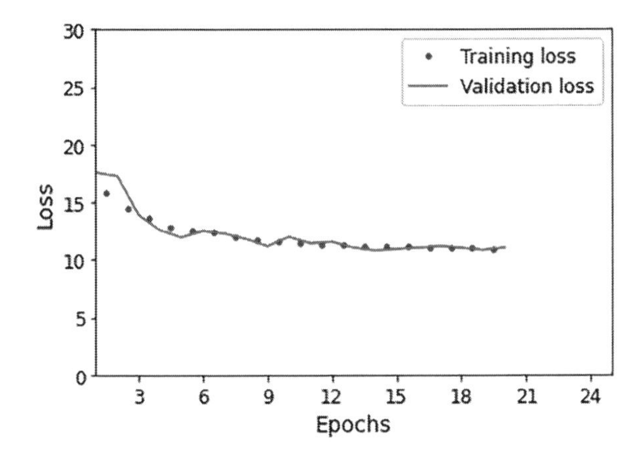

FIGURE 14.10 DLSTM model loss curve.

In DRNN, training loss and validation loss are nearly accurate for 50 epochs of the trained model. From Figure 14.11, it shows that validation loss and training loss are nearly the same when the model reaches the number of epochs. So the loss is also minimum as per mean absolute percent error (MAPE). SLSTM has a single hidden layer which shows the difference in the forecasting accuracy. DLSTM shows good results, which depends on the data given, the number of hidden layers, and accuracy related to model. When the LSTM memory cell has the capability of long-term dependencies of time sequence inputs, the accuracy of the prediction sequence and thus the results will be improved. The actual and predicted prices using SLSTM, DLSTM, and DRNN are shown in Figure 14.12a–c.

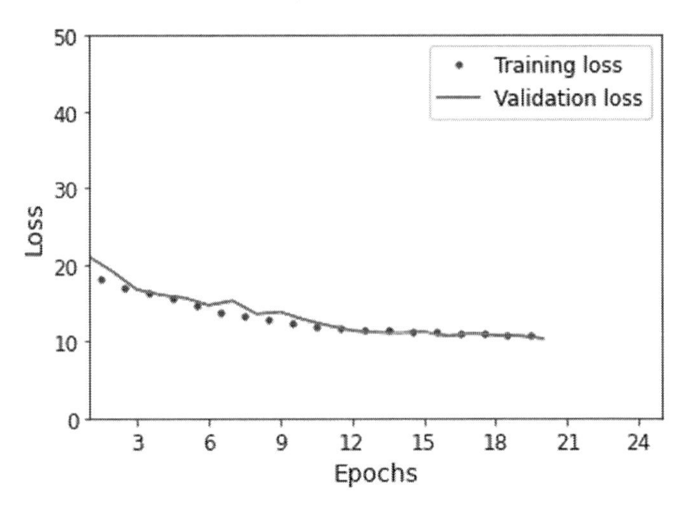

FIGURE 14.11 DRNN model loss curve.

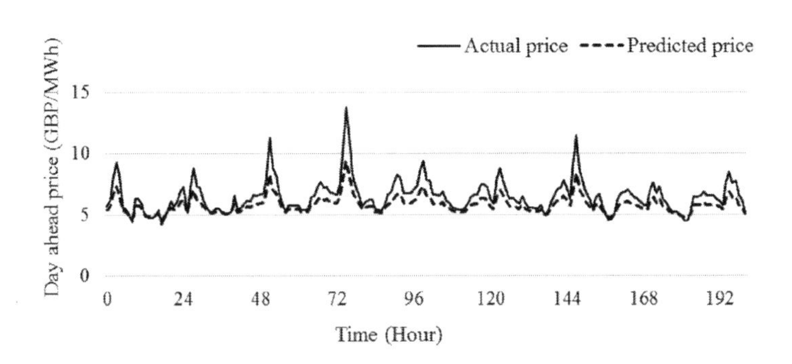

FIGURE 14.12 (a–c) Comparison between actual and predicted electricity prices using SLSTM, DLSTM, and DRNN, respectively.

(Continued)

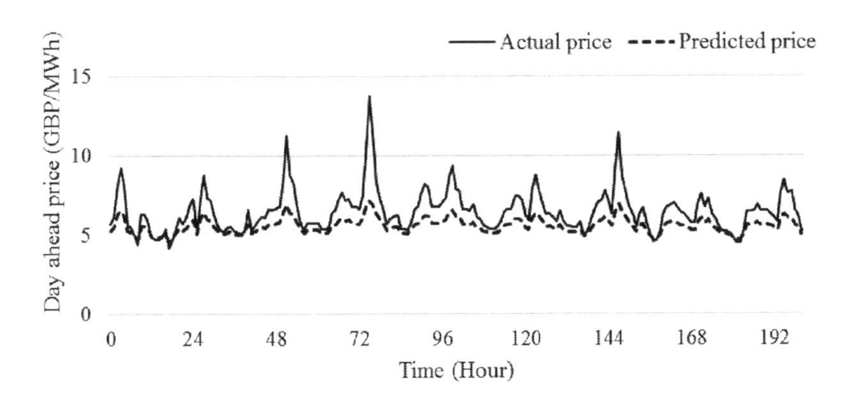

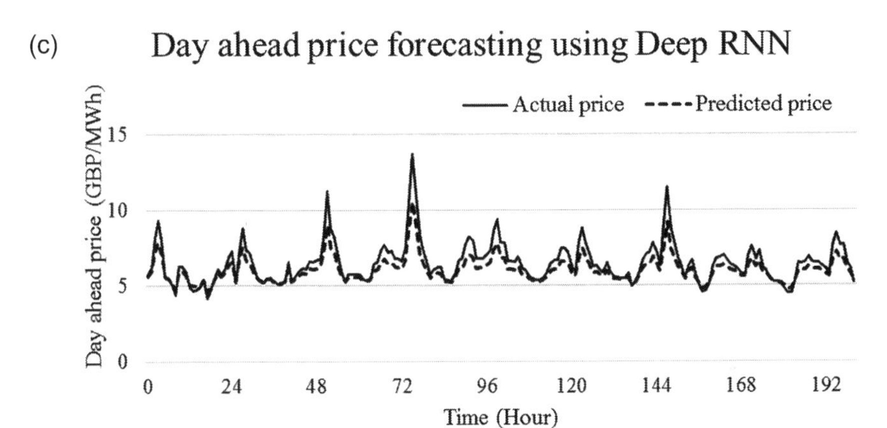

FIGURE 14.12 (*CONTINUED*) (a–c) Comparison between actual and predicted electricity prices using SLSTM, DLSTM, and DRNN, respectively.

The above plot shows the prediction for first 2,000 samples of day-ahead data and prediction of data. Comparing the performance and the prediction results using DLSTM, SLSTM over RNN method has resulted in much accuracy of the predicted results. RNN, SLSTM, and DLSTM networks are modelled, and their performance is validated using MAPE to provide accurate results. Table 14.2 express the evaluation of MAPE error in the day in advance price forecasting.

From Table 14.2, it shows that RNN when compared to SLSTM and DLSTM has shown excellent results with MAPE achieving to 7.49%.

TABLE 14.2
Comparison of Average MAPE Price Forecasting

Hour of Day 01/01/13	Price GBP/MWh	Temperature	Coal Price	Oil Price	Uranium Price	Natural Gas
00–01	37.76	48.48	95.59	111.11	43.5	3.351
01–02	30.01	48.3	95.59	111.11	43.5	3.351
02–03	21.98	46.4	95.59	111.11	43.5	3.351
03–04	18.06	45.4	95.59	111.11	43.5	3.351
04–05	18.06	44.44	95.59	111.11	43.5	3.351
05–06	22.09	42.52	95.59	111.11	43.5	3.351
06–07	22.09	41.12	95.59	111.11	43.5	3.351
07–08	26.08	40.16	95.59	111.11	43.5	3.351
08–09	30.08	39.54	95.59	111.11	43.5	3.351
09–10	33.18	41.37	95.59	111.11	43.5	3.351
10–11	43.19	43.2	95.59	111.11	43.5	3.351
11–12	50.4	44.65	95.59	111.11	43.5	3.351
12–13	50.19	45.3	95.59	111.11	43.5	3.351
13–14	44.92	45.37	95.59	111.11	43.5	3.351
14–15	48.02	45.25	95.59	111.11	43.5	3.351
15–16	54.95	44.39	95.59	111.11	43.5	3.351
16–17	85.83	42.65	95.59	111.11	43.5	3.351
17–18	50.8	42.35	95.59	111.11	43.5	3.351
18–19	44.79	41.9	95.59	111.11	43.5	3.351
19–20	44.7	42.41	95.59	111.11	43.5	3.351
20–21	43.05	41.76	95.59	111.11	43.5	3.351
21–22	31.6	41.32	95.59	111.11	43.5	3.351
22–23	37.53	41.32	95.59	111.11	43.5	3.351
23–00	34.96	40.11	95.67	112.47	43.25	3.233

14.5 CONCLUSION

This study suggested RNN-based method to forecast the 24-hour ahead electricity market prices because of its capability to cross-extend input lags and recall past pattern details in time sequence. The electricity price was predicted based on the Nordic pool data containing 60,000 samples to train and test, which consisted of multivariable input features. The 24-hour ahead future prices are predicted in a sequence manner by using RNN, SLSTM, and DLSTM methods, and the approach was validated using MAPE. The most accurate method for predicting 24-hour ahead future price for Nordic pool is RNN over other methods.

REFERENCES

1. S. Prabhakar Karthikeyan, I. Jacob Raglend, and D. P. Kothari. 2013. A review on market power in deregulated electricity market. *Electrical Power and Energy Systems*, 48, 139–147.
2. T. Hong. 2014. Energy forecasting: past, present, and future. *Foresight*, Winter, 43–48.

3. R. Weron. 2006. *Modeling and Forecasting Electricity Loads and Prices: A Statistical Approach*. Chichester: Wiley.
4. A. J. Conejo, M. A. Plazas, R. Espinola, and A. B. Molina. 2005. Day-ahead electricity price forecasting using the wavelet transform and ARIMA models. *IEEE Transactions on Power Systems*, 20(2), 1035–1042.
5. N. M. Pindoriya, S. N. Singh, and S. K. Singh. 2008. An adaptive wavelet neural network based energy price forecasting in electricity markets. *IEEE Transactions on Power Systems*, 23(3), 1423–1432.
6. L. Wu and M. Shahidehpour. 2010. A hybrid model for day-ahead price forecasting. *IEEE Transactions on Power Systems*, 25(3), 1519–1530.
7. J. P. S. Catalão, H. M. I. Pousinho, and V. M. F. Mendes. 2011. Hybrid wavelet-PSO-ANFIS approach for short-term electricity prices forecasting. *IEEE Transactions on Power Systems*, 26(1), 137–144.
8. G. Aneiros, J. Vilar, and P. Raña. 2016. Short-term forecast of daily curves of electricity demand and price. *International Journal of Electrical Power & Energy Systems*, 80(Sep), 96–108.
9. F. Saadaoui. 2017. A seasonal feed forward neural network to forecast electricity prices. *Neural Computing and Applications*, 28(4), 835–847.
10. J. Nowotarski and R. Weron. 2018. Recent advances in electricity price forecasting: a review of probabilistic forecasting. *Renewable and Sustainable Energy Reviews*, 81(1), 1548–1568.
11. G. Marcjasz, B. Uniejewskiab, and R. Weron. 2019. On the importance of the long-term seasonal component in day-ahead electricity price forecasting with NARX neural networks. *International Journal of Forecasting*, 35, 1520–1532.
12. I. P. Panapakidis and A. S. Dagoumas. 2016. Day-ahead electricity price forecasting via the application of artificial neural network based models. *Applied Energy*, 172, 132–151.
13. C. R. Knittel, M. R. Roberts, C. Knittel, and M. Roberts. 2005. An empirical examination of restructured electricity prices. *Energy Econoomics*, 27(5), 791–817.
14. D. Wang, H. Luo, O. Grunder, Y. Lin, and H. GuoMulti. 2017. Step ahead electricity price forecasting using a hybrid model based on two-layer decomposition technique and BP neural network optimized by firefly algorithm. *Applied Energy*, 190, 390–407.
15. J. Zhang, Z. Tan, S. Yang.2012. Day-ahead electricity price forecasting by a new hybrid method. *Computers and Industrial Engineering*, 63, 695–701.
16. E. Raviv, K. E. Bouwman, and D. van Dijk. 2015. Forecasting day-ahead electricity prices: utilizing hourly prices. *Energy Economics*, 50, 227–239.
17. M. Filomena Teodoro, M. A. P. Andrade, E. Costa e Silva, et al. 2018. *Energy Prices Forecasting Using GLM*. New York: Springer Science and Business Media LLC.
18. P. Kiran, K. R. M. Vijaya Chandrakala, and T. N. P. Nambiar. 2017. Multi-agent based systems on microgrid – a review. *International Conference on Intelligent Computing and Control (I2C2)*, Coimbatore.
19. P. Kiran and K. R. M. Vijaya Chandrakala. 2020. Variant Roth-Erev reinforcement learning algorithm-based smart generator bidding as agents in electricity market. *Advances in Intelligent Systems and Computing*, 1048, 981–989.
20. R. Weron. 2014. Electricity price forecasting: a review of the state-of-the-art with a look into the future. *International Journal of Forecasting*, 30(4), 1030–1081.
21. N. Singh and S. R. Mohanty. 2015. A review of price forecasting problem and techniques in deregulated electricity markets. *Journal of Power and Energy Engineering*, 3, 1–19.
22. S. Takiyar, K. G. Upadhyay, and V. Singh. 2015. Fuzzy ARTMAP and GARCH-based hybrid model aided with wavelet transform for short-term electricity load forecasting. *Energy Science & Engineering*, 4, 14–22.

23. I. R. Widiasari, L. E. Nugoho, Widyawan, and R. Efendi. 2018. Context-based hydrology time series data for a flood prediction model using LSTM. *5th International Conference on Information Technology, Computer, and Electrical Engineering (ICITACEE).*
24. https://www.nordpoolgroup.com/Market-data1/Dayahead/Area-Prices/ALL1/Monthly/?view=table.

15 MEMS Devices and Thin Film-Based Sensor Applications

Ashish Tiwary and Shasanka Sekhar Rout
GIET University

CONTENTS

15.1 INTRODUCTION

Thin film technology is a crucial key factor especially in the field of manufacturing various sensors and actuators. It defines the properties of a microsystem with its miniaturized sizes in form of either structural or sacrificial layers. Various researchers have investigated the role of thin films in the manufacturing of microelectronics with the help of IC (integrated circuit) fabrication techniques. It pushes the technology and the performance of these devices at different platforms, which further widens the applications and is useful in developing many miniaturized microsystems. With the development in thin film growth techniques, it is used in so many industrial applications and kindles the interest of academicians in their research towards the testing, design, and fabrication of thinfilm MEMS (micro-electromechanical system) devices that provide a light in solving issues. Thin films reflect the critical aspects of the MEMS device by their well-defined properties such as optical, electrical, mechanical, and structural. By doing so, it is quite possible to obtain semiconductors or insulator silicon-based thin films by tuning the electrical conductivity along with other parameters. These films enable the development of new sensors and devices with their broadened applications in miniaturized systems, and is best to use thinfilm technology. At the micro-level, the miniaturized devices can sense the environmental parameters very precisely controlling the system, and they can operate (actuate) also on macro-level. Many widened application areas have the glimpse of MEMS technology such as Bio-MEMS; this technology is generally useful to patients on the lookout for health conditions, which can be monitored keenly through the Bio-MEMS devices. An understanding of normal sensory processing can help in addressing biological issues towards the identification of any virus presence or delivery of the drugs; the heart conditions, neurological disorders (epilepsy), and many more are extremely analysed in a rigorous way. Similarly, other application area imprints the miniaturized technology useful in the fabrication of optical MEMS. It's a very essential and prominent tool in controlling the optical signals and switching it in the communication system networks. MEMS empowers the functionality, constancy, and performance of these miniscule devices concisely holding up the minimum cost of the instrument.

This chapter renders a broader outline of thin film-based microsystem technology with the related applications and discusses thin film processing techniques. In addition to the above, this chapter enlightens the fundamental scope of lithography and the art of fabrication. It can create very minute patterns ranging from few micrometres to nanometres.

15.1.1 Motivation and Background

The exploration made on the IC technology assembles both electronic and mechanical domains on a single platform to create a microsystem technology. Microsystem or MEMS focuses on human experiments prior to the invention made at the AT&T Bell Laboratories in the late 1940s. The invention continues with the development of a composed semiconductor device such as a transistor, which plays a vital role in amplifying or increasing electrical signals and power energy. Furthermore, Gordon Earle Moore observed in 1965 that the number electronic components incorporated on a single IC chip almost doubled every year. The prediction made by Gordon Earle Moore on an expeditious development in the increase in the number of transistors doubled in every one or one and a half years of span had a widespread impact on semiconductor industries, followed by Moore's law [1]. This is a remarkable achievement of originality and establishment of the technology in the past several decades. The process begins with the transistor; new electromechanical devices have been developed by reducing structures to a miniaturized framework.

MEMS devices penetrate into an idea of manufacturing tiny integrated devices that put an effect on larger volume that really aspires various individuals. The MEMS history emphasizes its differences, diversity, market challenges, and developed applications [2].

15.1.2 MEMS Devices: Sensors and Actuators

15.1.2.1 Energy Domains and Transducer

MEMSs are considered as microchips when combined with very miniscule movable mechanical parts fabricated using IC techniques [3]. Electronic circuitry is often connected and integrated with micro-structured mechanical components to yield satisfactory outcomes, such as cantilevers, micro-pumps, mechanical relays, combs, membranes, and channels. MEMSs, on the other hand, behave as sensors, retrieve environmental information, and actuate the gathered data to the control system, which makes a drastic change in the environmental conditions as per requirements.

There are six major energy domains that exist:

I. Electrical domain
II. Mechanical domain
III. Chemical domain
IV. Radiative domain
V. Magnetic domain
VI. Thermal domain.

15.1.2.2 Micro-Electromechanical System (MEMS) Development

MEMS technology ought to be considered as microchips combined with very miniscule movable mechanical parts. Conceptually, a micro device is designed with a 3-D solid model that can be firstly developed using some 3-D modelling tools [4]. Various simulation tools are available to visualize the designs and implement the same to get fruitful results:

i. COMSOL multiphysics
ii. Coventer
iii. Ansys
iv. SUGAR
v. MATLAB.

MEMS devices are built on a platform using ICs, and précised mechanical machining fabricates various microsensors and microactuators. Through Finite element method

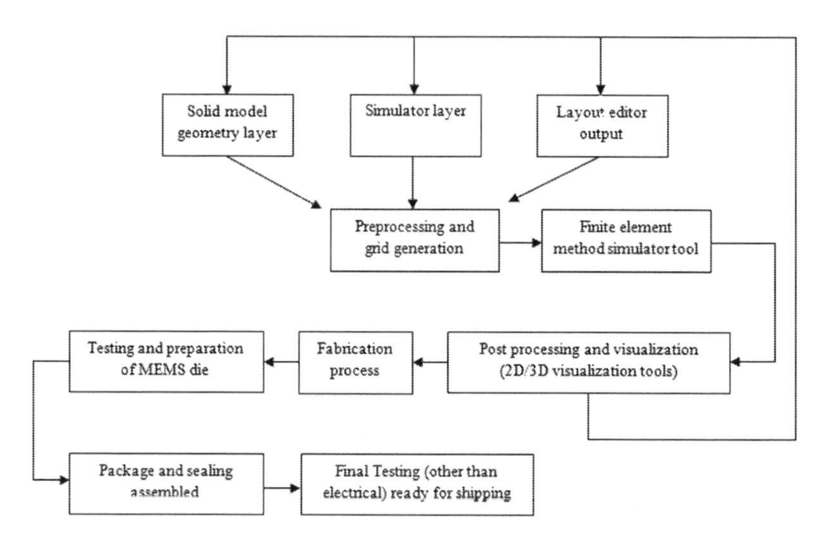

FIGURE 15.1 MEMS development.

(FEM) analysis, a breakthrough can be achieved, which enables rapid and efficient growth in manufacturing with mathematically accurate tools. These tools provide an optimization option for the researcher to develop an advanced application from the currently existing one [5], which is shown in Figure 15.1.

15.1.3 Photolithography

In modern semiconductor manufacturing, photolithography is a process that uses optical or UV light beam to pattern the photomask on the surface of a silicon substrate using photoresist materials.

Photolithography involves the following steps:

i. Superficial cleaning of wafer
ii. Barrier oxide layer formation
iii. Deposit of photoresist (spin coating)
iv. Soft bake
v. Photomask align
vi. Exposure
vii. Develop
viii. Hard bake
ix. Pattern transfer
x. Stripping
xi. Wafer cleaning and contaminants.

15.1.3.1 Superficial Cleaning of Wafer

In the first step, if any traces of disinfectants, ionic, and contaminants exist on the wafers, usually they are being cleaned up by some wet chemicals such as hydrogen peroxide (H_2O_2), trichloroethylene, acetone, or methanol [6].

15.1.3.2 Barrier Oxide Layer Formation

After cleaning, a very common step before spinning on a resist is the formation of a thin oxide layer on a wafer surface that is then subjected to thermal heating between 900°C and 1,200°C, which serves as a mask for the subsequent etching.

15.1.3.3 Deposit of Photoresist (Spin Coating)

A photoresist, sensitive to the ultraviolet radiation, is firmly placed on the wafer surface. Then, the substrate or wafer is held tightly on a spinner vacuum coater, and the resist is spread over the surface with a uniform thickness. Henceforth, the spinner is allowed to rotate at a very highly controlled speed ranging from few rpm to 8,000 rpm with a time limit of 15–30 seconds. As a result, a desired thickness of the film (1–10μm) is obtained, which depends on the viscosity of the resist material.

15.1.3.4 Soft Bake

The photoresist contains around 15% of solvent with additional built-in stress after spin coating. So, wafers are soft-baked at 75°C–100°C to evaporate the coating solvent and remove the built-in stress. By doing so, a strong fix of the resist layer is ensured onto the wafer surface. Commercially, hot plates are faster, are more controllable, and don't trap solvents.

15.1.3.5 Photomask Align

Once the soft baking process is done, a photomask (a glass plate) is placed with a metal film pattern on the one side aligned onto the resist surface directly. Here, UV lamp produces high-intensity UV radiations and causes the metal film pattern to transport to the exposed wafer surface. Basically, three aligners primarily used for the exposure process are mentioned in Figure 15.2, which are as follows:

i. Contact aligner
ii. Proximity aligner
iii. Projection aligner.

15.1.3.6 Exposure

In the simplest form, the exposure system, so-called UV lamp, lights up resist-coated wafer through a photomask, which is sensitive to the UV radiation. The foremost aim of the exposure approach is to transfer light to the wafer (substrate) to have a proper intensity and well-defined directionality, and to possess spectral characteristics and uniformity throughout the process [7]. As a result, an image can be achieved and patterned with greater resolution without any abnormalities.

15.1.3.7 Develop

High-energy radiations such as UV rays, electron beam, and X-rays are applied to a light-sensitive photoresist where it gets exposed completely. A developing solution such as sodium hydroxide (NaOH) or tetramethylammonium hydroxide (TMAH) is used to remove loosely bound photosensitive polymers. Generally, in the case of positive photoresist, the portion strike by the light becomes soluble and dissolved. In the case of negative photoresist, the portion strike by the light will not be dissolved and stay. The development process is shown in Figure 15.3 with both positive and negative photoresist.

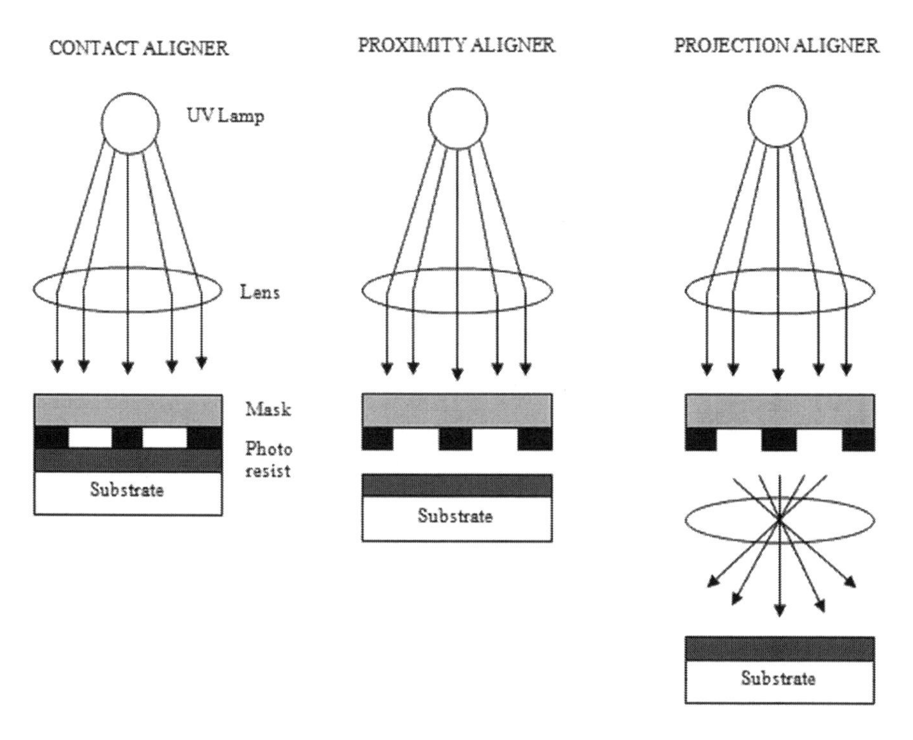

FIGURE 15.2 Mask alignment process.

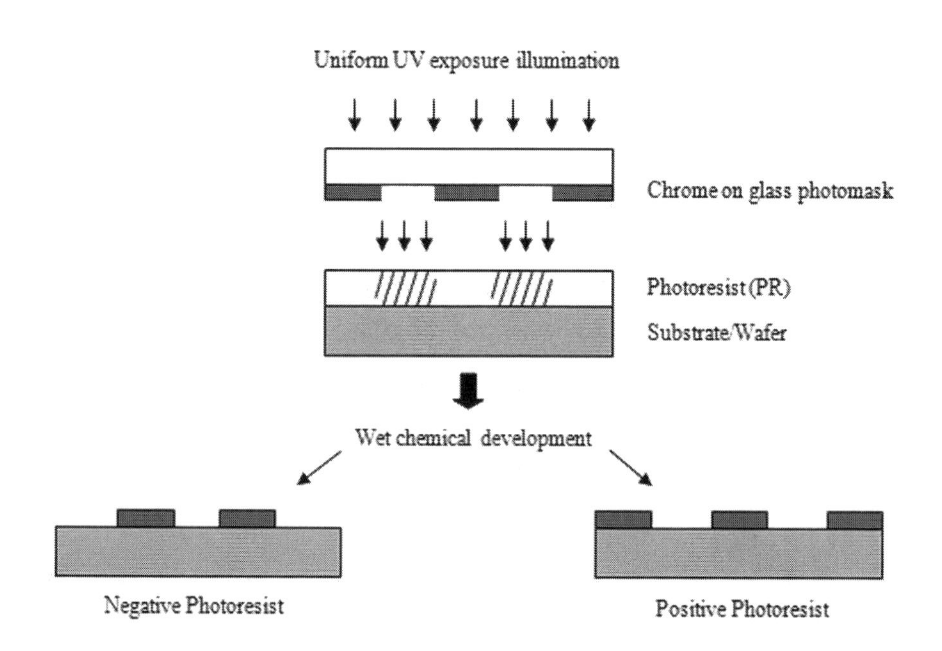

FIGURE 15.3 Development process.

15.1.3.8 Hard Bake

After development process, the photoresist is required to be baked again at an elevated temperature for a longer time period. This step is known as hard baking, to remove the persistent solvents on the wafer. Hard baking is aimed at strengthening the adhesion of the photoresist to the wafer even stronger than before. However, it improves some of the features such as thermal, chemical, and physical stabilities of the developed structure of photoresist, which are proven to be the most essential parameters. Depending upon the nature of the material, the hard bake process will extend further.

15.1.3.9 Pattern Transfer

With the help of lithography process, the subsequent patterns are now printed on the photoresist layer. These patterns need to be moved efficiently to the underlying substrate. It can be done through a few approaches as mentioned below:

i. Etching
ii. Deposition
iii. Ion implantation.

Etching is the most common pattern transfer approach. A uniform layer of the material to be patterned is deposited on the substrate. After the completion of the etching process, the left resist is taken away and leaves the desired etched pattern into the deposited layer.

15.1.3.10 Stripping

In lithography, photoresist stripping method is considered to be almost the last step in achieving the product physically. Basically, stripping phenomenon momentarily depends on wet and dry stripping. Acetone is one of the best examples of an organic stripper. Laboratories are meant for performing the experiments where a layer of dirt (scum) in the form of residues is coated onto the wafer surface. Commercially, to avoid the formation of scum, phenol-based organic strippers are used at a very high temperature and to progress with the semiconductor processing.

15.1.3.11 Wafer Cleaning and Contaminants

The most essential step in the preparation of wafers is to clean different types of contaminated particles available on the substrate such as dust, skin flakes, bacteria, solvents, chemicals, atmosphere, moisture, photoresist residue, and metallic. It's a deep concern to the user about the electronics device failing sooner than expected.

15.1.4 Role of Thin Films in Sensor

Thin film sensors are developed with more sensitivity, and there are challenges that are encountered during fabrication. The thickness of thin films holds a crucial role in the development of various sensor applications. This technology emerges into creating and developing primarily for building up the ICs and brings a revolutionary change in the field of microelectronics, communication, optical, deposit of all kinds, and energy harvesting strategies [8]. Due to the thin nature of films, they can be deformed more easily and are more sensitive to changes than bulk materials, and the minute changes in a film's conductivity and/or resistivity are more apparent.

15.1.4.1 Types of Thin Film Sensor

Depending upon the requirements in almost all fields, the composition and the properties of the thin films are varied to have many sensor applications [9]. There are different types of sensors where thin films are incorporated:

i. Gas sensors
ii. Strain sensors
iii. Heat flux sensors
iv. Humidity sensors
v. Corrosion sensors.

15.1.5 Categorization of Thin Film Deposition

Solid-state technology acquired the process of conversion from mechanical parts to semiconductor, which provided a scope for the inclusion of thin film deposition process. The film deposition technique should be either a physical or chemical process with different liquid or gas phases. Thin film deposition techniques have been categorized into two types: physical and chemical.

15.1.5.1 Thin Film Deposition Processes

MEMS devices are fabricated similar to the process used in manufacturing ICs. Generally, fabrication requires some processes such as lithography, etching, plating, and thermal oxidation of semiconductors, ion implantations, and deposition. Moreover, the deposition of thin film layers requires being more concise focusing on the development of thick layers of about 100 μm or ranging to a few nanometres. Apparently, for MEMS devices manufacturing, thin films carry different activities such as thickness layers of metals, insulators, and semiconductors along with the deposition processes [10].

Various thin film deposition techniques are used, which include:

i. Spin film technique
ii. Oxide layer growth (thermal)
iii. Chemical vapour deposition (CVD)
iv. Physical vapour deposition (PVD)
v. Electrodeposition process.

The main objective of deposit of thin films is to classify the deposition techniques used in the fabrication of microsystems. It seeks the recent development by virtue of meaningful research done globally through any of these deposition processes, which is being contingent on the material used, thickness of the film, and the designed structure being fabricated. Thin films have different functionalities during the fabrication of microsystems, which rely on the definite thickness and verified composition for the application within the device. Some distinctive layers are defined in Table 15.1.

TABLE 15.1
Type of Layers and Their Utilizations

Type of Layers	Utilization
Structural layer	Used to deposit a silicon dioxide by LPCVD process
Sacrificial layer	Sandwiched between structural layers for the separation of mechanical parts
Conductive layer	Electrical breakdown occurs in metal layers
Insulating layer	Formation of barrier in electrical components
Protective layer	Used for the prevention of corrosion through the layer or device
Etch stop layer	Used to stop etching process placed underneath the etched material
Mask layer	A photoresist film is patterned on a thick material

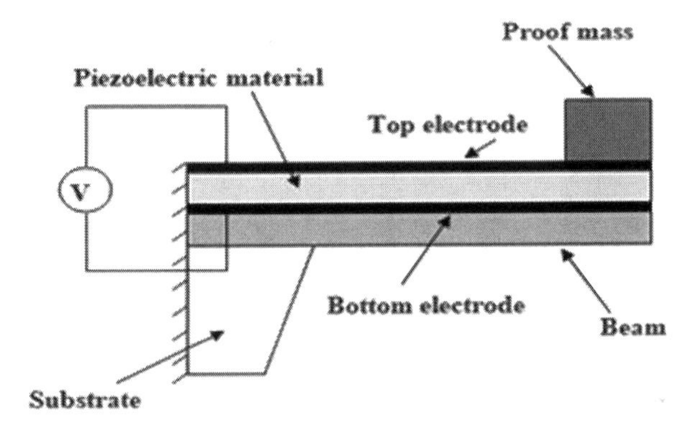

FIGURE 15.4 Unimorph cantilever.

Then, during fabrication of MEMS devices, different thin film layers are used for different applications.

15.1.5.2 Deposition Overview for MEMS

Figure 15.4 shows a schematic of a cantilever beam loaded with a piezoelectric material sandwiched between two top and bottom electrodes. Here, a proof mass is placed at the tip of beam where certain pressure is applied and results in the vibration of the piezoelectric material providing the desired output, i.e., voltage. Generally, thin films are required to deposit in a layer of some thickness through deposition techniques. Various deposition processes for microsystems are as follows:

i. Spin-on film
ii. Thermal oxidation
iii. CVD
iv. PVD
v. Electrodeposition.

15.1.5.2.1 Spin-on Film

It is the process of coating a few amounts (mL) of liquid that is dispensed or placed on a wafer surface or substrate through a spinning process at a rotational speed of about 1,000 rpm. The film thickness is dependent on the liquid density and its rotating velocity as shown in Figure 15.5. Now, the liquid is spun onto the wafer surface due to the centrifugal force, thus resulting in uniformity of the resist coating throughout the substrate. Spin-on film deposition occurred primarily beneficiary for resist material and spin-on glass (SOG).

15.1.5.2.2 Thermal Oxidation

In the process of fabricating miniature devices, thermal oxidation is considered as a chemical process, which gives rise to the uniform growth of highly graded quality silicon dioxide (SiO_2) layer in an ambient with elevated temperature onto the silicon substrate. Thermal oxidation process creates a huge difference in deposition of the silicon dioxide layer from many other existing deposition techniques without the occurrence of any reaction with the surface molecules. In IC technology, the silicon dioxide is used as an insulator material for different purposes:

i. Used in semiconductor fabrication techniques
ii. Used as an insulator in IC technology
iii. A 3-D device in microsystem
iv. Gate dielectric in electrical components
v. Ultrathin layer in CMOS devices
vi. MEMS sensor as a sacrificial layer.

15.1.5.2.2.1 Thermal Oxidation Process In this oxidation process, a silicon substrate at a very high temperature exposed to oxygen which eventually chemically oxidizes causes a formation of silicon dioxide (SiO_2) layer on the silicon surface. The grown layer of silicon dioxide (SiO_2) provides a measurement of its thickness due to the controlled temperature, process time consumption, and available amount of oxygen present in the surrounding.

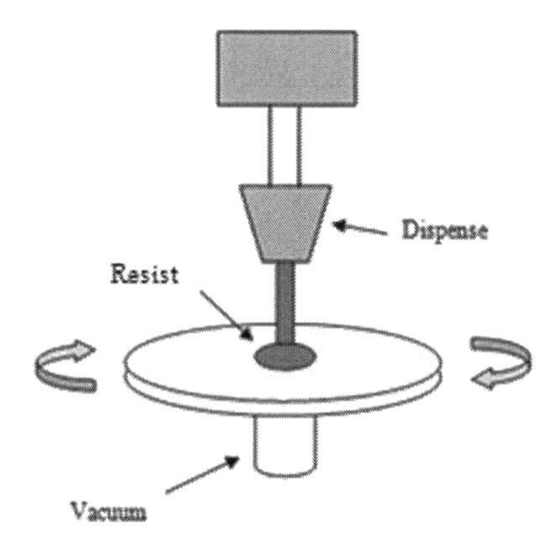

FIGURE 15.5 Photoresist film deposition on a substrate.

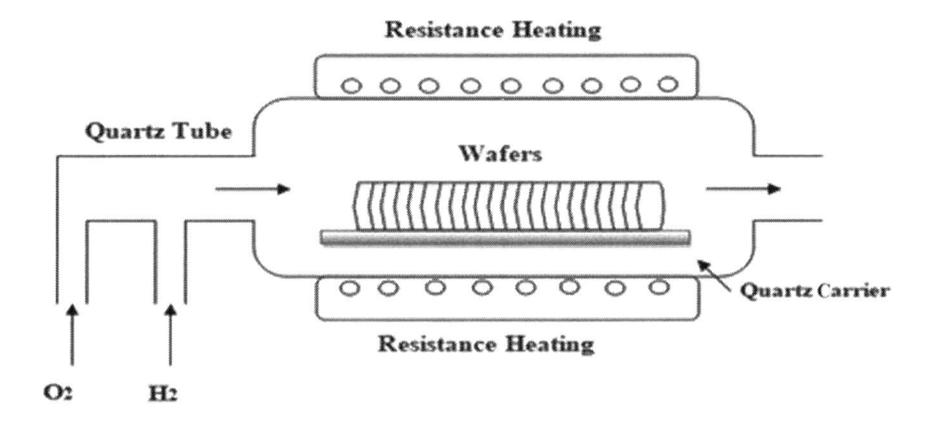

FIGURE 15.6 Thermal oxidation furnace.

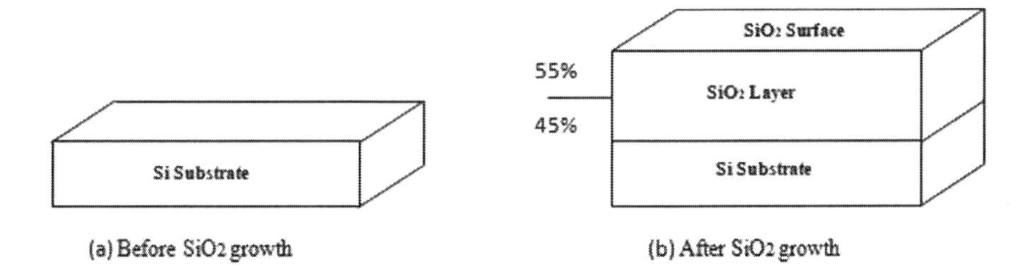

FIGURE 15.7 Thermal oxidation.

Usually, steam as an oxidizing agent is used in this process, known as wet oxidation. On the other hand, pure oxygen is used for the process, considered as dry oxidation [11]. A model was proposed by Deal and Grove in 1965 known as a linear parabolic model, which is a thorough representation of the oxide growth on silicon semiconductor. Here, a thermal oxidation furnace is shown in Figure 15.6, in which the process of thermal oxidation is carried out through some methods.

At first, silicon wafers are inserted inside the furnace comprising high temperature ranging from 800°C to 1,300°C through resistance heating process. A quartz carrier is available to hold the silicon wafers tightly inside the chamber. Then, from the quartz tube, a source of oxygen and hydrogen gas is allowed to pump inside the hot furnace. As a result, the oxygen molecules accelerate high and react with the silicon substrate or wafers to form a silicon dioxide (SiO_2) layer, also known as sacrificial layer. The higher the temperature and humidity, the faster the reaction. According to the Deal–Grove model, the reaction between oxygen and substrate takes place at the interface of the silicon oxide. Consequently, the thickness of the silicon layer is absorbed around 45% of the entire oxide thickness with the oxide growth depending upon the densities and molecular weights of silicon and silicon dioxide. The interface of the silicon oxide shifts deep inside the silicon substrate during the reaction as shown in Figure 15.7.

Thermal oxidation based on heating is broadly classified into two categories:

i. Hydrothermal wet oxidation
ii. Dry oxidation.

The hydrothermal wet oxidation process is treated as a process where the dissolved elements present in the water get associated with the oxygen, which acts as an oxidizing agent. The entire system is maintained at certain pressure to prevent excessive liquid evaporation.

$$\text{Si}(\text{solid}) + 2\text{H}_2\text{O}(\text{vapor}) \rightarrow \text{SiO}_2(\text{solid}) + 2\text{H}_2(\text{gas}).$$

On the other hand, the dry oxidation uses dry oxygen that is pushed directly to the chamber having high temperature, which physically reacts with the solid silicon material to form the silicon dioxide (SiO_2) layer.

The grown silicon dioxide layer is useful in the manufacturing of transistor gates and capacitor devices.

$$\text{Si}(\text{solid}) + \text{O}_2(\text{gas}) \rightarrow \text{SiO}_2(\text{solid}).$$

To achieve a faster oxide growth, H_2O is much more preferable than O_2. Dry oxidation is used to grow thin layers of nanometre size and possess a well-defined control over the thin oxides growth.

15.1.5.2.3 Chemical Vapour Deposition

CVD is referred as a process technology mostly utilized in the deposition of films to obtain high-quality refined solid materials with speedy performance. In this regard, it has not only restricted to involve in solid thin film coating but also produces bulk materials, powders, various composite materials, and fibres [12]. CVD is basically defined as one of the deposition techniques in which a solid material is deposited on a hot surface chemically reacted to convert into vapour phase.

15.1.5.2.3.1 Chemical Vapour Deposition Process The CVD system comprises three subsystems:

a. Pumping reactants inside the heating chamber
b. Gas exhaust from the vent system
c. Resistance heating source.

In Figure 15.8, sliced wafers are placed on a wafer stand inside a reaction chamber. As per the requirement, temperatures and pressure inside the chamber are set and programmed. Selection is done for the reactants (inert gases) to introduce to the system. The selected gases are allowed to move onto the placed wafer surface, which causes a reaction between them at an elevated temperature. The gas molecules further react chemically and form a solid thin film layer onto the wafer surface, termed as adsorption process. The entire procedure that occurs at high

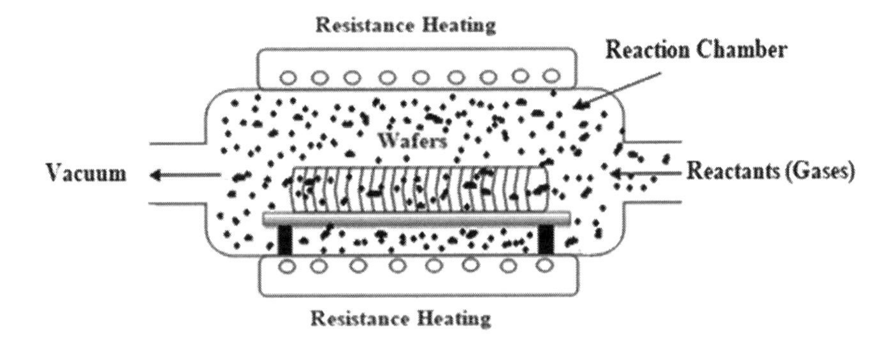

FIGURE 15.8 CVD chamber.

temperature produces by-products inside the reaction chamber that are expelled through a vacuum outlet as exhaust. This results in achieving the film layer thickness, which depends on certain parameters such as pressure, temperature, and high reactants concentration.

The most common CVD techniques used for the MEMS fabrications are as follows:

- LPCVD (low pressure CVD)
- APCVD (atmospheric pressure CVD)
- PECVD (plasma-enhanced CVD).

15.1.5.2.4 Physical Vapour Deposition

PVD technique is considered as a process that describes a vacuum deposition technique which supports the fabrication of various films and coating of the material surface. It utilizes certain methods in which the desired material is vaporized, changing its state from liquid level to gas level. PVD has been utilized a lot in industries, thus giving rise to many consumers' applications with an enormous advent of technology. There are two major types of PVD processes that exist and are used in microsystem fabrications:

i. Sputtering
ii. Evaporation.

15.1.5.2.4.1 Sputtering Sputtering is a kind of phenomenon which ejects the molecules from the surface of the solid materials bombarded by the high-energy particles of the plasma or gas tends to accelerate the ions. The sputtering process initiated when a substrate is placed in a vacuum chamber coated with metallic and polymeric thin films that come in contact with the argon gas (inert gas). The sputter deposition is shown in Figure 15.9.

The main objective of sputtering is to use the plasma energy which is partially ionized in nature on the target surface (cathode) and bring down all individual atoms of the specific material to get deposited on the substrate. Here, a substrate is mounted

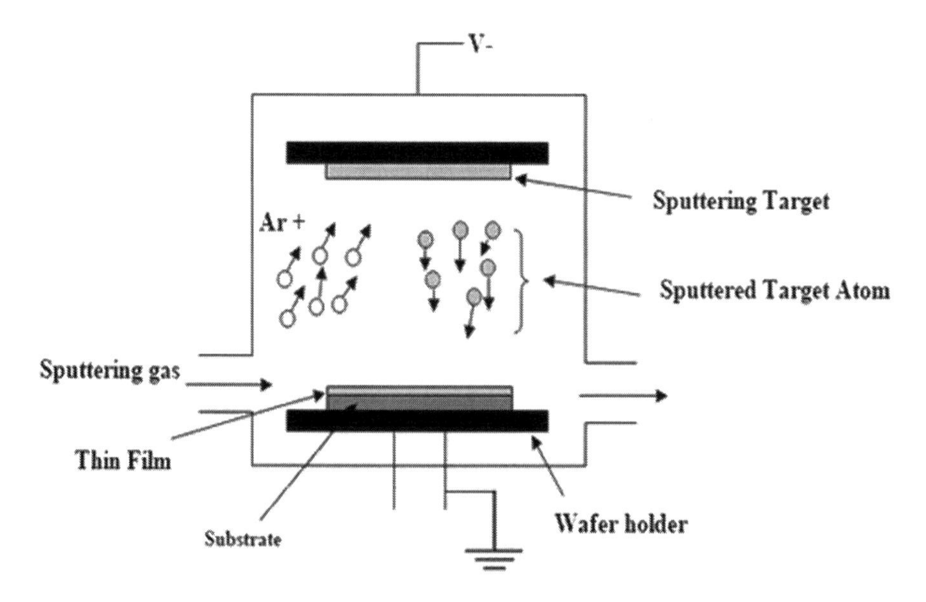

FIGURE 15.9 Sputter deposition: an overview.

inside of the vacuum chamber comprising inert gas, i.e., argon gas. Using this gas, plasma is created by a RF power source (pulsed DC). This causes some of the gas molecules to lose their electrons and become a positive ion. These ions travel towards the target (cathode), which is a negative end and ionized heavily. The ionized molecules hit the target material by bombarding at a high energy. This energy is transferred to breakdown the target atoms by converting them into vapour phase, which in turn carry enough energy to be projected onto the substrate. Several types of sputtering processes are available:

i. Direct current (DC) sputtering
ii. Radio frequency (RF) sputtering
iii. Magnetron sputtering
iv. High power impulse magnetron sputtering (HIPIMS) sputtering.

15.1.5.2.4.2 Evaporation In PVD evaporation, a thin film material is used as the source which tends to convert its solid state to vapour state when subjected to high thermal energy. The evaporation process is carried out under a high vacuum zone, ensuring reduction in the collision amid atoms or molecules which results in vapour expansion to occupy a large volumetric space inside the chamber, coating all surfaces, including the substrate. Once on the substrate (or any surface), the vapour condenses forming the desired thin film. In order to progress with the process of evaporation, a wafer (substrate) is chosen and kept inside the vacuum chamber tightly held by a wafer holder. Facing the substrate, a source material is placed from which vaporization occurs later [13]. Initially, a high temperature is applied to the source material which specifically changes from solid state to liquid and finally to vapour

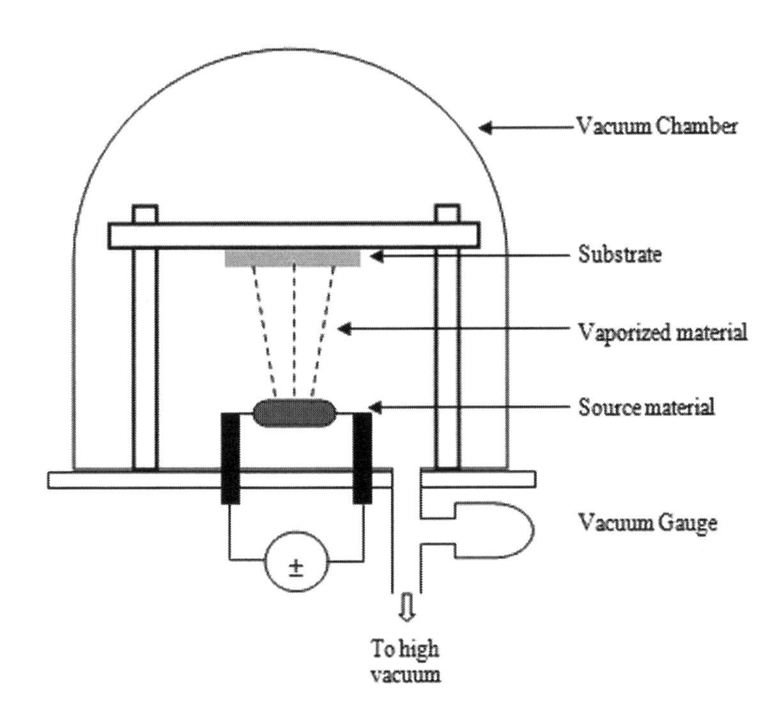

FIGURE 15.10 Thermal evaporation process.

state and evaporates. The evaporation is associated with the expansion of particles at the atomic or molecular level in order to fill the volume of the chamber, as shown in Figure 15.10.

15.1.5.2.5 Electrodeposition (also Known as Electroplating or Electroforming)

Electrodeposition is a process of coating a thin metallic layer on a conductive object mostly by the action of electrical current also known as electroplating, or to coat and fill a micro-sized cavity with metal (electroforming) [14]. It is the most suitable deposition technique in the world of nanostructure, in which thin films are produced abundantly. Now, it plays a vital role in the metallic deposition applied to the semiconductor materials.

Electroplating makes its use in different application areas such as in bicycle to prevent rust, tin cans, design of ornaments (gold and silver), in ICs for conduction, and LPG stoves. During the fabrication of electronic circuits, in order to have good conductivity among the components on the board, thin film metallic lines are prepared using metals such as copper, gold, platinum, and nickel. The film thickness ranges from less than 1 to 100 μm. The Lithography galvanformung abformung (LIGA) process uses electroforming for the construction of devices with very high aspect ratios, i.e., ratios of 100:1 or greater.

Figure 15.11 illustrates the typical electrodeposition process in which a current source in the form of a battery or any low-voltage dc source is used to provide a necessary electric current. The counter electrode and the wafer are immersed in

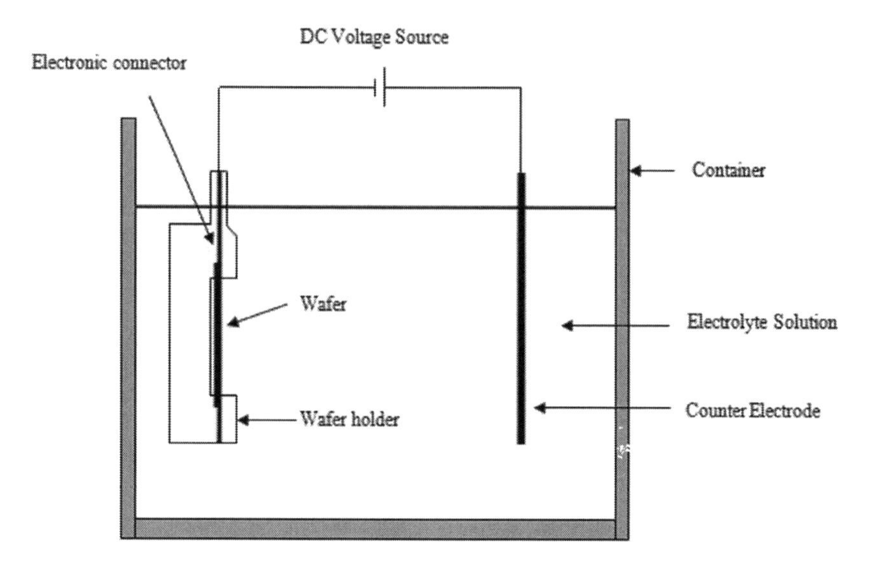

FIGURE 15.11 Electrodeposition set-up.

the electrolyte solution connected to the two ends of the DC voltage source [15]. The wafer to be coated with metallic layer is connected to the negative terminal of the power source through an electrical connector and becomes a negative electrode (cathode). The counter electrode behaves as a positive electrode (anode), which is connected to the positive terminal of the power source, thus completing the entire electric circuit arrangements. When the direct current is supplied to the electrolyte, a chemical reaction occurred at the electrode terminals placed at some distance. The anode terminal releases metal ions and the ions are transferred to the wafer (cathode) end. The chemical substance carries the ions from one electrode to another depending upon their polarities. The transportation of charged particles will continue as long as supply exists. Faraday's law of electrolysis states that the amount of electricity used is directly proportional to the material deposited on the electrode. Thus, the wafer will have a very thin metallic deposition until the process ends.

15.1.6 Technological Development in the Field of Thin Films

Technological development in thin films enables new devices and in the manufacturing of sensors. Materials like ceramics, metals, or polymers are used in thin film technology that could be applied on the substrate directly or in some other means. However, the fundamental materials that are chosen for the wafer (substrate) materials are silicon, steel, and glass. These substrate materials allow various deposition phenomena and enhance the substrate properties customized further to meet the requirements of manufacturing specific application-oriented devices [16]. Here, thin film technologies are applicable to flat-made substrates and those having many-faceted geometrical patterns. More focus is needed in miniaturization to build low-cost instruments. Thin film technology help in developing and constructing different devices with the deposition process for many applications usually by lithography tools and etching. Furthermore, the level of packaging MEMS devices also helps to produce the best thin film devices. As a result, thin film techniques render an innovative advancement in the field of science and technology with the leading-edge research work.

15.1.7 Future Perspective of Microelectromechanical Systems (MEMS)

MEMS or microsystem is a kind of technology that produces miniature device which incorporates sensors and actuators on a single chip with multifunctional characteristics. It can be restructured and down scale to very miniscule dimensions less than the size of a human hair. Generally, MEMS is a leading-edge mechanism used in various fields such as automotive, biomedical, and electronics, and serves a better option for the preparation of sensors in a bulk manner lowering down the cost efficiently. It drives a newer age of technology transforming to the digitization. In the present era, semiconductor industries find the opportunity to explore the technology globally. It advances more than 70% in manufacturing the devices at consumer and automotive end products almost grow double within a past few decades. In IC market, devices such as radio frequency MEMS, resonators, filters, and image sensors are integrated, which imprints the accountability of microsystem.

Big companies such as Google, TCS, and ADANI are drafting their business solution towards this critical technology. In general, MEMS is powering with new application and action in goods transportation, security system, health, housing the projects like smart city and industry. With the advent of MEMS technology, sensors outcome the data that can be processed in a qualitative manner, and also the information collected about the mankind and the surrounding has a huge impact in all aspects.

Lab-on-chip is a kind of biochip which gathers useful information during the reaction occurred among DNA, and biochemical reagents on testing surface like glass and plastic plates. It is clear that the term 'lab-on-chip' is to place the entire process of a laboratory onto a single chip-lab-on-chip, which has been developed for many applications and is now used for medical diagnosis of SARS, leukaemia, breast cancer, dipolar disorder, and several infectious diseases. With the help of MEMS, these devices now work with less power loss and can be integrated with other electronics devices, and can enhance the performance and reliability of these products while decreasing the size and the cost related to these devices. From the above given data, it can be concluded that MEMS technology has changed various markets and also transforming them currently like the automotive market, IT peripherals, telecommunication, medical, electronics, industry process control, and household.

15.1.8 Summary

This chapter mainly focuses on the persistent role of a thin film technology in device application based on MEMS technology. Almost each domain scales down with the miniaturization of devices and their cost by successfully employing thin film upgraded technology. This technology provides an effective ambience of research opportunities running through the verifiable facts applied in various challenging environments. Perhaps, predictions have already been made to resolve the issues currently reflected in the science world so far from the diversified fields of micro-/nanotechnology. At present, thin film devices are widely used in the field of electronics, medical, environmental predictor sensors, defence sectors, and many more yet to step in for the technological development of the society.

Therefore, the rapid and consistent growth of MEMS technology becomes the key factor for the achievement of miniature devices associated with a lot of application areas satisfying the criteria that enhance the usage of thin films coated over more devices. This uplifts the development of newer devices globally accepted with collaborative research.

REFERENCES

1. Schaller, R. R. 1997. Moore's law. Past, present, and future. *IEEE Spectrum*. 34(6): 52–59.
2. https://www.yumpu.com/en/document/view/4673421/emerging-markets-for-microfluidic-applications-i-micronews
3. Longden, A. C. 1900. Physical review. *Series I*. 11: 40.
4. Madou, M. J. 1997. *Fundamentals of Microfabrication*. Boca Raton, FL: CRC Press.
5. https://iopscience.iop.org/article/10.1088/1742-6596/34/1/041/pdf
6. https://www.azosensors.com/article.aspx?ArticleID=1431
7. Sitti, M., et al. 2016. Stretchable, skin-mountable, and wearable strain sensors and their potential applications: a review. *Advanced Functional Materials*. doi: 10.1002/adfm.201504755
8. https://www.intechopen.com/books/thin-film-processes-artifacts-on-surface-phenomena-and-technological-facets/
9. Eranen, S. 2015. Thin films on silicon. *Handbook of Silicon Based MEMS Materials and Technologies*. 124–205. doi:10.1016/b978-0-323-29965-7.00006-3
10. Creighton, J. and Ho, P. 2001. Introduction to chemical vapor deposition (CVD). *ASM International*. 1–11.
11. Bryzek, J., Peterson, K., and McCulley, W. 1994. Micromachines on the march. *IEEE Spectrum*. 31(5): 20–31. doi:10.1109/6.278394
12. Ehrfeld, W. and Lehr, H. 1995. Deep x-ray lithography for the production of three-dimensional microstructures from metals, polymers and ceramics. *Radiation Physics and Chemistry*. 45(3): 349–365.
13. Schmaljohann, F., et al. 2015. Thin film sensors for measuring small forces. *Journal of Sensors and Sensor Systems*. 4: 91–95. doi: 10.5194/jsss-4-91-2015
14. Reiner, L. and Bavarian, B. 2007. Thin film sensors in corrosion applications. *CORROSION 2007*, 11–15, Nashville, Tennessee.
15. Schneider, K., et al. 2015. VOx thin films for gas sensor applications. *Procedia Engineering*.doi:10.1016/j.proeng.2015.08.1009
16. Herner, S. B. 2018. Application of thin films in semiconductor memories. *Handbook of Thin Film Deposition*. 417–437. doi:10.1016/b978-0-12–812311-9.00013-x

16 Structural, Optical, and Dielectric Properties of Ba-Modified $SrSnO_3$ for Electrical Device Application

Aditya Kumar, Bushra Khan, and Manoj K. Singh
University of Allahabad

Upendra Kumar
Banasthali Vidyapith

CONTENTS

16.1 INTRODUCTION

Perovskite oxide has a general chemical formula ABO_3, where A and B are the cations whose valence state lies from +1 to +3, and +3 to +5, respectively [1]. The valence state of A and B has been selected in such a way that the total ionic charge must be equivalent to +6 and compensated by ionic charge by oxygen [2]. The alkaline earth-based stannates ($MSnO_3$) compounds belong to the giant dielectric materials, which are used in electronics industry for thermally stable capacitor applications [3,4]. $SrSnO_3$ (strontium stannate) is widely investigated as a humidity sensor application [5], similar to other investigated stannates, and also matched with the other appropriate oxides in terms of capacitive sensors for the detection of carbon dioxide and other gases [6]. $SrSnO_3$ mixed with different concentrations of WO_3 exhibits higher gas sensitivity for temperature >773 K and works as a capacitive-type sensor to detect NO and gasoline [7]. The Hall measurements taken on $BaSnO_3$ explain the existence of defects in the material [8]. Moreover, the optical band gap of stannates group has been observed to be in the range of 3.0–4.5 eV [9]. The semiconducting behavior of materials has been widely acceptable for gas-sensing applications. The solid solutions of titanate in alkaline earth stannates are extensively used for barrier layer and thermally stable capacitor applications [10,11]. The high temperature used in the preparation of ceramics leads to larger crystallite and the formation of different grain sizes [12]. The capacitive behavior of the ceramics is tuned by varying crystallite size and grain size and by doping of different kinds of homovalent and heterovalent substitution [13]. However, sintering of the ceramics increases the density above 90% and results in better performance of capacitive behavior [14]. In comparison, the loosely sintered particles or thin films provide better response for the application of the sensor. As the particle size or grain size has been found to be lower, it increases the reactivity of surface area, which makes easier percolation of gas particle into bulk [10,15]. So one of the motives in the modification of sensor material is to increase the porosity of the system.

To use the ceramics for barrier layer capacitor, the processing parameters play a crucial role in materials synthesis. During sintering process, with approaching sintering temperature, the materials should oxidize, while reoxidation process takes place on cooling from the firing temperature into the product. Since during the reoxidation process of grain, the time is not sufficient, and therefore, only the grain boundaries reoxidize and grow into insulator. This results in a barrier layer at the grain/grain boundary interfaces, which results in the high value of dielectric constant [16].

Another approach to synthesize the barrier layer capacitor with high dielectric constant is the modification of grain by doping of homovalent substitution. The substitution of Ti^{4+} at Sn^{4+} site of $BaSnO_3$ has been investigated as a ferroelectric relaxor and thermally stable capacitor [17,18]. At the Ba site of $BaSnO_3$, Ca replacement is also investigated as an option for thermally stable capacitor and gas sensor applications [19]. The basic idea behind the application of perovskite is the modification of grains by creating microheterogeneity in samples. In case of homovalent substitution, the higher ionic radii as well as lower ionic radii displaced at host lattice sites create lattice strain, as well as microheterogeneity in sample, which results into small nanopolar regions. These nanopolar regions also result to high value of dielectric constant [20]. Based on the literature, the Ba^{2+} has been selected as dopant at Sr^{2+} site of $SrSnO_3$ for this chapter. The difference in the ionic radius of host and dopant is within the range of ±15%, according to Hume-Rothery criterion. Therefore, Ba can be used as substituted for Sr site in $SrSnO_3$.

In this chapter, Ba-doped $SrSnO_3$ samples were synthesized using sol–gel chemical route followed by calcination at 1073 K and sintering at 1173 K. X-ray diffraction (XRD) analysis was used to identify the phase, and the crystallite size and lattice strain of the samples were further calculated from XRD data using Williamson–Hall (W-H) plot. Furthermore, the optical and dielectric properties of the samples have been thoroughly investigated in this chapter.

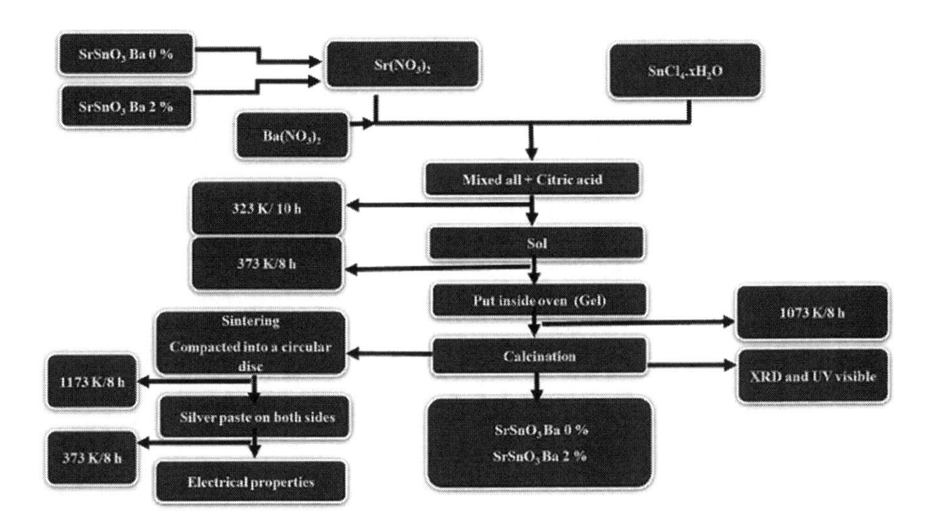

FIGURE 16.1 Flowchart of the steps involved for the synthesis of samples.

16.2 EXPERIMENTAL PROCEDURE

The nitrate of strontium ($Sr(NO_3)_2 \cdot 4H_2O$), barium $Ba(NO_3)_2$ (Alfa aesar, 99%), and chloride of tin (iv) ($SnCl_4 \cdot xH_2O$, Alfa aesar, 98%) were used as starting materials for this work. Citric acid ($C_6H_8O_7$) and deionized (DI) water were used as solvents for these materials, while polyvinyl alcohol (PVA) was used as binder.

In a typical procedure, stoichiometric amounts of strontium nitrate tetra hydrate, barium nitrate, and tin (iv) chloride hydrate were separately dissolved in 10 mL DI water. Next, these solutions were mixed into a solution at room temperature. Then, citric acid was added into the solution and the solution is stirred at 323 K for 10 hours to obtain the gel. Then, the obtained gel was dried in oven at 373 K for 8 hours. The gel was then directly transferred into muffle furnace and calcined at 1073 K for 8 hours. The furnace was cooled to room temperature and a white fine powder is obtained at the end of calcination process. These fine powders were compacted into a disk of thickness 1–2 and 10 mm in diameter by using a hydraulic cylindrical press. These pellets were put up at 1,173 K for 10 hours for sintering. Finally, these sintered pellets were coated with conductive silver paste on both the surfaces. Then, the pellets were dried at 373 K for 2 hours to remove moisture and projected for characterizations. Different protocol was adopted for preparing $SrSnO_3$, except for adding stoichiometric concentrations of barium nitrate in solutions. The flowchart of steps involved in the synthesis of sample is shown in Figure 16.1.

16.3 CHARACTERIZATION TECHNIQUES

The phase formation and structural determination of these samples have been analyzed by Bruker D8 advance (U.S.A.) X-ray diffractometer using Cu-Kα ($\lambda = 1.5418$ Å) radiation. Data of all these samples were collected in $20 \leq 2\theta \leq 80$ with a step size of 0.05° and in the presence of 40 kV and 30 mA of the accelerating voltage and applied current, respectively. Cary 4000 UV-Vis spectrophotometer has been used to record spectra in reflectance mode between the 200 and 800 nm wavelength range. Alfa-A high-frequency impedance analyzer Nova-Control was used to record data of the electrical measurements for these synthesized samples.

16.4 RESULTS AND DISCUSSIONS

In the present case, the samples $Sr_{1-x}Ba_xSnO_3$ with $x = 0.00$ and 0.02 were abbreviated as SBS0 and SBS2, respectively, during discussion.

16.4.1 Phase Determination Using X-ray Diffraction (XRD)

Phase formation and crystal structure determination of samples have been carried out using XRD technique. Figure 16.2 depicts the powder XRD profile of the samples. The entire diffraction patterns are well sharp, which indicates a the perfect crystalline nature of sample. The XRD pattern has been well matched to the theoretical crystallographic open database (COD) file no. 1533387, which belongs to the orthorhombic crystal structure [21]. The most intense peak observed for plane (200) at angular position 25.2° for sample SBS0 has been shifted towards lower angle at 25.1° for sample SBS2. The shifting of the most intense peak towards lower angle might be related to the higher ionic radii of Ba^{2+} (1.61 Å) compared to that of Sr^{2+} (1.44 Å) [22]. Further, in order to determine the lattice parameters of samples, **Unit Cell** software has been used. The X-ray density for samples is calculated using the following formula:

$$\rho = \frac{nM}{N_a V},$$

where, n, M, N_a, and V are the number of atoms/unit cell, atomic weight for desired compositions, Avogadro number, and volume of unit cell, respectively. The values obtained for the samples of the such as lattice parameters, volume, and density are

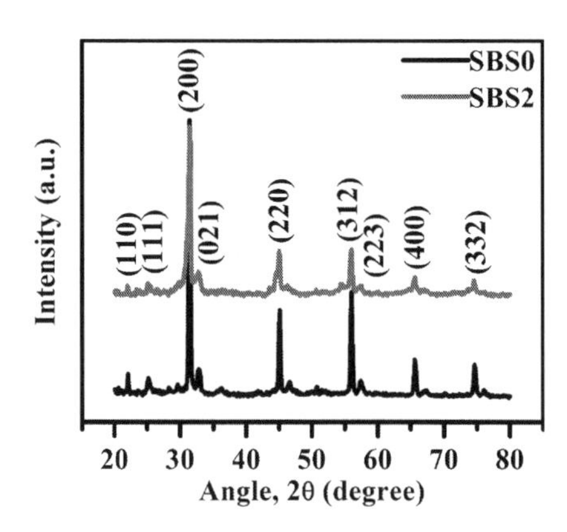

FIGURE 16.2 Powder X-ray diffraction (XRD) of the prepared samples.

TABLE 16.1
Lattice Parameters, Volume, Density, Crystallite Size, Micro-strain, and Optical Band Gap of Samples

Lattice Parameters	SBS0	SBS2
a (Å)	5.67933(5)	5.68112(3)
b (Å)	5.72829(4)	5.75486(8)
c (Å)	7.90664(7)	7.88200(2)
V (Å^3)	257.2256	257.6945
ρ (gm/cm^3)	6.56	6.58
Debye–Scherrer		
Crystallite size (nm)	25.4	32.07
W-H plot		
Crystallite size (nm)	34.62	43.89
Strain	2.80×10^{-3}	7.75×10^{-3}
Optical band gap	3.88 eV	3.97 eV

given in Table 16.1. From Table 16.1, it is noted that for Ba-incorporated samples, the values of lattice parameters and volume are found to be higher than undoped that might be due to the larger ionic radii of Ba^{2+}.

16.4.2 Determination of Crystallite Size

The broadening that occurred in XRD peak is associated with smaller crystalline size and contribution of micro-strain. The crystallite size of the samples is mathematically determined by the following formula [20]:

$$D = \frac{0.9\lambda}{\beta\cos\theta}, \tag{16.1}$$

where λ is the wavelength of Cu Kα radiation, β is the full width at half maxima (FWHM), and θ is the angular position of Bragg diffraction peak. Crystallite size for samples has been calculated for the most intense peak (200) and is indexed in Table 16.1. From Table 16.1, it was noticed that the crystallite size was observed to be increased with the addition of Ba relative to undoped. Since the value of crystallite size was not accurate in the present case due to the presence of micro-strain in sample. Therefore, it is important to separate the value of micro-strain from the XRD peak to calculate the exact value of crystallite size. A well-known relation has been used to determine the micro-strain present in the sample known as W-H plot described in the literature. According to this relation, the total width of XRD peak can be written as the sum of the contribution that occurs due to smaller crystallite size (β_D) and due to micro-strain (β_ε).

$$\beta = \beta_D + \beta_\varepsilon. \tag{16.2}$$

The values of β_D from Debye equation, and β_ε as $4\varepsilon \tan\theta$, were substituted into equation (16.2) and found to be [23]

$$\beta = \frac{0.9\lambda}{D\cos\theta} + 4\varepsilon\tan\theta \tag{16.3}$$

$$\beta\cos\theta = \frac{0.9\lambda}{D} + 4\varepsilon\sin\theta \tag{16.4}$$

The W-H plot for all the samples was generated using equation (16.4) and is shown in Figure 16.3. The linear relationship between $\beta\cos\theta$ and $4\sin\theta$ suggests that the straight line has slope ε and its intercept on *y*-axis is $\frac{0.9\lambda}{D}$. The intercept and slope value were used to determine the sample crystallite size with micro-strain, and are recorded in Table 16.1. From Table 16.1, it is again found that the size of the crystallite follows the same pattern as that obtained from the Debye equation. However, it is noted that the value obtained from the W-H plot is greater than the value obtained from Debye–Scherrer. It is found from equations (16.1) and (16.4) that the crystallite size is inversely proportional to the FWHM of XRD peak. Since the contribution of micro-strain is derived from the overall peak width, the average width is considered to be less than the overall width. Therefore, the value of crystallite size obtained after micro-strain correction is larger. The micro-strain of sample is found to be larger than undoped, which might be related to the larger ionic radii of Ba^{2+} and higher lattice volume of Ba-doped sample.

16.4.3 Study of Optical Properties

The UV-V is absorption spectroscopy has been used to study the optical properties of the samples. The absorption spectra of the prepared sample are shown in Figure 16.4a. It is found that in the UV region, the sample displays strong absorption,

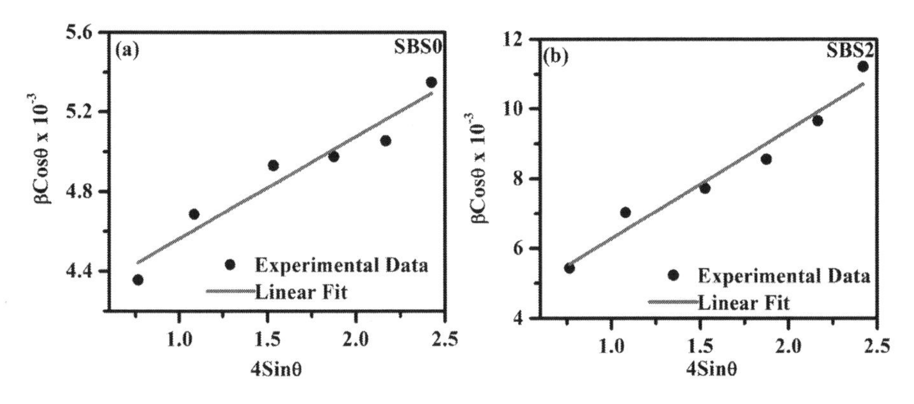

FIGURE 16.3 Williamson–Hall (W-H) plot generated using equation (16.4) from XRD data of samples (a) SBS0 and (b) SBS2.

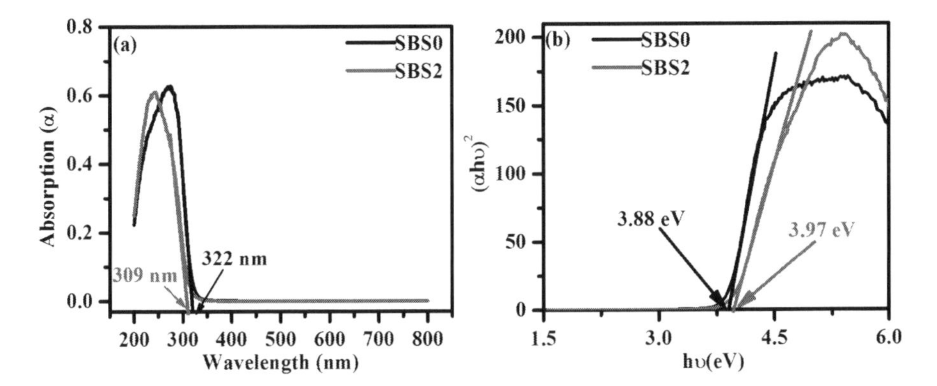

FIGURE 16.4 (a) Room temperature UV-visible spectrum of samples. (b) The Tauc plot generated using equation (16.5) for samples.

and translucent in the visible and near infrared region. The absorption band edge for the samples was determined using extrapolating linear curve of absorption to the wavelength axis. The value of band edge has been found to be 309 and 322 nm for samples SBS0 and SBS2, respectively. The absorption edge spectra clearly define the radiation absorbed between the highest occupied molecular orbital (HOMO) and the lowest unoccupied molecular orbital (LUMO), and corresponding energy known as band gap (E_g) [24]. The value of band gap is directly related to the wavelength of absorption by mathematical formula $E_g = hc/\lambda$ and calculated to be 3.86 and 4.02 eV for samples SBS0 and SBS2, respectively. With Ba incorporation in $SrSnO_3$ at Sr site, the band gap is found to be increased. The band gap of $SrSnO_3$ resulted due to overlapping of the 5s orbital of Sn and 2p orbital of O, and as Ba incorporated a repulsive force between the overlapped orbital acts which results in a slight increase in band gap [25]. Also, the amount of absorption in UV region was found to be decreased by Ba substitution, and such materials have their applications in filters and sensors for UV radiation.

Tauc's equation expresses dependency on photon energy in high-energy-absorption region. The absorption coefficient (α) and direct band gap (E_g) material are given according to the Tauc relationship [24]:

$$(\alpha h\nu)^m = B\left(h\nu - E_g\right), \tag{16.5}$$

where B, $h\nu$, and m are the independent energy constant, incident photon energy, and index parameter, respectively. The index parameter m decides the nature of electronic transition responsible for the optical absorption. Based on the literature, there are four values possible for m, namely, 2, 3/2, 1, and 1/2 which represent the direct allowed, direct forbidden, indirect forbidden, and indirect allowed transition, respectively. The $SrSnO_3$ is found to be a direct band gap semiconductor [9], so here Tauc plot was generated for direct allowed transition and is illustrated in Figure 16.4b. The optical band gap has been determined by extrapolating the linear curve of Tauc plot to energy axis. The experimental optical band gap was observed to be 3.88 and 3.97 eV, respectively, for the SBS0 and SBS2 samples. Thus, it is observed that the calculated band gap of SBS0 and SBS2 is roughly equal to what we received from the absorption spectra. The present value of optical band gap for sample SBS0 is found to be smaller than the reported value, which might be ascribed to the larger crystallite size of sample obtained in the present investigation [9]. The value of band gap obtained in the present case reflects the wide band gap semiconducting nature of samples.

By utilizing these states as metastable state, the present material can be explored for various semiconductor device applications.

16.4.4 Dielectric Properties

The dielectric properties of the sintered samples were analyzed as a function of the frequency from 500 Hz to 100 kHz and within temperature 25°C–500°C. Figure 16.5a and b depicts a variation in real part of dielectric constant for samples SBS0 and SBS2, respectively. The variation in real part of dielectric constant shows two regions in both curves: one is below 10 kHz and second above 10 kHz. The dielectric properties of ceramic materials resulted due to contribution of four types of polarizations, namely, interfacial polarization, orientational polarization, ionic polarization, and electronic polarization [20].

Interfacial polarization arises due to the presence of local microheterogeneity, which has two different values of conductivity and results in a very large value of dielectric constant. Normally, this kind of polarization was found to be active from 1 mHz to 1 kHz.

Orientational polarization arises due to the orientation of dipole (same kind of charge with opposite in nature placed apart formed a dipole) in the presence of small AC electric field. This kind of polarization was found to be active in the frequency range of 1 kHz to 1 MHz.

Ionic polarization is the process in which the ionization of atoms takes place (separation of charges resulting in the ionization of the atom) in the presence of small

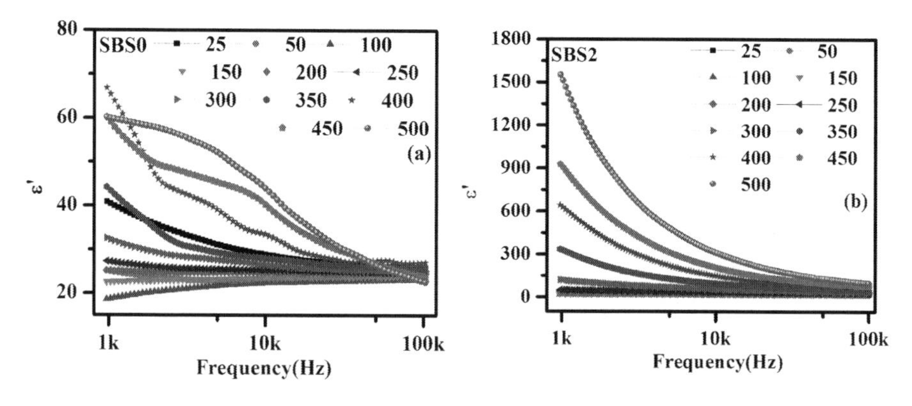

FIGURE 16.5 Variation of dielectric constant with frequency at different temperatures of (a) SBS0 (b) SBS2. The temperature shown in the box is represented by degree Celsius.

AC electric field. This kind of polarization is found to be active in the frequency range of 1 MHz to 1 GHz.

At last, the electronic polarization resulted due to motion of electrons in the presence of AC electric field. This polarization is found to be active in the frequency range higher than 1 GHz.

Based on the available frequency range, only two kinds of polarization were found to be active in the present sample, namely, interfacial polarization and orientational polarization. Since the present sample has been synthesized at 1,200°C, at this temperature the possibility of the oxygen loss can't be ruled out. The presence of oxygen vacancy can be understood as follows:

$$O_o \rightarrow \frac{1}{2}O_2(g) + V_o^{\cdot\cdot} + 2e'. \quad (16.6)$$

The presence of these electrons reduced the valence state of Sn^{4+} to Sn^{2+}, and can be understood by the following equation:

$$Sn^{4+} + 2e' \rightarrow Sn^{2+}. \quad (16.7)$$

In perovskite structure, the Ba occupied at Sr site (dodecahedral site), oxygen vacancy at face-centered (fcc) site, and Sn^{2+} at body-centered site. The occupation of Ba, Sn^{2+}, and oxygen vacancy is completely random process, so the micro-regions having different atoms at their respective site create microheterogeneities with a variation in their conductivity, which results in interfacial polarization [10]. At lower frequency, interfacial polarization is mainly dominant over others so it results in a large dielectric constant. However, the presence of Sn^{2+} at Sn^{4+} acts as the negative defect and is denoted by ${Sn^{2+}}_{Sn^{4+}}''$, and oxygen vacancy $V_o^{\cdot\cdot}$ acts as the positive defect which is situated far away from each other and forms an electric dipole (${Sn^{2+}}_{Sn^{4+}}'' - V_o^{\cdot\cdot}$) in the presence of small AC field. With increasing frequency, the interfacial polarization ruled out and other three polarizations are present (mostly dominant one is orientation polarization). Therefore in the present case, the orientation polarization results in a lower value of dielectric constant.

The effect of Ba on dielectric constant has observed by comparing the values of dielectric constant at the lowest frequency room temperature (25°C) and highest temperature (500°C). The dielectric constant for sample SBS0 was found to be 41, and 60 while for SBS2 it is 28, and 1,551 at temperatures 25°C and 500°C, respectively. From Table 16.1, it is observed that with the incorporation of Ba, the volume of unit cell becomes expanded, and therefore, the polarization becomes difficult at room temperature, while it becomes easier at higher temperature, so it results in a higher dielectric constant at 500°C [24].

Figure 16.6a and b depicts the variation in tangent loss with frequency at different temperatures. Both curves illustrate a similar trend: it shows a higher value of tangent loss below 10 kHz and a smaller value beyond 10 kHz, which is similar to dielectric constant. Hence, the tangent loss is defined as [26]

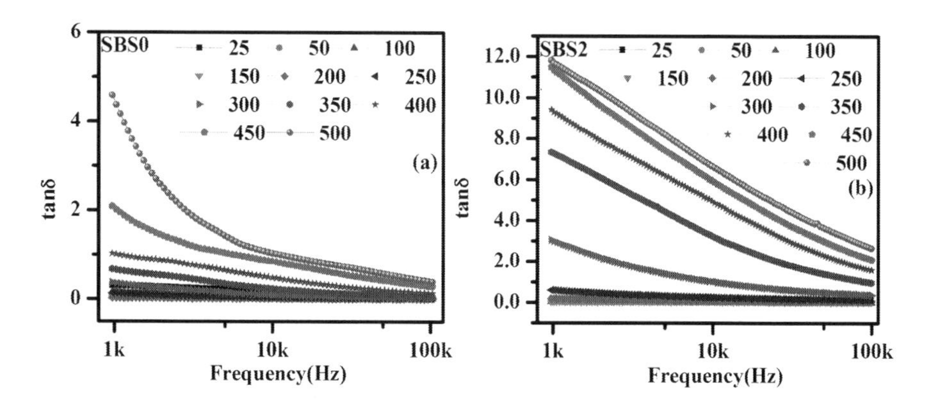

FIGURE 16.6 Variation of tangent loss with frequency at different temperatures of (a) SBS0 (b) SBS2. The temperature shown in the box is represented by degree Celsius.

$$\tan\delta = \frac{\varepsilon''}{\varepsilon'},$$

where ε'' is dielectric constant's imaginary part related to conduction of charge carrier, and ε' is dielectric constant's real part related to polarization of charge carrier. At lower frequency, the time required to hopping the charges between degenerated sites of Sn, i.e., Sn^{4+}/Sn^{2+}, is enough; therefore, it results in a higher value of tangent loss. Moreover, with increasing frequency, the time required for hopping the charge between Sn^{4+}/Sn^{2+} and/or for orientation of dipole $\left(Sn^{2+}{}_{Sn^{4+}}{}'' - V_o^{\cdot\cdot}\right)$ becomes less, thus resulting in a smaller value. To see the effect of Ba on tangent loss, again the value of dissipation factor has been compared at the lowest frequency on room temperature (25°C) and the highest temperature (500°C). The tangent loss for sample SBS0 is found to be 0.30 and 4.58, while for SBS2 it is 0.10 and 11.82 at the temperatures of 25°C and 500°C, respectively. As the volume of unit cell expands, the distance between dipole increases so the conduction of mobile charge carrier becomes difficult, thus resulting in a smaller value. However, as temperature increases, the thermal energy has been increased and the hopping of charge carrier becomes easy, thus resulting in a large value.

The value of dielectric constant 44 and tangent loss in the range of 0.30–0.66 for sample SBS0 while 50 and 0.10–0.60 for sample SBS2 made it a potential candidate for thermally stable capacitor and barrier layer capacitor application. Moreover, the increase in temperature makes it a potential candidate for sensor application.

16.5 CONCLUSIONS

The undoped and Ba-doped $SrSnO_3$ samples were successfully synthesized using sol–gel method followed by calcination at 1,073 K and sintering at 1,173 K. The phase formation of samples has been studied using XRD analysis, and it is found that the samples are crystallized into orthorhombic structure. Ba doping has improved the values of the lattice parameters, volume, and sample density. An UV-Vis absorption spectroscopy was used to study the optical properties of the samples. The samples have shown an intense absorption in UV region, while the magnitude of absorption gets reduced with Ba incorporation. Tauc plot was used to evaluate the optical band gap of samples and found to be 3.88 and 3.97 eV, which may be due to the higher value of lattice strain and the presence of defects in sample. The optical band gap of sample reflects the semiconducting behavior of sample and makes the present materials as UV filter, UV detector, and semiconductor device application. The dielectric properties and tangent loss of samples as a function of frequency and temperature were analyzed. The dielectric constant and tangent loss of samples decreased with increasing frequency, while the dielectric constant and tangent loss of samples increased with increasing temperature. Further, the value of the dielectric constant increased with Ba due to the expansion of unit cell and the presence of microheterogeneity in sample. The dielectric constant value is observed to be 44 and tangent loss in the range of 0.30–0.66 for sample SBS0, while 50 and 0.10–0.60 for sample SBS2, which make them possible candidates for thermally stable capacitor applications for barrier layer capacitor. Furthermore, the temperature change makes it a possible choice for the deployment of sensors.

ACKNOWLEDGMENT

AK thankfully acknowledges the University Grant Commission (UGC) of India for providing UGC research fellowship.

REFERENCES

1. Pradhan, S.; Roy, G. S. Study the Crystal Structure and Phase Transition of $BaTiO_3$ – A Pervoskite. *Researcher*, **2013**, *55* (33), 63–67. https://doi.org/10.7537/marsrsj050313.10.
2. Schorr, S. The Crystal Structure of Kesterite Type Compounds: A Neutron and X-Ray Diffraction Study. *Solar Energy Materials and Solar Cells*, **2011**. https://doi.org/10.1016/j.solmat.2011.01.002.
3. Azad, A. M.; Hon, N. C. Characterization of $BaSnO_3$-Based Ceramics Part 1. Synthesis, Processing and Microstructural Development. *Journal of Alloys and Compounds*, **1998**. https://doi.org/10.1016/S0925-8388(98)00370-3.
4. Azad, A. M.; Shyan, L. L. W.; Pang, T. Y.; Nee, C. H. Microstructural Evolution in $MSnO_3$ Ceramics Derived via Self-Heat-Sustained (SHS) Reaction Technique. *Ceramics International*, **2000**. https://doi.org/10.1016/S0272-8842(00)00005-5.
5. Shimizu, Y. Humidity-Sensitive Characteristics of La^{3+}-Doped and Undoped $SrSnO_3$. *Journal of The Electrochemical Society*, **1989**. https://doi.org/10.1149/1.2096854.
6. Azad, A. M. Solid-State Gas Sensors: A Review. *Journal of The Electrochemical Society*, **1992**. https://doi.org/10.1149/1.2069145.
7. Ishihara, T.; Fujita, H.; Nishiguchi, H.; Takita, Y. $SrSnO_3$-WO_3 as Capacitive-Type Nitrogen Oxide Sensor for Monitoring at High Temperature. *Sensors and Actuators, B: Chemical*, **2000**. https://doi.org/10.1016/S0925-4005(99)00416-5.
8. Ostrick, B.; Fleischer, M.; Meixner, H. High-Temperature Hall Measurements on $BaSnO_3$ Ceramics. *Journal of the American Ceramic Society*, **2005**. https://doi.org/10.1111/j.1151-2916.1997.tb03102.x.
9. Singh, D. J.; Xu, Q.; Ong, K. P. Strain Effects on the Band Gap and Optical Properties of Perovskite $SrSnO_3$ and $BaSnO_3$. *Applied Physics Letters*, **2014**, *104* (1), 1–5. https://doi.org/10.1063/1.4861838.

10. Singh, S.; Singh, P.; Parkash, O.; Kumar, D. Synthesis, Microstructure and Electrical Properties of Ti Doped $SrSnO_3$. *Advances in Applied Ceramics*, **2007**. https://doi.org/10.1179/174367607X202573.
11. Ansaree, M. J.; Kumar, U.; Upadhyay, S. Solid-State Synthesis of Nano-Sized $Ba(Ti_{1-x}Snx)O_3$ Powders and Dielectric Properties of Corresponding Ceramics. *Applied Physics A: Materials Science and Processing*, **2017**, *123* (6), 1–12. https://doi.org/10.1007/s00339-017-1047-6.
12. Kumar, U.; Ansaree, M. J.; Verma, A. K.; Upadhyay, S.; Gupta, G. Oxygen Vacancy Induced Electrical Conduction and Room Temperature Ferromagnetism in System $BaSn_{1-x}Ni_xO_3$ ($0 \leq x \leq 0.20$). *Materials Research Express*, **2017**, *4* (11), aa9416. https://doi.org/10.1088/2053-1591/aa9416.
13. Jiang, S. P. A Review of Wet Impregnation - An Alternative Method for the Fabrication of High Performance and Nano-Structured Electrodes of Solid Oxide Fuel Cells. *Materials Science and Engineering A*, **2006**. https://doi.org/10.1016/j.msea.2005.11.052.
14. Callister, W. D. Materials Science and Engineering: An Introduction (2nd Edition). *Materials & Design*, **2003**. https://doi.org/10.1016/0261-3069(91)90101-9.
15. Wu, X. H.; Wang, Y. De; Liu, H. L.; Li, Y. F.; Zhou, Z. L. Preparation and Gas-Sensing Properties of Perovskite-Type $MSnO_3$ (M=Zn, Cd, Ni). *Materials Letters*, **2002**. https://doi.org/10.1016/S0167-577X(02)00604-3.
16. Kumari, U. S.; Suresh, P.; Rao, A. V. P. Solid-State Metathetic Synthesis of Phase Pure $BaSnO_3$ and $BaZrO_3$. *International Research Journal of Pure & Applied Chemistry*, **2013**, *3* (4), 347–356.
17. Singh, S.; Singh, P.; Parkash, O.; Kumar, D. Structural and Relaxor Behavior of $(Ba_{1-x}La_x)(Ti_{0.85}Sn_{0.15})O_3$ Ceramics Obtained by Solid State Reaction. *Journal of Alloys and Compounds*, **2010**, *493* (1–2), 522–528. https://doi.org/10.1016/j.jallcom.2009.12.148.
18. Ansaree, M. J.; Kumar, U.; Upadhyay, S. Structural, Dielectric and Magnetic Properties of Particulate Composites of Relaxor ($BaTi_{0.85}Sn_{0.15}O_3$) and Ferrite ($NiFe_2O_4$) Synthesized by Gel-Combustion Method. *Journal of Electroceramics*, **2018**, *40* (3), 257–269. https://doi.org/10.1007/s10832-018-0127-0.
19. Kumar, A.; Singh, B. P.; Choudhary, R. N. P.; Thakur, A. K. A.C. Impedance Analysis of the Effect of Dopant Concentration on Electrical Properties of Calcium Modified $BaSnO_3$. *Journal of Alloys and Compounds*, **2005**, *394* (1–2), 292–302. https://doi.org/10.1016/j.jallcom.2004.11.012.
20. Kumar, U.; Ankur, K.; Yadav, D.; Upadhyay, S. Synthesis and Characterization of Ruddlesden-Popper System $(Ba_{1-x}Sr_x)_2SnO_4$. *Materials Characterization*, **2020**, *162* (July 2019), 110198. https://doi.org/10.1016/j.matchar.2020.110198.
21. Udawatte, C. P.; Kakihana, M.; Yoshimura, M. Low Temperature Synthesis of Pure $SrSnO_3$ and the $(Ba_xSr_{1-x})SnO_3$ Solid Solution by the Polymerized Complex Method. *Solid State Ionics*, **2000**. https://doi.org/10.1016/S0167-2738(99)00306-9.
22. Shannon, R. D.; Prewitt, C. T. Revised Values of Effective Ionic Radii. *Acta Crystallographica Section B Structural Crystallography and Crystal Chemistry*, **1970**. https://doi.org/10.1107/s0567740870003576.
23. Prabhu, Y. T.; Rao, K. V. X-Ray Analysis by Williamson-Hall and Size-Strain Plot Methods of ZnO Nanoparticles with Fuel Variation. *World Journal of Nano Science and Engineering*, **2014**, *04* (01), 21–28.
24. Kumar, U.; Upadhyay, S. Structural, Microstructure, Optical, and Dielectric Properties of $Sr_{1.99}M_{0.01}SnO_4$ (M: La, Nd, Eu) Ruddlesden–Popper Oxide. *Journal of Materials Science: Materials in Electronics*, **2020**, *31* (7), 5721–5730. https://doi.org/10.1007/s10854-020-03140-0.

25. Deepa, A. S.; Vidya, S.; Manu, P. C.; Solomon, S.; John, A.; Thomas, J. K. Structural and Optical Characterization of $BaSnO_3$ Nanopowder Synthesized through a Novel Combustion Technique. *Journal of Alloys and Compounds*, **2011**, *509* (5), 1830–1835. https://doi.org/10.1016/j.jallcom.2010.10.056.
26. Upadhyay, S.; Parkash, O.; Kumar, D. Dielectric Relaxation and Variable-Range-Hopping Conduction in $BaSn_{1-x}Cr_xO_3$ System. *Journal of Electroceramics*, **2007**, *18*, 45–55. https://doi.org/10.1007/s10832-007-9007-8.

17 Fabrication and Characterization of Nanocrystalline Lead Sulphide (PbS) Thin Films on Fabrics for Flexible Photodetector Application

Kinjal Patel, Jaymin Ray, and Sweety Panchal
Uka Tarsadia University

CONTENTS

17.1 INTRODUCTION

Photodetectors are an important member of the optoelectronic device family. The primary function of photodetectors is to convert light energy into an accurate electrical signal in the form of photovoltage or photocurrent. There is always a demand for extravagant performance of photodetectors among industrial and scientific communities. The major areas that employ photodetectors include fire detection, motion detection, night vision monitoring, hazardous environment monitoring, and optical telecommunication [1–5]. Currently, the market of photodetectors is dominated by the crystalline silicon material and its derivatives. In addition, photodetectors are also developed on solid fixed substrates, which have many applications in our day-to-day life, viz. sensors in digital camera, fire sensors, portable lux meter in mobile phones, portable distance measurement, and temperature measurement [6–8].

In recent days, we can observe that there is a trend of miniaturization for various optoelectronic devices, including photodetectors, which results in the utilization of flexible substrates for sensing elements, to increase the flexibility and reduce the cost of the devices. In addition, the weight of the devices is also efficaciously reduced, which opens the new door for lightweight, flexible, and wearable photodetectors. A variety of flexible substrates are readily available in the market, which have different characteristics such as softening point, inertness, degree of bending, and electrical conductivity. Subsequent to this, recent trends in the fields of scientific research on photodetectors, such as paper/fibre-based devices, have shown potential advantages including low cost, high sensitivity, foldable nature, and easy disposability. In order to achieve all the said advantages, the materials that are used as sensing elements must have a wide range of sensing characteristics within a small amount of consumption. However, small, flexible, and supersensible photodetectors rely on the generation of free carriers (electron and hole pairs) by absorbing the light photon. An extreme sensitivity is required to minimize the loss of photo-generated carriers, especially at the defect sites. In comparison with mono- and polycrystalline materials, semiconducting nano-sized or nanocrystalline (NC) materials are better candidates for controlling the defect chemistry. In addition, optical absorption phenomena can be modified by controlling the size and shape of NCs, thus reducing the quantum mechanical coupling effect from thousands of atoms [9,10]. Hence, absorption and emission of NCs can be easily controlled. On the other hand, chemical-based methods were mostly employed to synthesize NCs, which resulted in a significant reduction in the device fabrication cost. Hence, NCs are feasible to integrate conventional Si-based electronics with other flexible paper/plastic/fabric-based substrates.

Various material groups were studied in the form of NCs for flexible photodetectors. The most analysed material is carbon nanotube (CNT), as it has challenging optical and electrical properties due to its multidimensional attributes [11]. CNT can be acted or used as either a semiconductor or a metallic depending upon its chirality. Both of these forms can produce photocurrent depending on the illumination [12]. Zhang et al. have prepared a flexible single-wall CNT photodetector on a low thermal conductive material, viz. polyimide, by printing technique [13]. The detector demonstrated a stable photoresponse and a greater mechanical strength. Apart from CNT, in the modern era, dozens of nanostructured materials are utilized for flexible photodetectors. Single-crystalline silicon nanowires (NWs) on polyethylene terephthalate (PET) substrates were utilized to prepare a transparent flexible photodetector [14]. Owing to their higher carrier mobility, higher intrinsic carrier concentration, and direct band gap, III–V NWs were considered as better candidates for the sensible and flexible photodetectors. GaSb NW [15] and GaN/InGaN NW [16] are the most studied materials for flexible photodetectors. In the case of metal oxides, ZnO [17,18], Ga_2O_3 [19], Zn_2GeO_4, and $In_2Ge_2O_7$ [20] are extensively used for high-performance flexible UV detectors. Besides these, metal chalcogenide materials also demonstrate better performance for stable and highly sensible flexible NC photodetectors, such as CdS, PbS, SnS, Sb_2S_3, and In_2S_3 [21,22].

The most recent and freshly emerged option for flexible photodetectors is perovskite NW that is made of hybrid organic–inorganic materials. Direct band gap, high visible light absorption coefficient, and higher carrier mobility are the basic properties of perovskite materials, which indicates their potential towards utilization in highly sensitive and stable photodetectors [23]. Besides this, stability, lifetime, and flexibility are the most concerned limitations of perovskite detectors [24,25]. Despite all convinced and promising research outputs regarding flexible photodetectors, there are still a lot of scope to grow. In this context, fabrication and characterization of NC PbS-based flexible photodetectors are discussed. These photodetectors are prepared using different fibres that are easily available in the market.

Lead sulphide (PbS) has novel properties such as tunable band gaps (due to the 18 nm Bohr radius) and ease of solution processing, which makes it a suitable material for sensing application [18]. Several reports intimate the ease of preparation procedure for NC thin films and particles to fabricate flexible photodetectors. Mamiyev and Balayeva [26] investigated to prepare the NC PbS in MA/octene-1 copolymer matrix at a temperature of 80°C. XRD reveals a cubic structure having a crystalline size around 10–15 nm. The observed band gap is 0.41 eV, which indicates the formation of nanostructures. Patil et al. in 2006 [27] prepared PbS by using a successive ionic layer-by-layer adsorption and reaction (SILAR) method. The deposition was carried out at room temperature on a glass substrate. Wang et al. [28] synthesized PbS materials using a solid-state reaction. They have used surfactant $C_{18}H_{37}O(CH_2CH_2O)_{10}H$ (abbreviated as $C_{18}EO_{10}$) to prepare NC PbS. They investigated the role of this surfactant at low temperatures. XRD revealed that the prepared PbS has a grain size of 10–15 nm. The surfactant allows preparation of NCs at room temperature having a uniform size and shape with a high yield. Zhu et al. [29] used the sono-chemical irradiation to the ethylene diamine precursor solution containing solution of sulphur element and lead acetate in air atmosphere. The XRD study showed that the crystalline size of the prepared material is around 20 nm. They employed the same method to prepare HgS particles too. Chen et al. [30] synthesized a PbS NC material using a simple chemical method, and then the material was surface-modified with dialkyldithiophosphate (DDP). The purity of the prepared PbS was determined by X-ray photoelectron spectroscopy (XPS) and transmission electron microscopy (TEM), which also show the formation of pure PbS with a crystalline size of ~5 nm. Kruis et al. [31] prepared nanocrystals of PbS using the nucleation and aggregation processes under a controlled furnace. A gas-phase synthesis method was employed at normal pressure to prepare sub-20 nm, crystalline, quasi-spherical, and monodisperse PbS particles. The heating temperature of the furnace was set around 700°C to control the formation of particles.

In this work, PbS powder and its films on a flexible fabric-based substrate prepared via chemical root were characterized by XRD, FTIR (Fourier transform infrared spectroscopy), SEM, optical, and electrical methods. As flexibility reduces the cost of the devices, it will surely change the dimension of the electronic components in future.

17.2 EXPERIMENTAL

In order to prepare a defect-free NC-based photodetector, confirmation of size and shape of the prepared material is necessary. This work was executed in two parts of experiments: one is PbS NC material preparation, and the other is NC PbS thin film on a flexible substrate. In both cases, different sources of sulphur were used in order to check the best response towards the light. A PbS NC material has been prepared by ball milling followed by the solid-state reaction. PbS NC films on the flexible substrate were deposited by the dip (soak) coating method.

17.2.1 Ball Milling

Ball milling was used for mixing appropriate salts in the metal jar with zircon balls at room temperature. The time and speed of milling were fixed for all four samples, i.e. 20 minutes and 600 rpm, respectively. We have used lead acetate [$Pb(C_2H_3O_2)_2$] as a lead source, and four different components were used as sulphur sources, viz. Na_2S (sodium sulphide), CH_4N_2S (thiourea), $Na_2S_2O_3$ (sodium thiosulfate), and C_2H_5NS (thioacetamide). As per the atomic stoichiometry of PbS, 1:1 ratio of lead and sulphur sources was taken for ball milling. After 20 minutes of milling, a black colour material was obtained. Then, the drying process was carried out in two steps. The first step is to dry in air at 80°C for 1 hour, and the second step is to dry in air at 200°C for 3 hours. Figure 17.1 shows the actual photographs of the prepared PbS material using four different sulphur sources, and Table 17.1 shows the sample codes of the prepared PbS material.

Visual inspection of the dried material, as shown in Figure 17.1, indicates the various forms of granular nature. The sodium sulphide and thioacetamide-based material appeared as a pure black fine-textured powder. Both were easily collected after the drying process. The sodium thiosulfate-based dried powder was light brown in colour and granular in texture. However, the thiourea-based dried powder was in black mirror-polished, shiny, and crunchy texture, and hence, it requires further manual milling. After the required filtering, the prepared materials were characterized by XRD to confirm the crystalline structure and by FTIR to assure the purity.

17.2.2 XRD Analysis of Ball-Milled NC PbS Material

XRD profile of the prepared PbS materials using different sulphur sources is shown in Figure 17.2. The influence of the sulphur source is clearly observed in the form of the intensity of the crystalline planes and the peak width.

(a)

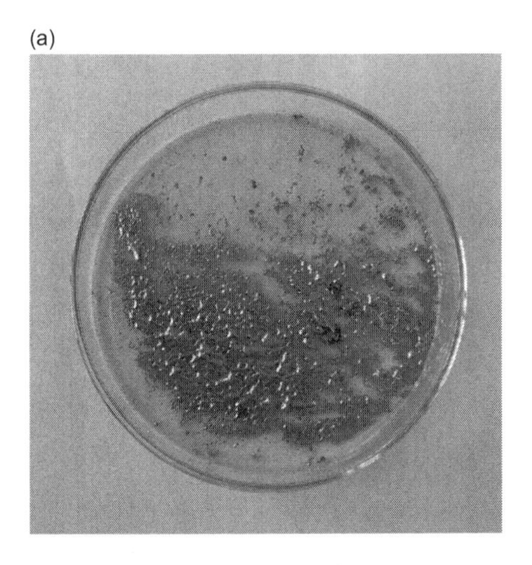

(b)

(c)

FIGURE 17.1 Dried PbS material using different sulphur sources: (a) CH_4N_2S (thiourea), (b) Na_2S (sodium sulphide), (c) C_2H_5NS (thioacetamide), and (d) $Na_2S_2O_3$ (sodium thiosulfate).

(Continued)

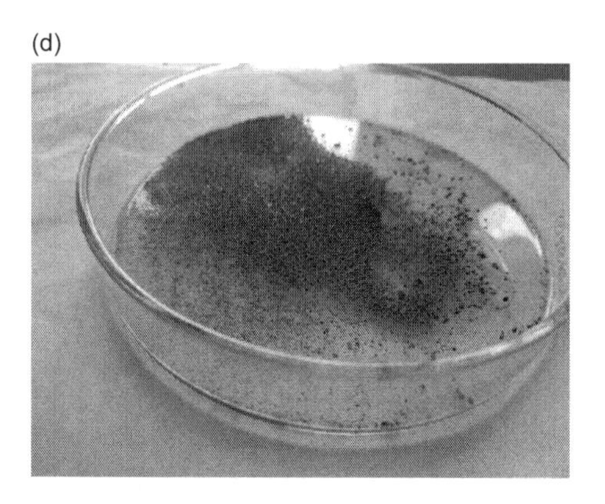

FIGURE 17.1 (CONTINUED) Dried PbS material using different sulphur sources: (a) CH_4N_2S (thiourea), (b) Na_2S (sodium sulphide), (c) C_2H_5NS (thioacetamide), and (d) $Na_2S_2O_3$ (sodium thiosulfate).

TABLE 17.1
Sample Code for Prepared PbS Materials Using Four Different Sulphur Sources

Sample Code	Used Sulphur Source
P1	CH_4N_2S (thiourea)
P2	Na_2S (sodium sulphide)
P3	C_2H_5NS (thioacetamide)
P4	$Na_2S_2O_3$ (sodium thiosulfate)

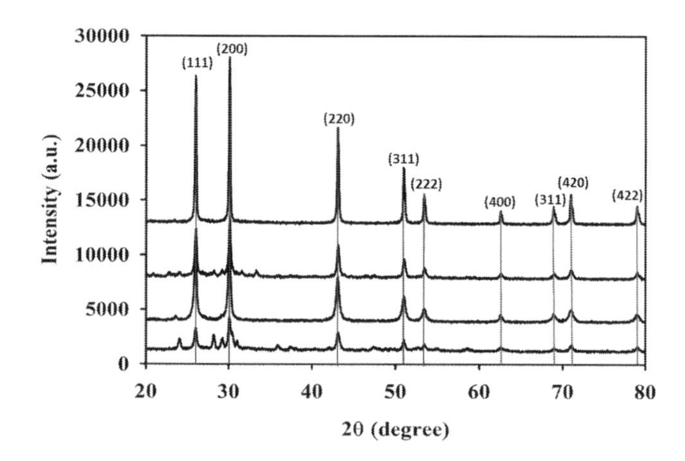

FIGURE 17.2 XRD pattern of the lead sulphide material using four sulphur sources: (from top to bottom) CH_4N_2S (thiourea), Na_2S (sodium sulphide), C_2H_5NS (thioacetamide), and $Na_2S_2O_3$ (sodium thiosulfate).

XRD pattern reveals the formation of PbS phase in all samples. The peaks corresponding to (111), (200), (220), (311), (222), (400), (420), and (422) planes of PbS represent the cubic phase. The interplanar distance and the lattice constant match well with the values reported in JCPDS Card No. 05-0592, which can be attributed to cubic phase and confirmation of PbS powder. The average crystalline size D was calculated by the following Scherrer formula:

$$D = 0.9\lambda/\beta\cos\theta,$$

where λ is the wavelength of X-ray (1.54 nm) used, β is the fill width half maxima of peak, and θ is the Bragg angle. The average crystalline size of PbS powder is estimated to be from 26 to 47 nm. The strain is related to a lattice "misfit" that relies on the preparation conditions of the film materials. The strain was calculated by the following formula:

$$\varepsilon = \beta\cos\theta/4,$$

where θ is the Bragg angle and β is the full width half maxima.

Furthermore, the dislocation density can be calculated by the following relation:

$$\rho = 1/D^2.$$

The length of dislocation lines per unit volume of crystal is normally defined as dislocation density. It can be attributed to the crystallographic defect present in the crystal structure. All the calculated structural parameters are given in Table 17.2.

XRD plot of sodium thiosulfate (sample P1)-based PbS indicates a couple of unidentified peaks, which may be the feature of the remaining unwanted organic/inorganic compound. However, XRD plot of sodium sulphide (sample P2)-based PbS doesn't show any unwanted peaks of any organic or inorganic compound, but the crystallinity is poor. Thioacetamide (sample P3)-based PbS spectra show a similar behaviour to sample P2. In comparison with samples P1, P2, and P3, thiourea-based PbS (sample P4) has a better crystalline structure with an average crystalline size of about 38 nm.

TABLE 17.2
Structural Parameters of PbS NC Materials Using Different Sulphur Sources

Sample Code	Crystalline Size (D) nm	Micro-Strain (ε) × 10^{-2}	Dislocation Density (ρ) L/m^2
P1	47	4.34	4.3×10^{14}
P2	26	7.24	1.4×10^{15}
P3	29	7.26	1.2×10^{15}
P4	38	5.55	7.6×10^{14}

17.2.3 FTIR (Fourier Transform Infrared Spectroscopy)

In order to confirm the presence of organic or inorganic molecules, FTIR was used. FTIR spectra of the prepared PbS materials using four different sulphur sources are shown in Figure 17.3. Certain functional groups in a compound absorb definite

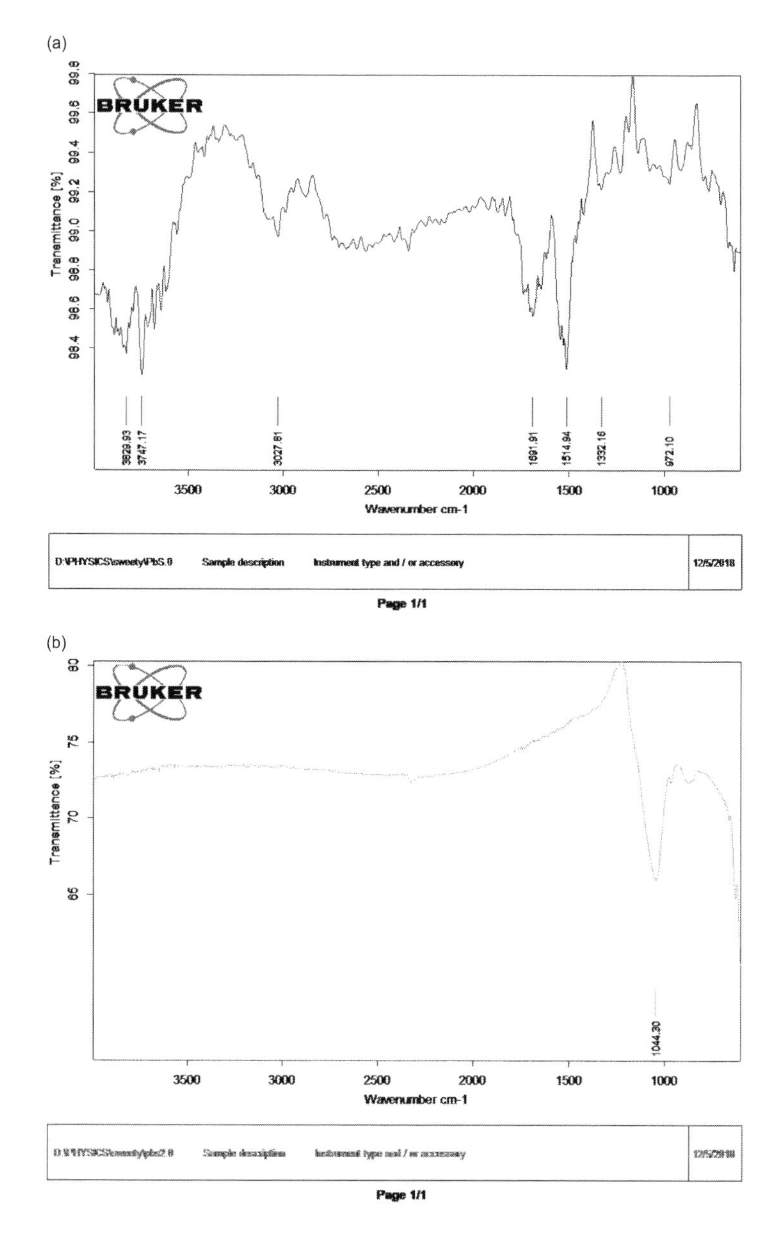

FIGURE 17.3 FTIR spectra of the prepared PbS materials using four different sulphur sources: (a) Na_2S (sodium sulphide), (b) C_2H_5NS (thioacetamide), (c) $Na_2S_2O_3$ (sodium thiosulfate), and (d) CH_4N_2S (thiourea).

(*Continued*)

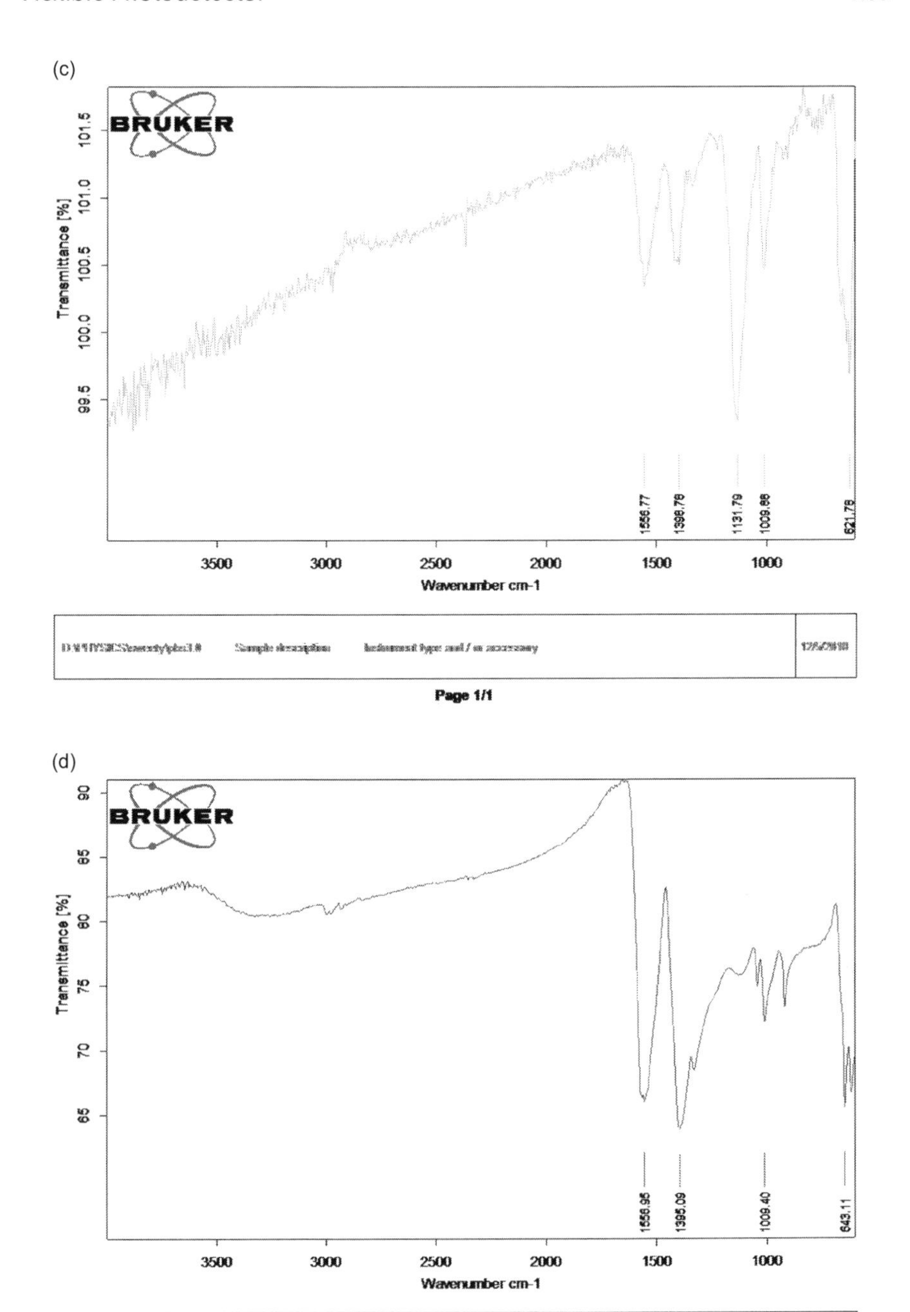

FIGURE 17.3 (CONTINUED) FTIR spectra of the prepared PbS material using four different sulphur sources: (a) Na_2S (sodium sulphide), (b) C_2H_5NS (thioacetamide), (c) $Na_2S_2O_3$ (sodium thiosulfate), and (d) CH_4N_2S (thiourea).

frequencies, and hence, the compound can be easily identified. In all FTIR spectra, the strong peak near 1,550 cm^{-1} corresponds to the long alkyl chain. The peaks found at 650 to 625 cm^{-1} correspond to P=S band.

17.3 FABRICATION AND CHARACTERIZATION OF A FLEXIBLE DETECTOR

17.3.1 Choice of Substrate

In this work, we have used dip coating method for depositing PbS thin film on the flexible substrate. We have used different types of fabrics available in the normal textile market. In order to prepare a flexible photodetector, there are certain criteria for the selection of flexible substrate: it sustains a temperature above 150°C, and it sustains flexibility even after heating and the porous structure on both sides. Based on these criteria, four types of flexible substrates were chosen, which were easily available in the market. The basic surface properties of these selected fabrics and their common names are shown in Table 17.3.

Oil painting hard canvas is static, and a polished fabric is generally used for the supporting purpose, as shown in Figure 17.4a. The mesh of the fabric is very fine and non-stretchable. While mono canvas or needlepoint canvas is normally stable, weave canvas is generally used in many types of needlework or stiches (Figure 17.4b). Usually, it is available in the counts of mono 10, 12, 13, 14, 16, and 18. The numbers point out the size of the holes. Linen fabrics have a mismatched surface feature due to the combination of thin and thick threads (Figure 17.4c), which makes its surface slightly rough. We have used the Edinburgh linen of 36 count. Satin has smooth and shiny surface having a tighter weave. The threads of it were difficult to handle after using them for some time, due to their snagging and surface texture difference (Figure 17.4d).

17.3.2 Dip or Soaking of PbS NC Films

Thin films of PbS NCs were deposited on all four types of fabrics, as mentioned in Figure 17.3, using the simple dip coating method. In this method, the dipping time varies as it requires time to soak the precursor solution into it. The precursor solution was prepared by dissolving lead sulphide and thiourea in methanol at 1:1 molar ratio. The ball milling process and its results, in the preparation of NC PbS using four different sulphur sources, suggest that the thiourea-based PbS shows a better

TABLE 17.3
Type of Fabrics Used and Their Surface Features

Paper No.	Type of Fabric	Surface Features
1	Oil painting hard Canvas	Smooth
2	Mono Canvas	Very rough
3	Linen	Slight rough
4	Satin	Smooth

crystalline structure with a pure chemical composition. Hence, in this study, only thiourea was used as the sulphur source with lead sulphide in order to prepare the precursor solution for a flexible detector. The substrates for the flexible detector, i.e. fabrics (shown in 3), were cut in the size of 1 cm × 0.5 cm, dipped into the precursor solution, and then placed in oven for thermolysis for 20 minutes. The oven was set at a temperature of 200°C and controlled by a PID (SELEC 500) temperature controller. The sustainability of all four types of fabric cloths was first checked at thermolysis temperature. The result shows that oil painting hard canvas was not affected

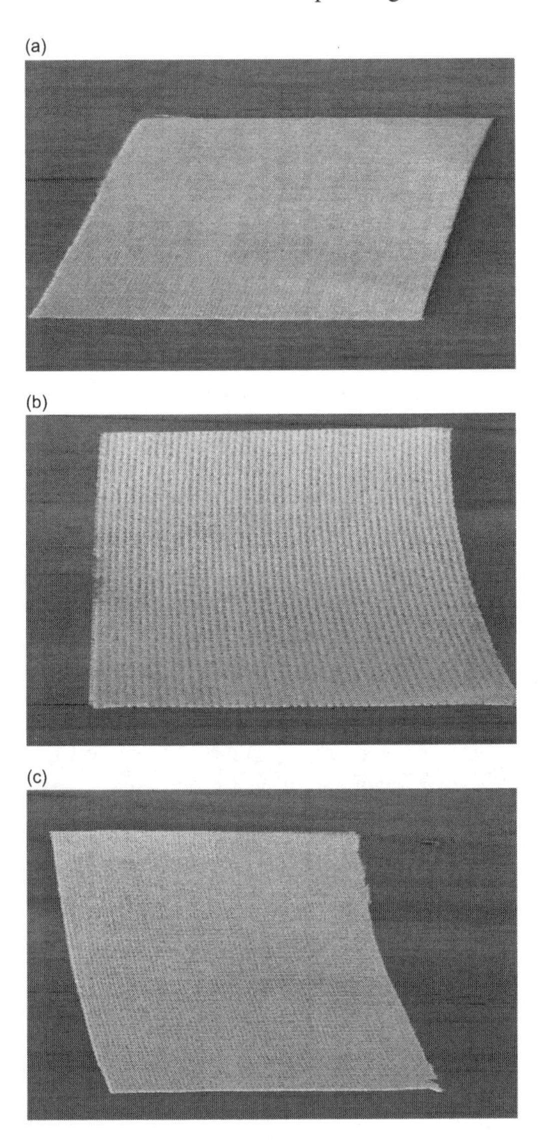

FIGURE 17.4 Types of fabric cloths used for preparation of a flexible photodetector.

(*Continued*)

(d)

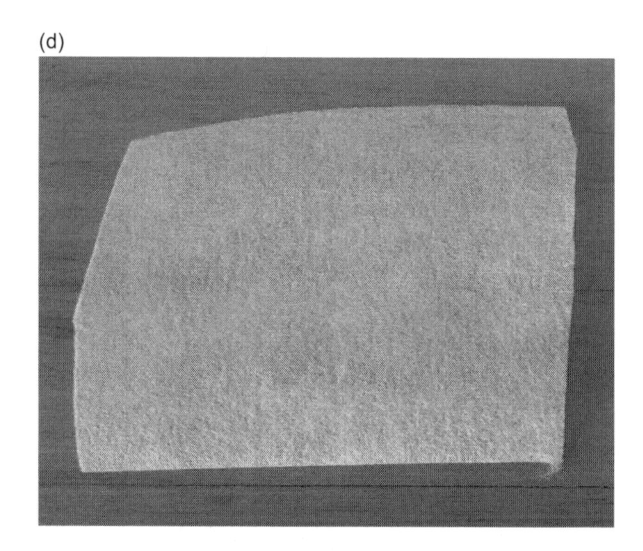

FIGURE 17.4 (CONTINUED) Types of fabric cloths used for preparation of a flexible photodetector.

by the thermolysis temperatures, whereas some others were shrunken or lost their flexibility. At the end, oil painting hard canvas was used as the substrate to prepare a flexible photodetector.

The dipping time varied, viz. 48, 72, 96, and 108 minutes. The reason behind varying the dipping time is the texture-dependent soaking capability of the fabric. After thermolysis, the prepared samples were characterized by XRD, SEM, and UV-VIS spectrophotometer in order to analyse the grown crystal structure, surface morphology, and optical band gap, respectively. In order to measure the photoconductivity of the prepared samples, two types of contact materials were used: one is silver paste, and the other is graphite paste. The tungsten halogen lamp (80 W) was used as the light source, and 3-3/4 Digits Digital LCR multi-meter (Metravi make) was used to measure the photoconductivity at room temperature.

17.3.3 XRD Analysis of the PbS NC Films on Fabrics

XRD patterns of the prepared dip-coated fabric having different dipping times are shown in Figure 17.5. For a better analysis, the sample of oil painting hard canvas was analysed by XRD first, as a reference. The XRD plot reveals that as the dipping time increases, the intensity of the peak increases, which indicates an improvement in crystallinity, which is in turn a function of the dipping time. More dipping time can increase the adsorption of the precursor solution. In addition, the characteristic peak of the base, i.e. fabric, decreases as the dipping time increases. The sample dipped for 48 minute doesn't show any relevant peaks, which are not found in the comparative plot. The highest crystalline size was 14 nm, and it was observed for 108-minute dipping time.

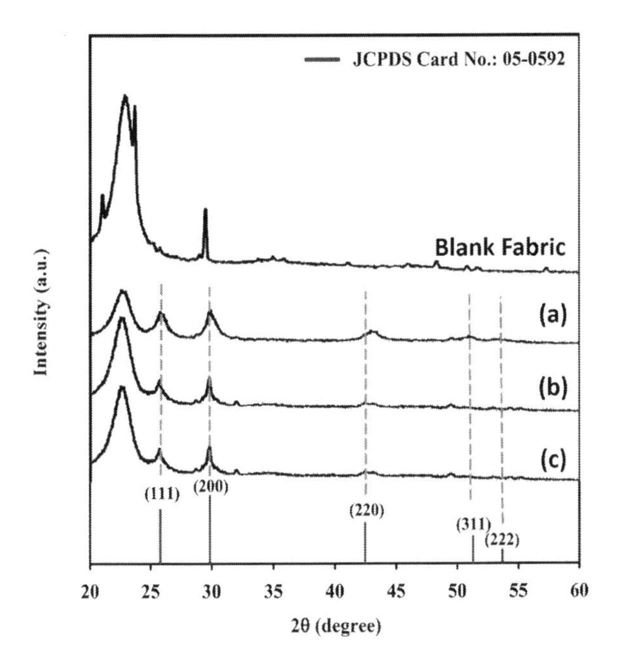

FIGURE 17.5 XRD spectra of blank fabric and dipped NC PbS on fabrics for (a) 72 minutes, (b) 92 minutes, and (c) 108 minutes.

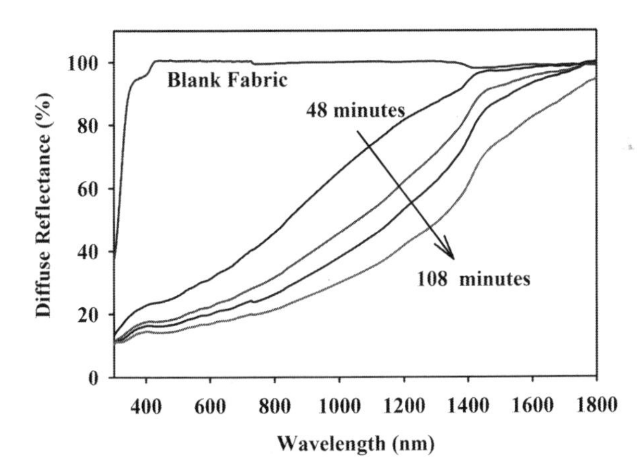

FIGURE 17.6 Diffuse reflectance spectra of blank and dipped NC PbS on fabrics for 48, 72, 92, and 108 minutes.

17.3.4 UV-Vis Analysis of PbS NC Films on Fabrics

Diffuse reflectance was measured in order to calculate the band gap of the prepared samples, as shown in Figure 17.6. The blank fabric's reflectance was the highest (nearly 100% in the visible region), as it is pure white in colour. In the case of the prepared sample, the reflectance decreased as the dipping time increased. But the absorption edge (around 1,410 nm, i.e. ~0.9 eV) remained unchanged.

17.3.5 SEM Analysis of PbS NC Films on Fabrics

All the prepared flexible photodetectors show a uniform coating of PbS material. SEM topography of the blank fabric and 108-minute dipped PbS-coated fabric (2,500 ×, 8,000 × and 20,000 ×, resolution) is shown in Figure 17.7. The SEM topography clearly indicates the deposition of PbS on the fibre of the fabric. 20,000 × resolution image demonstrates the consistent deposition of PbS films.

17.3.6 Photoconductivity Measurement

Photoconductivity of the prepared flexible photodetector was measured using two types of contacts (silver paste and graphite paste). The base resistance of the silver paste was in the range of 40–50 Ω. On the other hand, graphite paste was prepared using 2H, 2B, HB, and H grades. Amongst them, the base resistance of 4H grade is the lowest, viz. 80–90 Ω. Table 17.4 shows the photoconductivity data, i.e. ΔR – difference between dark and light resistance, of the all prepared samples having different dipping times. Figures 17.8 and 17.9 show the graphical behaviour of ΔR for both contacts, i.e. silver paste and graphite paste, respectively. When comparing the ΔR of both contacts, the graphite paste shows unevenness and higher resistance of the sample. The samples having silver paste contact show a significant improvement up to 96 minutes; afterwards, the ΔR reduces. The reduction in the ΔR may be attributable to the overgrowth or loosely bound PbS material.

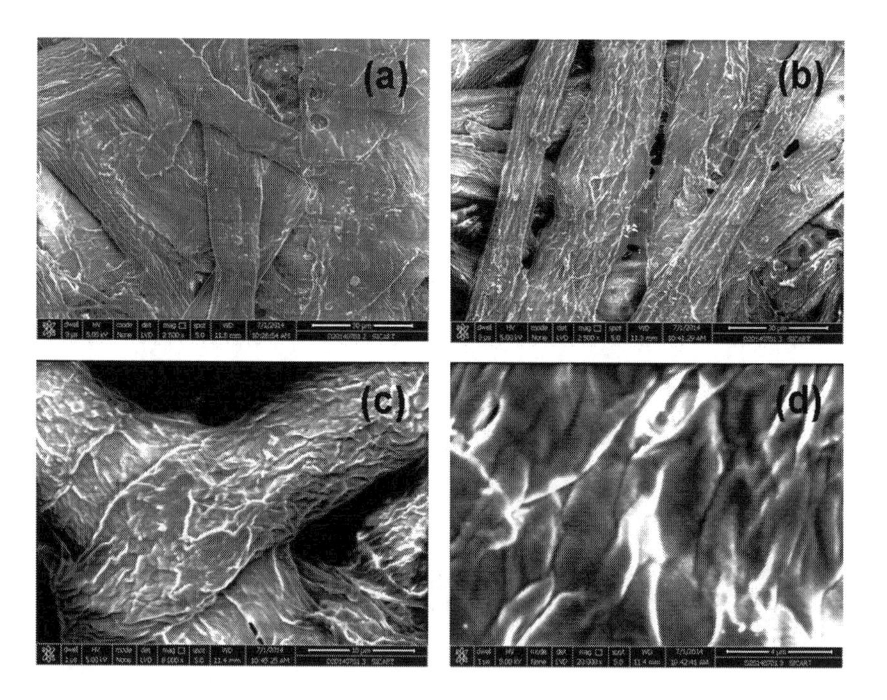

FIGURE 17.7 SEM images of (a) uncoated fabric, (b) PbS-deposited fabric (108 minutes), (c) magnified images at 8,000× and (d) 20,000×.

TABLE 17.4
Photoconductivity Parameters of the Flexible PbS Photodetector

	Silver Paste			Graphite Paste		
Dipping Time (minute)	Resistance – With Light – R_L (MΩ)	Resistance – Without Light – R_D (MΩ)	ΔR (MΩ)	Resistance – With Light – R_L (MΩ)	Resistance – Without Light – R_D (MΩ)	ΔR (MΩ)
48	103.4 ± 3	128.8 ± 4	25 ± 4	70.8 ± 2	152.8 ± 3	82 ± 2
72	104.8 ± 5	127.9 ± 6	23 ± 5	90.9 ± 3	120.8 ± 4	30 ± 3
96	93.5 ± 3	127.1 ± 3	33 ± 4	84.2 ± 4	128.4 ± 4	44 ± 3
108	98.0 ± 2	107.2 ± 2	9 ± 2	67.6 ± 3	99.2 ± 5	30 ± 4

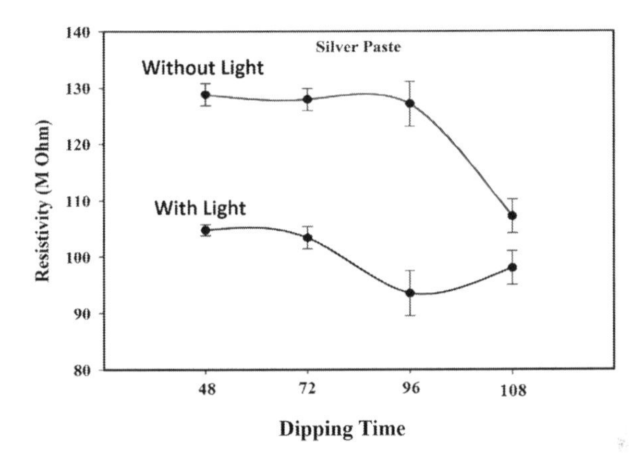

FIGURE 17.8 Change in resistivity of the PbS NC flexible photodetector having silver paste contact.

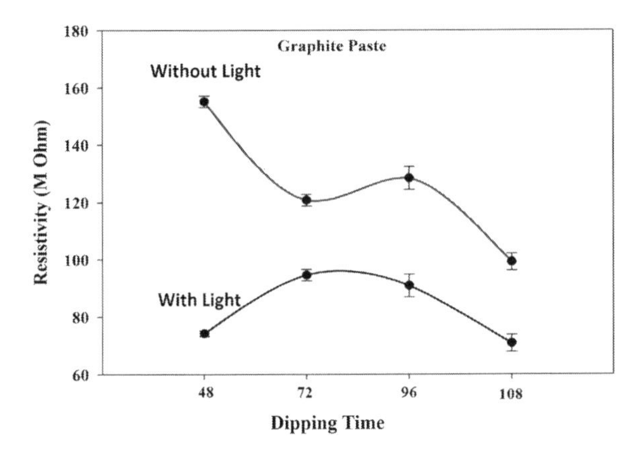

FIGURE 17.9 Change in resistivity of the PbS NC flexible photodetector having graphite contact.

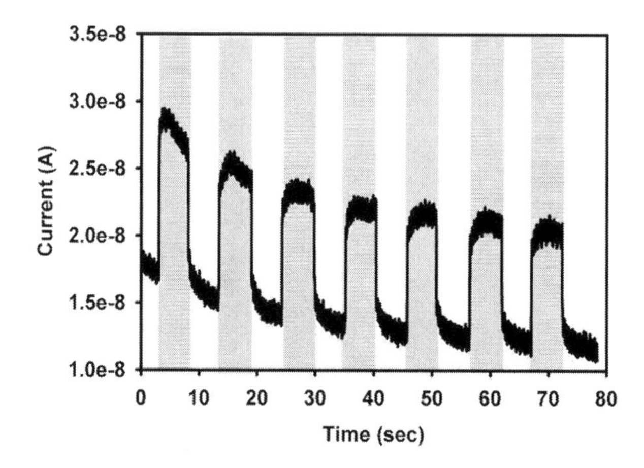

FIGURE 17.10 Rise and decay response of the prepared PbS NC (96-minute-dipped) flexible detector having silver paste contact.

Repeatability of photoresponse is the basic characteristic and requirement of any photodetector. In this analysis, time-dependent photoresponse was measured for more than 70 seconds. The rise and decay of the photocurrent were measured at an interval of 5 second illumination and 5 second dark. Figure 17.10 shows the rise and decay response of the prepared PbS NC (96-minute-dipped) flexible detector having silver paste contact.

17.4 CONCLUSION

PbS NC materials and films were successfully prepared using four different sulphur sources. After optimizing the sulphur source, the NC PbS films were deposited on four different fabric cloths using dip coating method. In this method, the dipping time varied from 48 to 108 minutes as it measured the adsorption of precursor solution into the fabric. XRD confirms the pure NC PbS cubic phase. FTIR and optical properties indicate the presence of the consummate PbS material with its substantial features. SEM topography confirms the full coverage of the PbS material on the whole fabric cloth. The photoconductivity measurements are utilized to identify the photoresponse of the NC PbS coating onto the fabric.

ACKNOWLEDGEMENT

The authors are thankful to B. U. Patel Research Promotion Scheme of Uka Tarsadia University (UTU/RPS/1260/2018) for the financial assistance.

REFERENCES

1. Koppens F. H. L., Mueller T., Avouris P., et al. (2014). Photodetectors based on graphene, other two-dimensional materials and hybrid systems. *Nat. Nanotechnol.*, 9, 780.

2. Li J., Niu L., Zheng Z., Yan F. (2014). Photosensitive graphene transistors. *Adv. Mater.*, 26(31), 5239.
3. Sun Z., Chang H. (2014). Graphene and graphene-like two-dimensional materials in photo detection: mechanisms and methodology. *ACS Nano*, 8(5), 4133.
4. Buscema M., Island J. O., Groenendijk D. J., et al. (2015). Photocurrent generation with two-dimensional van der Waals semiconductors. *Chem. Soc. Rev.*, 44, 3691.
5. Xie, C., Mak, C., Tao, X. M., Yan, F. (2017). Photodetectors based on two dimensional layered materials beyond graphene. *Adv. Funct. Mater.*, 27, 1603886.
6. Choi C., Choi M. K., Liu S., et al. (2017). Human eye-inspired soft optoelectronic device using high-density MoS_2-graphene curved image sensor array. *Nat. Commun.*, 8, 1664.
7. Yang Y., Gao W. (2019). Wearable and flexible electronics for continuous molecular monitoring. *Chem. Soc. Rev.*, 48, 1465.
8. Ali A., Shehzad K., Guo H., et al. (2017). High-performance, flexible graphene/ultra-thin silicon ultra-violet image sensor. *IEEE International Electron Devices Meeting*, IEEE, 8.6.1–8.6.4.
9. Sargent E. H. (2012). Colloidal quantum dot solar cells. *Nat. Photonics*, 6, 133.
10. Sukhovatkin V., Hinds S., Brzozowski L., Sargent E. H. (2009). Colloidal quantum-dot photodetectors exploiting multiexciton generation. *Science*, 324, 1542.
11. Richter M., Heumüller T., Matt G. J., Heiss W., Brabec C.J. (2017). Carbon photodetectors: the versatility of carbon allotropes. *Adv. Energy Mater.*, 7(10), 1601574.
12. Barkelid M., Zwiller V. (2014). Photocurrent generation in semiconducting and metallic carbon nanotubes. *Nat. Photonics*, 8(1), 47.
13. Zhang S., Cai L., Wang T., et al. (2017). Fully printed flexible carbon nanotube photodetectors. *Appl. Phys. Lett.*, 110, 123105.
14. Hossain M., Sandeep Kumar G., Prabhava S. N. B., et al. (2018). Transparent, flexible silicon nanostructured wire networks with seamless junctions for high-performance photodetector applications. *ACS Nano*, 12(5), 4727.
15. Luo T., Liang B., Liu Z., et al. (2015). Single-GaSb-nanowire-based room temperature photodetectors with broad spectral response. *Sci. Bull.*, 60, 101.
16. Zhang H., Dai X, Guan N., et al. (2016). Flexible photodiodes based on nitride core/shell p–n junction nanowires. *ACS Appl. Mater. Interfaces*, 8, 26198.
17. Xia F., Mueller T., Lin Y. M., Valdes-Garcia A., Avouris P. (2009). Ultrafast graphene photodetector. *Nat. Nanotechnol.*, 4, 839.
18. Zheng Z., Gan L., Zhang J., Zhuge F., Zhai T. (2017). An enhanced UV-Vis-NIR and flexible photodetector based on electrospun ZnO nanowire array/PbS quantum dots film heterostructure. *Adv. Sci.*, 4(3), 1600316.
19. Wang S., Sun H., Wang Z., et al., (2019). In situ synthesis of monoclinic β-Ga_2O_3 nanowires on flexible substrate and solar-blind photodetector. *J. Alloys Compd.*, 787(30), 133.
20. Liu Z., Huang H., Liang B., et al. (2012). Zn_2GeO_4 and $In_2Ge_2O_7$ nanowire mats based ultraviolet photodetectors on rigid and flexible substrates. *Opt. Express*, 20, 2982.
21. Xie X., Shen G. (2015). Single-crystalline In_2S_3 nanowire-based flexible visible-light photodetectors with an ultra-high photoresponse. *Nanoscale*, 7, 5046.
22. Graham R., Miller C., Oh E., Yu. D., (2011), Electric field dependent photocurrent decay length in single lead sulfide nanowire field effect transistors. *Nano Lett.*, 11(2), 717.
23. Chen Q., De Marco N., Yang Y., et al. (2015). Under the spotlight: The organic-inorganic hybrid halide perovskite for optoelectronic applications. *Nano Today*, 10(3), 355.
24. Xu X., Zhang X., Deng W., et al. (2018). Saturated vapor-assisted growth of single-crystalline organic-inorganic hybrid perovskite nanowires for high-performance photodetectors with robust stability. *ACS Appl. Mater. Interfaces*, 10(12), 10287.
25. Asuo I. M., Fourmont P., Ka I., et al. (2019). Highly efficient and ultrasensitive large-area flexible photodetector based on perovskite nanowires. *Small*, 15(1), 1804150.

26. Mamiyev Z. Q., Balayeva N. O. (2015). Preparation and optical studies of PbS nanoparticles. *Opt. Mater.*, 46, 522–525.
27. Patil R. S., Pathan H. M., Gujar T. P., Lokhande C. D., (2006). Characterization of chemically deposited nanocrystalline PbS thin films. *J. Mater. Sci.*, 41(17), 5723–5725.
28. Wang W., Liu Y., Zhan Y., Zheng C., Wang G. (2001). Synthesis and characterization of aragonite whiskers by a novel and simple route. *J. Mater. Chem.*, 11, 1752–1754.
29. Zhu J., Liu S., Palchik O., Koltypin Y., Gedanken A. (2000). A novel sonochemical method for the preparation of nanophasic sulfides: synthesis of HgS and PbS nanoparticles. *J. Solid State Chem.*, 153(2), 342–348.
30. Chen S., Liu W., Yu L. (1998). Preparation of DDP-coated PbS nanoparticles and investigation of the antiwear ability of the prepared nanoparticles as additive in liquid paraffin. *Wear*, 218, 153–158.
31. Kruis F. E., Nielsch K., Fissan H. (1998). Preparation of size-classified PbS nanoparticles in the gas phase. *Appl. Phys. Lett.*, 73, 547.

18 Effect of Stiffness in Sensitivity Enhancement of MEMS Force Sensor Using Rectangular Spade Cantilever for Micromanipulation Applications

Monica Lamba, Himanshu Chaudhary, and Kulwant Singh
Manipal University Jaipur

CONTENTS

18.1 INTRODUCTION

The MEMS market is expected to witness a CAGR of 6.34% for the forecast period of 2020–2025 due to its increasing demand over the past few years in various fields of application, including electronics such as wearable devices, smartphones, tablets, digital cameras portable navigation devices and media players, and gaming consoles [1].

MEMS sensors are mainly categorized depending upon their type, application, and geography, as indicated in Figure 18.1. Among all MEMS sensors, the force sensor shares a big portion of the total revenue. Microscale force sensors are required to protect small-scale structures as they efficiently measure forces in micro-Newton ranges and can be utilized as force feedback in various fields, including minimally invasive surgeries, material science, lifescience, and mechanobiology [2–6], as indicated in Figure 18.2.

The force feedback improves the speed and accuracy in performing micromanipulation tasks [7,8]. MEMS sensor mainly consists of mechanical and sensing structures. MEMS mechanical structures include diaphragms, cantilever beams, and electrostatic motors; however, microcantilever beams are the most preferred due to their flexibility, versatility, high sensitivity, and low cost [9,10]. Microcantilever beams convert the applied force into displacement. The sensing structure of the MEMS force sensor is mainly categorized into electrical and optical force sensors, as shown in Figure 18.3. In this chapter, the focus will be on the piezoresistive sensing mechanism falling under the category of electrical-type MEMS force sensor owing to its small size, high resolution, low phase lag, low cost, high sensitivity, high dynamic range, easy fabrication, and easy integration.

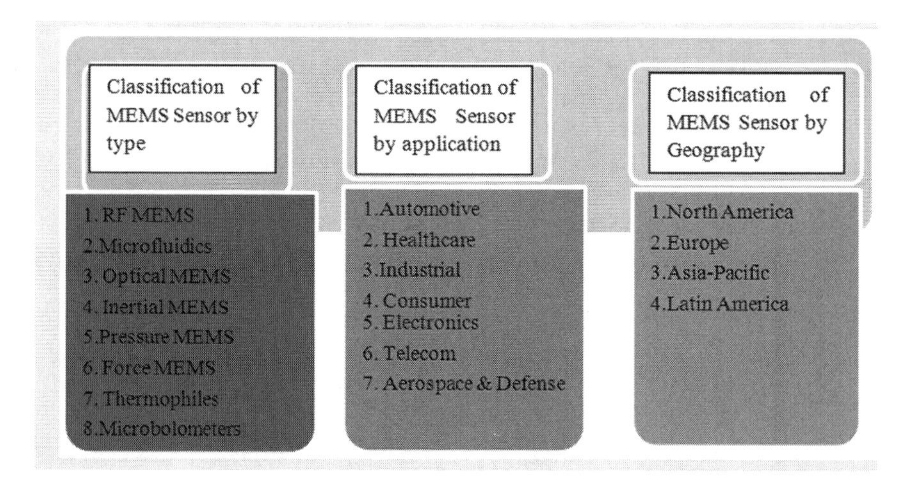

FIGURE 18.1 Classification of MEMS sensors.

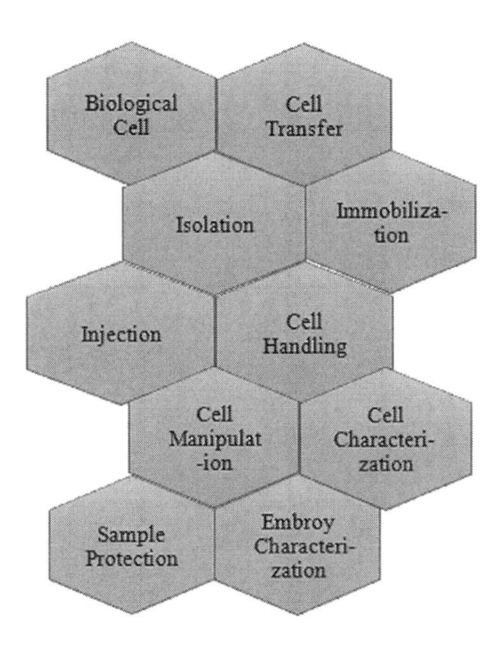

FIGURE 18.2 Applications of MEMS force sensors.

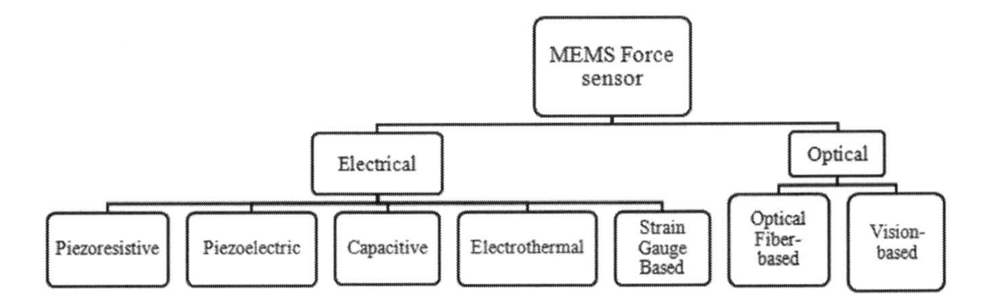

FIGURE 18.3 Different types of sensing mechanisms in MEMS force sensors.

To analyze the performance of the sensor, sensitivity is one of the key parameters that need to be investigated. There are various factors, as shown in Figure 18.4, which influence the sensitivity. The sensitivity of piezoresistive microcantilever-based force sensor can be enhanced by varying the sensors' design parameters, which include the dimension of the cantilever [11] and design, as reported by Ansari and Cho [12] and Zhang et al. [13], cantilever material, as reported by Wee et al. [14] and Nordström et al. [15], and by varying piezoresistor dimension and location on microcantilever, as discussed by Goericke and King [16]. The cantilever and piezoresistors are arranged in such a manner that higher stress is induced inside the piezoresistors to enhance the sensitivity of the designed sensor. Stress-concentrated region is also one of the techniques of sensitivity enhancement, as mentioned in references [17–21].

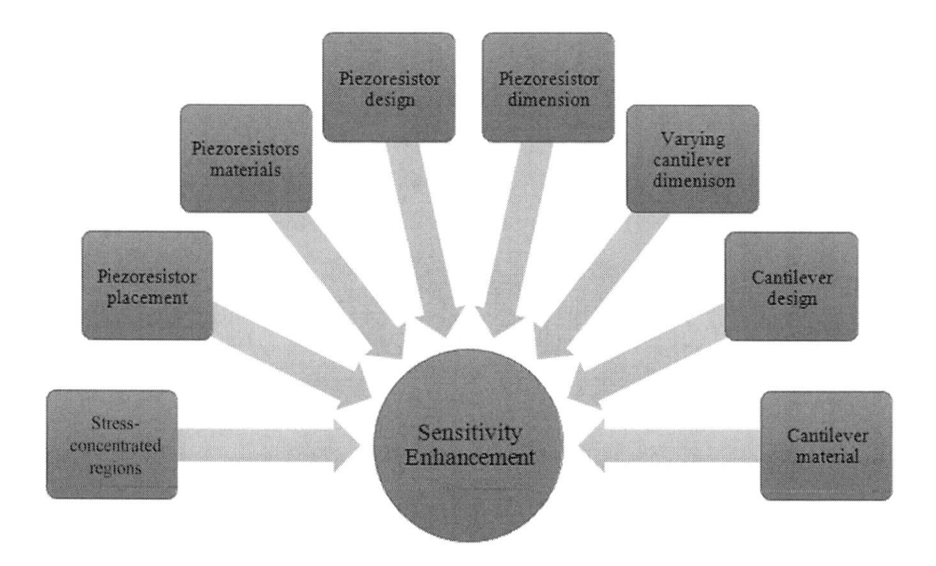

FIGURE 18.4 Various factors affecting sensitivity enhancement in piezoresistive microcantilever-based force sensor.

Moreover, the above-mentioned studies related to the sensitivity enhancement of piezoresistive microcantilever-based force sensor were restricted to the geometrical parameters of cantilevers, piezoresistors, and their placement along with their material properties. To the best of the authors' knowledge, the correlation between the stiffness and electrical sensitivity has not been reported yet. Therefore, there exists a need to analyze the effect of stiffness on a piezoresistive force sensor. In this pursuit, an investigation has been undertaken in this study by considering a unique design of a rectangular microcantilever with a rectangular spade as a mechanical structure along with different combinations of substrate and piezoresistor materials. Finite element analysis has been performed using COMSOL Multiphysics 5.3a software to examine maximum displacement and electrical sensitivity of a sensor using a varied combination of flexible and non-flexible materials operated in the range of 1–10 μN.

18.2 THEORETICAL ANALYSIS AND MATHEMATICAL EQUATIONS

18.2.1 Basic Operating Principle

When low-magnitude forces in the micro-Newton range are to be sensed for micromanipulation application, the sensor needs to be highly efficient to sense these forces effectively. The force exerted by these micromanipulation tasks on the tip of the microcantilever is indicated in the form of displacement which is further converted into an electrical signal by piezoresistive sensing mechanism. The displacement of the microcantilever due to applied force is calculated by equation (18.1) [22].

$$z = \frac{4(1-v)\sigma L^2}{Et^2} \tag{18.1}$$

where z is the displacement of the microcantilever which depends upon the Poisson's ratio (ν), stress (σ), Young's modulus of elasticity (E), and the geometrical parameters length (L) and thickness (t).

18.2.2 Mechanical Sensing Structure

The microcantilever is one of the versatile, simple, and flexible sensing structures most widely used in physical, chemical, and biological MEMS sensors. To enhance the bending stress at the joint of the fixed and moving parts of the microcantilever, circular, triangular, or rectangular-shaped spade can be connected at the free end of the microcantilever. However, for this study, a rectangular spade is taken into consideration because it provides more area under investigation than other geometrical shapes, as shown in Figure 18.5. The total length of the microcantilever is defined as L, which is the sum of the cantilever's length l_c and spade length l_S. The width of the cantilever and rectangular spade is represented as b_c and b_S. The thickness of the spade is greater than the thickness of the microcantilever such that $b_S = \lambda b_c$. The thickness of the cantilever and rectangular spade is kept the same.

The maximum deflection and corresponding maximum bending stress of rectangular spade microcantilever structure given by Euler–Bernoulli beam theory for a microcantilever with a uniformly distributed load applied at the top surface is indicated in equation (18.2), defining the longitudinal stress at a distance c above the neutral axis as a function of x from the fixed end [23].

$$\sigma = \frac{6w(L-x)^2 c}{bt^3} \tag{18.2}$$

From equation (18.2) it can be easily perceived that bending stress developed on the microcantilever depends on the geometrical dimensions of the microcantilever and is independent of the mechanical properties of the material. By increasing the length and reducing the thickness, longitudinal stress is enhanced after a point reduction in the thickness of the cantilever, which increases the fabrication cost and the structural reliability of the sensor.

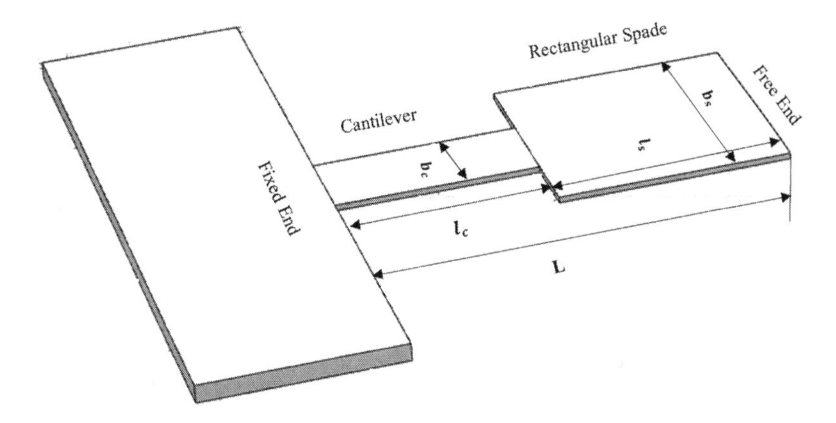

FIGURE 18.5 Rectangular spade microcantilever structure.

18.2.3 Modes of Operating of Rectangular Spade Microcantilever Structure

Based on the principle of translating the recognition event into mechanical motion, the modes of the cantilever are mainly categorized into three, namely, static, dynamic, and heat mode. In static mode, deflection of the microcantilever due to stress within can be measured by various sensing mechanisms such as optical, capacitive, and piezoresistive. In dynamic mode, because of external actuation, microcantilever oscillates at its resonance (natural) frequency. Variation in load or mass can shift the resonance frequency of the microcantilever. In heat mode, the bending of the microcantilever is caused by a change in temperature by taking advantage of the bimetallic effect. In this study, for sensing low-magnitude force in micro-Newton range, a rectangular spade microcantilever structure is operated in static mode utilizing a piezoresistive sensing mechanism.

18.2.4 Deflection Detection Method: Piezoresistive Sensing

Force is measured by detecting the deflection of the rectangular spade microcantilever structure via a piezoresistive sensing mechanism. It engrosses the piezoresistors on the microcantilever in Wheatstone bridge configuration in such a manner that out of four equally valued resistors at least one should be in a high-stress region to sense the mechanical deflection of the microcantilever, which is indicated through a change in resistance ΔR further calibrated into voltage. The bending stress of the cantilever is linearly associated with the piezoresistive effect. The ratio of change in resistance ΔR (with stress) with respect to original resistance (without stress) R is expressed as follows [24,25]:

$$\frac{\Delta R}{R} = \sigma_l \pi_l \tag{18.3}$$

where σ_l is the longitudinal stress developed on rectangular spade microcantilever structure, and π_l is the piezoresistive coefficient of the material under consideration. The Wheatstone bridge is shown in Figure 18.6 indicating the supply voltage V_{in} applied between terminals A and B, and V_{out} is obtained across terminals C and D. The output voltage V_{out} of quarter Wheatstone is indicated below defining R_1, R_2, R_3, and R_4 as four equally valued resistances. When no force is applied on the microcantilever, Wheatstone bridge is in a balanced condition, but shifts to an unbalanced condition when uniformly distributed force is applied on the top side of the microcantilever. The voltage of the quarter Wheatstone bridge under unbalanced condition is given by the following equation calculated by simplifying Figure 18.6 into Figure 18.7.

$$V_{AC} = \frac{R_4}{R_2 + R_4} V_{CC} \tag{18.4}$$

$$V_{BC} = \frac{R_2}{R_1 + R_2} V_{CC} \tag{18.5}$$

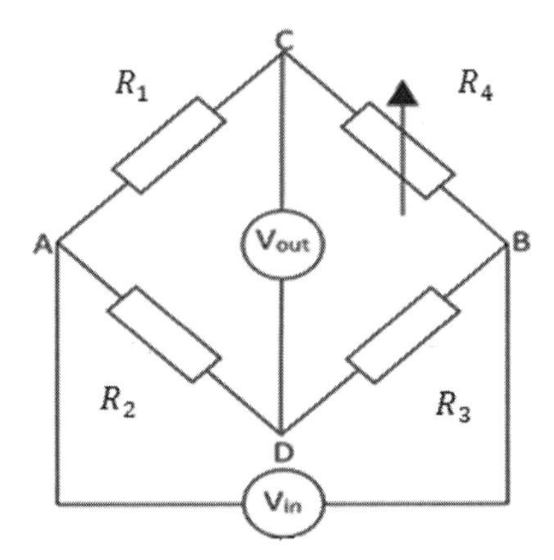

FIGURE 18.6 Quarter Wheatstone bridge configuration.

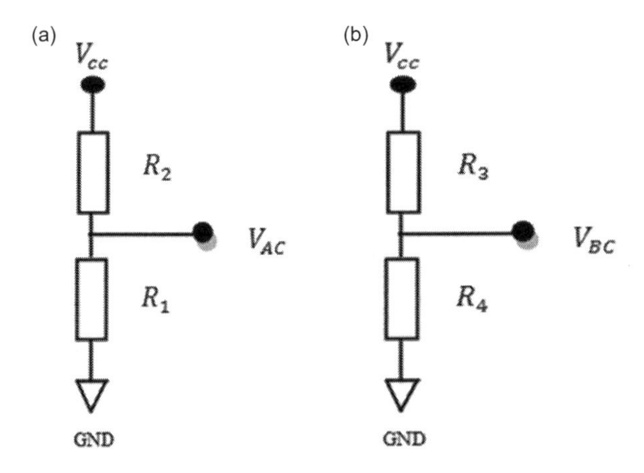

FIGURE 18.7 Voltage divider configuration of Wheatstone bridge for solving the circuit: (a) voltage between A and C terminal and (b) voltage between B and C terminal.

$$V_{\text{out}} = V_{AC} - V_{BC} \tag{18.6}$$

$$\begin{aligned}
&= \frac{R_4}{R_2 + R_4} V_{CC} - \frac{R_2}{R_1 + R_2} V_{CC} \\
&= V_{CC}\left(\frac{R_4}{R_2 + R_4} - \frac{R_2}{R_1 + R_2} \right) \\
&= V_{CC}\left(\frac{R_1R_4 + R_2R_4 - R_2R_2 - R_2R_4}{(R_2 + R_4)(R_1 + R_2)} \right) \\
V_{\text{out}} &= V_{CC}\left(\frac{R_1R_4 - R_2R_2}{(R_2 + R_4)(R_1 + R_2)} \right)
\end{aligned} \tag{18.7}$$

As $R_1 = R_2 = R_3 = R_4 = R$, at unbalanced condition of Wheatstone bridge $R_4 = R + \Delta R$, and the output voltage is given by the following equations:

$$V_{AC} = \frac{R + \Delta R}{2R + \Delta R} V_{CC} \tag{18.8}$$

$$V_{BC} = \frac{R}{R + R} V_{CC} \left(\frac{V_{CC}}{2} \right) \tag{18.9}$$

$$V_{\text{out}} = \frac{R + \Delta R}{2R + \Delta R} V_{CC} - \frac{R}{R + R} V_{CC} \left(\frac{V_{CC}}{2} \right) \tag{18.10}$$

$$= V_{CC} \left(\frac{(2(R + \Delta R) - 2R - \Delta R)}{2(2R + \Delta R)} \right)$$

$$= V_{CC} \left(\frac{(2R + 2\Delta R - 2R - \Delta R)}{(4R + 2\Delta R)} \right)$$

$$= V_{CC} \left(\frac{(\Delta R)}{(4R + 2\Delta R)} \right)$$

$$= V_{CC} \left(\frac{\Delta R}{4R(1 + \frac{\Delta R}{2R})} \right)$$

As non-linearity component $\frac{\Delta R}{2R} \ll 1$

$$V_{\text{out}} \approx V_{CC} \frac{\Delta R}{4R} \text{ as } \frac{\Delta R}{2R} \ll 1 \tag{18.11}$$

18.2.5 Sensitivity of MEMS Force Sensor

The performance of the sensor is judged based on the sensitivity of the designed sensor. On increasing the sensitivity of the sensor, its performance increases. The sensitivity (S) of the MEMS force sensor is defined as the ratio of change in voltage to the change in applied force, and is calculated using equation (18.12)

$$S = \frac{\Delta V}{\Delta F} \tag{18.12}$$

For obtaining the output voltage corresponding to the applied force, the supply voltage of 3.3V is applied at the sensor. The unit of the measured sensitivity is milli-volt per micro-Newton (mV/μN).

18.3 MODELING OF RECTANGULAR SPADE MICROCANTILEVER

Here, analytical modeling of rectangular spade-based microcantilever is evaluated to determine the deflection and bending stress in rectangular spade microcantilever caused by uniformly distributed force applied by microbots. The width of the spade and cantilever is denoted by b_s and b_c, whereas the length of the spade and cantilever is denoted by l_c and l_s. The relation between b_s and b_c shows that $b_s = \lambda b_c$ where $\lambda \geq 1$. The thickness of the cantilever and spade is the same, t. Because of the uniformly distributed force applied on the top side of the cantilever and spade, the deflection δ_x at distance x from the free end is indicated in the free body diagram shown in Figure 18.8.

Deflection in the micro cantilever region is $\delta_c\,(0 \leq x \leq l_c)$ and in the spade region is $\delta_s\,(l_c \leq x \leq L)$, and total deflection δ_x is obtained by the addition of δ_c and δ_c [26].

$$\delta_{\max} = \delta_c(0 \leq x \leq l_c) + \delta_s(l_c \leq x \leq L) \tag{18.13}$$

As, $\delta_c = \left(\dfrac{q_c l_c^4}{8EI_c}\right)_c$ and $\delta_c = \left(\dfrac{q_s l_s^4}{8EI_s}\right)_s$ equation (18.13) is written as:

$$\delta_{\max} = \left(\frac{q_c l_c^4}{8EI_c}\right)_c + \left(\frac{q_s l_s^4}{8EI_s}\right)_s \tag{18.14}$$

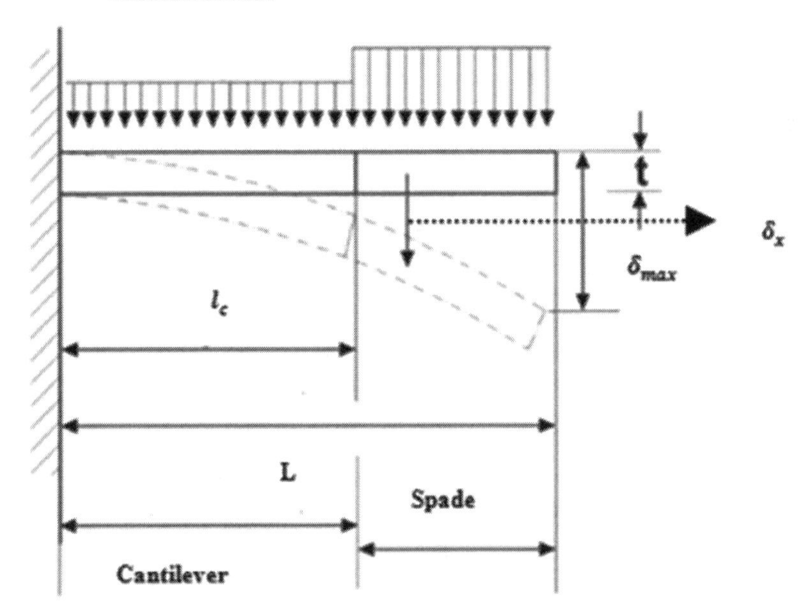

FIGURE 18.8 Free body diagram of the rectangular cantilever with a rectangular spade due to uniformly distributed force.

where q is the uniformly distributed load, L is the length of the cantilever, and E and I are the Young's modulus of elasticity and moment of inertia, respectively. As $q_c = \frac{F}{l_c}$, $q_s = \frac{F}{l_s}$, $I_c = \frac{b_c h_c^3}{12}$, and $I_s = \frac{b_s h_s^3}{12}$, so equation (18.14) is modified as follows:

$$\delta_{\max} = \left(\frac{3Fl_c^3}{2Eb_c t^3}\right)_c + \left(\frac{3Fl_s^3}{2Eb_s t^3}\right)_s \tag{18.15}$$

$$\delta_{\max} = \frac{3F}{2Et^3}\left(\frac{l_c^3}{b_c} + \frac{l_s^3}{b_s}\right) \tag{18.16}$$

The maximum bending stress at the fixed end of the microcantilever is $\sigma_{\max}$ which is the sum of bending stress due to the microcantilever and spade σ_c and σ_s is calculated as

$$\sigma_{\max} = \sigma_c(0 \le x \le l_c) + \sigma_s(l_c \le x \le L) \tag{18.17}$$

The maximum stress developed at the joint of fixed and moving part of the cantilever is calculated as

$$\sigma_{\max} = \frac{3F}{t^3}\left(\frac{l_c}{b_c} + \frac{l_s}{b_s}\right) \tag{18.18}$$

Using equations (18.16) and (18.18), maximum deflection and maximum stress can be analyzed.

18.4 DEVICE DESIGN

In this chapter, the piezoresistive rectangular spade MEMS force sensor is taken into consideration. The top view of the piezoresistive rectangular spade microcantilever is shown in Figure 18.9 indicating rectangular spade, cantilever, contact probes, and piezoresistors. The substrate material used for developing a force sensor is made up of silicon or PDMS (Young's Modulus = 750 kPa), which freely deflects with uniformly distributed applied force. The length of the cantilever is $l_c = 700$ µm and breadth is $b_c = 300$ µm. The dimension of rectangular spade having $l_s = 800$ µm with breadth $b_s = 900$ µm is attached at the tip of the cantilever so that more stress can be developed at the joint of the fixed and moving part of the microcantilever. This thickness to the fixed part of the sensor is higher than the thickness of the cantilever and rectangular paddle. The thickness allotted to the cantilever and the paddle is $t_c = t_s =$ 20 µm. Even thickness of the cantilever and spade provide mechanical strength to the sensor. To avoid the effect of temperature, four equal dimensions piezoresistors made of graphene or polysilicon are connected in a quarter Wheatstone bridge configuration. The dimensions of all the piezo resistors are kept the same as $l_{\text{piezo}} = 1{,}460$ µm, $b_{\text{piezo}} = 20$ µm, and $t_{\text{piezo}} = 1$ nm to attain an identical value of resistance. One of the

(a)

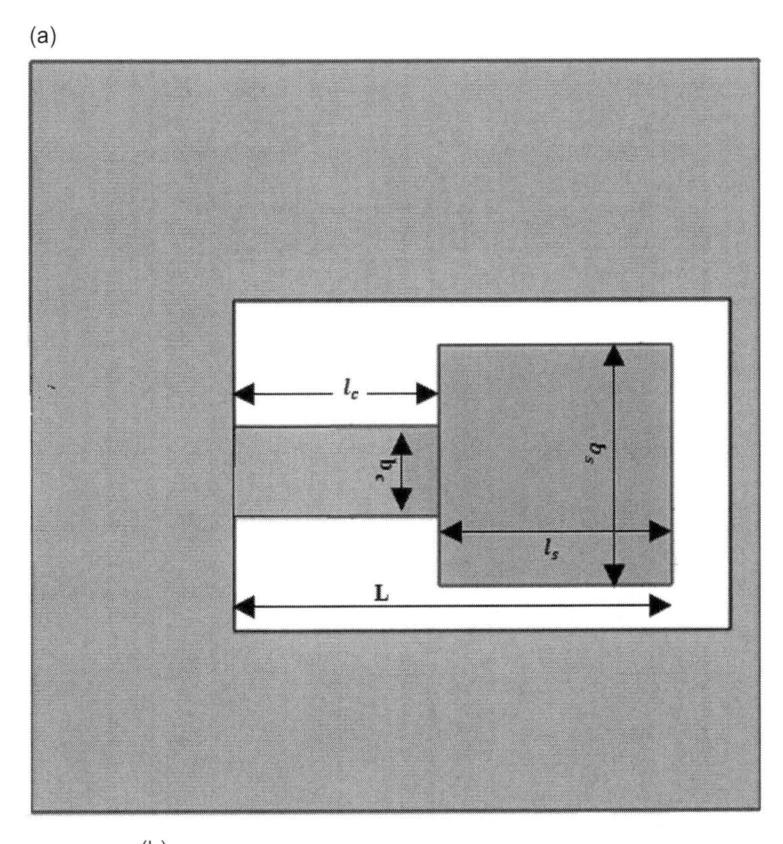

(b)

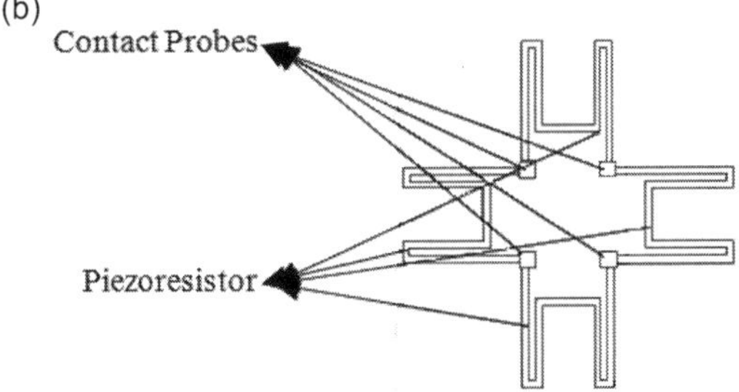

FIGURE 18.9 Top view of the MEMS force sensor (a) rectangular cantilever with a rectangular spade and (b) piezoresistors connection with contact pads.

resistors of the Wheatstone bridge is placed on the moving part of the cantilever to experience maximum stress, which further affects the electrical sensitivity of the designed sensor. The remaining three resistors are placed on the fixed part of the cantilever so that they cannot experience any variations.

The properties of the material under consideration for simulation are indicated in Table 18.1, along with definitions of the material properties of silicon, polysilicon, PDMS, graphene, and gold, which are considered for the simulation of the designed

TABLE 18.1
Material Properties Under Consideration for Force Sensor

	Materials				
Properties and Its Units	**Silicon (Substrate)**	**Polysilicon (Piezoresistors)**	**PDMS (Substrate)**	**Graphene (Piezoresistors)**	**Gold (Metal Contacts)**
Density (kg/m^3)	2,329	2,320	970	2,250	19,300
Young's modulus (Pa)	170 e^9	169 e^9	750 e^3	1 e^{12}	70 e^9
Poisson's ratio (unit less)	0.28	0.22	0.49	0.456	0.44
Coefficient of thermal expansion (1/K)	2.6 e^{-6}	2.9 e^{-6}	9 e^{-4}	2.6 e^{-6}	14.2 e^{-6}

sensor. It mainly includes density, Young's modulus of elasticity, poison's ratio, and coefficient of thermal expansion.

18.5 RESULTS AND DISCUSSION

Simulation of a piezoresistive rectangular microcantilever with a rectangular spade is carried out in COMSOL Multiphysics 5.3a software by considering four different combinations of substrate-piezoresistor materials. Different combinations of materials are chosen to analyze the effect of stiffness in the sensor on displacement, displacement sensitivity, and electrical sensitivity.

18.5.1 Displacement and Displacement Sensitivity

The simulated model of the designed force sensor shown in Figure 18.10 indicates the displacement of microcantilever under uniformly distributed force. The maximum displacement is at the free end of the rectangular spade.

The designed force sensor is simulated for different substrate-piezoresistor materials by considering a combination of silicon-polysilicon, silicon-graphene, PDMS-polysilicon, and PDMS-graphene. Of these four combinations, silicon-polysilicon and silicon-graphene are considered in Category 1, whereas PDMS-polysilicon and PDMS-graphene are in Category 2, as tabulated in Table 18.2.

Figure 18.11a and b indicates the force-displacement characteristics of Category 1 and 2 materials. From Figure 18.11a, it can be observed that the displacement sensitivity of Category 1 materials is the maximum in silicon-graphene compared to silicon-polysilicon. Figure 18.11b shows that the displacement sensitivity of Category 2 materials is the maximum in PDMS-graphene as compared to PDMS-polysilicon. Substrate materials of Category 2 are more flexible than substrate materials in Category 1, due to which displacement sensitivity of Category 2 materials is much higher than Category 1.

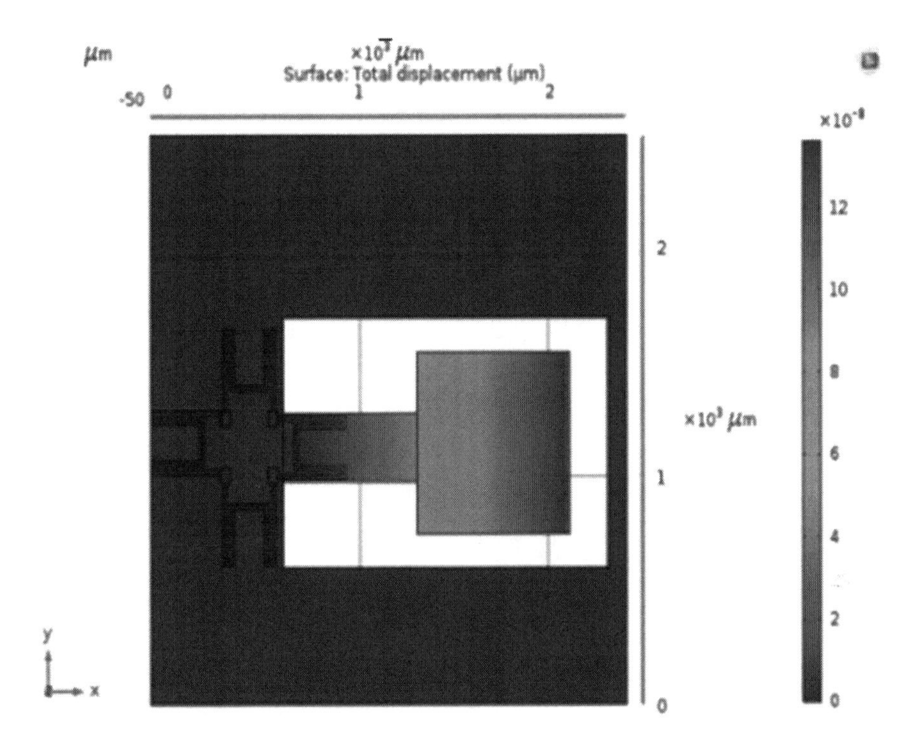

FIGURE 18.10 Displacement developed in a simulated model of rectangular spade MEMS force sensor.

TABLE 18.2
Categorization of the Substrate-Piezoresistor Materials for the Simulated Force Sensor

Category	Substrate-Piezoresistors	Type of Materials	Type of Sensor
1	Silicon-polysilicon	Non-flexible–non-flexible	Fully non-flexible
	Silicon-graphene	Non-flexible–flexible	Partially flexible
2	PDMS-polysilicon	Flexible–non-flexible	Partially flexible
	PDMS-graphene	Flexible–flexible	Fully flexible

18.5.2 Stress and Electrical Sensitivity

Figure 18.12a indicates that the maximum stress is observed at the joint of fixed and moving parts of the microcantilever. One of the piezoresistors is placed in the maximum stress region to detect this stress. Developed stress is sensed by the piezoresistive sensing mechanism connected in Wheatstone bridge configuration and generates an electrical potential across the contact pads of the piezoresistors, as shown in Figure 18.12b.

Figure 18.13 aindicatesthe force-voltage characteristics of Category 1 materials, which shows that electrical sensitivity is higher in silicon-graphene

(a)

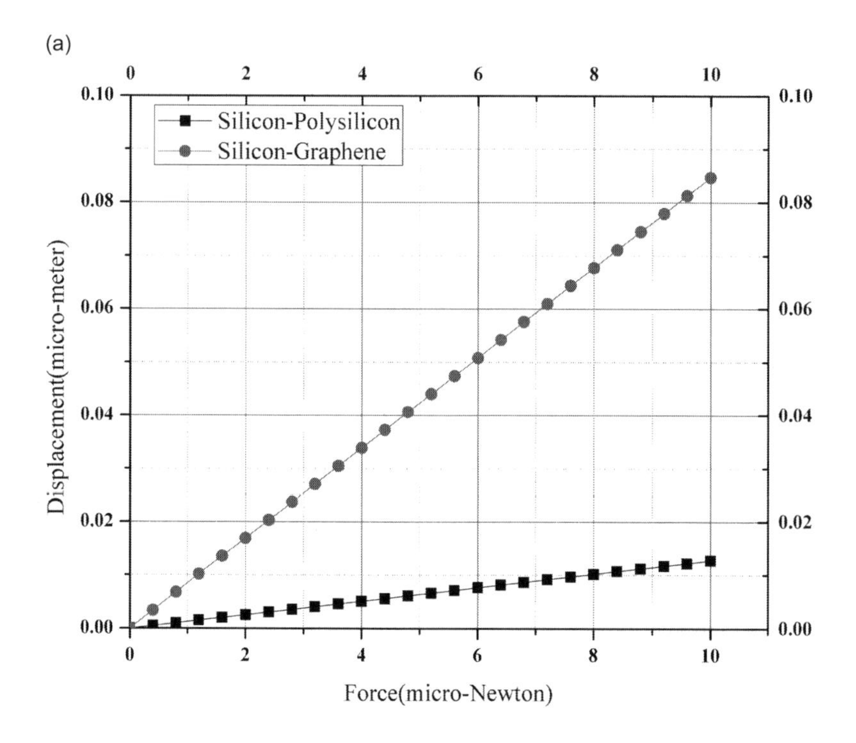

(b)

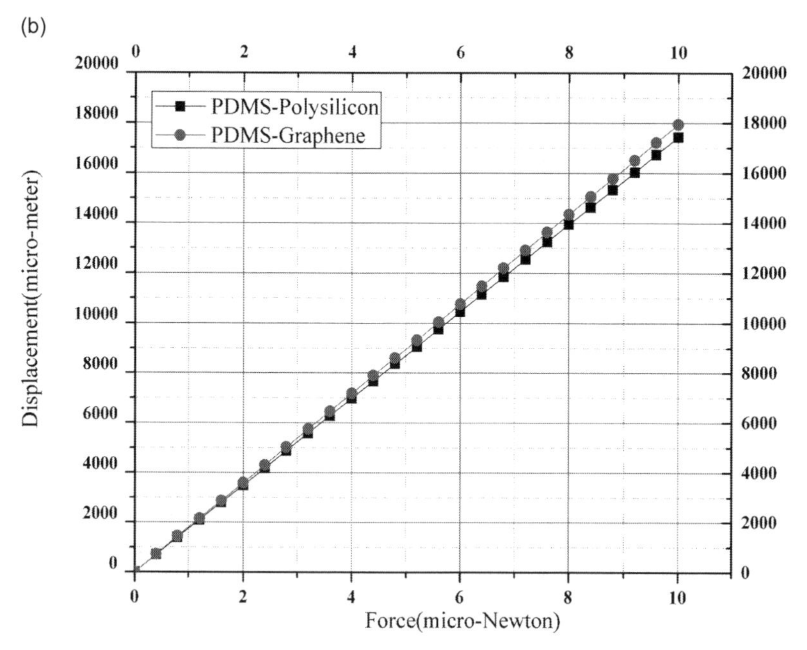

FIGURE 18.11 Comparison of displacement sensitivity of rectangular cantilever rectangular spade force sensor for (a) silicon-polysilicon and silicon-graphene and (b) PDMS-polysilicon and PDMS-graphene.

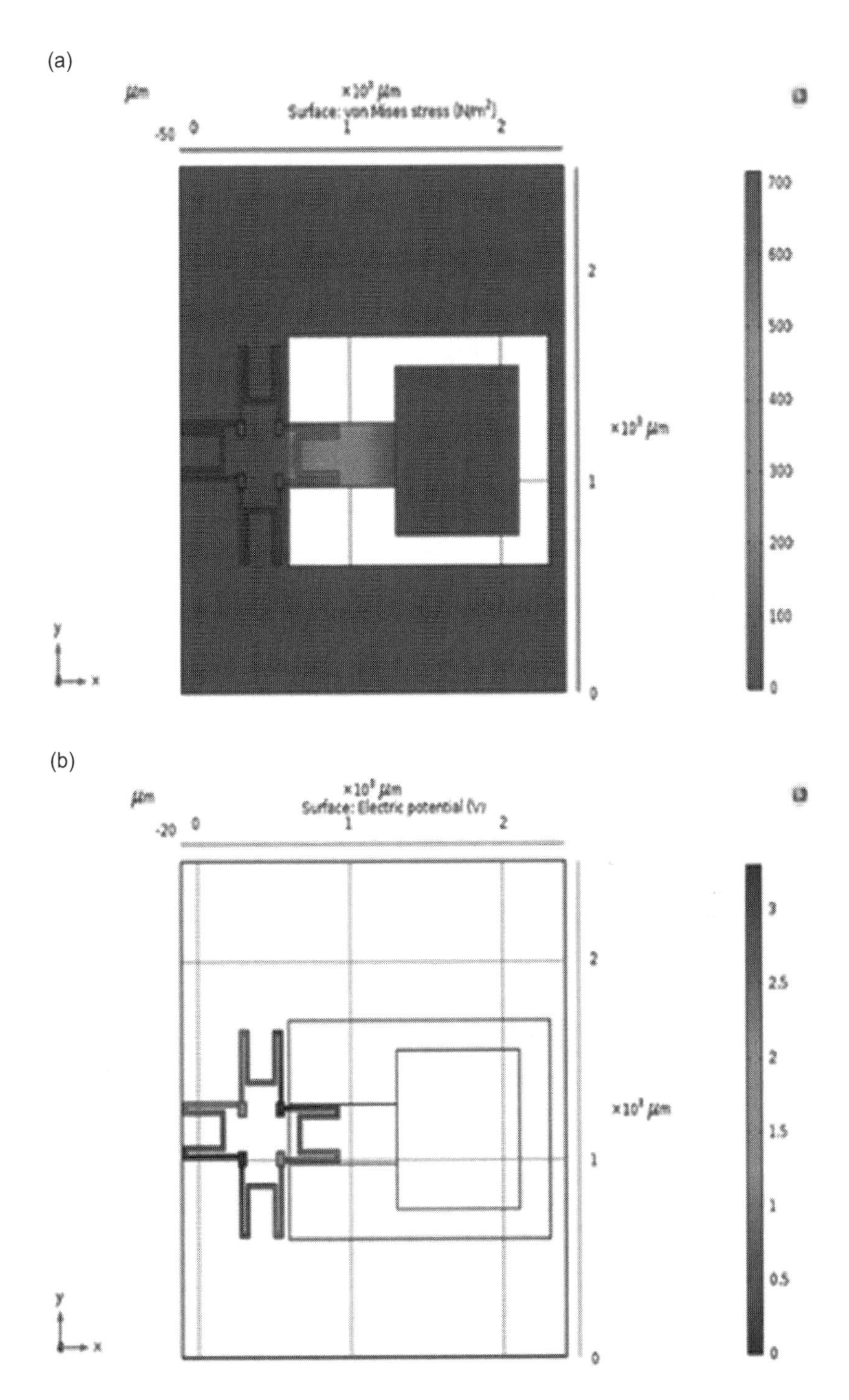

FIGURE 18.12 Designed MEMS force sensor (a) Von-Mises stress and (b) electrical potential developed across piezoresistors.

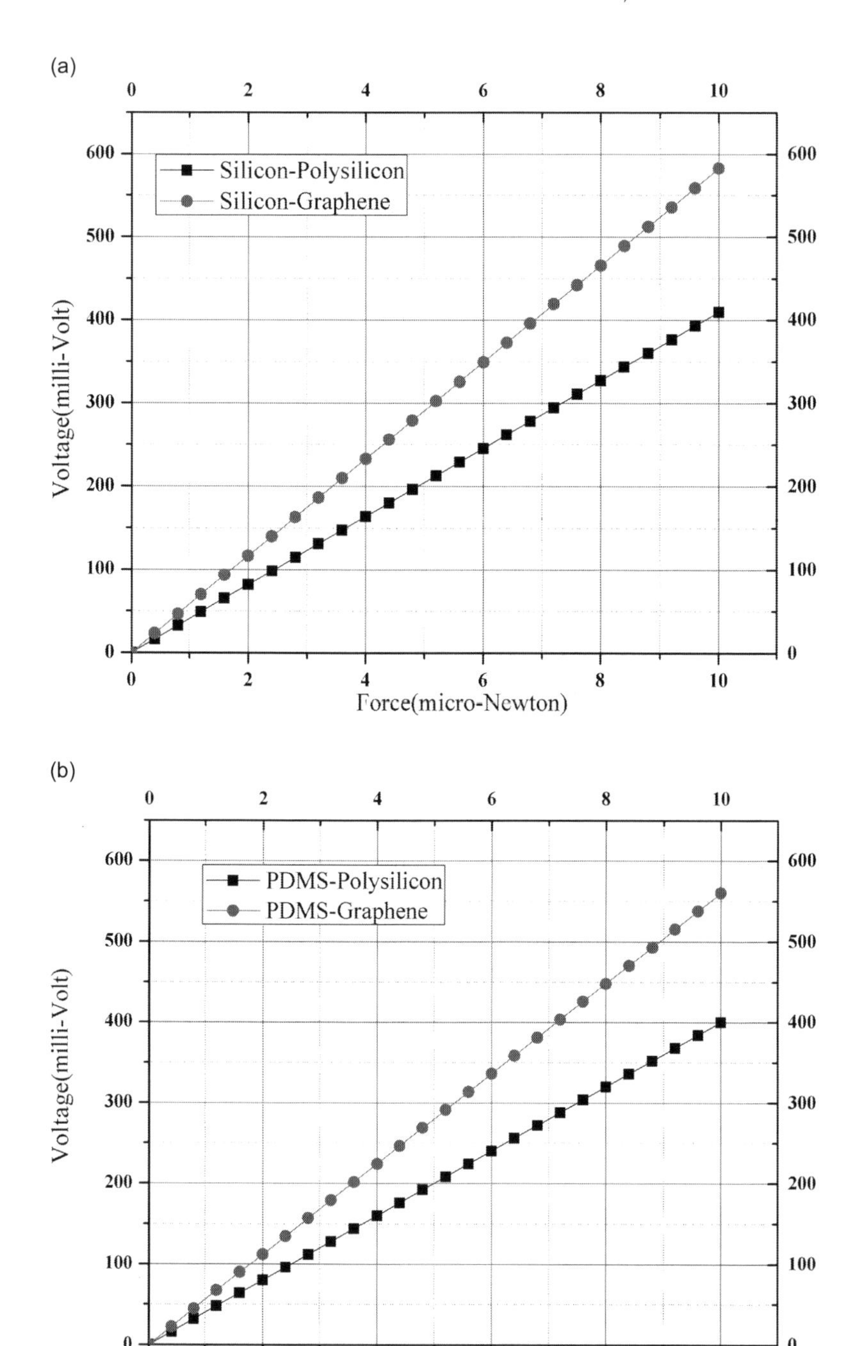

FIGURE 18.13 Comparison of electrical sensitivity of rectangular cantilever rectangular spade force sensor for (a) silicon-polysilicon and silicon-graphene and (b) PDMS-polysilicon and PDMS-graphene.

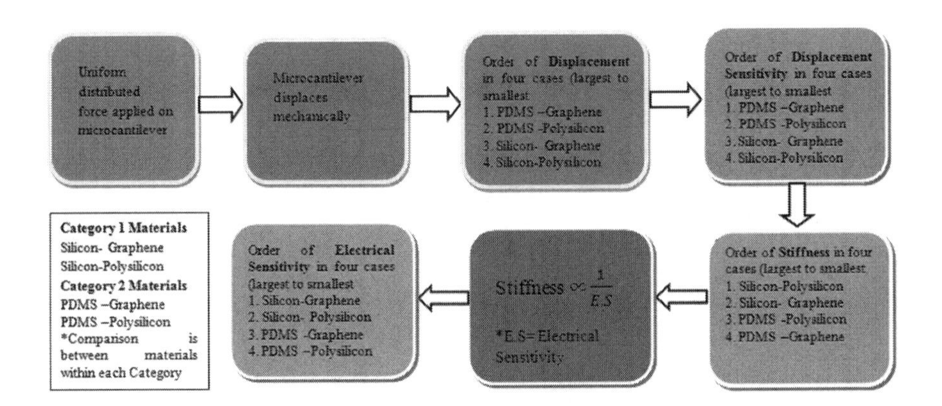

FIGURE 18.14 Block diagram indicating the effect of stiffness in the sensitivity of the designed MEMS force sensor.

(58.245 mV/μN) compared to silicon-polysilicon (40.938 mV/μN), whereas in Category 2 PDMS-graphene has higher electrical sensitivity (56.021 mV/μN) compared to PDMS-polysilicon (40.006 mV/μN).

18.5.3 Effect of Stiffness on Electrical Sensitivity

Figure 18.14 explains the role of stiffness in affecting the electrical sensitivity of the sensor. Initially, when uniformly distributed force is applied on the microcantilever, it displaces from its original position and causes fully flexible material (PDMS-graphene) to have the maximum displacement and displacement sensitivity compared to fully non-flexible material (silicon-polysilicon). Stiffness is defined as the ratio of the applied force to the corresponding displacement, and has an inverse relation with the electrical sensitivity of the sensor. Therefore, electrical sensitivity is the maximum in silicon-graphene compared to other combinations.

18.5.4 Application-Specific Operating Range

The geometrical model of the designed sensor is considered analogous for different combinations. The operating range of the sensor is from 1 to 10μN, which makes it suitable for a vast range of applications, including cell manipulation, embryo characterization, microbotics, cell handling, immobilization, etc., as shown in Figure 18.15.

18.6 CONCLUSIONS

This study has established a direct relationship between the stiffness of materials used in rectangular spade cantilever and electrically sensitive piezoresistive MEMS force sensor. For this analysis, four different combinations of substrate-piezoresistor materials of the MEMS force sensor are simulated individually using finite element analysis. From the analyzed results, it can be concluded that when uniformly distributed force is applied to the moving part of the cantilever, stress and the induced

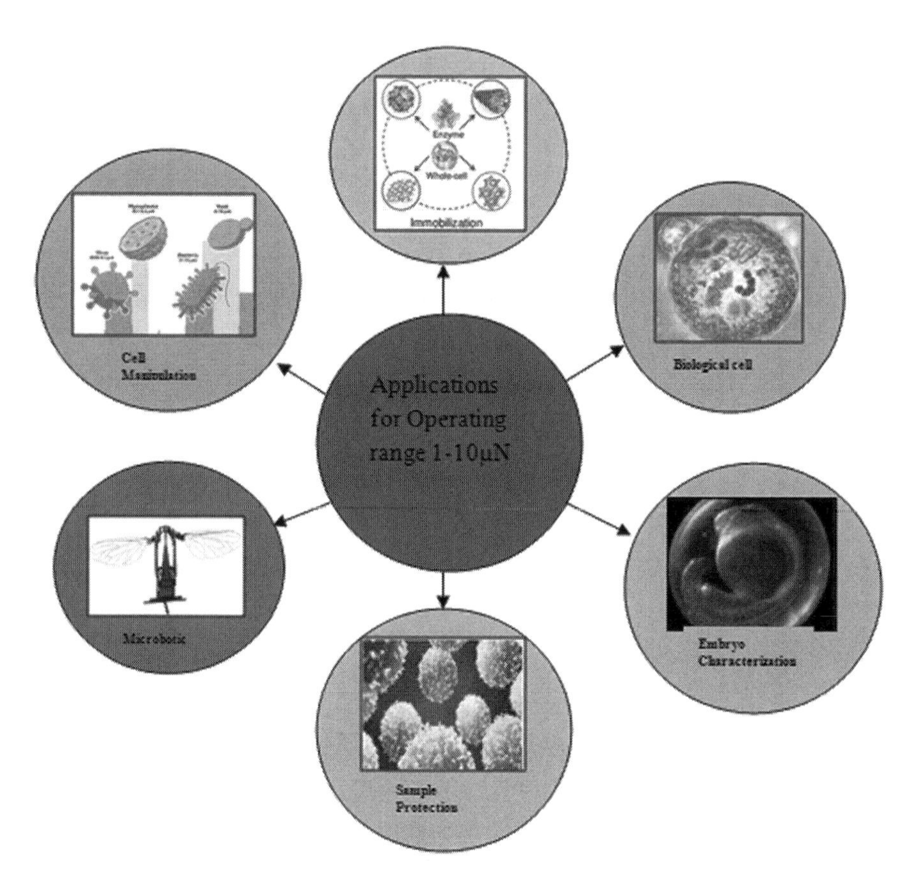

FIGURE 18.15 Applications of the rectangular spade MEMS force sensor.

stiffness increases the electrical sensitivity of the sensor. Materials considered for this analysis are divided into two categories. The findings of the Category 1 materials revealed that the electrical sensitivity of silicon-graphene is 23.8% higher than silicon-polysilicon, whereas in Category 2 materials, electrical sensitivity of PDMS-graphene was 22.4% higher than PDMS-polysilicon. However, overall, observed electrical sensitivities of the different combination under consideration for this range (1–10µN) of MEMS force sensor makes it suitable for cell transferring, isolation, injection, cell manipulations, etc.

REFERENCES

1. DUBLIN, March 19, 2020 /PRNewswire/The MEMS Market - Growth, Trends, and Forecast (2020-2025) report has been added to Research and Markets.com's offering.
2. Sun Y., Nelson B. J.(2007). MEMS capacitive force sensors for cellular and flight biomechanics. *Biomedical Materials*, 2(1), S16.
3. Rebello K. J. (2004). Applications of MEMS in surgery. *Proceedings of the IEEE*, 92(1), 43–55.

4. Ku S., Salcudean S. E. (1996). Design and control of a teleoperated microgripper for microsurgery. *Proceedings of IEEE International Conference on Robotics and Automation, IEEE*, 889–894.
5. Praprotnik J., Ergeneman O., Chatzipirpiridis G., Weidlich A., Blaz S., Pané S. &Nelson B. J. (2015). An array of 2D magnetic micro force sensors for life science applications. *Procedia Engineering*, 120, 220–224.
6. Noda K., Hashimoto Y., Tanaka Y. & Shimoyama I. (2009). MEMS on robot applications. *TRANSDUCERS 2009 – 2009 International Solid-State Sensors, Actuators and Microsystems Conference, IEEE*, 2176–2181.
7. Duc T.C., Creemer J. F. & Sarro P. M. (2006). Piezoresistive cantilever for nano-Newton sensing in two dimensions. *19th IEEE International Conference on Micro Electro Mechanical Systems, IEEE*, 586–589.
8. Boudaoud M. (2014). Regnier S. An overview on gripping force measurement at the micro and nano-scales using two-fingered microrobotic systems. *International Journal of Advanced Robotic Systems*, 11(3), 45.
9. Gopinath P.G., Mastani S. A. & Anitha V. R. (2014). MEMS microcantilevers sensor modes of operation and transduction principles. *International Journal of Computer Engineering Research*, 4(2), 2250–3439.
10. Vashist S. K.(2007). A review of microcantilevers for sensing applications. *Journal of Nanotechnology*, 3, 1–8.
11. Ansari M.Z., Cho C., Choi W., Lee M., Lee S. & Kim J. (2013). Improving sensitivity of piezoresistive microcantilever biosensors using stress concentration region designs. *Journal of Physics D: Applied Physics*, 46(50), 505501.
12. Ansari M.Z., Cho C. (2009). Deflection, frequency, and stress characteristics of rectangular, triangular, and step profile microcantilevers for biosensors. *Sensors*, 9(8), 6046–6057.
13. Zhang G., Zhao L., Jiang Z., Yang S., Zhao Y., Huang E., Wang X. & Liu Z. (2011). Surface stress-induced deflection of a microcantilever with various widths and overall microcantilever sensitivity enhancement via geometry modification. *Journal of Physics D: Applied Physics*, 44(42), 425402.
14. Wee K.W., Kang G. Y., Park J., Kang J.Y., Yoon D.S., Park J. H. & Kim T. S. (2005). Novel electrical detection of label-free disease marker proteins using piezoresistive self-sensing micro-cantilevers. *Biosensors and Bioelectronics*, 20(10), 1932–1938.
15. Nordström M., Keller S., Lillemose M., Johansson A., Dohn S., Haefliger D., Blagoi G., Havsteen-Jakobsen M. & Boisen A. (2008). SU-8 cantilevers for bio/chemical sensing; fabrication, characterisation and development of novel read-out methods. *Sensors*, 8(3), 1595–1612.
16. Goericke F. T. & King W.P. (2008). Modelling piezoresistive microncantilever sensor response to surface stress for biochemical sensors. *IEEE Sensors Journal*, 8, 1404–1410.
17. Bashir R., Gupta A., Neudeck G.W., McElfresh M. & Gomez R. (2000). On the design of piezoresistive silicon cantilevers with stress concentration regions for scanning probe microscopy applications. *Journal of Micromechanics and Microengineering*, 10(4), 483.
18. Yang M., Zhang X., Vafai K. & Ozkan C.S. (2003). High sensitivity piezoresistive cantilever design and optimization for analyte-receptor binding. *Journal of Micromechanics and Microengineering*, 13(6), 864.
19. Yang S.M. & Yin T.I. (2007). Design and analysis of piezoresistive microcantilever for surface stress measurement in biochemical sensor. *Sensors and Actuators B: Chemical*, 120(2), 736–744.
20. Mohammed A.A., Moussa W.A. & Lou E. (2009). Optimization of geometric characteristics to improve sensing performance of MEMS piezoresistive strain sensors. *Journal of Micromechanics and Microengineering*, 20(1), 015015.

21. Wahid K.A., Lee H.W., Shazni M. A. & Azid I. A. (2014) Investigation on the effect of different design of SCR on the change of resistance in piezoresistive micro cantilever. *Microsystem Technologies*, 20(6), 1079–1083.
22. Sader J.E. (2001). Surface stress induced deflections of cantilever plates with applications to the atomic force microscope: Rectangular plates. *Journal of Applied Physics*, 89(5), 2911–2921.
23. Doll J. C., Park S. J. & Pruitt B. L. (2009). Design optimization of piezoresistive cantilevers for force sensing in air and water. *Journal of Applied Physics*, 106(6), 064310.
24. Du L., Zhao Z. & Pang C. (2007). Design and fabrication MEMS-based micro solid state cantilever wind speed sensor. *2007 International Conference on Information Acquisition, IEEE*, 336–340.
25. Lamba, M., Mittal, N., Singh, K. &Chaudhary, H. (2020). Design analysis of polysilicon piezoresistors PDMS (Polydimethylsiloxane) microcantilever based MEMS Force sensor. *International Journal of Modern Physics B*, 34(09), 2050072.
26. Agrawal V.K., Patel R., Boolchandani D. & Rangra K.(2018). Analytical modeling, simulation, and fabrication of a MEMS rectangular paddle piezo-resistive microcantilever-based wind speed sensor. *IEEE Sensors Journal*, 18(18), 7392–7398.

19 Successive Ionic Layer Adsorption and Reaction Deposited ZnS-ZnO Thin Film Characterization

Sampat G. Deshmukh
S. V. National Institute of Technology
SKN Sinhgad College of Engineering

Rohan S. Deshmukh
SKN Sinhgad College of Engineering

Ashish K. Panchal and Vipul Kheraj
S. V. National Institute of Technology

CONTENTS

19.1 INTRODUCTION

Recently, Takuya Kato from Solar Frontier (SF) reported a recorded efficiency of 22.9% for thin-film polycrystalline solar cells based on Cu(In, Ga)(Se, S)$_2$ (CIGSSe) with 10-nm thick CdS as the first buffer layer for the size of 1 cm^2. In 2016, SF also reported a conversion efficiency of 22.0% for small area cells by omitting the CdS buffer layer [1], which is considerably higher than 16.4% previously stated by Bjorkman et al. [2] for SLG/Mo/CuInGaSe$_2$/ZnO, S/ZnO/ZnO:Al structure. However, a major hurdle in the field of PV solar cells is the availability of constituents, production cost, and toxicity of Cd with respect to environmental concerns. In fact, highly abundant elements are good candidates to reduce the production cost; alternative non-toxic buffer layers and growth techniques have been proposed for this technological momentous field.

To overcome this, in the field of photoelectric conversion applications, some ternary or binary compounds whose constitutes are abundant in the Earth's crust are potential can didates, for example, (Zn, Mg)O, Zn(O, S, OH)$_x$ [1], Zn(O, S) [3], ZnS/Zn(S, O), Zn(OH, S), Zn(OH, Se) [4], SnO$_2$/ZnO [5], ZnO/ZnS [6–8], Bi$_2$S$_3$/ZnO [9] and ZnS/ZnO [10]. At this juncture, among semiconductors, ZnS and ZnO members of II–VI group with a wide bandgap value of 3.8 and 3.2 eV, respectively, are attracting additional attention from the scientific community in view of their photoconductivity, photosensitivity, photocatalyst, or photoelectric conversion [4,5,7,8]. Therefore, the facile fabrication of the ZnS-nanoparticle/ZnO-nanoflower structure is currently of crucial importance. In the past, chemical bath deposition [4], chemical precipitation method [5], magnet sputtering [7], hydrothermal [11], and chemical sulfidation [12] have been involved in the deposition of ZnS/ZnO heterostructures. Furthermore, special morphologies, such as hollow dumbbells, nanowires, and nanorods, have been deposited by physical and chemical methods [8,11,13]. To the best of the authors' knowledge, less attention has been given to the synthesizing of ZnS/ZnO films by simple, inexpensive, and eco-friendly successive ionic layer adsorption and reaction (SILAR) route. The SILAR technique requires low-cost chemicals. Moreover, a large number of substrates and areas can be deposited in a single run with a legitimate plan of the substrate holder. In this technique, cationic and anionic precursors are isolated and the cyclic substrate is immersed in both precursors such that a heterogeneous chemical reaction occurs on the surface of the substrate, forming a film with no precipitation in the precursor solution. Deshmukh and Kheraj reported a detailed comparison of physical and wet chemical synthesis techniques [14].

In this work, we file the deposition of the ZnS-nanoparticle/ZnO-nanoflower structure film via a simple and low-cost SILAR method at 300K. An aqueous alkaline precursor containing Zn^{2+} and S^{2-} was employed in the deposition of the ZnS-nanoparticle/ZnO-nanoflower structure. In addition, these synthesized films were characterized for their optical, structural, morphological, and wettability properties by employing various characterization techniques.

19.2 EXPERIMENTAL DETAILS

19.2.1 Chemical Deposition of ZnS-Nanoparticle/ZnO-Nanoflower Thin Films

In the present study, the ZnS-nanoparticle/ZnO-nanoflower composite thin films were prepared via the SILAR process using commercially available glass as a substrate. All chemical reagents were analytical grade and were utilized for the preparation of ZnS-nanoparticle/ZnO-nanoflower composite thin films without further purification. Zinc chloride was obtained from Merk Ltd. Sodium sulfide, acetone, and aqueous NH_3 solution (25%) were obtained from SD Fine Chem Ltd. For the synthesis of the composite thin films, a commercially available glass substrate of size 75 mm × 25 mm × 1.45 mm was cleaned by following a process described in an earlier report [15]. Primarily, chemical SILAR deposition of ZnS-nanoparticles at room temperature on a pre-cleaned substrate was completed as per an earlier reported technique [15]. For this deposition, 0.1 M precursor solutions of $ZnCl_2$ and Na_2S were utilized as a supply of Zn^{2+} and S^{2-} ions, respectively. Figure 19.1a presents a pictorial representation of the SILAR process. Here, ZnS-nanoparticle thin films of 20, 40, and 60 SILAR cycles were formed, as described earlier [15]. These films were white in shading and were well-adherent to the glass substrate.

Furthermore, chemical deposition of the ZnO-nanoflower by SILAR was done onto the ZnS-nanoparticle thin films, which were previously deposited on the glass substrate. An appropriate amount of $ZnCl_2$ precursor (0.1 M) was prepared by dissolving zinc chloride with aqueous NH_3 in distilled water under constant stirring using a magnetic stirrer. Further, aqueous NH_3 was added for dissolving the curd-like precipitate. Thus, drop-by-drop aqueous NH_3 was uninterruptedly added until the precursor solution turned colorless and transparent. The deposition cycle consists of three steps, as presented in Figure 19.1b. The previously ZnS-nanoparticle-coated glass substrate was dipped in the cationic precursor for 10 s, an optimized time, to adsorb Zn^{2+} ions. Then, this substrate was dipped in hot H_2O for 10 s to form ZnO-nanoflower on the ZnS-nanoparticle film. After that, the substrate was dipped into distilled H_2O for 10 s, allowing the removal of the lightly attached ZnO and other unreacted constituents. The distilled water was changed every 10 cycles. The deposition of 40 cycles on all formerly prepared ZnS-nanoparticle films was done. Lastly, substrates were air-dried in the atmosphere at 300K. The ZnS-nanoparticle/ZnO-nanoflower films formed were tagged as *R*1, *R*2, and *R*3 correspondingly and were used for further characterizations.

19.2.2 Characterization of ZnS-Nanoparticle/ZnO-Nanoflower Films

The ZnS-nanoparticle/ZnO-nanoflower-structured thin films *R*1, *R*2, and *R*3 deposited on a glass substrate by the SILAR technique were characterized using various techniques. The thickness of *R*1, *R*2, and *R*3 thin films was calculated by weight difference procedure utilizing sensitive microbalance. The film density was assumed as mean bulk density of zinc oxide (5.675 g/cc) and zinc sulfide (3.98 g/cc). The crystal structure and surface morphologies of ZnS-nanoparticle/ZnO-nanoflower films were

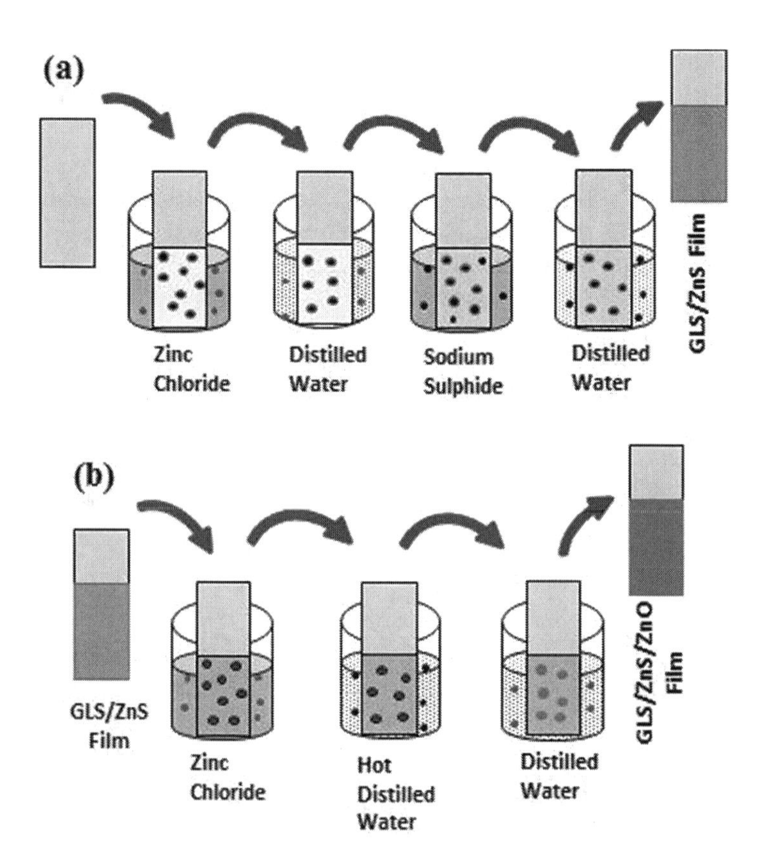

FIGURE 19.1 Scheme for the deposition of (a) ZnS-nanoparticle and (b) ZnS-nanoparticle/ZnO-nanoflower composite films.

investigated using X-ray diffraction (XRD) (Model: Ultima IV Rigaku) and scanning electron microscope (SEM) (JBM-6360A: Model) techniques, respectively. The chemical composition of the prepared *R*1, *R*2, and *R*3 films was evaluated by the EDAX analysis. Raman spectra of ZnS-nanoparticle/ZnO-nanoflower films were studied at room temperature (Model: Bruker RFS27-Stand-alone). The UV-VIS study of these films was performed on a UV-VIS spectrophotometer (Model: Varian) equipped with a thermostat cell compartment at 300K. The wettability of *R*1, *R*2, and *R*3 films was measured by the water contact angle (WCA) measurement (Rame Hart Inc.).

19.3 RESULTS AND DISCUSSION

19.3.1 Film Formation of ZnS-Nanoparticle/ZnO-Nanoflower Structure

The formation of ZnS-nanoparticle by the SILAR technique at 300 K on a commercially available glass substrate was explained earlier [15]. When the substrate was dipped into the 0.1 M $ZnCl_2$ precursor for an optimized duration of 10 s, Zn^{2+} ions were adsorbed on the substrate, owing to the attractive forces among the substrate surface and ions. Yildirim et al. [16] reported that these forces can be chemical attraction, cohesive, or Van der Waals.

$$ZnCl_2 + 2NH_4OH_{(aq)} \rightarrow 2NH_4^+ + 2Cl^- + Zn(OH)_{2(s)} \tag{19.1}$$

$$Zn(OH)_{2(s)} \leftrightarrow \left[Zn^{2+} + 2H^+ + 2O^{2-}\right]_{(s)} \tag{19.2}$$

$$Zn(OH)_{2(s)} + NH_4OH_{(aq)} \rightarrow Zn^{2+} + OH^- + NH_4OH \tag{19.3}$$

After dipping in H_2O for an optimized duration of 10 s, the substrate was dipped in 0.1 M Na_2S precursor for 15 s to adsorb S^{2-} ion, eventually forming the ZnS-nanoparticles.

$$Na_2S \rightarrow 2Na^+ + S^{2-} \tag{19.4}$$

$$Zn^{2+} + S^{2-} \rightarrow ZnS_{(s)} \tag{19.5}$$

The peeling of the film from the substrate was done after 60 SILAR cycles. The obtained ZnS-nanoparticle films were white in shading and were well-adherent to the glass substrate.

Further, the glass/ZnS-nanoparticle film was immersed in 0.1 M zinc chloride precursor for 10 s to adsorb Zn^{2+} ions as per equations (19.1) and (19.2).

However, when the substrate was dipped in hot distilled H_2O for 10 s, the formation of ZnO-nanoflower on ZnS-nanoparticle film occurred.

$$Zn(OH)_{2(s)} \rightarrow ZnO_{(s)} + H^+ + OH^- \tag{19.6}$$

High temperature reinforced the formation of ZnO, as described by Yildirim et al. [16] and Rakhshani et al. [17], as well as the adherence of the ZnO-nanoflower on glass/ZnS-nanoparticle substrate.

19.3.2 Thickness of ZnS-ZnO Thin Film

For any material, film thickness is a key factor in understanding its properties. In this study, a weight difference method was used to measure the thickness of the prepared *R*1, *R*2, and *R*3 samples using the following equation:

$$t = \frac{m}{\rho \times A} \tag{19.7}$$

where,

m = mass of the deposited ZnS/ZnO material on the substrate (g)
ρ = density of the deposited material
A = surface area of the ZnS/ZnO film (cm^2)

At this juncture, the material density was proposed as a mean of bulk density of ZnS (ρ = 3.98 g/cc) [3] and ZnO (ρ = 5.675 g/cc) [18]. In chemically deposited films, the accurate thickness measurement was not possible due to their rough morphology, non-uniformity, porosity, and edge-tapering effects. The thickness of *R*1, *R*2, and *R*3 was found to be 755, 932, and 1590 nm respectively. The deposited ZnS-nanoparticle/ZnO-nanoflower structured films were found to be whitish, uniform, and well-adherent to the surface of the glass substrate.

19.3.3 Structural Studies

Figure 19.2 presents the XRD pattern of the ZnS-nanoparticle/ZnO-nanoflower structured *R*1, *R*2, and *R*3 thin films deposited on the glass substrate. From the XRD profile of *R*1, *R*2, and *R*3, the preferential growth of (102), (110), and (116) planes at 2θ values 28.51°, 47.50°, and 56.48° confirms the hexagonal phase of ZnS-nanoparticles (JCPDF # 89-2191). In contrast, additional well-resolved growth of (100), (002), (101), (102), (110), (103), (112), and (201) planes at 2θ values 31.68°, 34.35°, 36.18°, 47.50°, 56.48°, 62.76°, 67.80°, and 68.97°, respectively, approves the hexagonal structure of ZnO-nanoflowers (JCPDF # 79-0207). The formation of the ZnS-nanoparticle/ZnO-nanoflower can be attributed to the overlapping of the diffraction peaks of ZnO indexed to (102) and (110) with the (110) and (116) planes of ZnS. No impurity peaks were identified, confirming the establishment of the pure phase of ZnS-nanoparticle/ZnO-nanoflower. The intensity of diffraction peaks increases with film thickness, which might be credited to the crystallinity of *R*1, *R*2, and *R*3 thin films [16,19]. Table 19.1 provides the correlation between the observed 2θ values of *R*1, *R*2, and *R*3 films and standard values. The intense and broad diffraction peaks in Figure 19.2 confirm that *R*1, *R*2, and *R*3 are nanocrystalline in nature [20].

Further, the particle size of *R*1, *R*2, and *R*3 were determined using the well-known Debye-Scherrer relation [21],

$$D = \frac{0.9 \times \lambda}{\beta \cdot \cos\ \theta} \tag{19.8}$$

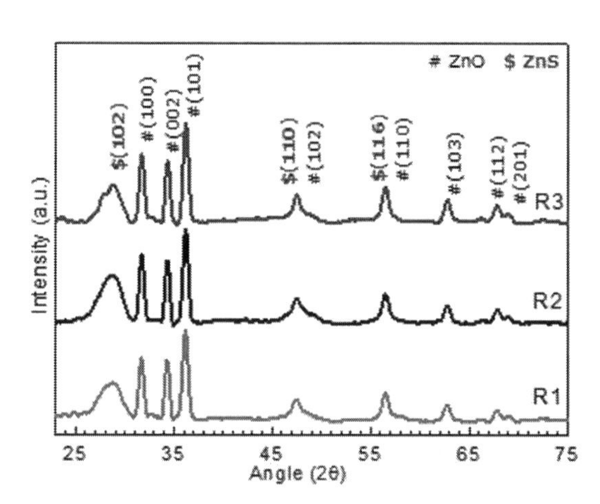

FIGURE 19.2 X-ray diffraction (XRD) pattern of ZnS-nanoparticles/ZnO-nanoflowers structured *R*1, *R*2, and *R*3 thin films.

TABLE 19.1
Correlation of Observed 2*θ* Values of *R*1, *R*2, and *R*3 Films with Standard Values

	Plane (hkl) ZnS			Plane (hkl) ZnO							
Angle 2*θ*	**102**	**110**	**116**	**100**	**002**	**101**	**102**	**110**	**103**	**112**	**201**
*R*1	28.490	47.399	56.490	31.688	34.409	36.160	47.399	56.490	32.783	67.813	68.892
*R*2	28.580	47.513	56.499	31.605	34.339	36.176	47.513	56.499	62.764	67.845	68.922
*R*3	28.517	47.505	56.489	31.685	34.352	36.185	47.505	56.489	62.763	67.802	68.970
Standard	28.571	47.529	56.386	31.699	34.382	36.182	47.459	56.463	62.760	67.805	68.924

where,

λ = wavelength of the X-rays used (in this case, $\lambda = 1.5405$ Å)
θ = Bragg diffraction angle (radian)
β = full-width-at-half-maximum (FWHM)

The determined particle size for ZnS-nanoparticles was found to be between 8.41 and 10.14 nm and for ZnO-nanoflowers was between 20.33 and 26.23 nm. Furthermore, the dislocation density (δ) of *R*1, *R*2, and *R*3 film was determined using the following equation [21],

$$\delta = \frac{1}{D^2} \tag{19.9}$$

where D is the particle size.

Stresses were the major hurdles in the development of promising structural properties. Stress may arise from geometric discrepancy on the interface between the substrate and crystalline lattice. These stresses can be the source of microstrain (ε) in the films, which are calculated using the following equation [21],

$$\varepsilon = \frac{\beta \times \cos\ \theta}{4} \tag{19.10}$$

The achieved values of δ, ε, and D for *R*1, *R*2, and *R*3 are listed in Table 19.2. Lesser values of microstrain and dislocation density show a minor degree of lattice imperfections [20], and, consequentially, establishment of high-quality ZnS-nanoparticle/ZnO-nanoflower structure of *R*1, *R*2, and *R*3 thin films.

19.3.4 Raman Spectroscopy Studies

Raman spectroscopy study is essential in condensed matter physics to investigate the crystalline nature as well as rotational and vibrational phonon modes in nanomaterials. Figure 19.3 reveals the Raman spectra of ZnS-nanoparticles/ZnO-nanoflower structured *R*2 and *R*3 thin films. The hexagonal structure of *R*2 and *R*3 belongs to

TABLE 19.2
Achieved Values of δ, ε, and D of R1, R2, and R3 Thin Films

	Plane (hkl) ZnS			Plane (hkl) ZnO								
	102			100			002			101		
Sample ↓	δ nm^{-2} ($\times 10^{-3}$)	ε ($\times 10^{-3}$)	D (nm)	δ nm^{-2} ($\times 10^{-3}$)	ε ($\times 10^{-3}$)	D (nm)	δ nm^{-2} ($\times 10^{-3}$)	ε ($\times 10^{-3}$)	D (nm)	δ nm^{-2} ($\times 10^{-3}$)	ε ($\times 10^{-3}$)	D (nm)
*R*1	9.732	3.419	10.14	1.453	1.321	26.23	1.649	1.407	24.63	2.217	1.632	21.24
*R*2	11.310	3.687	9.40	2.414	1.703	20.35	2.420	1.705	20.33	2.216	1.632	21.24
*R*3	14.140	4.122	8.41	1.672	1.417	24.46	1.813	1.476	23.48	2.216	1.632	21.24

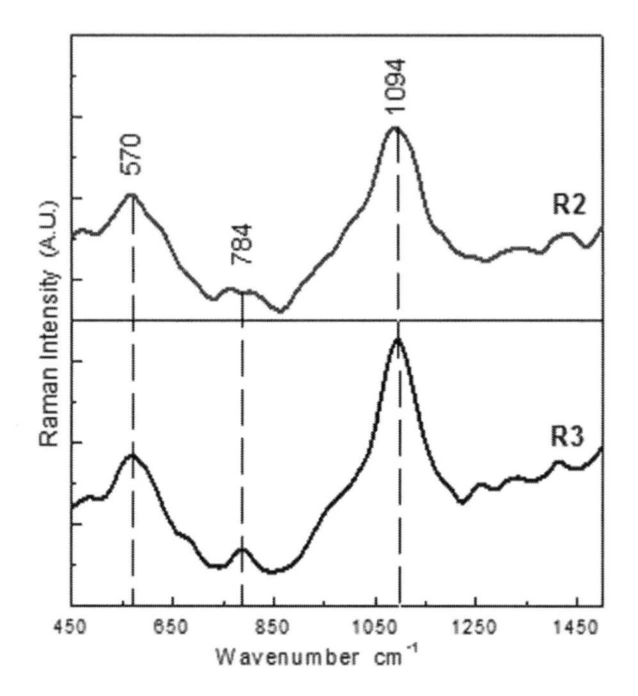

FIGURE 19.3 Raman spectra of ZnS-nanoparticles/ZnO-nanoflowers structured *R*2 and *R*3 thin films.

the space group of $^4C_{6V}$ with four atoms per unit cell [22,23]. For *R*2 and *R*3, the predicted optical phonon modes by group theory are expressed as follows:

$$\Gamma_{\text{opt}} = A1 + 2B1 + E1 + 2E2 \tag{19.11}$$

Among these optical phonons, *E*1, *E*2, and *A*1 are Raman active, whereas *B*1 (low) and *B*1 (high) are generally silent [24]. The polar *E*1 and *A*1 phonon modes are divided into two components: (i) longitudinal optical (LO), and (ii) transverse optical (TO) because of their polar symmetry. In addition, the two frequency components,

E2 (high) and E2 (low), belong to non-polar phonon E2. Therefore, ZnS-nanoparticle/ZnO-nanoflower structure has A1 (LO), A1 (TO), E1 (LO), E1 (TO), E2 (high), and E2 (low) Raman-active phonon modes [6]. In this study, the phonon peaks at 784 and 1094 cm^{-1} can be attributed to second and third LO phonons, which are inconsistent with the earlier reported Raman spectra of ZnS by Gode [25]. Additionally, the peaks observed at 1094 and 570 cm^{-1} may be allotted to E1 (LO) and A1 (LO) phonon mode of ZnO-nanoflower [26,27]. The second and third-order LO phonon of ZnO-nanoflowers indicate a shift toward the lower energy, consistent with earlier reports of Xu [26] and Tong [27] for ZnO, which can be ascribed to the confinement effect of optical phonons [22]. This confirms that the ZnS-nanoparticle/ZnO-nanoflower structure of *R*2 and *R*3 has high purity crystalline nature.

19.3.5 Surface Morphological Studies

The surface morphology study assumes incredible efforts to communicate the surface profile and the nature of the films. The SEM images of *R*1, *R*2, and *R*3 ZnS-nanoparticles/ZnO-nanoflower structured thin films are displayed in Figure 19.4a–c. *R*1, *R*2, and *R*3 films are homogeneous pinhole-free, fully covered, and adherent to the glass substrate. A careful observation confirms the uniform distribution of ZnS tightly packed nanospherical particles. After the deposition of ZnO on the ZnS-nanospherical particle, the development of flower-like structures was observed, as shown in Figure 19.4a–c for *R*1, *R*2, and *R*3. This flower-like structure was developed as an additional layer was deposited. These nanoflowers have a diameter of the order of 2–4 μm involving the vertical growth of multilayer pedals. The magnification image of ZnO nanoflowers for *R*2 and *R*3 was presented in Figure 19.4b and c, respectively. Every nanoflower contains a major vertical round rod with a diameter

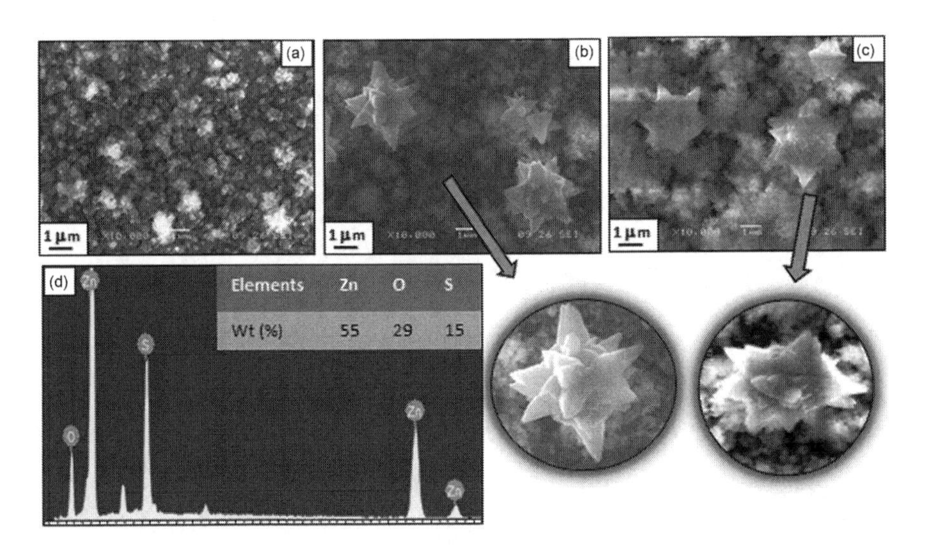

FIGURE 19.4 Scanning electron microscope (SEM) images of *R*1 (a), *R*2 (b), and *R*3 (c), and EDAX of *R*3 (d) thin films.

of ~400 nm with multilayer pedals. These pedals exhibit the tapering feature with a tip size of ~50–60 nm. All these results are in acceptable concurrence with XRD, as discussed in Section 19.3.3.

Figure 19.4d depicts the representative EDAX spectrum of R3 ZnS-nanoparticle/ZnO-nanoflower thin film. It shows the presence of only three Zn, S, and O elements. The inset of Figure 19.4d exhibits the quality proportion of Zn, O, and S elements as 55:29:15, which is in good stoichiometric agreement of Zn(O, S) [3,28].

19.3.6 Wettability Studies

The wetting of solid with water, wherein air is the immediate medium, depends on the relation between the interfacial tensions (solid/air, water/air, and water/solid). The proportion between these tensions defines the contact angle (CA) among a water droplet on a certain surface. A CA of 0° and 180° corresponds to complete wetting and non-wetting properties, respectively. Both super-hydrophilic (CA $\leq$ 10°) and super-hydrophobic (CA $\geq$ 150°) material surfaces are desirable for practical applications [21]. A wettability study was conducted to examine the interaction between water and ZnS-nanoparticle/ZnO-nanoflower structure. Figure 19.5 presents the photo pictures of the WCA measurement for *R*1, *R*2, and *R*3 thin films at room temperature. The WCA is probably influenced by the surface morphology, chemical composition, and uniformity of the thin films [29]. In the present case, the WCA was 36.9°, 44.5°, and 61.4° for *R*1, *R*2, and *R*3 samples of ZnS-nanoparticles/ZnO-nanoflower structure, respectively.

Figure 19.5 reveals that all thin-film surfaces are hydrophilic in nature as WCA is less than 90°. The CA of *R*1, *R*2, and *R*3 thin films increased with an increase in SILAR cycles. The increase in WCA of *R*1, *R*2, and *R*3 thin films may be due to (i) vertically oriented nanoflowers, the air is trapped in the crevices between the nanoflower and water droplet; (ii) little enhancement in ZnS/ZnO nanoflowers and grain size; and (iii) nanocrystalline nature of the films that are expected to possess very high surface energy [29].

19.3.7 Optical Studies

The transmission (*T*) spectra of a typical ZnS-nanoparticle/ZnO-nanoflower *R*1, *R*2, and *R*3 thin films within the wavelength region 340–525 nm is presented in Figure 19.6. *R*1, *R*2, and *R*3 films have low absorbance within the visible region of the spectrum. Figure 19.6 reveals that all films are transparent to light in the visible sector.

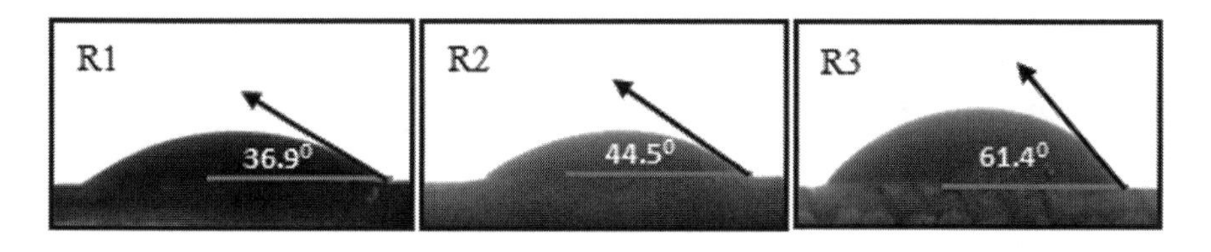

FIGURE 19.5 Measurement of water contact angles (WCAs) of *R*1, *R*2, and *R*3 thin films.

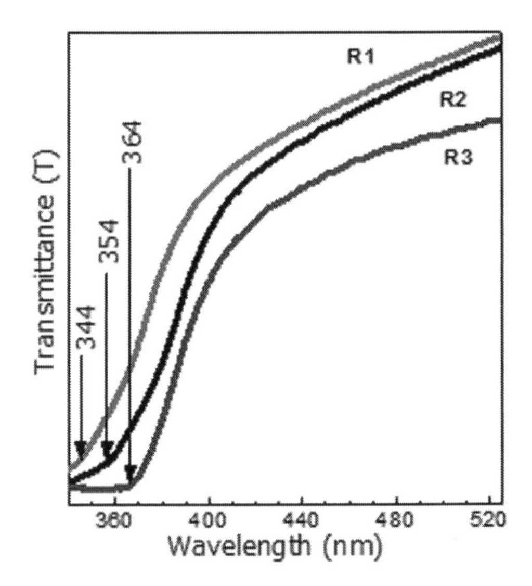

FIGURE 19.6 Variation of transmittance (T) against wavelength (nm) of $R1$, $R2$, and $R3$ thin films.

The decrease in transmittance occurs with an increase in film thickness, as described by Mesa et al. [30]. The direct optical bandgap of all ZnS-nanoparticle/ZnO-nanoflower $R1$, $R2$, and $R3$ films were determined using the following relation [15],

$$E_g = \frac{1240}{\lambda_c} \tag{19.12}$$

For $R1$, $R2$, and $R3$ thin films, the cut-off wavelengths (λ_c) were detected as 344, 354, and 364 nm and corresponding bandgap value estimated as 3.6, 3.5, and 3.4 eV, respectively. These assessed energy bandgap values are consistent with the previous reports of the bulk hexagonal phase of ZnS (3.8 eV) and ZnO (3.2 eV) reported by Sookhakiann et al. [31] and Lokhande et al. [3]. The decrease in the bandgap of ZnS-nanoparticle/ZnO-nanoflower $R1$, $R2$, and $R3$ thin films could be attributed to (i) increase in the film thickness with SILAR cycles, and (ii) improvement in the crystallinity (Figure 19.2).

19.4 CONCLUSION

In this study, an environment-friendly, inexpensive SILAR synthesis route was adopted to deposit ZnS-nanoparticle/ZnO-nanoflower structure at room temperature. The XRD studies revealed the hexagonal phase of ZnS-nanoparticle/ZnO-nanoflower thin films. For the ZnS/ZnO nanocomposite, the LO phonon mode was confirmed by the Raman study. The energy bandgap of the prepared ZnS-nanoparticle/ZnO-nanoflower structure was found to be 3.6 to 3.4 eV. The SEM images of the prepared structure showed the formation of ZnO-nanoflowers on ZnS-nanospherical particles. The WCA of ZnS/ZnO nanocomposite confirmed the hydrophilic nature as the contact angle was below 90°. This prepared ZnS-nanoparticle/ZnO-nanoflower structure may be used as a building block for nanoscale electronics and optoelectronic devices.

ACKNOWLEDGEMENT

One of the authors (SGD) is grateful to Shrimati Vimal G. Deshmukh for her consistent inspiration and guidance. The authors are thankful to PAH Solapur University, Solapur; Shivaji University, Kolhapur; and SP Pune University, Pune for providing characterization facility.

REFERENCES

1. T. Kato. (2017). Cu(In, Ga)(Se, S)2 solar cell research in Solar Frontier: Progress and current status. *Jpn. J. Appl. Phys.*, 56, 04CA02.
2. C. Platzer Bjorkmana, T. Torndahl, D. Abou-Ras, J. Malmstrom, J. Kessler. (2006). Zn(O, S) buffer layers by atomic layer deposition in Cu(In, Ga)Se_2 based thin film solar cells: Band alignment and sulfur gradient. *J. Appl. Phys.*, 100(4), 044506.
3. C. D. Lokhande, H. M. Pathan, M. Giersig, H. Tributsch. (2002). Preparation of $Zn_x(O, S)_y$ thin films using modified chemical bath deposition method. *Appl. Surf. Sci.*, 187, 101–107.
4. M. Bar, A. Ennaoui, J. Klaer, T. Kropp, N. Allsop, M. C. Lux-Steiner. (2006). Formation of a ZnS/Zn(S, O) bilayer buffer on $CuInS_2$ thin film solar cell absorbers by chemical bath deposotion. *J. Appl. Phys.*, 99, 123503.
5. L. Zheng, Y. Zheng, C. Chen, Y. Zhan, X. Lin, Q. Zheng, K. Wei, J. Zhu. (2009). Network structured SnO_2/ZnO heterojunction nanocatalyst with high photocatalytic activity. *Inorg. Chem.* 48(5), 1819–1825.
6. R. Zamiri, D. M. Tobaldi, H. A. Ahangar. (2014). Study of far infrared optical properties and photo catalytic activity of ZnO/ZnS hetero-nanocomposite structure. *RSC Adv.*, 4, 35383–35389.
7. X. Gu, S. Zhang, Y. Zhao, Y. Qiang. (2015). Band alignment of ZnO/ZnS heterojunction prepared through magnetron sputtering and measured by X-ray photoelctron spectroscopy. *Vacuum*, 122, 6–11.
8. J. Schrier, D. O. Demchenko, L. W. Wang. (2007). Optical propeties of ZnO/ZnS and ZnO/ZnTe heterostructures for photovoltaic applications. *Nano Letts.*, 7(8), 2377–2382.
9. S. Balachandran, M. Swaminathan. (2013). The simple, template free synthesis of a Bi_2S_3–ZnO heterostructure and its superior photocatalytic activity under UV-A light. *Dalton Trans.*, 42(15), 5338–5347.
10. J. Yan, X. Fang, L. Zhang, Y. Bando, U. K. Gautam, B. Dierre, T. Sekiguchi, D. Golberg. (2008). Structure and cathodoluminescence of individual ZnS/ZnO biaxial nanoblet heterostures. *Nano Lett.*, 8(9), 2794–2799.
11. P. Guo, J. Jiang, S. Shen, L. Guo. (2013). ZnS/ZnO heterojunction as photoelectrode: Type II band alignment towards enhanced photoelectrochemical performance. *Int. J. Hydrogen Energy*, 38(29), 13097–13103.
12. A. K. Giri, C. Charan, A. Saha, V. K. Shahi, A. B. Panda. (2014). An amorperometric cholesterol biosensor with excellent sensitivity and limit of detection based on an enzyme-immobilized microtubular ZnO@ZnS heterostructure. *J. Mater. Chem. A*, 2, 16997–17004.
13. X. Yu, G. Zhang, H. Cao, X. An, Y. Wang, Z. Shu, Fei Hua. (2012). ZnO@ZnS hollow dumbbells-graphene composites as high-performance photo catalysts and alcohol sensors. *New J. Chem.*, 36, 2593–2598.

14. S. G. Deshmukh, V. Kheraj. (2017). A comprehensive review on synthesis and characterizations of Cu_3BiS_3 thin films for solar photovoltaics. *Nanotechnol. Environ. Eng.*, 2(1), 15.
15. S. G. Deshmukh, A. Jariwala, A. Agarwal, C. Patel, A. K. Panchal, V. Kheraj. (2016). ZnS nanostructured thin-films deposited by successive ionic layer adsorption and reaction. *AIP Conf. Proc.*, 1724, 020033.
16. M. Ali Yildirim, A. Ates. (2010). Influence of film thickness and structure on the photo-response of ZnO films. *Optics Commun.*, 283, 1370–1377.
17. A. E. Rakhshani, A. Bumajdad, J. Kokaj. (2007). ZnO films grown by successive chemical solution deposition. *Appl. Phys. A*, 89, 923–928.
18. K. V. Gurav, V. J. Fulari, U. M. Patil, C. D. Lokhande, O.-S. Joo. (2010). Room temperature soft chemical route for nano fibrous wurtzite ZnO thin film synthesis. *Appl. Surf. Sci.*, 256, 2680–2685.
19. V. Senthamilselvi, K. Ravichandran, K. Saravanakumar. (2013). Influence of immersion cycles on the stoichiometry of CdS films deposited by SILAR technique. *J. Phys. Chem. Solids*, 74, 65–69.
20. M. M. Salunkhe, K. V. Khot, S. H. Sahare, P. N. Bhosale, T. Bhave. (2015). Low temperature and controlled synthesis of $Bi_2(S_{1-x}Se_x)_3$ thin films using a simple chemical route: Effect of bath composition. *RSC Adv.*, 5(70), 57090–57100.
21. S. G. Deshmukh, S. J. Patel, K. K. Patel, A. K. Panchal, V. Kheraj. (2017). Effect of annealing temperature on flowerlike Cu_3BiS_3 thin films grown by chemical bath deposition. *J. Electron. Mater.*, 46(10), 5582–5588.
22. R. Zamiri, H. Abbastabar Ahangar, D. M. Tobaldi, A. Rebelo, M. P. Seabra, M. Shabani, J. M. F. Ferreira. (2014). Fabricating and characterising ZnO–ZnS–Ag2S ternary nanostructures with efficient solar-light photocatalytic activity. *Phys. Chem. Chem. Phys.*, 16, 22418–22425.
23. Q. Xiong, O. Jinguo Wang, L. C. Reese, V. Lew Yan, P. C. Eklund. (2004). Raman scattering from surface phonons in rectangular cross-sectional w-ZnS nanowires. *Nano Lett.*, 4 (10), 1991–1996.
24. J. Fan, T. Li, H. Heng. (2016). Hydrothermal growth of ZnO nanoflowers and their photocatalyst application. *Bull. Mater. Sci.*, 39(1), 19–26.
25. F. Gode. (2011). Annealing temperature effect on the structural, optical and electrical properties of ZnS thin films. *Physica B*, 406, 1653–1659.
26. N. Xu, Y. Cui, Z. Hu, W. Yu, J. Sun, N. Xu, J. Wu. (2012). Photoluminescence and low-threshold lasing of ZnO nanorod arrays. *Optics Express*, 20(14), 14857–14863.
27. Y. Tong, L. Dong, Y. Liu, D. Zhao, J. Zhang, Y. Lu, D. Shen, X. Fan. (2007). Growth and optical properties of ZnO nanorods by introducing ZnO sols prior to hydrothermal process. *Mater. Letts.*, 61, 3578–3582.
28. A. Grimm, D. Kieven, I. Lauermann, M. C. Lux-Steiner, R. Klenk. (2012). Zn(O, S) layers for chalcoyprite solar cells sputtered from a single target. *EPJ Photovoltaics*, 3, 30302.
29. S. G. Deshmukh, V. Kheraj, K. J. Karande, A. K. Panchal, R. S. Deshmukh. (2019). Hierarchical flower-like Cu_3BiS_3 thin film synthesis with nano-vacuum chemical bath deposition technique. *Mater. Res. Express*, 6(8), 084013.
30. F. Mesa, W. Chamorro, M. Hurtado. (2015). Optical and structural study of In_2S_3 thin films grown by co-evaporation and chemical bath deposition (CBD) on Cu_3BiS_3. *Appl. Surf. Sci.*, 350, 38–42.
31. M. Sookhakiann, Y. M. Amin, W. J. Basirun, M. T. Tajabadi, N. Kamarulzaman. (2014). Syntheis, structural, and optical properties of type-II ZnO-ZnS core-shell nanostructure. *J. Luminescence*, 145, 244–252.

20 State of Art for Virtual Fabrication of Piezoresistive MEMS Pressure Sensor

Samridhi and Parvej Ahmad Alvi
Banasthali Vidyapith

CONTENTS

20.1 INTRODUCTION

Technology is the chief agent in today's modern era which depends profoundly on electronics and digitalization. To make more efficient, compatible, and portable electronic gadgets, scientists and researchers are focusing on reducing size to strengthen device performance and responsivity. Microelectromechanical systems (MEMS) bring to attention the mechanical constructions of micro sizes executing an electronically controlled preset function. The one associated with MEMS is a piezoresistive pressure sensor. Piezoresistivity is a concept based on the Wheatstone bridge, in which the resistance of a material varies when external mechanical stress is applied. On applying external pressure to the diaphragm, the piezoresistors on the diaphragms that are connected to metallic terminals transduce the input signal into an output electrical signal [1]. Micromachined silicon piezoresistive pressure sensors have caught the public eye due to their extensive application in the consumer market [2–4]. They have proven to be very accurate and more reliable, use less power, are robust, perform a quick analysis, have a wider range of applications, and are economically inexpensive. These sensors have been widely used in a variety of applications such as aerospace [5], robotics [6], and petrochemicals for pressure measurement [7].

Diaphragm and piezoresistors are the crucial components of a piezoresistive pressure sensor. When external pressure is applied to the diaphragm, stress is generated, and it becomes important to locate the resistors where stress is the maximum to achieve the maximum sensitivity [8]. In addition, to understand the physical mechanics involved in the piezoresistive pressure sensor, it is necessary to understand the fabrication process associated with the MEMS pressure sensor.

The rapid developments in IC technology in recent years have reduced the dimension of the sensors with increasing complexity in the fabrication of VLSI chips. In device modeling, with the help of the best model of a device, one can design a device with selected specifications based on the set of device parameters. In turn, the overall performance of VLSI devices is determined by process conditions. Consequently, it is essential to understand, characterize, and optimize the system steps involved in device fabrication. As the dimension of the device reduces, the fabrication step becomes essential as the shape of the metal film deposited has an instantaneous effect on the performance of the final device and circuit. Once the models are developed and coded in a complete computer application, the device parameters may be efficiently anticipated without going through the actual process and fabrication steps. Alvi et al. reported the case of a V-shaped cavity sealed by the CVD technique [9]. Malhaire and Barbier deposited a polysilicon thin film using the low chemical vapor deposition (LPCVD) method and the cavity etching was done by KOH solution [10].

In this chapter, we present a detailed stepwise fabrication process of the piezoresistive pressure sensor using the SILVACO tool. A polycrystalline thin film as a sensing element has been deposited by an electron beam physical vapor deposition technique on a square diaphragm. A layout of the pressure sensor is presented along with the fabrication methodology.

20.2 SILICON-BASED MEMS DEVICES

Silicon-based materials are still the most commonly used materials in the integration of MEMS devices, which are integrated into the identical environment and provide additional potential to combine electronics with mechanical devices. They offer the batch fabrication process for mechanical devices such as microstructures with a micron size, sensors, and actuators. MEMS devices take the advantage of the mechanical property of silicon rather than its electrical property. Single-crystal silicon has a cubic crystal structure with a lattice constant of $a = 0.543$ nm, as shown in Figure 20.1; therefore, it exhibits anisotropic property, which is obvious due to its mechanical property along with Young's modulus which can be used as a wafer. The properties of the wafer depend upon its orientation and the impurities added to it. The doping concentration of the impurity has a direct effect on the electrical property but not on the mechanical property. Silicon is a semiconductor of the IVth group to create p-type semiconductor dopants of group III (boron), which are added to create positively charged mobile charges. To create n-type semiconductor, dopants of group V (phosphorous) are added to create negatively charged mobile charges.

Silicon, amorphous silicon, and polysilicon exhibit piezoresistive property. Therefore, the resistivity of such materials are impacted by external pressure. The relative change in resistivity depends on the stress components parallel and

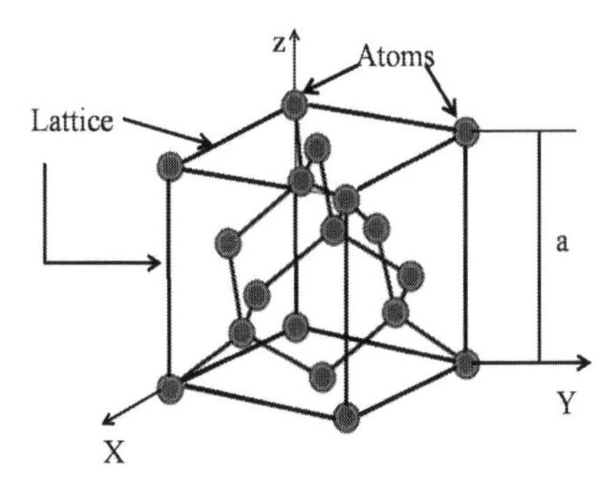

FIGURE 20.1 Single-crystal silicon structure.

perpendicular to the direction of the resistor [11,12]. This might be due to dopant concentration and crystallographic orientation. Out of the three components of silicon, the p-polysilicon has the advantage of being used as piezoresistors for MEMS devices. Polysilicon piezoresistors are placed on the diaphragm and isolate each other by the oxide layer. This isolation is maintained at a high temperature of 300°C, thus reducing the risk of current leakage [13].

20.3 SOFTWARE DETAILS

Today, most of the development in the field of electronics is done by computer modeling. The process associated with device modeling and simulation is known as TCAD (technology-computer-aided design). The use of such software reduces both time and cost. This chapter aims to create a standard structure that can be easily integrated within the framework of SILVACO 2D or 3D simulator. SILVACO is an electronic design automation software, and provides a platform for modeling and simulation of devices. This interactive software has unique features such as Deckbuild, Tony Plot, Maskview, and Optimizer. In this study, we worked on the DevEdit tool. It is a commanding module for structure modeling, enhancing, and remeshing. It can be used to create a device on the mesh and edit the same device [14]. Some early results related to the simulation of the proposed devices have been presented in [15–18].

20.4 DESIGNING

In this section, an organized approach has been discussed to design a polysilicon piezoresistive pressure sensor. A layout of the designed pressure sensor has been shown in Figure 20.2 based on the Wheatstone bridge. The sensitivity of the pressure sensor depends on four sensing elements and is fabricated from pressure-variable resistance. It measures stress within a thin crystalline diaphragm such that the resistance of two oppositely located resistors increases, whereas the resistance of the other two resistors decreases, generating a potential difference. It is important to locate the position of resistors located on a diaphragm to get the maximum potential difference.

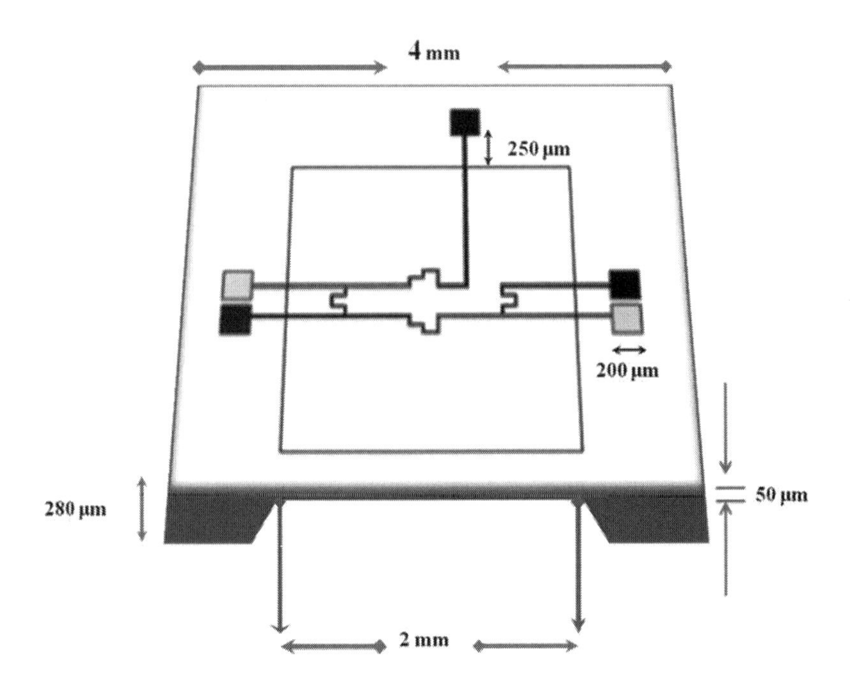

FIGURE 20.2 Schematic view of the MEMS pressure sensor.

The dimension and design parameters of the designed sensor have been shown in Table 20.1. The connections, resistors, and contact pads are created on the diaphragm. The substrate and diaphragm are made up of single-crystal Si-silicon. Lightly doped p-polycrystalline silicon are used as piezoresistors, and a metal line made of gold has been used to connect the resistors with the contact pads. To create a diaphragm, TMAH (tetra-methyl-ammonium hydroxide, $(CH_3)_3N(OH)C_6H_5)$) back etching has been done to obtain an angle of 54.74°.

TABLE 20.1
Dimensions of the Designed Pressure Sensor

Parameters	Values
Chip size	4 mm × 4 mm
Size of diaphragm	2 mm
	2 mm
Value of resistors	1–2 KΩ
Length of resistors	400 µm
Width of resistors	10 µm
Dimension of contact pads	200 µm × 200 µm
Contact pad location	250 µm (away from the diaphragm edge)
Diaphragm Thickness	50 µm

20.5 FABRICATION PROCESS

The fabrication process has been shown in pictorial detail in Figure 20.3. A p-type silicon wafer of 2″ diameter having <100> plane has been selected, and the process starts with cleaning of the wafer, which is subjected to RCA (Radio Corporation of America) according to the following sequence:

1. Silicon wafer is loaded into the furnace at a high temperature of 1,100°C for oxidation.
2. After oxidation, the E-beam process is conducted for the deposition of the polycrystalline thin film.
3. Doping of boron in a polysilicon thin film for piezoresistive.
4. A mask-based process, photolithography, PLG-I is conducted for patterning (9×9 array).
5. Patterning of piezoresistors is done with the help of PLG-II.
6. Deposition of gold is performed for metal lines connection and contact pads.
7. PLG-III is done for the transfer of patterns for the formation of the diaphragm.
8. TMAH etching for diaphragm formation.
9. PLG-IV to etch gold selectively.
10. Deposition of PECVD SiO_2 for passivation.
11. PLG-V is carried out for the opening of the contact pad and PECVD SiO_2 is etched out.

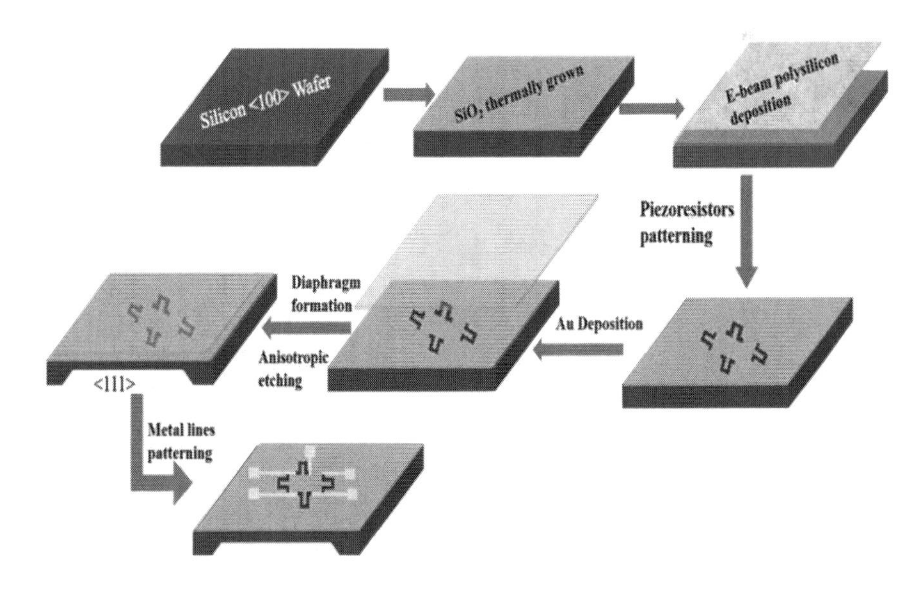

FIGURE 20.3 Schematic illustration of the fabrication process of the piezoresistive pressure sensor.

20.6 FABRICATION METHODOLOGY THROUGH SILVACO SOFTWARE

Step 1. In SILVACO software, click on the DevEdit tool, where the window appears as:

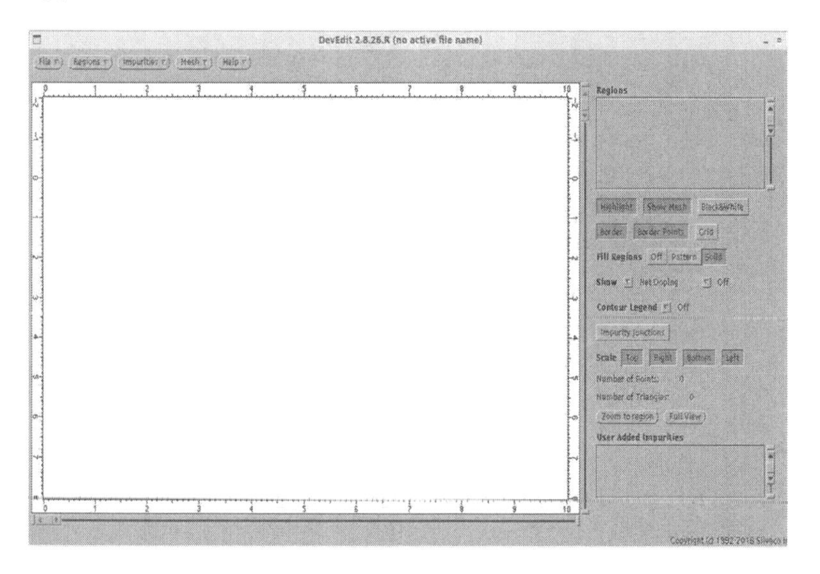

Step 2. In the region menu, select add region, and then select the region according to the requirement. Select the material from the material tab. As we have chosen silicon wafer in direction of <100>, the window appears as:

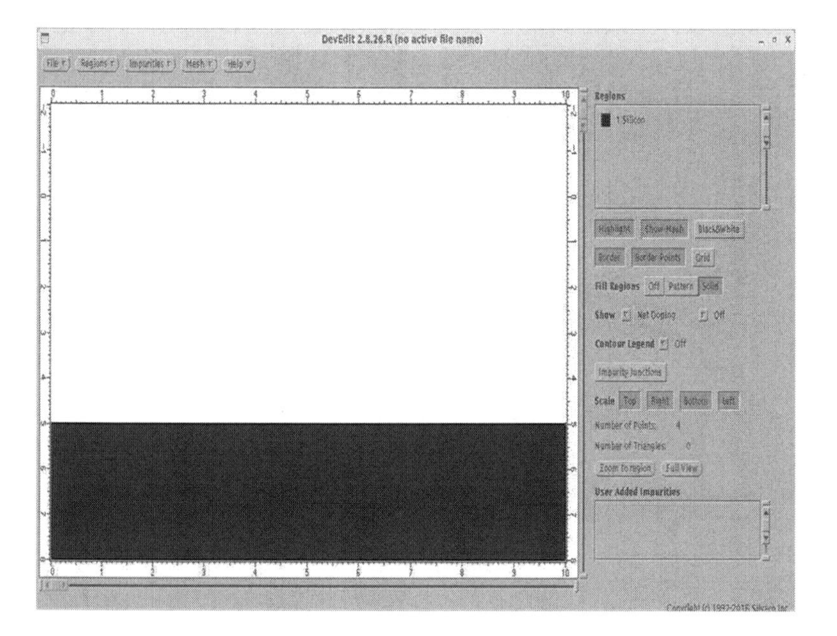

Step 3. For thermally grown SiO_2, Repeat step 2. Click on the region icon and add region, and then select the material from the material tab. In the add/point tab, select preference for color selection.

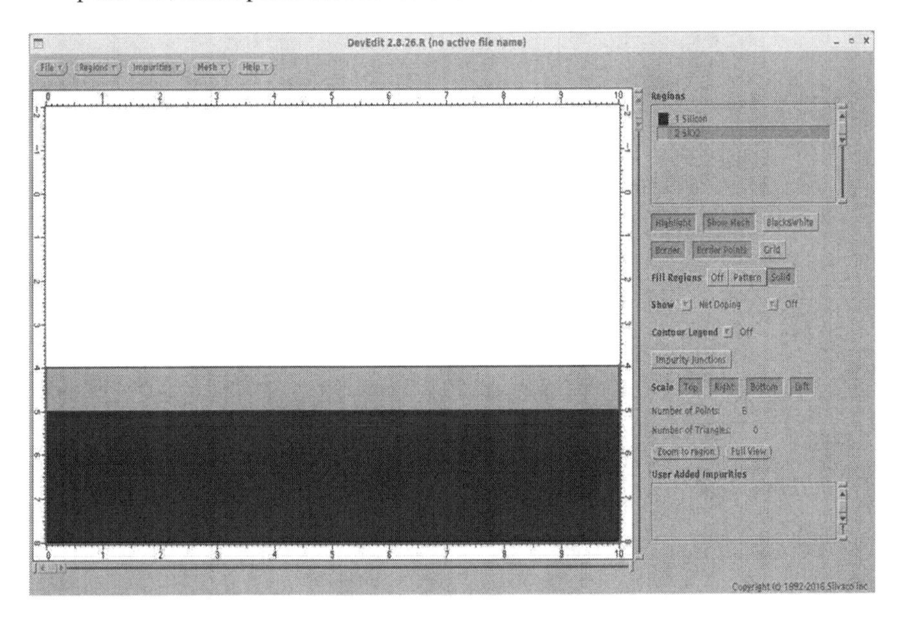

Step 4. The E-beam method has been used to add the polysilicon; repeat the same step as above.

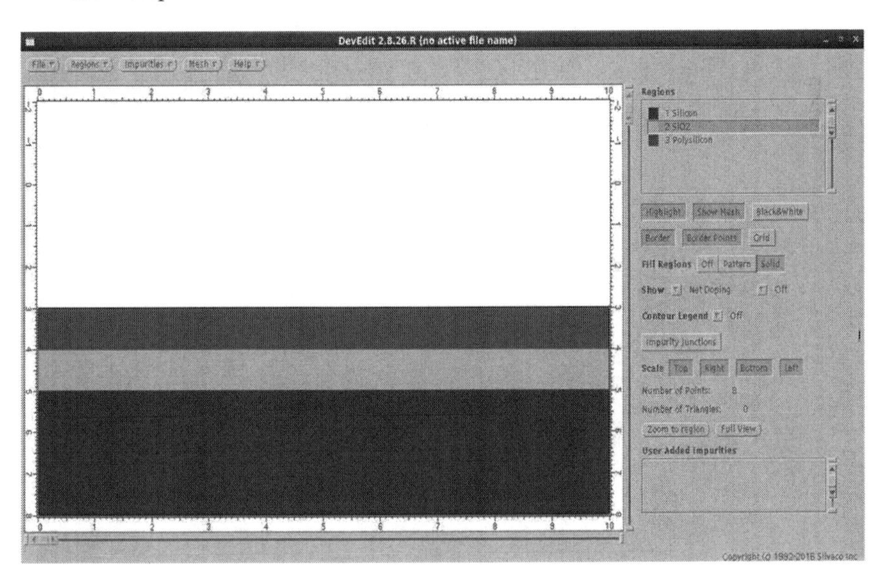

Step 5. For delination of piezoresistors using PLG-II, select the region to be etched. Window appears on the right side; in the regions section, select the selected region to be etched. Right-click on that and delete the region. Finally, the selected region has been deleted and the structure appears as:

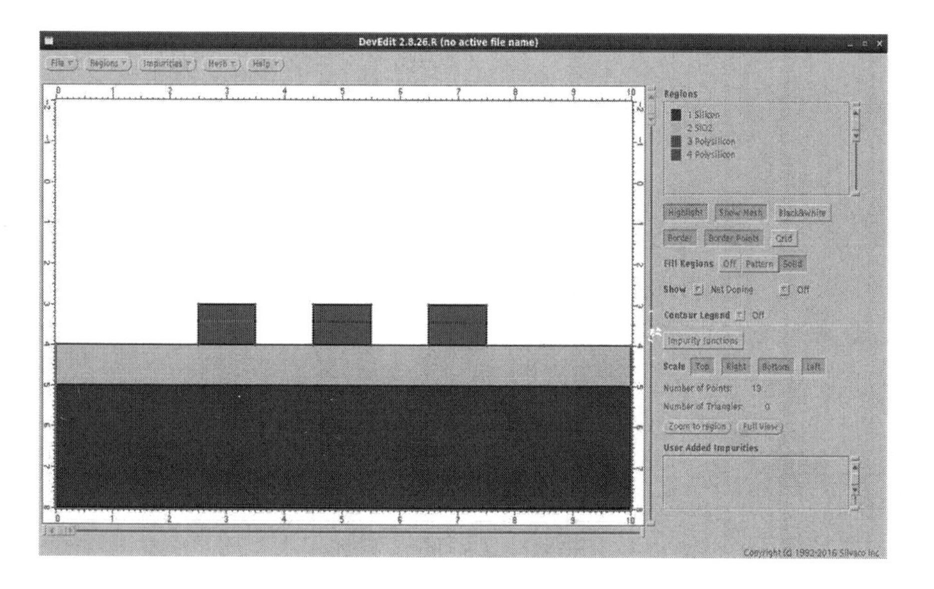

Step 6. Metalization of gold for contact lines and contact pads.

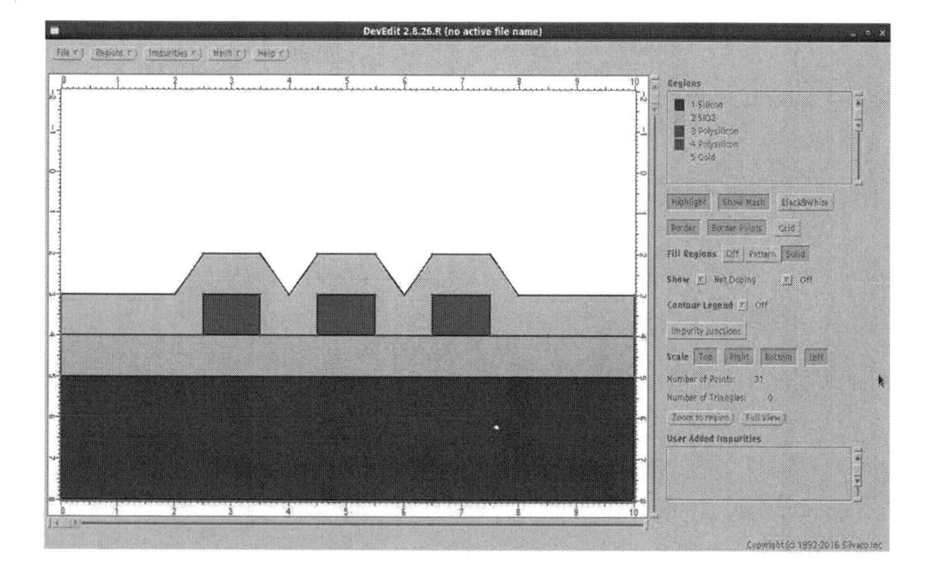

Step 7. Wet anisotropic etching is used for the formation of the diaphragm by TMAH back etching. In the region icon, select add region, and then mention the points for etching to form an Si-based diaphragm in <111> direction. KOH is known for higher anisotropy between <100> and <111> plane. It is considered as the best anisotropic etchant.

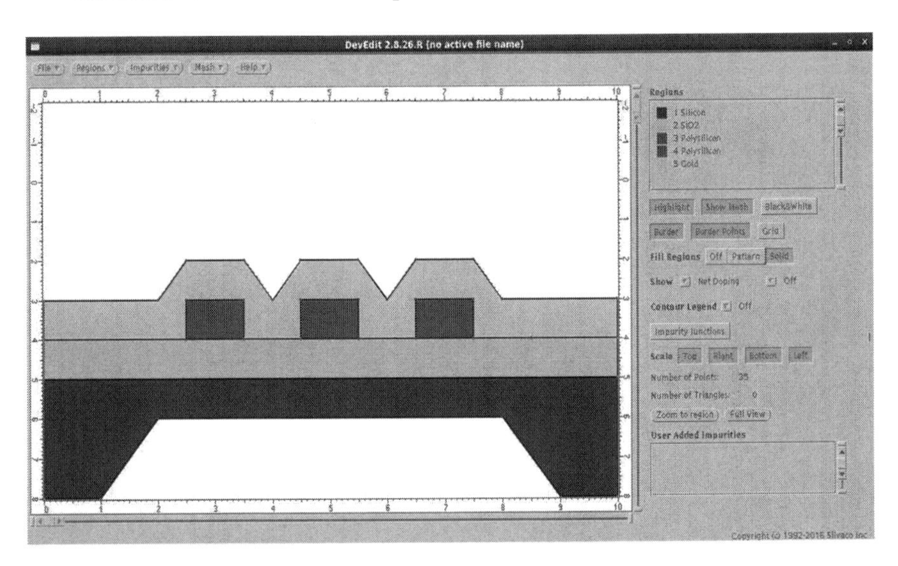

Step 8. Gold is etched selectively such that metalline formation can take place; PLG-IV is conducted to connect the metal lines with piezoresistors.

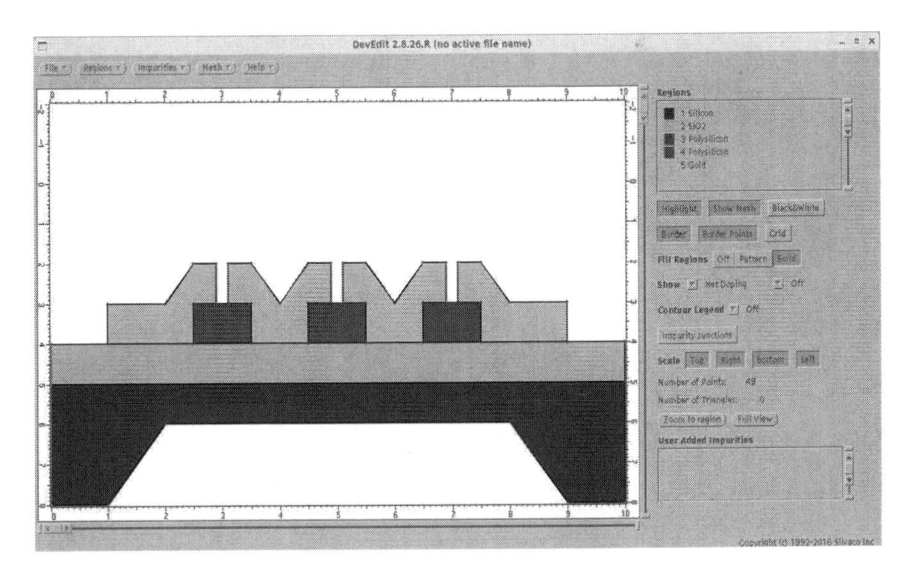

Step 9. Deposition of PECVD SiO_2 for passivation.

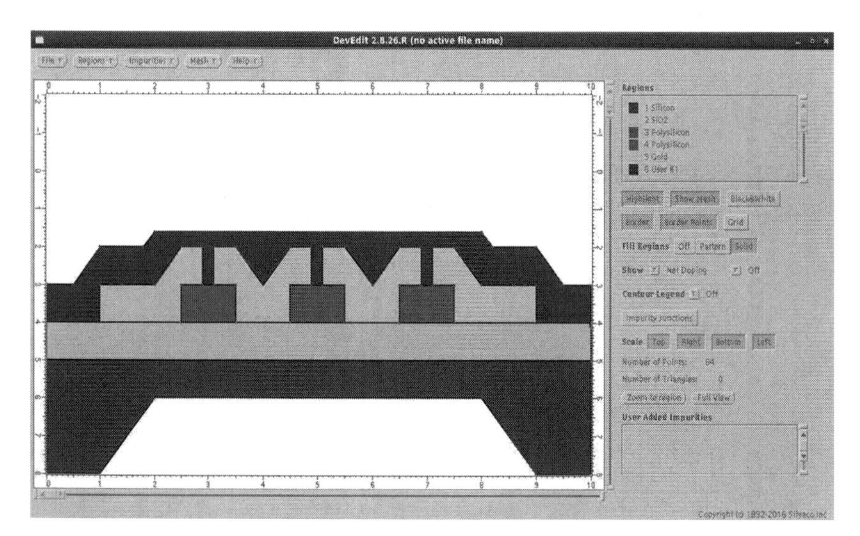

Step 10. PECVD SiO_2 is etched out and PLG-V is carried out for the opening of the contact pad.

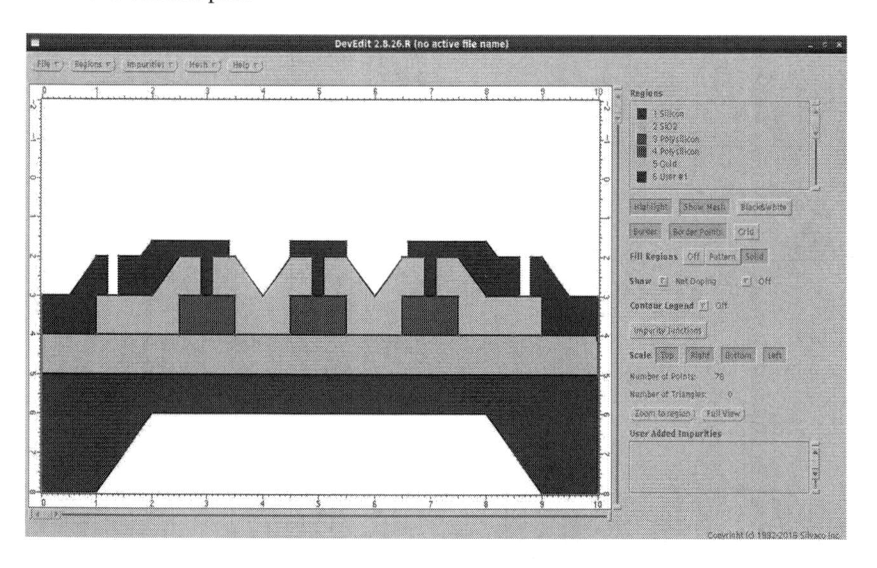

20.7 CONCLUSION

A detailed stepwise fabrication process of the piezoresistive pressure sensor has been illustrated within the framework of the SILVACO 2D or 3D simulator. DevEdit is a commanding module for structure modeling, enhancing, and remeshing. The fabrication process followed is important to investigate as it provides complete information about the process to be followed during fabrication.

ACKNOWLEDGEMENT

Authors are very thankful to the DST Government of India, New Delhi for providing facilities to Banasthali Vidyapith under the CURIE programme.

REFERENCES

1. Samridhi, Kumar M., Dhariwal S., Singh K., and Alvi P. A. (2019). Stress and frequency analysis of silicon diaphragm of MEMS based piezoresistive pressure sensor. *International Journal of Modern Physics B*, 33, 1950040.
2. Eaton W. P. and James S. H. (1997). Micromachined pressure sensors: Review and recent developments. *Smart Materials and Structures*, 6(5), 530.
3. Pramanik C., Saha H., and Gangopadhyay U. (2006). Design optimization of a high performance silicon MEMS piezoresistive pressure sensor for biomedical applications. *Journal of Micromechanics and Microengineering*, 16(10), 2060.
4. Bistué G., Elizalde J., García-Alonso I., Olaizola S., Castano E., Gracia F. J., and García-Alonso A. (1997). A micromachined pressure sensor for biomedical applications. *Journal of Micromechanics and Microengineering*, 7(3), 244.
5. Moghaddam M. K., Breede A., Brauner C., and Lang W. (2015). Embedding piezoresistive pressure sensors to obtain online pressure profiles inside fiber composite laminates. *Sensors*, 15(4), 7499–7511.
6. DeanJr R. N. and Luque A. (2009). Applications of microelectromechanical systems in industrial processes and services. *IEEE Transactions on Industrial Electronics*, 56(4), 913–925.
7. Niu Z., Zhao Y., and Tian B. (2014). Design optimization of high pressure and high temperature piezoresistive pressure sensor for high sensitivity. *Review of Scientific Instruments*, 85(1), 015001.
8. Zhu B., Zhang X., Zhang Y., and Fatikow S. (2017). Design of diaphragm structure for piezoresistive pressure sensor using topology optimization. *Structural and Multidisciplinary Optimization*, *55*(1), 317–329.
9. Alvi P.A., Akhtar J., Lal K. M., Naqvi S. A. H., and Azam A. (2008). Design and fabrication of micromachined absolute micro pressure sensor. *Sensors & Transducers*, 96(9), 1–7.
10. Malhaire C. and Barbier D. (2003). Design of a polysilicon-on-insulator pressure sensor with original polysilicon layout for harsh environment. *Thin solid films*, 427(1–2), 362–366.
11. Samridhi, Sharma M., Singh K., Kumar S., and Alvi P. A. (2020). Analytical study of graphene as a novel piezoresistive material for MEMS pressure sensor application. *Journal of Nano- and Electronic Physics*, 12(2), 02001.
12. Samridhi, Singh K., and Alvi P. A. (2019). Finite element analysis of polysilicon based MEMS temperature-pressure sensor. *Materials Today: Proceedings*, doi: 10.1016/j.matpr.2019.11.024.
13. Samridhi, Singh K., and Alvi P.A. (2020). Influence of the pressure range on temperature coefficient of resistivity (TCR) for polysilicon piezoresistive MEMS pressure sensor. *Physica Scripta*, doi: 10.1088/1402-4896/ab93e7.
14. Hossain F. M., Nishii J., Takagi S., Ohtomo A., Fukumura T., Fujioka H., ..., and Kawasaki M. (2003). Modeling and simulation of polycrystalline ZnO thin-film transistors. *Journal of Applied Physics*, 94(12), 7768–7777.
15. Samridhi, Kumar M., and Singh K. (2019). Stress analysis of dynamic silicon diaphragm under low pressure. *American Institute of Physics Conference Series*, 2115(3), 030464.

16. Alvi P. A., Lourembam B. D., Deshwal V. P., Joshi B. C., and Akhtar J. (2006). A process to fabricate micro-membrane of Si_3N_4 and SiO_2 using front-side lateral etching technology. *Sensor Review*, 26(3), 179–185.
17. Samridhi, Sharma M., Singh K, and Alvi P. A. (2019). Comparative study of displacement profile for circular and square silicon diaphragm. *In American Institute of Physics Conference Series*, 2100(2), 020132.
18. Samridhi, Kumar M., Singh K., and Alvi P. A. (2019). Finite element analysis of circular silicon diaphragm. *IOP Conference Series: Materials Science and Engineering*, 594(1), 012045.

21 Role of Aqueous Electrolytes in the Performance of Electrochemical Supercapacitors

Prakash Chand
National Institute of Technology

CONTENTS

21.1 INTRODUCTION

Over the decades, there has been steady development in digital cutting-edge technology along with portable consumer electronic appliances, such as laptops, cellular phones, digital cameras, pulsed light dynamo, and alternative power determinant for computer flashback, which have become an essential part of our everyday life. The rapid increase in electronic devices encourages extensive curiosity on economical, lightweight, environmentally favorable, intact, and high energy density battery materials for both commercial and environmental applications. Thus, to accomplish the requirement for cutting-edge digital applications and for basic requirements of human life, advanced electrochemical energy storage and exchange equipment are required. Currently, energy storage materials have attracted enormous interest and research curiosity owing to the increasing apprehension about the continual evolution of energy. Energy conversion and storage is a big challenge in the modernized world. To overcome these problems, the need for portable energy storage devices is increasing. The most adequate and practicable technology for electrochemical energy transformation and storage are fuel cells, batteries, and supercapacitors, which are resourceful and often worn out in numerous applications [1–3]. Among assorted energy storage equipment, supercapacitors or ultracapacitors are adequate, vigorous, and continual energy storage devices, overcoming the inconsistency among batteries and conventional capacitors and deliver specific capacitance of six to nine orders of degree greater, high energy density, high power density, small equivalent series resistance, and prolonged charge-discharge lifespan cycles than conventional capacitors [4,5]. Supercapacitors and ultracapacitors are identical. The disparity in the taxonomy can be attributed to the Europeans and the Americans. Europeans term the device as supercapacitor whereas Americans recognize it as ultracapacitors. With rapid charge and discharge rates and elongated life of supercapacitors, they are extensively used for realistic applications in portable user electronic equipment such as cell phones, computers, flashlights, memory cards, and digital devices, as well as facilitate the prospect of diverse varieties of electric vehicles, power military operations, such as navigate missile automation and eminently perceptive marine warheads, space applications, and everyday handheld electronics. Current advancements in nanoscience and nanotechnology have offered the opportunity to design new energy storage materials for the next-generation high performance supercapacitors with superior specific capacity and elongated cycle life. Owing to their elevated peculiar surface area, nanomaterials provide abundant active spots for electrochemical reactions, short diffusion pathways, and high freedom for volume transition throughout the charging-discharging mechanism to cultivate the structural integrity of the electrode. Supercapacitors play an indispensable function in accompanying batteries or fuel cells in their energy storage operation by furnishing alternative power requirements to shield in contrast to power suspension. Supercapacitors are electrochemical energy storage systems that are expected to overcome the energy problems because of their high specific power density and rapid charge-discharge processes (in a fraction of seconds, and longer life cycle and stability).

Figure 21.1 illustrates the Ragone plot that demonstrates the relationship between power and energy density for diverse electrochemical energy storage systems.

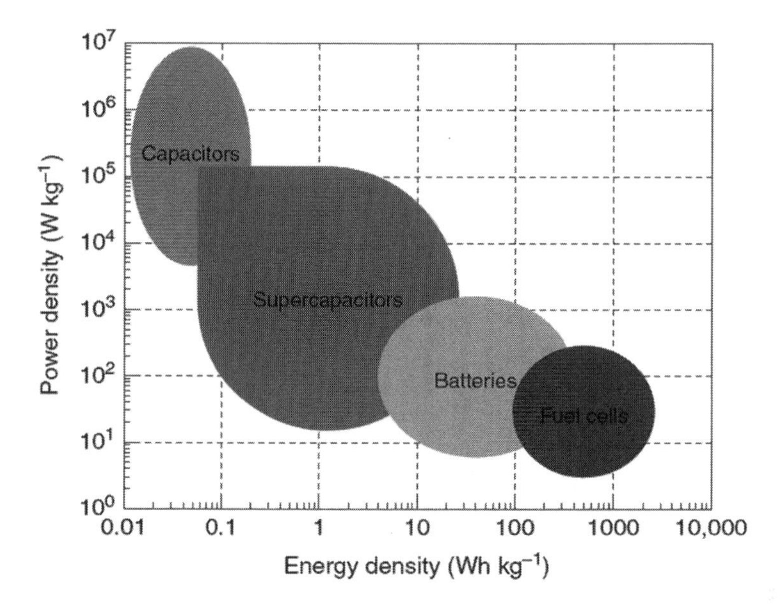

FIGURE 21.1 Ragone diagram showing variation in diverse energy depot systems of capacitor, batteries, fuel cells, and supercapacitor. (Reproduced with permission from Ref. [6], © Wiley 2020.)

From the Ragone contrive as depicted in Figure 21.1, it is clear that batteries endure less power density but possess high energy density, although conventional capacitors provide high power but miniature energy density. Therefore, supercapacitors as an electrochemical capacitor (EC) have drawn more consideration than batteries and fuel cells owing to their fast storage prospect and higher power density, flexible operating temperature, long life cycle, and stability. Beyond low energy density, supercapacitor pertaining to electrode materials and electrolytes embrace crucial impending and has bridged the inconsistency among the battery and conventional capacitor due to their unique characteristics [7]. Hence, at present, supercapacitors are treated as the most significant next-generation energy storage devices because of their fast charging and discharging processes. Consistent with a rechargeable battery, the critical demand of supercapacitors is their insufficient energy density.

21.1.1 Taxonomy of Supercapacitors

Active electrode materials and electrolytes need to accomplish the storage efficacy of supercapacitor devices (SCDs). Therefore, on the grounds of active electrode material and their charge accumulator mechanism, supercapacitors are generally categorized into three groups: (i) electrochemical double-layer capacitors (EDLCs) (ii) pseudocapacitor, and (iii) hybrid supercapacitors. EDLCs are broadly carbon occupying electrodes with high conductivity and stability that can accumulate energy through electrostatically adsorbing charges at the interface of the electrodes and electrolyte. Pseudocapacitors use rapid, reversible redox reaction to accomplish their energy accumulation process by electrochemical reactions at the surface interface. In contrast, hybrid capacitors integrate the EDLCs (twofold layer) and the pseudocapacitors (Faradaic reactions) to accumulate energy electrostatically as well as electrochemically [8,9]. Figure 21.2 shows a schematic diagram for the classification for supercapacitors.

21.1.1.1 Electrochemical Double-Layer Capacitors (EDLCs)

EDLCs accumulate charge electrostatically at the interface of broad surface area electrodes, for instance, carbon in a liquid electrolyte. In EDLCs, ionic charges soak up in the electrolyte whereas electric charges mount up in the electrode shaping a layer known as the electric twofold layer. The majority of the current EDLCs are made up of two extremely absorbent carbon-based electrodes in an electrolyte isolated by an insulative. The charge accumulation process in EDLCs is substantially reliant during interface charge segregation involving the electrolyte and electrode material, for example, activated carbon, and are merely electrostatic and non-Faradaic, where no oxidation–reduction reaction occurs and an extremely slight twofold layer is produced between the edge of the electrode and the electrolyte [10]. Hence, no transport of charge occurs from one electrode to the other. Because there is no chemical reaction, the transportation of ions in the electrolyte solution or electrons throughout the electrodes accounts for charging the accumulator. EDLCs could be completely indicted or liberated in an abrupt period through a large power density. Moreover, the

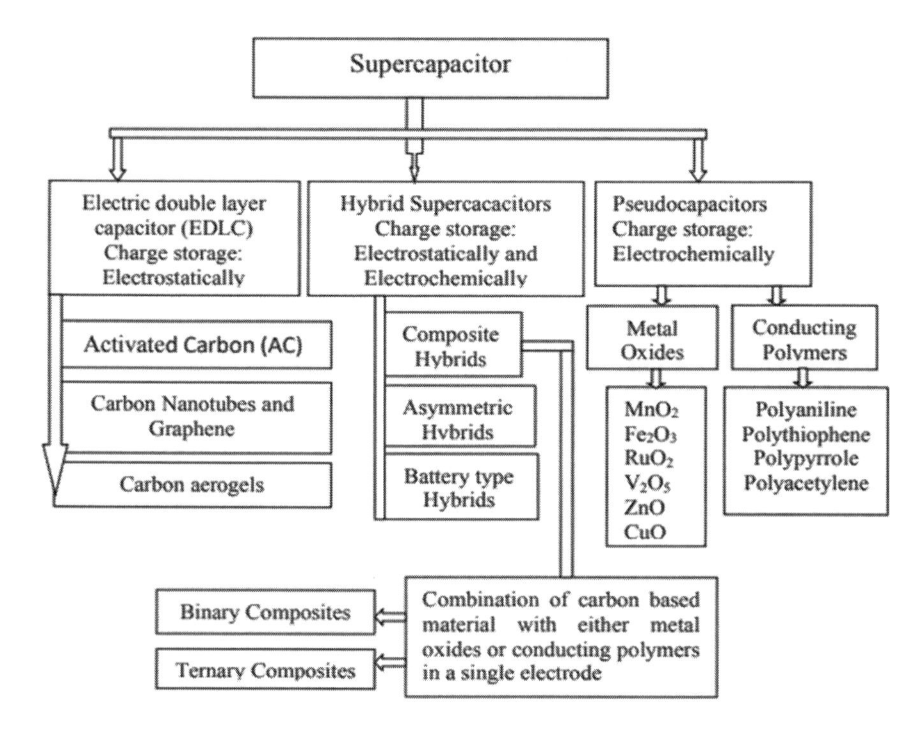

FIGURE 21.2 Classification of supercapacitors based on electrode material and the charge storage system.

charge storage mechanism does not embrace either chemical or physical alteration in the solid phase of the electrode; therefore, EDLCs have an elongated cycle life. The concentration of the electrolyte remains stable throughout the charging and discharging, which stores the energy in the EDLC. Carbon relies on materials, for instance, activated carbon, carbon nanotubes, graphene, and carbon aerogels are commonly utilized to manufacture EDLCs [11]. EDLCs have become promising aspirants for energy storage appliances involving electronics handy equipment, HEVs, and digital telecommunication devices, as well as renewable energy storage applications because of their huge power density and remarkable elongated lifespan cycle [12]. However, because they undergo from lower energy density in contrast to that of conventional batteries, their commercialization is limited. The energy density could be improved by a suitable selection of types of electrolytes as well as by building the electrode material with a large surface area.

21.1.1.2 Pseudocapacitors

In comparison to EDLCs, pseudocapacitor or redox supercapacitor or faradaic supercapacitor (FS) exploit rapid and reversible Faradaic reactions (oxidation–reduction reactions) for the charge accumulation, which occurs on the active electrode material and engrosses the route of the charge beyond the twofold layer. This mechanism is analogous to the charging and discharging progression in batteries, ensuing in Faradaic current transient throughout the supercapacitor cell. Pseudocapacitors store energy by electrochemical reactions at the interface. Pseudocapacitors exhibit high energy density in comparison to the EDLCs because the capacitance typically originates from the strong reversible redox (Faradaic) reactions at the electrode/electrolyte interface. Typically, pseudocapacitors are fabricated using materials enduring high redox reactions, such as conducting polymers like polythiophene, polyaniline, and polypyrrole and certain transition metal oxides/hydroxides like MnO_2, Fe_2O_3, RuO_2, V_2O_5, ZnO, and CuO, etc., which can store energy by rapid Faradaic redox reactions [13,14]. The aforementioned pseudocapacitive materials possess higher energy densities; hence, they exhibit greater energy storage in comparison to the EDLCs, possibly owing to the extremely reversible oxidation–reduction (Faradaic) reaction, which is in contrast to EDLCs. Such kinds of supercapacitors engross both the non-Faradaic (electric twofold layer) and Faradaic (oxidation–reduction) charge storage procedures. However, pseudocapacitors or FS frequently go through from somewhat inferior power density in contrast to EDLCs because the Faradaic process are typically sluggish than the non-Faradaic process. In addition, similar to batteries, FS frequently lack stability during cycling because oxidation–reduction reactions occur at the electrodes. As the Faradaic electrochemical reactions take place at or near the surface of the electrodes, they can be categorized into three kinds: (i) underpotential deposition, that is, absorption pseudocapacitive, (ii) redox pseudocapacitive, and (iii) intercalation pseudocapacitive [15].

21.1.1.3 Hybrid Capacitors

The power density and cycle lifespan for EDLC materials are superior to that of pseudocapacitor or FS, however, energy density and capacitance is higher for pseudocapacitors. Therefore, the combination of EDLCs and pseudocapacitors build a

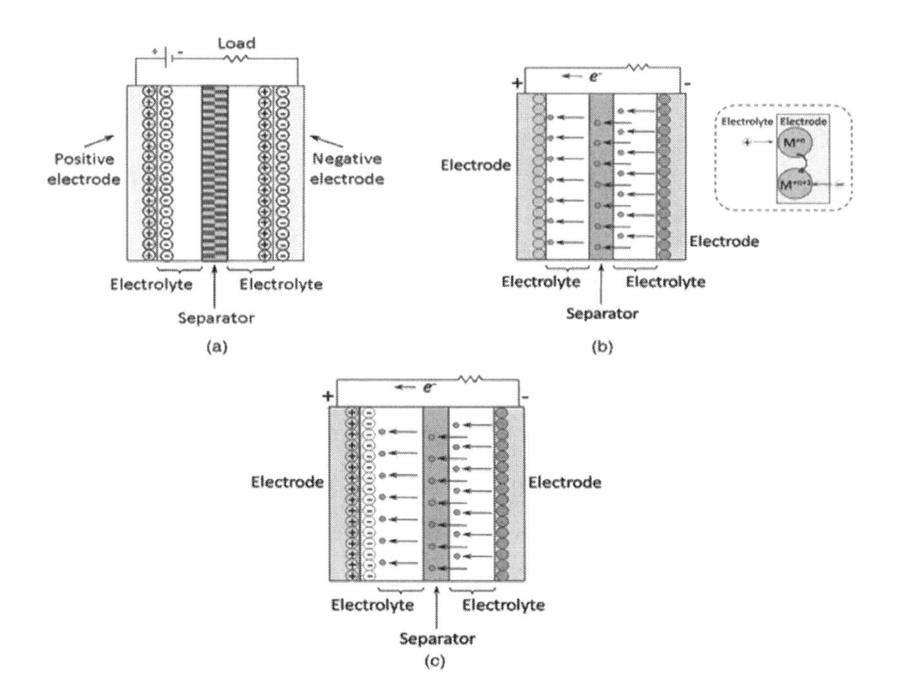

FIGURE 21.3 Schematic representation of supercapacitor taxonomy: (a) EDLCs type, (b) pseudocapacitor type, (c) hybrid capacitor type. (Reproduced with permission from Ref. [17], © American Society of Civil Engineers 2013.)

hybrid capacitor with higher specific capacitance in comparison with EDLCs and pseudocapacitive supercapacitors. As EDLC material (like activated carbon) and the pseudocapacitive material (such as conductive polymers, transition metal-based oxides, etc.) are produced using both types of material, it is called a hybrid capacitors [16]. Hybrid capacitor accumulates charges by either absorption–desorption oroxidation–reduction reactions, that is, electrostatically or electrochemically. The hybrid supercapacitor is further categorized into three groups, namely, composite hybrid, asymmetric hybrid, and battery type hybrid. The hybrid supercapacitor can accomplish extensive power and energy density along with superior cycling stability. Figure 21.3a–c shows the schematic diagram for the different types of supercapacitors.

21.2 BASIC REQUIREMENTS FOR DESIGNING ELECTROLYTES FOR SCDS

The electrochemical accomplishment of the SCDs depends upon various factors, for instance, nature of active electrode materials, nature of electrolyte and separator thickness, etc. Among these factors, electrolyte plays a critical role in deciding the comprehensive characteristics of the properties of the SCDs. The electrochemical performance of SCDs is defined by power density (P), energy density (E), specific capacitance value (C_s), equivalent series resistance (ESR), time constant (t), cyclic stability, charge-discharge capabilities, self-discharging, the time required for charging, and cost. The explicit surface vicinity of active electrode materials, the electrical conductivity of electrode material, porosity, crystallinity for deep diffusion of electrolyte, size and shape of redox-active species, and intrinsic properties of electrolyte are other important parameters which significantly influence the electrochemical properties of SCDs. Therefore, it is useful to exploit suitable amalgamation of electrode materials and electrolytes to accomplish higher energy and power density for SCDs. In general, three capabilities are adopted to estimate the electrochemical efficacy of SCDs, in particular, cyclic voltammetry (CV), galvanostatic charge/discharge (GCD), and electrochemical impedance spectroscopy (EIS).

To accomplish greater electrochemical performance for SCDs, the appropriate choice and preparation of active electrode materials and electrolytes with required features is indispensable, as the active electrode material and the electrolyte play an important function in establishing the electrochemical performance of SCDs. Electrolytes have been distinguished as one of the most dominant constituent elements in the functioning of the supercapacitor, which comprises EDLC, pseudocapacitor, and hybrid supercapacitor. The electrochemical property of supercapacitor can be estimated in relation to the electrochemical characteristics established during an amalgamation involving the electrode and the electrolyte materials. The interaction among the electrodes and electrolytes perform a crucial function in the extensive functioning of the supercapacitor. The diverse factors, for instance, power density, working temperature, ESR, self-discharge rate, and cycling stability of the supercapacitor are strongly influenced by the electrolyte in addition to the type and size of ion, the concentration of electrolyte, electrolyte and electrode interaction, and operating potential window (as depicted in Figure 21.4). The pseudocapacitance

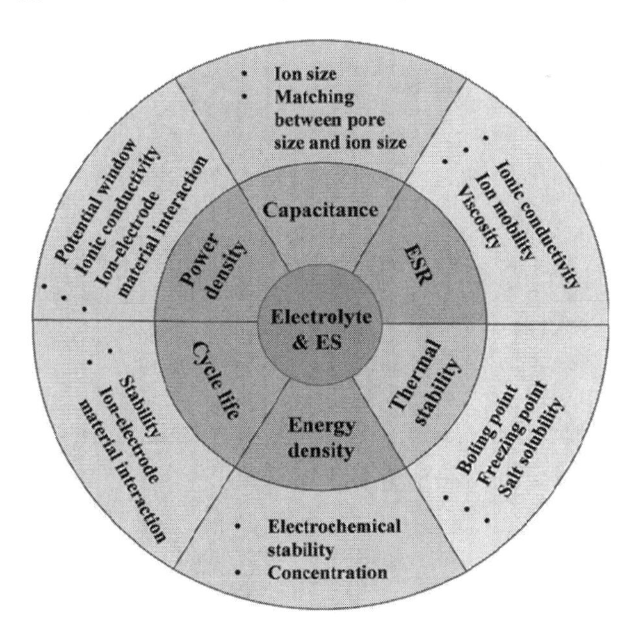

FIGURE 21.4 Effects of the electrolytes on the evolution of supercapacitors devices (SCDs). (Reproduced with permission from Ref. [20], © Royal Society of Chemistry 2015.)

values of the composite materials also depend on the nature as well as the type of electrolyte [18]. It is well recognized that the internal resistance of the supercapacitor is determined by the ionic conductivity of electrolytes. The electrolyte ion size must be equivalent or small from the pore extent of active electrode materials to acquire the highest specific capacity and extensive power density. In addition, the boiling point, viscosity, and freezing temperature of electrolyte influence the thermal constancy and the working temperature of a supercapacitor. Further, it is probably recognizable that the operating potential of the supercapacitor is connected to the electrochemical constancy of the electrolyte. Moreover, deterioration of the electrolyte is related to the aging and breakdown of the supercapacitor. For high electrochemical functioning of the supercapacitor, following properties of the electrolyte of supercapacitor are required: (i) abroad potential window, (ii) large ionic conductivity, (iii) extensive range of operating temperature, (iv) high electrochemical stability, (v) compatibility with the electrolyte materials, (vi) inferior volatility and flammability, (vii) low viscosity, (viii) low cost, and (ix) environmentally friendly [19,20]. It is hard to accomplish all the demands because each electrolyte has its benefits and shortcoming, and it is realistic to congregate all the above requirements with one electrolyte.

21.2.1 Equivalent Series Resistance (ESR)

ESR is a vital constraint for evaluating the power density of supercapacitor devices. Power density is directly related to the square of the operating potential (V) and diametrically related to the ESR of the electrochemical supercapacitor cell. Therefore, to develop both energy and power density of the supercapacitor device, it is necessary to raise the values of operating voltage (V) and at the same time reduce the values of ESR. The operating potential of the cell relies on the thermodynamic stability of electrolyte and electroactive materials of the electrode. Similar to the diverse electrochemical energy storage devices, an extensive value of ESR confines the charging-discharging rate and leads to a lesser value of power density. In general, ESR comprises different kinds of resistors involving inherent resistance of the active electrode materials and electrolyte solvent, mass transport resistance of the diffusion of the ions, and interaction resistance relating electroactive material and current collector [21]. Consequently, for a supercapacitor device to accomplish both superior energy and power density, it is necessary to build up an electrolyte with an extensive working potential while maintaining a minimum ESR or higher ion conductivity.

21.2.2 Electrolyte Conductivity, Ion Solvation, Mobility of the Free Ions, and Solvent Upshot

Electrolytes are a critical parameter and play a significant role in the overall performance of the supercapacitive system to have superior power as well as energy density, elongated cycle life, and security measure. To approximate the evolution of electrolyte, the ionic mobility and conductivities of the electrolytes are important factors. The conductivities of the aqueous electrolyte are normally higher in contrast to ionic and solid electrolytes owing to inferior vigorous viscosities of aqueous electrolytes. The conductivities (σ) of species (i) are fundamentally associated with the concentration of charge carriers (n_i), ionic mobility (μ_i), degree of the valence of the movable ion charges (z_i), and the elementary charge (e), expressed by the following equation [22]:

$$\sigma = \sum n_i \mu_i z_i e$$

The variables in exceeding the equation rely on the solvation effect, movement of the solvated ions, and lattice energy of the salt. Therefore, all components comprising solvent, additive, and salt affect the conductivities of the electrolyte. There are certain factors including the density of liberated charge transporters, that is, cation and anion and mobility of free ions, the solubility of salts in the solvent, cleavage of salts, and coupling of the ions of the decomposed salt, which directly or indirectly affect the conductivities of electrolytes. In addition, viscosities of the solvent and perspective range of electrostatic exchanges among freely or segregated ions that are resolute with the dielectric constant of the solvent also influence the conductance of the electrolytes. The solvent components such as dielectric constant and viscosity significantly influence the conductivity of electrolytes. Dielectric constant determines the cleavage of the salts, whereas viscosity determines the ionic mobilities of the salt [23]. Further, the solvent with a higher viscosity also has a higher dielectric constant [24]. Consequently, to get an appropriate solvent regime for the supercapacitive system, a solvent with lower values of viscosities and higher values of dielectric constant salt should be blended.

21.2.3 Salt Concentration Effect

The nature of electrolyte and their concentration used in the EC has significant impacts on their electrochemical performances. It is well known that the conductivities could be distinct for different salts in a similar solvent because of the correlation among the anions and cations and the volume variation of diverse anions [25–27]. The conductivities of the electrolytes are also different because of the salt density in a similar regime. The amount of free ions is largest at the petite density of the salts, as a result, the most favorable denseness of salt can too enhance the ionic conductivities. Moreover, as there is equilibrium among viscosity and free ions, the conductivities will be incredibly high. However, if the salt density is exceptionally higher in the solution, the anion and cation in the solution will amalgamate robustly with impartial ions, which decreases the figure of free ions as well as reduces the ionic conductivities of the electrolytes.

21.2.4 Electrochemical and Thermal Stability

Electrochemical and thermal constancy of the supercapacitor electrolyte are critical factors associated with the security and cycling lifespan of the supercapacitor appliances. The uppermost and lowest constraints of the oxidation–reduction reaction of the electrolyte, which are the manifestation of electrochemical windows. The galvanic and thermal constancy not just relies on the form of the electrolytes but also depends strongly on the consistency of the electrolyte with the electrode, the linkages of electrolyte and electrode mutually, and the thermal reliability of the electrolytes themselves [25]. The cycle life is one of the chief constraints and signs of stability for evaluating the overall performance of the supercapacitor device. However, the cyclic reliability is normally compressed owing to the idealistic electrochemical convertibility ensuing upon the correlations among the electrolyte ions and the electrodes material. The cycle lifespan of the supercapacitive device relies on various parameters, including the cell sort, active electrode material, electrolyte, charge-discharge rate, working potential, and working temperature. However, few electrolytes deteriorate under the charging-discharging process with the increase in working temperature and heat discharge; hence, it causes safety concerns for the supercapacitor devices. Further, the operating temperature can affect many characteristics of supercapacitors, such as the power density, performance during long lifespan cycle, ESR, and self-discharging rate. In particular, the electrochemical and thermal stability of supercapacitor rely on the category of the electrolyte, for instance, the density and the category of salt, any additive, working electrodes, as well as the specific characteristics of the solvent such as boiling temperature, viscosity, and freezing temperature [25].

21.3 ELECTROLYTES AND THEIR CLASSIFICATIONS

The electrolyte has been recognized as one of the vital components of SCDs, as well as active electrode material. Diverse kinds of electrolytes have been developed which can significantly advance with the electrochemical evolution of supercapacitor appliance. The electrolyte is a medium that conducts the electricity produced by the dissolution of the salt in an adequate ionizing solvent such as water. Typically, electrolyte exists within the separator and inside the active material layers. The electrolyte plays an important function in the development of the electrical double layer for EDLCs and redox reactions for a pseudocapacitor. Further, the electrolyte is an indispensable and extensive component in supercapacitors and play an incredibly vital function in transporting as well as compensating charges among the two electrodes. The key factors for an electrolyte are wide potential window to attain high energy density, higher ionic concentrations, small ohmic resistance, and small viscosities, which affect the power densities of the SCDs. Therefore, the electrolyte has a unique importance in SCDs as the energy and power density normally rely on the operating potential window of supercapacitor, which is determined by the electrolyte. A good quality electrolyte is described by an extensive voltage window, concentration of ions, low solvated ionic radii, small volatility, temperature coefficient, low resistivity, low viscosity, less toxicity, high electrochemical stability, and low cost [19]. In general, electrolytes are classified into three groups, viz., liquid electrolyte, solid-state electrolyte, and redox-active electrolyte [28]. Each electrolyte has its own merits and disadvantages. Among these, the liquid electrolyte can be further classified into two types, namely, aqueous electrolytes and non-aqueous electrolytes, of which the aqueous one is the most extensively employed in the literature because of its high ionic conductivity and excellent safety properties. Solid-state electrolytes are classified into three types and redox-active electrolytes are broadly divided into

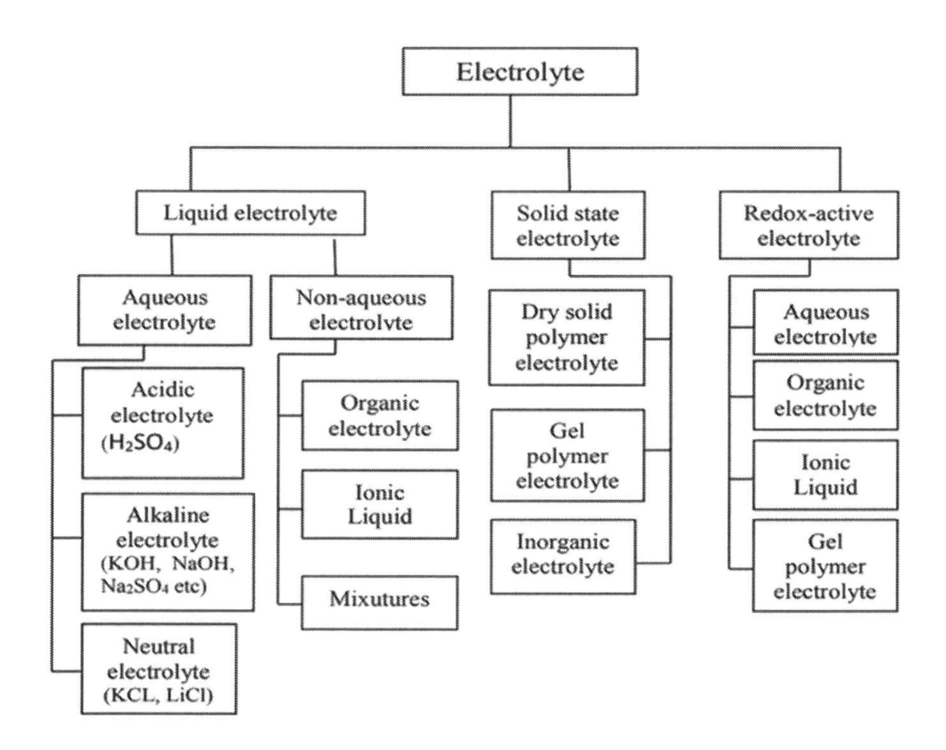

FIGURE 21.5 Classification of various electrolytes.

four types. Figure 21.5 demonstrates the classification of various electrolytes. In this chapter, the foremost emphasis is on discussing the effect of aqueous electrolytes on the evolution of the SCDs.

21.4 AQUEOUS ELECTROLYTES

Aqueous electrolytes possess a high concentration of ions with smaller ionic radius and low resistance and have been widely utilized in research owing to their simple treatment in the laboratories in contrast to organic electrolyte and ionic liquids, which involve refining processes. They produce charge/discharge rates owing to the comparatively extensive conductivity and low viscosity of concentrated solutions. Thus, SCDs fabricated using aqueous electrolytes illustrate superior capacitance and power compared to organic electrolytes. Further, aqueous electrolytes reveal high ionic conductivity (~10^{-3} Scm^{-1}) compared to organic and ionic electrolytes, which is supportive for reducing the ESR and leads to superior capacitance and power density of SCDs [29,30]. However, the smaller operating potential window of aqueous system restricts its performance in supercapacitors. However, the operating potential of organic electrolytes can reach up to 3 V. As the energy concentration of supercapacitor is directly related to the square of the cell potential, the organic electrolyte is more appropriate compared to other electrolytes. However, there are safety risks owing to flammability, highly toxic nature, and high cost, which restrict their commercial applications. The ionic electrolyte identified at ambient temperatures, molten salts and ionic liquids have many advantages such as non-toxicity, non-flammability, significant electrochemical and thermal stability and diverse amalgamation of selections of cation and anion. The feeble ionic conductivities at the ambient temperature of ionic liquid also limit its practical application. Hence, aqueous electrolytes are typically preferred compared to other electrolytes owing to their easy preparation, cost, non-hazardous nature, and high ionic conductivity [31,32]. Normally, aqueous electrolytes reveal enormously high conductivity, which is at least one magnitude higher in comparison to the organic or ionic liquids electrolytes. Further, to appraise the extensive concert of aqueous electrolytes, some distinctive criteria should be adopted into deliberation, for instance, the sizes of hydrated and bare ions, the flow of ions that change the ionic conductivities, and the specific capacity [33,34]. Aqueous electrolytes are further categorized into three groups based on the origin of solvent pH: (i) acidic electrolyte, (ii) alkaline electrolyte, and (iii) neutral electrolyte. Figure 21.5 demonstrates the classification of aqueous electrolytes. The most frequently utilized aqueous electrolytes are KOH, H_2SO_4, and Na_2SO_4 because of their higher ionic conductivities. The high ionic conductivities of aqueous electrolytes are favorable for reducing ESR that results in extensively high power density SCDs.

21.4.1 Acidic Electrolytes

Diverse acidic electrolytes are exploited in electrochemical capacitors, for example, HCl, KCl, K_2SO_4, and H_2SO_4, etc. However, the most frequently used acid electrolyte is H_2SO_4 because of its extremely higher ionic conductivities (~0.8 S cm^{-1} for 1 M H_2SO_4 at 25°C) [30]. Because the high conductivities of the H_2SO_4 electrolyte can reduce the value of ESR, it result in expansively extensive power density for SCDs. The value of conductivity relies on the concentration of the H_2SO_4electrolyte. The extremely higher or lower density of electrolyte induces a reduction in the value of ionic conductivities. According to a literature survey, it is noticed that the highest ionic conductivity of the H_2SO_4 electrolyte is accomplished at 1.0 M concentration at 25°C; hence, several studies use 1 M H_2SO_4 electrolytes solution, mainly for SCDs using carbon-based supercapacitors [19,35]. Recently, many research groups have employed aqueous electrolyte such as H_2SO_4 in carbon-based capacitors (EDLCs), pseudocapacitive supercapacitors, and hybrid supercapacitor because of the high value of ionic conductivity of H_2SO_4 electrolyte. The ESs have lower ESR and result in higher values of specific capacity for supercapacitor devices [19,35].

21.4.2 Alkaline Electrolytes

As many of the acidic electrolytes are not appropriate for analyzing the valuable cost-effective metal material as well as metal-based oxides, for instance, Co, Ni, Eu, etc., alkaline electrolytes are exploited instead of acidic electrolytes and have drawn considerable curiosity from the scientific community [19]. Among diverse alkaline electrolytes such as LiOH, NaOH, and KOH, the most typically used electrolyte is KOH because of its significantly extensive ionic conductivity. In general, the electrolyte characteristics such as concentration, types of ions, and the working temperature could influence the efficiency of electrochemical supercapacitors, for instance, alkaline electrolyte concentrations can influence the value of ESR, and hence the specific capacitance. However, there is one negative aspect, that is, corrosion at the electrode substrate surface when using a concentrated electrolyte which can result in the electrode materials shedding out upon the substrate. Thus, it is essential to optimize the electrolyte concentration for the general functioning of ESs. Joshi et al. [36,37] reported the electrochemical properties of $BiPO_4$ nanostructures for samples BP1 and BP2 synthesized via hydrothermal and facile microwave irradiation technique as electrode material for EC application (examined at ambient temperature) by CV, GCD, and EIS techniques. The CV, GCD, and EIS of $BiPO_4$ electrode material have been studied in 4M KOH and 2M KOH aqueous alkaline electrolytes, respectively, for the samples BP1 and BP2, as demonstrated in Figures 21.6a–c and 21.7 (a–c) for samples BP1 and BP2, respectively. The specific capacitance (C_s) value estimated from GCD for sample BP1 and BP2 were 446 Fg^{-1} at 1 A/g and 268 Fg^{-1} at 1 A/g, respectively. The Nyquist graph for the EIS is depicted in Figures 21.6c and 21.7c for the samples BP1 and BP2, which infer that the as-prepared electrode is partially polarizable.

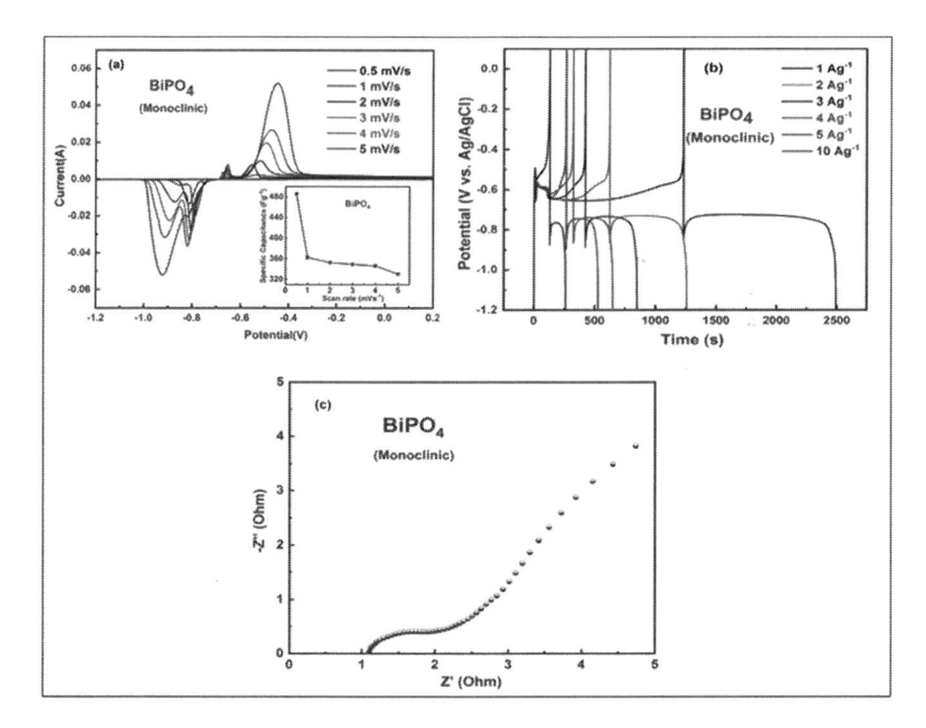

FIGURE 21.6 (a) Cyclic voltammetry (CV) curve of $BiPO_4$ nanostructures for sample BP1, (inset: Deviation of specific capacity with a sweep rate of $BiPO_4$ nanostructures), (b) galvanostatic charge/discharge (GCD) arc of $BiPO_4$ nanostructures for sample BP1, and (c) Nyquist plot for the EIS spectra of $BiPO_4$ nanostructures for sample BP1. (Reproduced with permission from Ref. [36], © Elsevier 2020.)

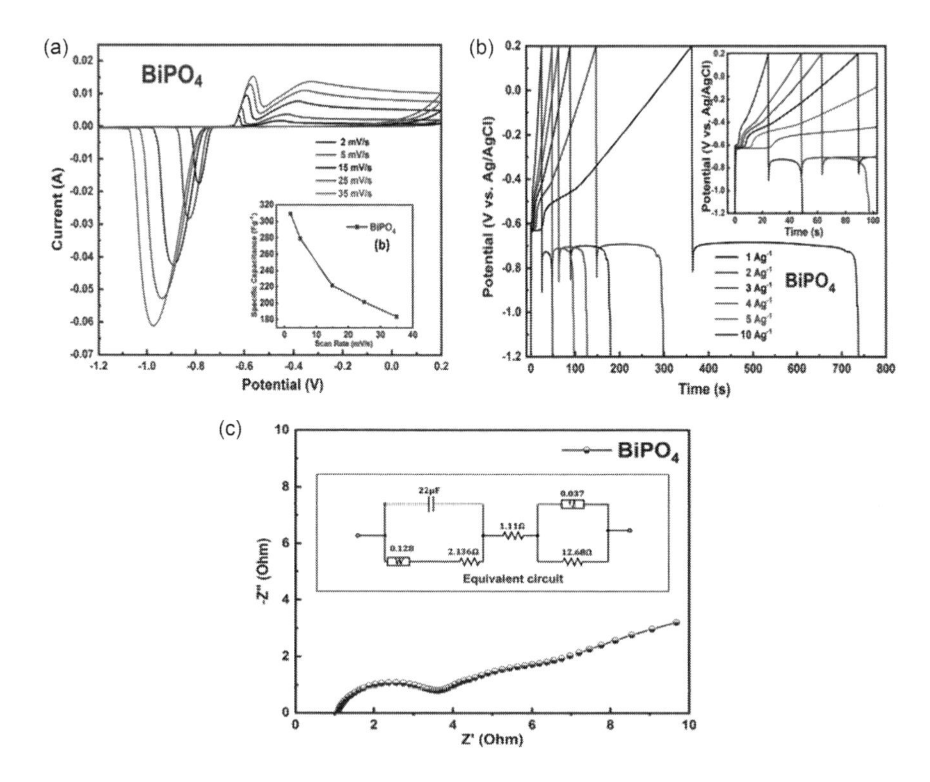

FIGURE 21.7 (a) Cyclic voltammetry (CV) curve of $BiPO_4$ nanostructures for sample BP2, (inset: Deviation of specific capacity with a sweep rate of $BiPO_4$ nanostructures), (b) Galvanostatic charge/discharge (GCD) curve of $BiPO_4$ nanostructures for sample BP2, and (c) Nyquist graph for the EIS of $BiPO_4$ nanostructures for sample BP2 (inset: An equivalent circuit). (Reproduced with permission from Ref. [37], © Elsevier 2020.)

The results show that $BiPO_4$ nanostructures prepared under optimal synthesis condition by hydrothermal and microwave technique is an alternative for cathode materials as electrode materials for energy storage applications.

21.4.3 Neutral Electrolytes

In contrast to alkaline and acidic electrolytes, neutral electrolytes have been considered extensively because of their higher potential window, safety, and inferior corrosion aspect. There are different kinds of neutral electrolytes, namely, LiCl, Li_2SO_4, $LiClO_4$, Na_2SO_4, NaCl, $NaNO_3$, KCl, K_2SO_4, KNO_3, $Ca(NO_3)_2$, and $MgSO_4$, which are typically exploited in ES studies. Among the diverse neutral electrolytes, Na_2SO_4 is an often employed neutral electrolyte and has shown potential electrochemical reactions for electrodes, particularly pseudocapacitance materials and hybrid capacitor. Further, neutral electrolytes-based uniform carbon electrochemical capacitors have been recognized as the utmost hopeful aspirant because of reduced ecological impact and superior energy density. However, a few salts, for instance, K_2SO_4, could not accomplish like a highly concentrated salt, particularly when it is utilized in low temperatures. Indeed, the outcome of a neutral electrolyte on the evolution of ESs also relies on the kind of electrolyte exploited. Some studies have revealed that the specific capacities of EDLC electrode material with neutral electrolytes are inferior compared to H_2SO_4 electrolyte or the KOH electrolyte [38,39]. The ESR of electrochemical capacitors improves with neutral electrolytes, and generally, it has lower ionic conductivities in comparison to H_2SO_4 and KOH electrolyte. However, compared to acidic and alkaline electrolytes, carbon-based electrochemical supercapacitors with neutral electrolyte could furnish better working potential window on account of having a broad electrolyte stable potential window (ESPW) [40]. Because a neutral electrolyte has lesser concentrations of H^+ and OH^- than acidic and alkaline electrolyte, a higher potential for hydrogen and oxygen development reactions could be anticipated, which results in an increased ESPW. Demarconnay et al. [41] demonstrate a superb lifespan cycle evolution with 10,000 recharging-discharging cycles at a larger cell potential of 1.6 V with 0.5 M Na_2SO_4 electrolyte for activated carbon-stemmed ECs.

21.5 CHALLENGES IN THE DEVELOPMENT OF SCDS ELECTROLYTES AND FUTURE RESEARCH DIRECTIONS

The aqueous electrolyte has been acknowledged as one of the critical elements in the electrochemical performance of SCDs. The main constraints in supercapacitors' evolution are to improve the lifespan cycle, energy and power density, and security concerns. The foremost aspects that influence both efficiencies as well as the realistic application of SCDs are extensive potential window, ionic conductivities, concentration of the ions, small solvated ionic radii, inferior volatility, temperature coefficient, small resistivity and viscosity, ion mobility, cost, and electrochemical and thermal stabilities of the electrolytes. The aqueous electrolytes possess a high concentration of ions with smaller ionic radius and low resistance and have been widely used, exhibiting superior capacitance and power. Despite extensive accomplishments taken in the domain of electrochemical capacitor electrolyte, there are diverse challenges in this field, still, which hamper the commercial applications of SCDs in the field of science and technology, for instance, smaller energy density and working voltage in aqueous electrolytes. Further, the energy and power density of the SCDs are instantly persuaded by the working cell potential of the electrolyte. In aqueous electrolytes, the ESPW values strongly rely on the anions and cations of a conductive salt. The electrolyte with extensive ESPW values can raise the cell potential of the electrochemical supercapacitors, and consequently, strengthen the energy density. In addition, the contaminant in electrolyte induces the pessimistic impact on ESPW which gives rise to a high self-discharge rate. Therefore, to improve ESPW values for better energy density, the purity of the electrolyte and optimization of the appropriate electrolytes is necessary. From other studies, it is not easy to distinguish the appropriate electrolytes since the metrics are typically accomplished according to distinct conditions. Thus, it is a large challenge to build ESs with superior energy densities.

Besides, the gravimetric capacity, energy, and power densities of the supercapacitor are usually claimed based on the mass of vigorous electrode material. Besides the electrode materials, the electrolyte also significantly contribute, which should not be disregarded. Further, to accomplish the high evolution of supercapacitor, the suitable consistency among the electrolyte-electrode material is especially important. The elemental perception of the reaction chemistry and consistency connecting electrode-electrolyte, as well as the interaction among new electrolyte and hybrid electrode materials can be enhanced by an exhaustive learning of theoretical and experimental study. Such a primary perceptive will not just present the route for upcoming novel electrolytes but also simplify the evolution of electrode material that is equivalent to other electrolytes. While evaluating the functioning of ESs, it is also obligatory to distinguish the volume and mass of the electrolytes. The improvement in standard techniques to appraise the concert of diverse electrolytes is also indispensable.

21.6 SUMMARY

In summary, ES is recognized as the green energy storage system for next-generation accepting significant performance and robustly contingent upon the electrolyte. The charge storage mechanisms in ESs rely on the adsorption/desorption of charges on the electrode–electrolyte interface, whereas for the pseudocapacitive material it is by fast oxidation–reduction reactions. This chapter offers a broad overview of the evolution and modern developments with respect to aqueous electrolyte for ESs of the reaction mechanism, requirements for designing better-quality electrolytes, aqueous electrolyte factors, including ionic conductivity, ion size, ion mobility, radii of bare and hydrate ion, viscosity, dielectric constant of the solution, ESPW, and type and molar concentration of the electrolytes, which impinges on ES accomplishment, for instance, specific capacity, cyclic constancy, energy and power densities, and challenges of ES electrolytes. Diverse improvements are still crucial to improve the functioning of ESs, for instance, in the selection of appropriate electrolytes, specific capacity, constancy, and cost-effectiveness. Although extensive accomplishments have been made in this domain, significant challenges remain to be addressed. The aqueous energy storage devices are encouraging applications in a wide range of energy storage because of their extensive ionic conductivity, safety, and cost-effectiveness. The competency of the supercapacitor not only relies on the active electrode materials but also on the appropriate electrolyte, which performs a vital function in the evolution of ESs. Hence, this study concludes that the selection of electrolytes is an essential constraint in accomplishing superior performance for next-generation energy storage applications.

ACKNOWLEDGMENT

The author is grateful to the Science and Engineering Research Board (SERB), Govt. of India for providing funding through research project No: SERB/F/10804/2017–18. The authors are also grateful to the Director, NIT Kurukshetra for providing the facilities in the Physics Department.

DECLARATION OF COMPETING INTEREST

The authors declare no competing financial interests.

REFERENCES

1. H. Pang, Y. Ma, G. Li, J. Chen, J. Zhang, H. Zheng, W. Du, Facile synthesis of porous ZnO-NiO composite micro polyhedrons and their application for high power supercapacitor electrode materials, *Dalton Transactions* 41 (2012) 13284–13291.
2. M. Winter, R.J. Brodd, What are batteries, fuel cells, and supercapacitors, *Chemical Reviews* 104 (2004) 4245–4269.
3. P. Simon, Y. Gogotsi, Materials for electrochemical capacitors, *Nature Materials* 7 (2008) 845–854.
4. J. Zhang, X. S. Zhao, On the configuration of supercapacitors for maximizing electrochemical performance, *ChemSusChem* 5 (2012) 818–841.
5. C. Liu, F. Li, L. P. Ma, H. M. Cheng, Advanced materials for energy storage, *Advanced Materials* 22 (2010) E28–E62.
6. B. K. Kim, S. Sy, A. Yu, J. Zhang, *Electrochemical Supercapacitors for Energy Storage and Conversion, Handbook of Clean Energy Systems*, John Wiley & Sons, Ltd. (2015), DOI: 10.1002/9781118991978.hces112.
7. L. L. Zhang, X. S. Zhao, Carbon-based materials as supercapacitor electrodes, *Chemical Society Reviews* 38 (2009) 2520–2531.
8. X. Wang, C. Yan, J. Yan, A. Sumboja, P. S. Lee, Orthorhombic niobium oxide nanowires for next-generation hybrid supercapacitor device, *Nano Energy* 11 (2015) 765–772.
9. C. Yuan, J. Li, L. Hou, J. Lin, G. Pang, L. Zhang, L. Lian, X. Zhang, Template - engaged synthesis of uniform mesoporous hollow $NiCo_2O_4$ sub-microspheres towards high-performance electrochemical capacitors, *RSC Advances* 3 (2013) 18573–18598.
10. Y. Li, G. Zhu, H. Huang, M. Xu, T. Lu, L. Pan, N, S dual doping strategy via electrospinning to prepare hierarchically porous carbon polyhedra embedded carbon nanofibers for flexible supercapacitors. *Journal of Materials Chemistry A* 7 (2019) 9040–9050.
11. S. L. Candelaria, Y. Shao, W. Zhou, X. Li, J. Xiao, J.-G. Zhang, Y. Wang, J. Liu, J. Li, G. Cao, Nanostructured carbon for energy storage and conversion, *Nano Energy* 1 (2012) 195–220.
12. F. Jaouen, E. Proietti, M. Lefevre, R. Chenitz, J. P. Dodelet, G. Wu, H. T. Chung, C. M. Johnston, P. Zelenay, Recent advances in non-precious metal catalysis for oxygen-reduction reaction in polymer electrolyte fuelcells. *Energy & Environmental Science* 4 (2011) 114–130.
13. Y. Wang, Y. Xia, Recent progress in supercapacitors: from materials design to system construction, *Advanced Materials* 25 (2013) 5336.
14. X. Chen, R. Paul, L. Dai, Carbon-based supercapacitors for efficient energy storage, *National Science Review* 4 (2017) 453–489.
15. V. Augustyn, P. Simon, B. Dunn, Pseudocapacitive oxide materials for high-rate electrochemical energy storage, *Energy Environmental Science* 7 (2014) 1597–1614.
16. V. C. Lokhande, A.C . Lokhande, C. D. Lokhande, J. H. Kim, T. Ji, Supercapacitive composite metal oxide electrodes formed with carbon, metal oxides and conducting polymers, *Journal of Alloys and Compounds* 682 (2016) 381–403.
17. M. Vangari, T. Pryor, L. Jiang, Supercapacitors: Review of materials and fabrication methods, *Journal of Energy Engineering* 139 (2013) 72–79.
18. A. González, E. Goikolea, J. A. Barrena, R. Mysyk, Review on supercapacitors: Technologies and materials, *Renewable and Sustainable Energy Reviews* 58 (2016) 1189–1206.

19. M. Z. Iqbal, S. Zakar, S. S. Haider, Role of aqueous electrolytes on the performance of electrochemical energy storage device, *Journal of Electroanalytical Chemistry* 858 (2020) 113793.
20. C. Zhong, Y. Deng, W. Hu, J. Qiao, L. Zhang, J. Zhang, A review of electrolyte materials and compositions for electrochemical supercapacitors, *Chemical Society Reviews* 44 (2015) 7484–7539.
21. Y. Shao, M. F. El-Kady, J. Sun, Y. Li, Q. Zhang, M. Zhu, H. Wang, B. Dunn, R. B. Kane, Design and mechanisms of asymmetric supercapacitors, *Chemical Reviews* 118 (2018) 9233–9280.
22. K. Xu, Nonaqueous liquid electrolytes for lithium-based rechargeable batteries, *Chemical Reviews* 104 (2004) 4303–4417.
23. G. Kamath, R. W. Cutler, S. A. Deshmukh, M. Shakourian-Fard, R. Parrish, J. Huether, D. P. Butt, H. Xiong, S. K. R. S. Sankaranarayanan, In silico based rank-order determination and experiments on non-aqueous electrolytes for sodium-ion battery applications, *Journal of Physical Chemistry C* 118 (2014) 13406–13416.
24. H. Che, S. Chen, Y. Xie, H. Wang, K. Amine, X. Z. Liao, Z. F. Ma, Electrolyte design strategies and research progress for room-temperature sodium-ion batteries, *Energy Environmental Science* 10 (2017) 1075–1101.
25. B. Pal, S. Yang, S. Ramesh, V. Thangadurai, R. Jose, Electrolyte selection for supercapacitive devices: A critical review, *Nanoscale Advances* 1 (2019) 3807–3835.
26. A. Bhide, J. Hofmann, A. Katharina Dürr, J. Janek, P. Adelhelm, Electrochemical stability of non-aqueous electrolytes for sodium-ion batteries and their compatibility with $Na_{0.7}CoO_2$, *Physical Chemistry Chemical Physics* 16 (2014) 1987–1998.
27. C. H. Wang, Y. W. Yeh, N. Wongittharom, Y. C. Wang, C. J. Tseng, S. W. Lee, W. S. Chang, J. K. Chang, Rechargeable $Na/Na_{0.44}MnO_2$ cells with ionic liquid electrolytes containing various sodium solutes, *Journal of Power Sources* 274 (2015) 1016–1023.
28. M. Kim, I. Oh, J. Kim, Effects of different electrolytes on the electrochemical and dynamic behavior of electric double-layer capacitors based on a porous silicon carbide electrode, *Physics Chemistry Chemical Physics* 17 (25) (2015) 16367–16374.
29. B. Pal, S. G. Krishnan, B. L. Vijayan, M. Harilal, C. C. Yang, F. I. Ezema, M. M. Yusoff, R. Jose, In situ encapsulation of tin oxide and cobalt oxide composite in porous carbon for high-performance energy storage applications, *Journal of Electroanalytical Chemistry* 817 (2018) 217–225.
30. M. Galinski, A. Lewandowski, I. Stepniak, Ionic liquids as electrolytes, *Electrochimica Acta*, 51 (2006) 5567–5580.
31. R. A. P. Jayawickramage, J. P. Ferraris, High-performance supercapacitors using lignin-based electrospun carbon nanofiber electrodes in ionic liquid electrolytes, *Nanotechnology* 3030 (2019) 155402. DOI: 10.1088/1361–6528/aafe95.
32. S. Madani, C. Falamaki, H. Alimadadi, S. H. Aboutalebi, Binder-free reduced graphene oxide 3D structures based on ultra-large graphene oxide sheets: High-performance green micro-supercapacitor using NaCl electrolyte, *Journal of Energy Storage* 21 (2019) 310–320.
33. M. Y. Kiriukhin, K. D. Collins, Dynamic hydration numbers for biologically important ions, *Biophysical Chemistry* 99 (2002) 155–168.
34. X. Liu, R. Tian, W. Ding, Y. He, H. Li, Adsorption selectivity of heavy metals by Naclinoptilolite in aqueous solutions, *Adsorption* 25 (4) (2019) 747–755.
35. M.S. Rahmanifar, M. Hemmati, A. Noori, M. F. El-Kady, M. F. Mousavi, R. B. Kaner, Asymmetric supercapacitors: An alternative to activated carbon negative electrodes based on earth-abundant elements, Materials Today Energy12 (2019) 26–36.
36. A. Joshi, P. Chand, V. Singh, Optical and electrochemical performance of hydrothermal synthesis of $BiPO_4$ nanostructures for supercapacitor applications, *Materials Today: Proceedings* (2020) doi: 10.1016/j.matpr.2020.02.716.

37. A. Joshi, P. Chand, V. Singh, Electrochemical and optical study of $BiPO_4$ nanostructures for energy storage applications, *Materials Today: Proceedings* (2020) doi:10.1016/j.matpr.2020.02.153.
38. G. Lota, K. Fic, E. Frackowiak, Carbon nanotubes and their composites in electrochemicalapplications, *Energy Environmental Science* 4 (5) (2011) 1592–1605.
39. H. Wu, X. Wang, L. Jiang, C. Wu, Q. Zhao, X. Liu, L. Yi, The effects of electrolyte on the supercapacitive performance of activated calcium carbide-derived carbon, *Journal of Power Sources* 226 (2013) 202–209.
40. K. Fic, G. Lota, M. Meller, E. Frackowiak, Novel insight into neutral medium as electrolyte for high-voltage supercapacitors, *Energy Environmental Science* 5 (2012) 5842–5850.
41. L. Demarconnay, E. R. Pinero, F. Beguin, A symmetric carbon/carbon supercapacitor operating at 1.6 V by using a neutral aqueous solution, *Electrochemistry Communications* 12 (2010) 1275–1278.

22 Graphene for Flexible Electronic Devices

S. Dwivedi
S.S. Jain Subodh P.G. (Autonomous) College

CONTENTS

22.1 INTRODUCTION

Flexible electronics [1,2] portray the combination of a thin passive wafer or substrate of, for example, metallic foil, textile or plastic, with active electrical components integrated over it so that the substrate turns conformally to highly curved surfaces on stretching along a direction [3,4]. Transparent electronics combine electronics of optically transparent materials (graphene or CNTs) with optically transparent substrate (glass) [5,6]. Similarly, the word "textile electronics" has been ascribed to integrated electronics fabricated on silk or other woven fabrics [7,8]. The various type of substrates for flexible electronics include polymers, thin substrates, paper, fabrics, and even thin metal foils as outlined above [9,10]. These substrates are cost-effective, good in thermal conduction, but resistant to electrical conduction, in addition to being flexible mechanically [11]. Robustness is a mechanical property that makes them potentially attractive for flexible device applications (Figure 22.1). These substrates do not consist of structural defects, possess high resistive properties to mechanical deformations to creep, and can resist high temperatures before melting, as well as possess a high glass transition temperature (T_g) [11]. The term "flexible" has a physical meaning that directly relates to the ability to suffer bending such that the strain, $=\frac{t}{r}$, developed on the application of bending force should not exceed yield strain $\sigma_{\text{Yield Strain}}$ of the film [12–14]. Deposition of μm-thin layer is performed, patterned for device structuration, bonded, and integrated over flexible platforms such that the substrate retains its flexible nature. It is necessary that the intra-layers of the electronic material are strongly bonded such that the strain (σ) formed as a result of bending does not surpass yield strain $\sigma_{\text{Yield Strain}}$ of the bilayer structure. In fact, durable flexible devices must possess the capability to withstand induced mechanical

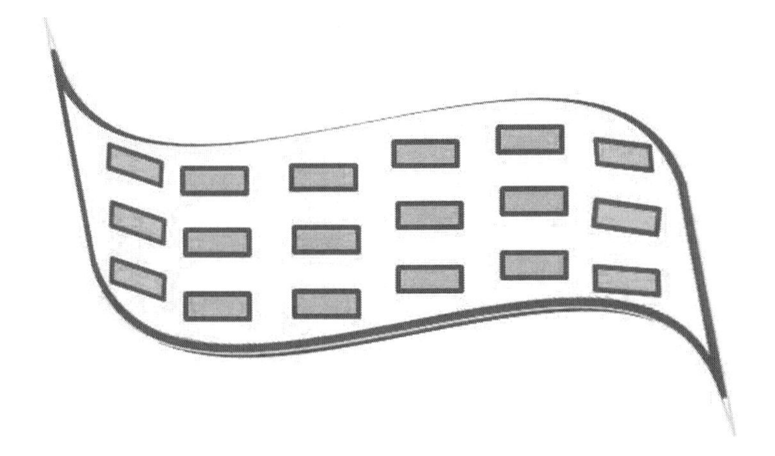

FIGURE 22.1 Printed flexible device displays no effect on bending.

deformation caused by stretching, bending, twisting, folding, and compressing, and should maintain structural integrity along with electronic performance. In regard to flexible characteristics, the following points should be focused on:

i. Bending status of flexible devices is characterized by bending angle, bending radius, and distance between the two ends of the bend.
ii. Structural design include choice of suitable flexible substrates, optimized architectural design of devices on flexible platforms, as thin as possible by reduction in thickness, and a neutral plane platform.
iii. Mechanical modeling for analysis of strain and deformation distributive pattern in the entire flexible energy storage devices.
iv. Experimental methods for the determination of different bending parameters to analyze flexible materials and devices.

Different types of polymer substrates are *poly*(tetrafluoroethylene) (PTFE) [15,16], kaptonz *poly*(imide) [17] and *poly*(ethylene terephthalate) (PET) [18], *poly*(dimethylsiloxane) (PDMS) [19], and cellulose paper composite-based substrates [20,21], which are routinely used for flexible device technology. Nathan et al. has pointed out physical characteristics of many polymer substrates separately [22]. Polyimide or Kapton, commonly called as thermal tape, is orange in color, possesses high thermal expansion coefficient, is costly, and chemically resistive along with a maximum deposition temperature of 250°C [22]. PET possesses a moderate coefficient of thermal expansion, is cost-effective, chemically resistive in nature, with a maximum bearable deposition temperature of 160°C [22]. Polyetherimide (PEI) is strong, brittle, and possesses a maximum tolerance of temperatures up to 180°C [22]. Polyetheretherketone (PEEK) has good chemical resistive properties with the maximum tolerance of temperatures up to 240°C [22]. These flexible substrates have been applied in a number of applications, including electronic skin, wearable electronic devices, portable devices for energy storage and harvesting, stretchable electronics technology, sensors fabricated on a flexible platform, high-end biotechnology devices, and logic devices.

22.2 MECHANICAL PROPERTIES OF FLEXIBLE SYSTEMS

Flexible energy storage and harvesting devices are extremely important for biomedical applications, long-life battery systems, and low-cost solar cells [12–14]. Mechanical stability of electrodes is a significant challenge in these flexible devices requiring them to be sufficiently thin and flexible. Consequently, on applying bending force on flexible platforms, deformation mechanism occurs and is resisted by an intrinsically produced stress. Considering the case of a thin film deposited over flexible substrate equivalent to a mechanical beam of radius "*r*," inner surface experiences compressive strain while outer surface sustains tensile strain [12–14]. In this device structure, a region devoid of uniaxial strain portrays a mechanically neutral plane. This mechanically neutral plane is placed in a position which is a function of thickness of each contributing layer and the Young's modulus. A mechanically neutral plane is defined as the plane that passes through the printed flexible device when mechanical deformation with minimal radius curvature is applied that results in no uniaxial bending strain [14].

It is this mechanically neutral plane that forms the point of crux so that deposition of film at this location develops an ultra-flexible platform. In case of a device structure consisting of a stack of thin films and electrodes, bending mechanics become slightly complex leading to the development of stress factor due to the difference in mechanical properties, such as elastic modulus. Figure 22.2 displays the bending mechanism in a thin film on a flexible substrate. Strain generated in the upper part of the curved surface is expressed as follows:

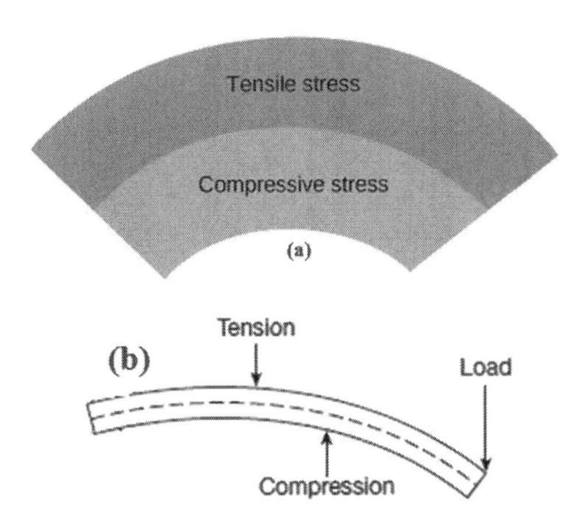

FIGURE 22.2 Bending mechanics of a thin film on a flexible substrate for describing bending mechanism.

$$\sigma = \frac{x}{r} \tag{22.1}$$

Here, x is the distance of the upper curved surface from the neutral plane and r is the cylindrical radius covered under the lower curved surface of the mechanical beam [12–14]. Equation (22.1) shows that the thin film endures minimum bending strain while sustaining its electrical performance at the threshold of bending radius. Mao et al. have given a mathematical expression for the distance of mechanically neutral plane to the upper curved convex surface as follows [14]:

$$x = \left[\frac{t_{\text{Thin Film}} + t_{\text{Substrate}}}{2}\right]\frac{\left(1 + 2\alpha + \delta\alpha^2\right)}{(1+\alpha)(1+\delta\alpha)} \tag{22.2}$$

Here, $t_{\text{Thin Film}}$ and $t_{\text{Substrate}}$ are the thicknesses of thin films and substrates, $\alpha = \frac{t_{\text{Thin Film}}}{t_{\text{Substrate}}}$ is the ratio of thickness of the thin film and substrate, and $\delta = \frac{Y_{\text{Thin Film}}}{Y_{\text{Substrate}}}$ expresses the ratio of Young's modulus of the thin film and substrate [14]. Hence, the strain of the upper curved surface is expressed as:

$$x = \left[\frac{t_{\text{Thin Film}} + t_{\text{Substrate}}}{2r}\right]\frac{\left(1 + 2\alpha + \delta\alpha^2\right)}{(1+\alpha)(1+\delta\alpha)} \tag{22.3}$$

In case of nearly similar Young's modulus of the thin film and substrate $Y_{\text{Thin Film}} \approx Y_{\text{Substrate}}$ so that ratio of these two quantities becomes $\delta \approx 1$. In this case, mechanically neutral plane lies along the same line coinciding with the line passing through the middle plane of the device structure. In that case, equations (22.2) and (22.4) are simplified as follows [14]:

$$x = \left[\frac{t_{\text{Thin Film}} + t_{\text{Substrate}}}{2}\right] \tag{22.4}$$

Extending the simplest case to a multi stacked structure, the constituent layers are theorized as a fused mechanical beam-type structure, then the length "x" of the top-surfaced bending curvature from the mechanically neutral plane is expressed as follows:

$$x = \frac{\sum_{j=1}^{n} \overline{Y}_j h_j \left[\left(\sum_{k=1}^{j} h_j\right) - \frac{h_j}{2}\right]}{\sum_{j=1}^{n} \overline{Y}_j\, h_j} \tag{22.5}$$

Here, n is the total number of constituent layers forming the multilayered structure, h_i is the thickness of jth layer in case of narrow-structured devices, and $\overline{Y}_j$ is Young's modulus of each component layer.

22.3 STRUCTURAL ARCHITECTURE AND DESIGN OF PRINTED FLEXIBLE DEVICES

Mechanical architectural design is the key factor in the fabrication of printed flexible energy storage and harvesting devices. The architecture should minimize the strain caused by the deposition of films and bending on flexible substrates. The threshold of deformation normally depends on the interplay of stress-sustaining capability of flexible material and acquired strain. In case of flexible electronic devices, mechanical neutral plane gets deviated from the plane lying along the mid-axis to the film plane, leading to generation of lesser strain. This is well-documented from equation (22.3) when $Y_{\text{Thin Film}} > Y_{\text{Substrate}}$ and $\delta \approx 1$ [14]. This is in contrast to the electronic materials deposited over silicon, diamond, gallium arsenide, or other similar non-flexible substrates. In such types of unilayered thin film or multilayer stacks, stress is developed in a film which severely limits their capability to sustain higher degree of mechanical deformation. Flexible electronics provide a pathway for printed devices with low strain and low Young's modulus of elasticity. Organic materials are intrinsically soft with obstructed charge-transport properties. Inorganic materials offer the possibilities of a wide range of applications with tailored appropriate architectural device design [23,24]. For suitable bending properties, electronic devices on non-flexible substrates operate efficiently for small radius of curvature, while printed flexible devices with similar architectural design do not show optimum performance in tensile bending configuration for the similar radius of curvature. To improve the bending performance of the device, the neutral plane can be moved to the rigid electronic material film possessing high flexural rigidity [14]. This condition can be expressed as:

$$Y_s t_s^2 = Y_e t_e^2 \tag{22.6}$$

Here, t_s and Y_s are the thickness and modulus of the substrate, respectively, while t_e and Y_e and are the thickness and modulus of encapsulation, respectively. In addition, large mismatch in modulus, proper lamination of devices, and untimely failure in typical multilayered device structures are the typical problems which need to be given due attention.

22.4 GRAPHENE AS POTENTIAL MATERIAL FOR STRETCHABLE ELECTRONICS

Graphene is a novel nanostructured material currently under investigation globally for application as transparent electrodes in flexible energy storage devices [5], logic devices [25], bioinspired devices [5], and sensing devices. For stretchable applications, large area deposition, high throughput, and cost-efficient production techniques of graphene are required [1,2,13]. In this case, chemical vapor deposition (CVD) in combination with transfer techniques along with chemical exfoliation combined with spray-assisted printing are the most productive techniques for graphene production [5]. CVD technique can be used for the nucleation and growth of high-quality graphene films over metallic substrates of Fe, Co, Ni, Cu, and Pt for growth of large-area and high-quality graphene films [5]. There are three stages of this growth process: (i) diffusion of carbon atoms into metallic lattice at an optimized temperature, (ii) nucleation of carbon seeds out of the thin metallic film in the thermodynamic process of cooling in a specific crystal orientation, and (iii) assembling of nucleated carbon nanostructures in a honeycomb lattice to form a blanket of graphene layer over metallic films.

Graphene patterning is an important step in the fabrication of stretchable electronic devices. Transparent electrodes of graphene or even heterostructures have been fabricated by direct patterned growth of graphene from nickel or polymer patterns. Plasma-enhanced chemical vapor deposition (PECVD) is another variant of CVD that is appropriate for low-temperature deposition, enabling the direct growth of graphene on polymer substrates [5,25]. However, the graphene grown by PECVD is not as high quality as CVD technique. Consequently, it is necessary to develop efficient transfer methods to produce graphene without defect formation and subsequently transfer it to a stretchable polymeric or other substrate. Among many transfer methods, *poly*(dimethylsiloxane) (PDMS) or *poly*(methyl methacrylate) (PMMA) is a solution deposited as a supporting layer [5]. This is followed by etching of metallic catalyst-cum-substrate layer or foil and subsequently transferring the polymer-encapsulated graphene over the target substrate. In the final step, the supporting polymer encapsulation layer is removed.

Two-dimensional (2D) graphene layers open up new technology platforms for stretchable electronics offering special properties of high mobility, high transmittance, and bending features. Graphene nanostructures possess carrier mobility as high as ~10^5 cm^2/V s on insulating substrates to 2.3×10^5 cm^2/V s in suspended systems, and have a current capacity of 10^9 A/cm^2 [26–28]. In another type of carbon nanostructures, semiconducting single-walled carbon nanotubes (SWCNTs) have a carrier mobility of ~80 $\times 10^3$ cm^2/V s [29]. Graphene was first prepared in form of small flakes by mechanical exfoliation [30]. However, electronic-grade graphene for applications in microelectronics and stretchable electronics can be prepared using CVD. CVD-fabricated graphene devices have lower field effect mobilities in comparison to the devices fabricated with mechanically exfoliated graphene or epitaxial graphene [5,30]. The cause underlies in the fact that there may be incorporation of defects and wrinkles in the deposited films along with significant electron scattering at the unusual grain boundaries. Another critical step in the fabrication of flexible device structure is transfer of deposited graphene without any degradation in quality from metal substrates to soft platforms or other polymer substrates. Wet etching is preferred for the detachment of graphene from parent substrates on which it is deposited [31]. Chemical etchants $FeCl_3$ or $(NH_4)_2S_2O_8$ are often used for peeling-off graphene layers from Cu substrates, while NaOH or KOH are used for the removal of sapphire substrates [31]. Polymer *poly*-methyl-methacrylate (PMMA) is an effective binder for holding graphene during wet etching [31]. However, few drawbacks associated are routine damages and contamination of graphene layer with residual materials, and hence does not turn up to be efficient for scaling-up of the fabrication process. Dry printing or stamping technique employs *poly*-dimethylsiloxane (PDMS) stamp for the transfer of graphene films from seed substrates to metallic films but suffers from problems of mechanical deformation [32]. Another technology for large-area graphene film is roll-to-roll (R2R) lamination process [5,33] for deposition on flexible substrates, as shown in Figure 22.3a. This technique uses a thermal release layer providing temporary support, enabling the fabrication of graphene layer on flexible substrates as large as 44 inch [39]. Graphene layer deposited on a Cu metallic foil is laminated with *poly*-ethylene co-vinyl acetate (EVA) [33]. Vinyl acetate acts as an auxiliary layer for providing support to the polymer film followed by etching of

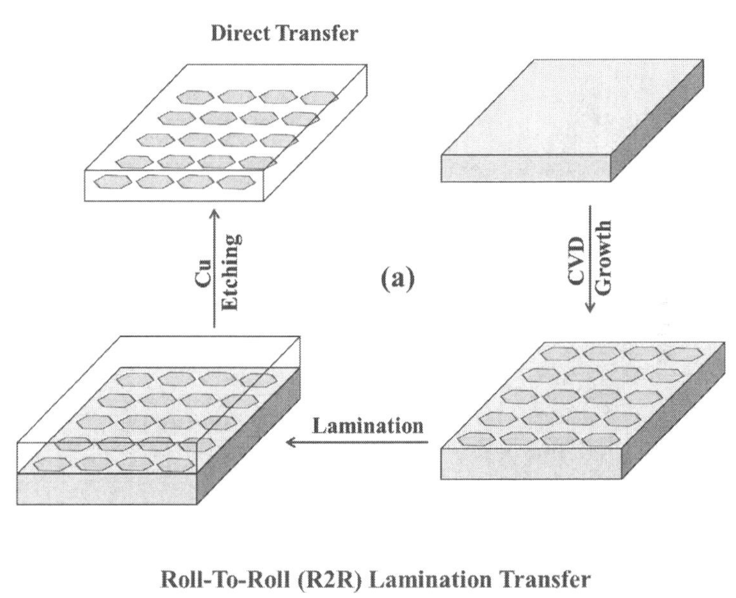

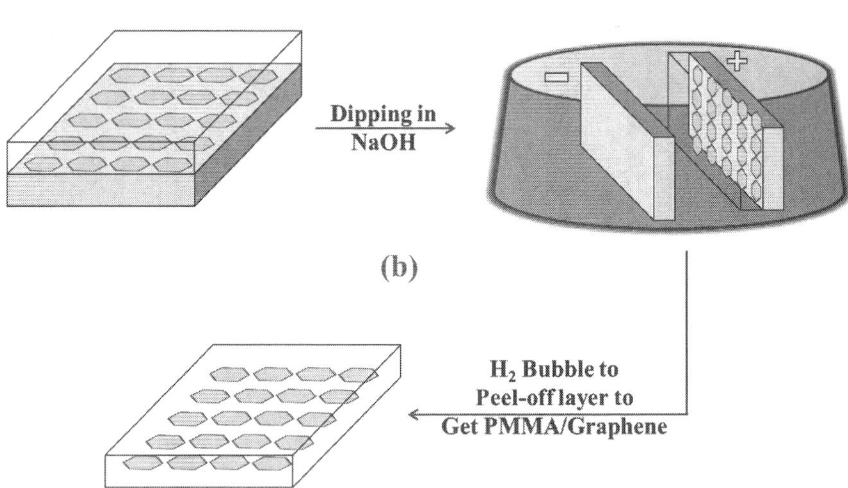

FIGURE 22.3 Schematic illustration of (a) roll-to-roll (R2R) lamination transfer technology, (b) bubbling technology, (c) chemical etchant technology, and (d) aligned transfer technology.

(Continued)

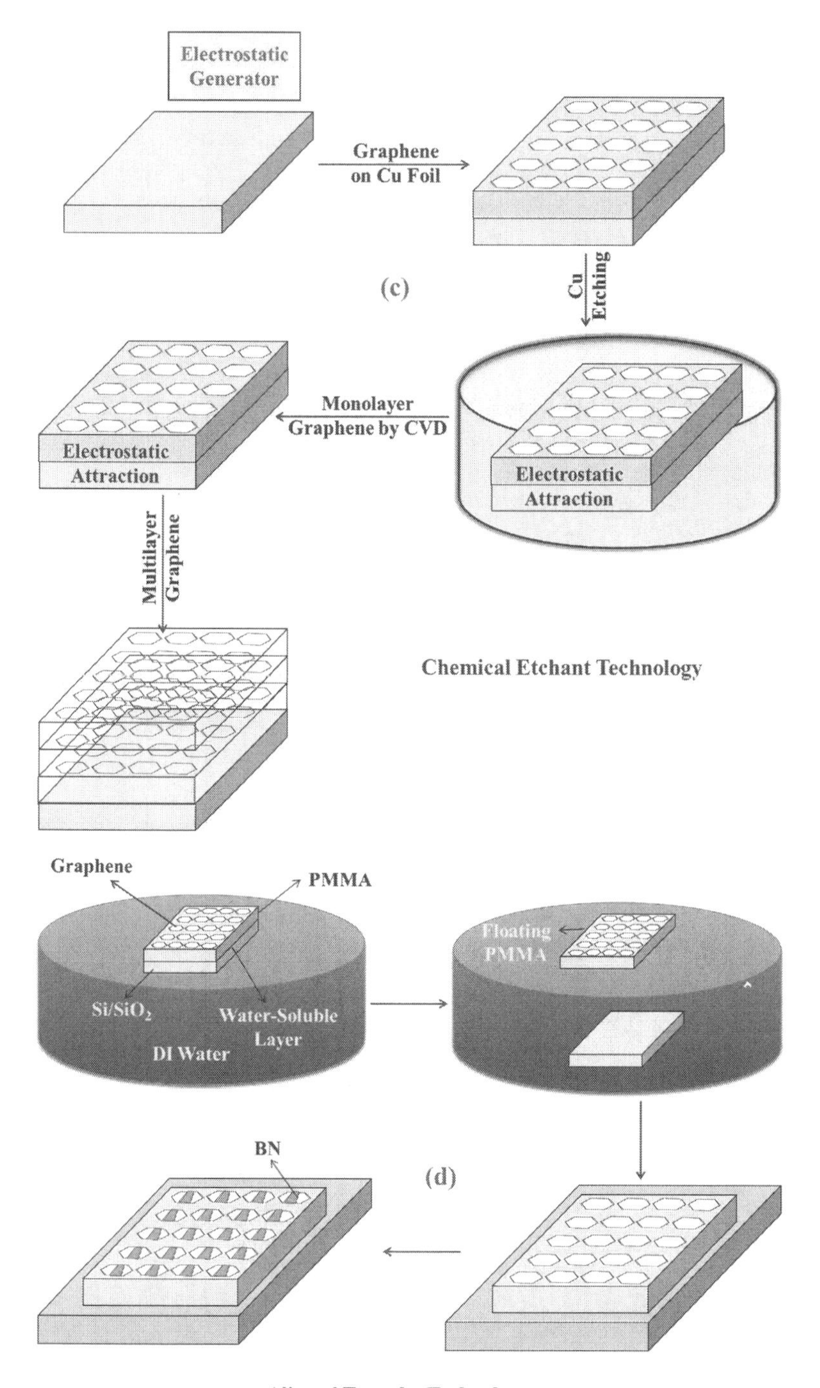

FIGURE 22.3 (CONTINUED) (c) chemical etchant technology, and (d) aligned transfer technology.

Cu thin-layered metallic substrate. The transferred mono-atomically thick graphene forms an almost uniform layer with variation in resistances lesser than 10% [28,33]. In addition, there may be organic contamination in few percentages from the thermal Kapton tape and incorporation of defects due to the transfer of graphene layers [28,33]. An extension of this transfer technology is the non-destructive bubbling method, as shown in Figure 22.3b. In this method, PMMA/graphene/metallic foil stack is dipped in NaOH solution for application as a cathode with a constant current source. In yet another method, clean-lifting transfer (CLT) technology employs electrostatic forces for transferring graphene to soft substrates but does not make use of covering layer of PMMA as shown in Fig. 3(c) [33,50]. The system consisted of an electrostatic generator kept from the substrate at a distance of one inch. An electric discharge is produced inside the electrostatic generator subsequently followed by a pressing mechanism which enabled uniform adherence of graphene molecules onto the substrate. After etching of Cu metallic thin foil, graphene film was transferred onto plastic substrate and rinsed with deionized (DI) water for removing the remaining etchant. However, this technique does not consider positioning of carbon film at a specific location on the substrate. In another transfer technique, 2D flakes can be aligned to the specific location as shown in Fig. 3(d) [33].

22.5 FLEXIBLE GRAPHENE FOR SOLAR CELL DEVICES

Silicon solar technology is low-cost, non-toxic, relies on low bandgap of Si [5], is an efficient converter of solar energy into power bank, and its cost of power is produced per watt. Graphene as a 2D material has potential for futuristic applications in lightweight, flexible, thinner solar cells for scaled-up roll-to-roll processing [5,28]. Applications of graphene include (undoped/doped) in the form of transparent conducting ultrathin electrode, as a junction layer based on Schottky non-linear barrier operating as hole collector, undoped/doped graphene in form of charge-transport functionalized layer/electrode for organic or perovskite solar cells, superthin 2D heterojunction solar cells, tandem and "hot-carrier" solar cells, and their integration into integrated circuitry for energy storage and harvesting devices [31]. At the heart of the operation of solar cells lies the fact that a potential energy barrier is formed at the interface between two differently polarized electronic materials. An electric field gets developed due to the separation of charge carriers at the interface due to absorption of light. The three types of junctions formed at the interface are homojunction, heterojunction, and Schottky junctions [31]. A homojunction of *p-n* junction occurs at the interface of identical semiconductors of equal bandgaps but with separate doping. A heterojunction is formed between two junctions of different types of semiconductors with dissimilar energy bandgaps. If the type of doping is the same in both the semiconductors, an isotype *n-n* or *p-p* heterojunction is formed. On the contrary, an anisotype *p-n* heterojunction is formed in which two oppositely doped semiconductor are joined together. In a heterojunction, the top layer has a high bandgap possessing high transparency for the illuminated optical radiation. The bottom layer has a low bandgap possessing the absorption capability of light. In the case of a Schottky junction, a metal-junction is fabricated so that the band bending occurs, leading to the thermal equilibrium of charge carriers. Graphene with zero bandgap on contacting with silicon semiconductor forms a Schottky junction-based solar cell [5,28,31]. Li et al. reported first graphene-Si Schottky junction-based solar cell on *n*-type Si [34]. In this fabrication process, graphene sheets were deposited by CVD on Ni metallic thin foils. These graphene flakes were subsequently dispersed onto pre-patterned Si/SiO_2 substrates forming a conformal coating on the exposed *n*-type Si substrate and interconnected with Au electrodes. There exists a difference of work function between graphene and Si, and the corresponding energy band diagram has been shown in Figure 22.4. Removal of electrons from *n*-type Si to graphene creates a depletion region devoid of charge carriers from the surface as well as at a certain distance below the surface of Si. In this process of junction formation, accumulation of positive charge carriers happens at the side of the semiconductor, which leads to band bending in an upward direction. Alignment of Fermi levels occurs on reaching equilibrium when the two junctions are contacted so that a built-in electric field

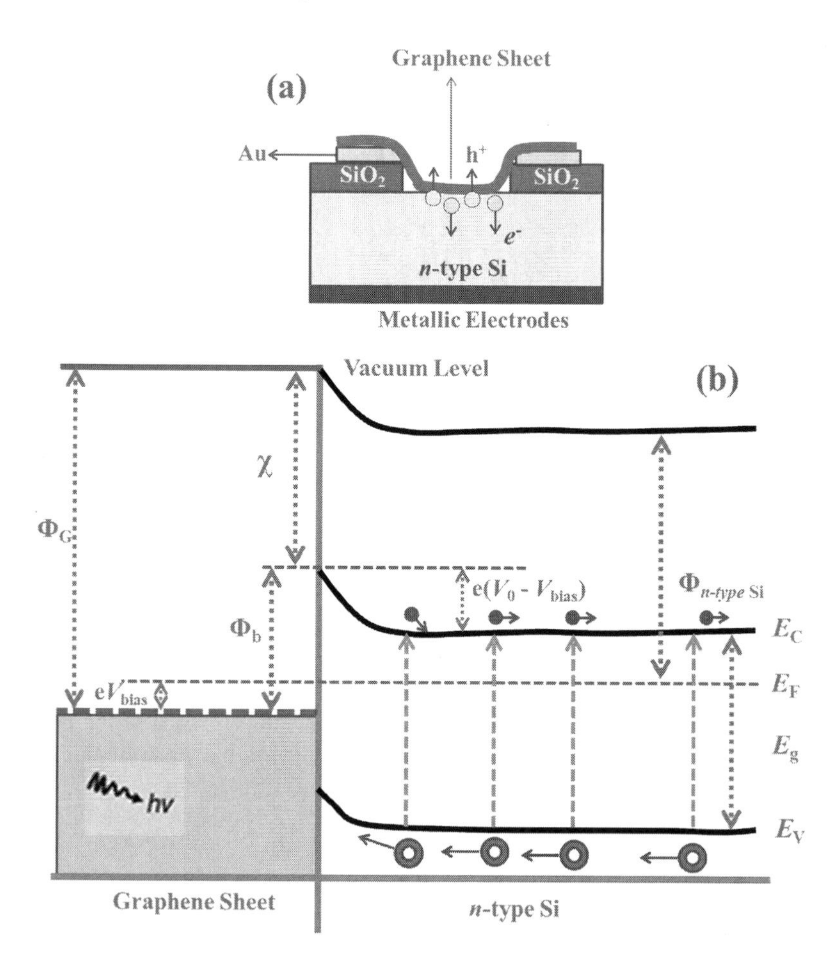

FIGURE 22.4 Energy diagram of graphene nanosheet/*n*-type Si Schottky junction in forward-biased mode.

(E_{bi}) is developed, which acts as a blockade or barrier for the movement of charge carriers across the two materials. In the case of Schottky junction-based solar cells, high temperature-based diffusion of foreign atoms is not required and the barrier height is much lower compared to built-in potential (V_{bi}) of a traditional *p-n* junction-based Si solar cell. On illumination of Si surface with photons of energy higher than the bandgap of Si, electron-hole pairs generated in *n*-type Si are segregated by the electric field. This leads to transport of holes into the side of graphene and electrons to the side of *n*-type Si. In case when incident radiation energy is lesser than the bandgap of Si but is greater than the Schottky barrier height, electron-hole pairs are produced in graphene on illuminating with light. A stream of holes is formed in graphene, and electrons cross the Schottky non-linear barrier with sufficient energy and are transported across the junction toward *n*-type Si. Das et al. have mentioned the formula for short short-circuit current (I_{sc}), that is, when voltage across solar cell is zero that, which is expressed as follows [31]:

$$I_{sc} = I_{ph} - I_0\left[e^{\frac{I_{sc}R_s}{V_T}} - 1\right] - \frac{I_{sc}R_s}{R_{sh}} \tag{22.7}$$

Here, I_{ph} is the photogenerated current, I_0 is the reverse saturation current, R_{sh} is shunted resistance, and R_s is series connection resistance of the solar cell. Schottky junction-based solar cells have the drawback of reverse saturation current of higher order because of thermionic emission as a result of low height of Schottky barrier. The reverse saturation current (I_0) is given as follows [31]:

$$I_0 = AA^*T^2e^{-\frac{\varphi_b}{kT}} \tag{22.8}$$

Here, A^* is the Richardson constant (≈112 A/cm/K^2 for *n*-type Si), A is the area of graphene/Si Schottky junction, T is absolute temperature, k is the Boltzmann constant, and φ_b is the Schottky barrier height.

In case the current flowing through the solar cell is zero, that is, if it is open-circuited, open-circuit voltage (V_{OC}) is obtained, which is given as follows [31]:

$$V_{OC} = \eta V_T \ln\left[1 + \frac{I_{ph}}{I_0}\right] \tag{22.9}$$

Here, η is the ideality factor of the diode and V_T is thermally infested voltage of 25 mV.

Some strategies for modification of solar cell performance are as follows,

i. Enhanced formation of electron-hole pairs on illumination of light and subsequent efficient absorption of photons to cause transition of electrons from ground state to the excited state;
ii. Free charge carriers formation by maximized separation of electron-hole pairs to flow across an external circuit.

Graphene plays a double role in the improved performance of graphene/Si heterojunction solar cell. Graphene works wonderfully as a transparent material allowing efficiently transmitted light to propagate into the semiconductor. It forms a non-linear Schottky junction in metal-semiconductor-based junction physics which develops effective segregation of electron-hole pairs and accumulation of charge carriers. On integration with silicon, graphene forms a highly bendable and stretchable platform that caters to both high optical transmission efficiency and low sheet resistance or high electrical conducting medium. Transmittance and sheet resistance of graphene vary as a function of number of layers. Both transmittance and sheet resistance are reduced with increase in the number of graphene layers. Reportedly, graphene possesses a sheet resistance of 10 Ω/sq on glass and 40–300 Ω/sq on polymer substrate poly-ethylene terephthalate (PET), which is significantly higher compared to commercially available ITO [35,36]. Transmission of graphene is 87% in the wavelength range of 350–800 nm, entailing a higher percentage of optical transparency [35,36]. CVD-grown graphene has a high sheet resistance in the range of 125–1,000 Ω/sq, limiting the operational efficiency of a solar cell. In practice, fabricated graphene is never inherent in nature, and hence, inherent conductivity at the Dirac point can never be realized at room temperature. This culminates in lower sheet resistance values in comparison to that calculated theoretically. The main drawbacks associated with stacking are the number of transfer processes, including enhancement of processing time, non-uniformity in thickness, and cost. This stacking of layer-by-layer doped graphene helps in lowering the sheet resistance value. Doping is an important factor that decreases the sheet resistance value of graphene nanoflakes. Sheet resistance is important to develop lower series resistance values and enhancement of fill factor of the solar cell. Doping of graphene is a technological process that can be divided into chemical modification, heteroatom doping, and electrostatic field tuning. Heteroatom doping involves doping of heteroatoms, such as boron, oxygen, sulfur, nitrogen, and phosphor, as well as substitution of carbon atoms or those atoms covalently with carbon atoms. Chemical modification by doping of atoms involves reaction with chemicals, such as NO_2, H_2O, and NH_3 [37]. Ultraviolet/ozone treatment for charge doping has been performed successfully with a major focus toward modification of resistance of graphene sheets by doping through chemical species. Co-doping is another method in which a permutation of two categories of *p*-dopants is used to modify the electrical properties of graphene [38]. Graphene has been coupled with one-dimensional metal nanowires to form hybrid networks to form transparent electrodes with low sheet resistance of ~33 Ω/sq and high transmittance of approximately 94% in the visible range of 550 nm [39], with enhanced light-trapping characteristics. For graphene doped using nitric acid (HNO_3) on silicon pillar array formation of a Schottky junction-based solar cell an efficiency of 3.55% has been reported. Device fabricated over stretchable ultrathin Si substrate produced an efficiency of 5.09% to 1.30% as a result of PMMA. By enhancing charge carriers of holes type by inducing *p*-dopants, sheet resistance can be reduced, enhancing the fill factor parameter of solar cell devices to 10.6%.

Graphene/silicon interface fabrication resembles a metal-semiconductor interfaced junction of the non-linear Schottky junction-type solar cell. Graphene shows metallic properties having zero bandgap energy of virgin graphene which develops a Schottky barrier. Graphene possesses transparent properties that allow high optical transmission to optically responsive indirect bandgap semiconductor Si. This transparent graphene allows absorption of high percentage of optical radiation onto active semiconductor Si to generate a large amount of electron-hole pairs. Difference in work function between graphene and Si separates and extracts the electron-hole charge carriers developing a built-in electric field (E_{bi}) [40]. In addition, graphene functions as a dynamic layer for the segregation of photo-produced carriers. On doping graphene with *n*-type or *p*-type foreign atoms, *n*-type impurity atoms pushed the Dirac points in the bandgap structure of graphene lie below the Fermi level (E_F), while *p*-type foreign atoms drives the Dirac point above E_F. Mohammed et al. have shown that the energy levels at *n*-type Si surface bend upside while energy levels bend downward for *p*-type Si when graphene semiconductor contact is made [41]. Reportedly, graphene/*n*-type Si junction has 0.52–0.67 eV of barrier height while graphene/*p*-type Si junction has barrier height of 0.61–0.73 eV at operating temperature of 300 K [31,41]. A drawback in Si solar cell device technology is the presence of a large number of dangling bonds on Si surface, which causes recombination of charge carriers. Consequently, surface of Si semiconductor needs to be passivated with an oxide layer to reduce the defects on the surface. Native oxide on *n*-Si presents blockade to hole movement which are collected in graphene acting as a reservoir. Holes start accumulating at the oxide-Si interface to tunnel across the barrier but recombine with the pool of electrons to form a recombination current.

Stretchable multifunctional logic devices are heavily envisaged for bioelectronics and wearable electronics [42,43]. Graphene is a carbon nanomaterial with distinguished Young's modulus because of the high degree of mechanical flexibility and strong atomic bonding [5]. Monoatomic thick honeycomb structure in 2D geometry with stretchable, flexible, and conformal characteristics make graphene the most suitable for a scalable commercial fabrication technology. Low temperature-enabled printing technology is used for the fabrication of graphene-based stretchable devices on not-so-usual substrates, for example, rubber balloon [5]. Graphene electronics printed on rubber substrates have degraded electrical properties because stretchable and porous substrates absorb different types of molecular species inducing significant scattering. Other potential applications of graphene include flexible biosensors, detachable graphene-based sensor attached to enamel of the dental for monitoring a patient's health, and graphene-based bioelectronics devices for high-resolution electrophysiological imprints of brain cell activity at the brain–machine interface.

REFERENCES

1. Huang, Siya, Liu, Yuan, Zhao, Yue, Ren, Zhifeng. "Flexible electronics: Stretchable electrodes and their future." *Advanced Functional Materials* 29 no. 6 (2019): 1805924(1–15).
2. Gu, Yiding, Zhang, Ting, Chen, Hao. "Mini review on flexible and wearable electronics for monitoring human health information." *Nanoscale Research Letters* 263 no. 14, (2019): 1–15.

3. Li, Kaiwei, Zhang, Nan, Zhang, Ting, Wang, Zhe, Chen, Ming, Wu, Tingting, Ma, Shaoyang, Zhang, Mengying, Zhang, Jing, U. S. Dinish, Shum, Perry, Ping, Olivo, Malini, Wei, Lei. "Formation of ultra-flexible, conformal, and nano-patterned photonic surfaces via polymer cold-drawing." *Journal of Materials Chemistry C* 6 no. 17 (2018): 4649–4657.
4. Maji, Debashish, Das, Debanjan, Wala, Jyoti, Das, Soumen. "Buckling assisted and lithographically micropatterned fully flexible sensors for conformal integration applications." *Scientific Reports* 5 no. 17776 (2016): 1–16.
5. Jang, Houk, Park, Yong Ju, Chen, Xiang, Das, Tanmoy, Kim, Min-Seok, Ahn, Jong-Hyun. "Graphene-based flexible and stretchable electronics." *Advanced Materials* 28 (2016): 4184–4202.
6. Nair, Raveendran Rahul, Blake, Peter, Grigorenko, Alexander, Novoselov, Konstantin Sergeevich, Booth, Timothy J., Stauber, Tobias, Peres, Nuno, Geim, Andre K. "Fine structure constant defines visual transparency of graphene." *Science* 320 no. 5881 (2008): 1308(1).
7. Wicaksono, Irmandy, Tucker, I Carson, Sun, Tao, Guerrero, A. Cesar, Liu, Clare, Woo M. Wesley, Pence, J. Eric, Dagdeviren. "A tailored, electronic textile conformable suit for large-scale spatiotemporal physiological sensing in vivo." *npj Flexible Electronics* 4 no. 5 (2020): 3225(1–15).
8. Wang, Gang, Hou, Chengyi, Wang, Hongzhi. "Stimuli-responsive electronic skins." In *Flexible and Wearable Electronics for Smart Clothing*, 29–48. Wiley-VCH Verlag GmbH & Co. KGaA, Germany, 2020.
9. Tao, Xiaoming. *Wearable Electronics and Photonics*, 136–154. Woodhead Publishing Series in Textiles, 2005.
10. Lu, Qing-Hua, Zheng, Feng. "Polyimides for electronic applications" In *Advanced Polyimide Materials*, Ed. Shi-Yong Yang, 195–255. Elsevier, 2018.
11. Schwartz, Mel. "*Smart Materials*". CRC Press, Taylor & Francis Group, London, UK.
12. Yu, Cunjiang, Jiang, Hanqing. "Forming wrinkled stiff films on polymeric substrates at room temperature for stretchable interconnects applications." *Thin Solid Films* 519 (2010): 818–822.
13. Wang, Bo, Bao, Siyuan, Vinnikova, Sandra, Ghanta, Pravarsha, Wang Shuodao. "Buckling analysis in stretchable electronics." *npj Flexible Electronics* 5 (2017): 1–9.
14. Mao, Lijuan, Meng, Qinghai, Ahmad, Aziz, Wei, Zhixiang. "Mechanical analyses and structural design requirements for flexible energy storage devices." *Advanced Energy Materials* 7 (2017): 1700535(1–19).
15. Schröder, Stefan, Strunskus, Thomas, Rehders, Stefan, Gleason, K. Karen, Faupel, Franz. "Tunable polytetrafluoroethylene electret films with extraordinary charge stability synthesized by initiated chemical vapor deposition for organic electronics applications." *Scientific Reports* 9 (2019): 2237(1–7).
16. Pakhuruddin, Mohd. Zamir, Ibrahim, Kamarulazizi, Aziz Abdul Azlan. "Properties of polyimide substrate for applications in flexible solar cells." *Optoelectronics and Advanced Materials – Rapid Communications* 7 no. 5–6 (2013): 377–380.
17. Fang, Yunnan, Hester, G. D. Jimmy, deGlee, M. Ben, Tuan, Chia-Chi, Brooke, D. Philip, Le, Taoran, Wong, Ching-Ping, Tentzeris, M. Manos, Sandhage, H. Kenneth. "A novel, facile, layer-by-layer substrate surface modification for the fabrication of all-inkjet-printed flexible electronic devices on Kapton." *Journal of Materials Chemistry C* 4 no. 29 (2016): 7052–7060.
18. Zardetto, Valerio, Brown, M. Thomas, Reale, Andrea, Carlo di Aldo. "Substrates for flexible electronics: A practical investigation on the electrical, film flexibility, optical, temperature, and solvent resistance properties." *Journal of Polymer Science Part B* 49 no. 9 (2011): 638–648.

19. Chen, Jing, Zheng, Jiahong, Gao, Qinwu, Zhang, Jinjie, Zhang, Jinyong, Omisore, Mumini Olatunji, Wang, Lei and Li, Hui. "Polydimethylsiloxane (PDMS) – Based flexible resistive strain sensors for wearable applications." *Applied Science* 8 (2018): 345(1–15).
20. Adly, Nouran, Weidlich, Sabrina, Seyock, Silke, Brings, Fabian, Yakushenko, Alexey, Offenhäuser, Andreas, Wolfrum, Bernhard. "Printed microelectrode arrays on soft materials: from PDMS to hydrogels." *npj Flexible Electronics* 2 no. 15 (2018): (1–9).
21. Barras, Raquel, Cunha, Inês, Gaspar, Diana, Fortunato, Evira, Martins, R., Pereira, Luis "Printable cellulose-based electroconductive composites for sensing elements in paper electronics." *Flexible and Printed Electronics* 2 (2017): 014006(1–13).
22. Nathan, Arokia, Ahnood, Arman, Cole, T. Matthew, Lee, Sungsik, Suzuki, Yuji, Hiralal, Pritesh, Bonaccorso, Francesco, Hasan, Tawfique, Garcia-Gancedo, Luis, Dyadyusha, Andriy, Haque, Samiul, Andrew, Piers, Hofmann, Stephan, Moultrie, James, Chu, Daping, Flewitt, J. Andrew, Ferrari, C. Andrea, Kelly, J. Michael, Robertson, John, Amaratunga, J. A. Gehan, Milne, I. William. "Flexible electronics: The next ubiquitous platform." *Proceedings of The IEEE* 100 (2012): 1486–1517.
23. Ouyang, Meng, Muisener, R. J., Boulares, Alya, Koberstein, Jeff. "UV-ozone induced growth of a SiO_x surface layer on a cross-linked polysiloxane film: Characterization and gas separation properties." *Journal of Membrane Science* 177 no. (1–2) (2000): 177–187.
24. Li, Junpeng, Liang, Jiajie, Li, Lu, Ren, Fengbo, Hu, Wei, Li, Juan, Qi, Shuhua, Pei, Qibing. "Healable capacitive touch screen sensors based on transparent composite electrodes comprising silver nanowires and a furan/maleimide diels-alder cycloaddition polymer." *ACS Nano* 8 no. 12 (2014): 12874–12882.
25. Ryu, Jaechul, Kim, Youngsoo, Won, Dongkwan, Kim, Nayoung, Park, Sung, Jin, Lee, Eun-Kyu, Cho, Donyub, Cho, Sung-Pyo, Kim, Jin, Sang, Ryu, Hee Gyeong, Shin, Hae-A-Seul, Lee, Zonghoon, Hong, Hee Byung, Cho, Seungmin. "Fast synthesis of high-performance graphene films by hydrogen-free rapid thermal chemical vapor deposition." *ACS Nano* 8 no. 1 (2014): 950–956.
26. Bolotin, Kirill, Sikes, K. J., Jiang, Zhifang, Klima, M., Fudenberg, G., Hone, James, Kim, Phaly, Stormer, H. L. "Ultrahigh electron mobility in suspended graphene." *Solid State Communication* 146 (2008): 351–355.
27. Mayorov, Alexander S, Gorbachev, Roman, Morozov, Sergey, Britnell, Liam, Jalil, Rashid, Ponomarenko, Leonid A, Blake, Peter, Novoselov, Kostya S, Watanabe, Kenji, Taniguchi, Takashi, Geim, Andre. "Micrometer-scale ballistic transport in encapsulated graphene at room temperature." *Nano Letters* 11 (2011): 2396–2399.
28. Chae, Hoon Sang, Lee Hee, Young. "Carbon nanotubes and graphene towards soft electronics." *Nano Convergence* 1 no. 15 (2014): 1–26.
29. Durkop, T., Getty, Stephanie A., Cobas, Enrique, Fuhrer, Michael. "Extraordinary mobility in semiconducting carbon nanotubes." *Nano Letter* 4 (2004): 35–39.
30. Urmimala, Maitra, Ramakrishna, H. S. S. Matte, Prashant, Kumar, Rao, C. N. R. "Strategies for the synthesis of graphene, graphene nanoribbons, nanoscrolls and related materials." *Chimia International Journal of Chemistry* 66 no. 12 (2012): 941–948.
31. Das, Sonali, Pandey Deepak, Thomas, Jayan, Roy, Tania. "The role of graphene and other 2D materials in solar photovoltaics." *Advanced Material* 1802722 (2018): 1–35.
32. Hur, Seung-Hyun, Park, O. Ok, Rogers, John A. "Extreme bendability of single-walled carbon nanotube networks transferred from high-temperature growth substrates to plastic and their use in thin-film transistors". *Applied Physics Letter* 86 no. 24 (2005): 243502(1–4).

33. Lee, Whan Seung, Mattevi, Cecilia, Chhowalla, Manish, Sankaran, Mohan R. "Plasma-assisted reduction of graphene oxide at low temperature and atmospheric pressure for flexible conductor applications." *The Journal of Physical Chemistry Letters* 3 no. 6 (2012): 772–777.
34. Li, X., Zhu, Hongwei, Wang, Kunlin, Cao, Anyuan, Wei, Jinquan, Li, Chunyan, Jia, Yi, Li, Zhen, Li, Xiao, Wu Dehai. "Graphene-on-silicon Schottky junction solar cells." *Advanced Materials* 22 no. 25 (2010): 2743–2748.
35. Minami, Tadatsugu. "Transparent conducting oxide semiconductors for transparent electrodes". *Semiconductor Science and Technology* 20 no. 4 (2005): S35–S44.
36. Granqvist, Claes Goran. "Transparent conductors as solar energy materials: A panoramic review." *Solar Energy Materials and Solar Cells* 91 no. 17 (2007): 1529–1598.
37. Bausi, Francesco, Schlierf, Andrea, Treossi, Emanuele, Schwab, Georg Matthias, Palermo, Vincenzo, Cacialli, Franceo. "Thermal treatment and chemical doping of semi-transparent graphene films." *Organic Electronics* 18 (2015): 53–60.
38. Liu, X., Zhang, W. X., Meng, J. H., Yin, Z. G., Zhang, L. Q., Wang, H. L., Wu, J. L. "High efficiency Schottky junction solar cells by co-doping of graphene with gold nanoparticles and nitric acid." *Applied Physics Letter* 106 no. 23 (2015): 233901(1–6).
39. Lee, Mi- Sun, Lee Kyongsoo, Kim, So-Yun, Lee Heejoo, Park, Jihun, Choi Kwang-Hyuk, Kim Han-K, Kim, Dae-Gon, Lee, Dae-Young, Nam, SungWoo, Park, Jang-Uung. "High-performance, transparent, and stretchable electrodes using graphene-metal nanowire hybrid structures." *Nano Letter* 13 no. 6 (2013): 2814–2821.
40. Perello, David J., Lim, Seong Chu, Chae, Seung Jin, Lee, Innam, Kim, Moon J, Lee, Young Hee, Yun, Minhee. "Thermionic field emission transport in carbon nanotube transistors." *ACS Nano* 5 no. 3 (2011): 1756–1760.
41. Mohammed, Muatez, Li, Zhongrui, Cui, Jingbiao, Chen, Tar-Pin. "Junction investigation of graphene/silicon Schottky diodes." *Nanoscale Research Letters* 7 no. 302 (2012): (1–6).
42. Jiao, Tianpeng, Wei, Dapeng, Liu, Jian, Sun, Wentao, Jia, Shuming, Zhang, Wei, Feng, Yanhui, Shi, Haofei, Du, Chunlei. "Flexible solar cells based on graphene-ultrathin silicon Schottky junction." *RSC Advances* 5 no. 89 (2015): 73202–73206.
43. Rogers, John A., Ghaffari, Roozbeh, Kim, Dae-Hyeong. "Stretchable bioelectronics for medical devices and systems." *Microsystems and Nanosystems*, XII, 314 Springer (2016): 93.

23 Flexible Microfluidics Biosensor Technology

Supriya Yadav, Mahesh Kumar, Kulwant Singh, Niti Nipun Sharma, and Jamil Akhtar
Manipal University Jaipur

CONTENTS

23.1 INTRODUCTION

In microfluidics, fluid flow is laminar and fluid is confined in a microchannel with dimensions at μm-to-nm scale; moreover, no turbulent mixing and high electric field is possible in microchannels [1]. Microfluidic devices offer the ability to work with a small sample volume with a shorter reaction time, with the possibilities of multiple reactions at a time. Therefore, microfluidics hold an entire lab into a single chip, that is, lab-on-a chip (LOC) [2]. Microfluidics are employed in many biological applications, for example, cell sorting [3], DNA sequencing on a chip [4], and capillary electrophoresis [5]. For the fabrication of microfluidic devices, various materials are being used, such as silicon [6–8], glass [9], polymer [10], and hydrogel [11]; however, because of their brittle nature, high cost, and lack of flexibility and compatibility with biological fluid, paper is being seen as an attractive substrate for the advancement of microfluidic devices in the future. Flexibility is a major concern for the fabrication of microfluidic devices [12]. Paper is thin, low-cost, lightweight with variety of thicknesses, easy to handle, portable, biocompatible, self-driven fluid flow, flexible, and offers simple fabrication technique without requirement of a clean room and capacity of minor loaded sample for analytical study [13]. In this chapter, we will discuss flexible microfluidic sensor as well as the microstructure of paper with selected properties and making of paper as a substrate for the fabrication of flexible microfluidic devices.

23.2 FLEXIBLE MICROFLUIDICS BIOSENSOR TECHNOLOGY

A biosensor is a device that can detect an analyte by generating signals proportional to the concentration of an analyte present in biological or chemical reaction occurring on the biosensor device [14]. Biosensors are employed in environmental monitoring, food quality control, drug discovery, and detection of pathogen, or we can say that biosensors are ubiquitous; the best examples of current biosensors is a home pregnancy kit [15] and glucose sensor [16]. Integration of microfluidics has had an impact on recent advances in biosensing technology. The microfluidic biosensors convert traditional laboratory analysis into a miniaturized lab-on-chip device with reduced sample volume, short reaction time, low power demand, low waste production, and short detection time compared to the regular detection methods. Hence, microfluidic biosensors, as shown in Figure 23.1, are portable, low cost, and provide high throughput assay with high reaction rate, which is also favorable for point-of-care testing. Figure 23.1 shows a point-of-care test, the medical diagnostic with one drop of body fluid is only possible by the use of microfluidic device system in any primary healthcare center, in doctor's clinic, in the field, at home, etc. Because of its miniaturized size, these microfluidic biosensor devices use minimum sample, reduce power requirement, improve reproducibility, improve accuracy and reliability, easy to handle, and cost-effective.

Microfluidic biosensors have shown promise for biomedical applications involving traditional material in microelectromechanical systems (MEMS) technology in combination with polymer-based microfluidic devices using complex fabrication processes. High cost and sensitivity is a major concern for the next stage in the development of microfluidic devices. The advanced development of microfluidic devices requires materials which are flexible, portable, and, most importantly, compatible with both chemical and biological fluids. As shown in Figure 23.2, fluid flows in a flexible microchannel demonstrate a pressure drop in the interface between the microchannel wall and fluid wall, and this deformation of pressure in interface proceeds fluid further in the microchannel. Fluid flow is also affected by the viscosity of the fluid as well as geometry and dimensions of the microchannel. Flexibility and elasticity are intrinsic properties of a material and are defined by Young's (elastic)

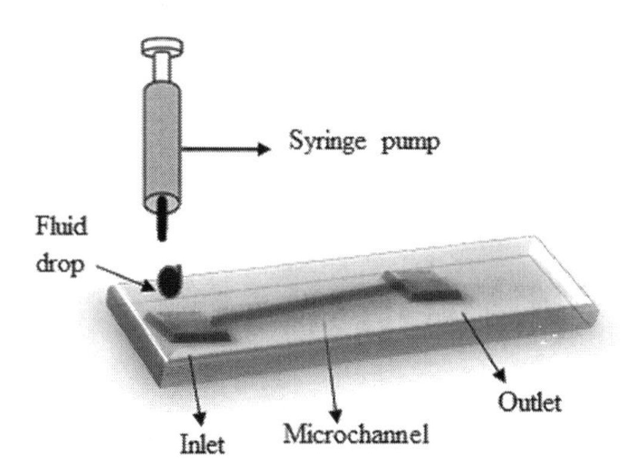

FIGURE 23.1 Schematic structure of a microfluidic device.

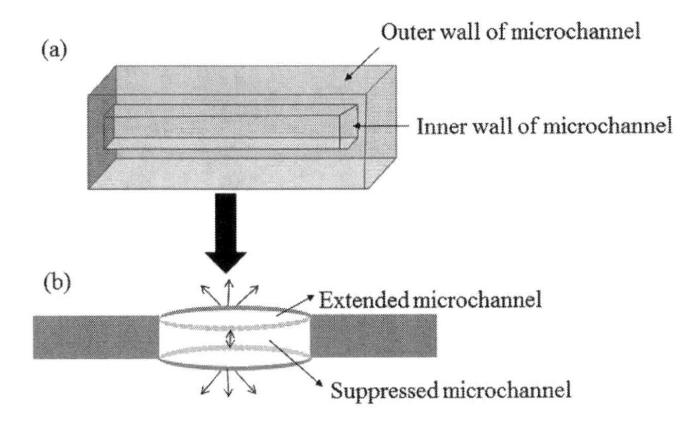

FIGURE 23.2 A basic flexible microchannel view (a) in original condition and (b) after bending.

modulus and shear modulus, respectively. Low values of Young's and shear modulus show high flexibility in materials. Flexibility or elasticity of any material might be attuned by their manufacturing process.

Paper-based flexible microfluidic biosensors are being pursued to overcome fabrication complexities and make them cost-effective, portable, and biodegradable. Paper-based flexible microfluidic biosensors suggest that cellulose is a smart material in biomedical applications for the detection of any bio-recognition elements, such as aptamers, antibodies, nucleic acids, cells, enzymes, bacteria, and viruses. Understanding of paper structures and their physical and chemical properties play an important role in designing micropatterns in microfluidic applications. Cellulose-based paper is cost-effective, easy to handle, flexible, simple to fabricate, and uses small sample volume [17]. Capillarity is the major mechanism of fluid flow in a fine area of cellulose-based paper from micro-meter scales without using any external force [18,19]. Capillarity occurs between the intermolecular forces of liquid molecule and the surrounding solid and vapor surfaces.

23.3 PAPER AS A SUBSTRATE FOR FLEXIBLE MICROFLUIDIC BIOSENSORS

In our society paper has been used for more than 2,000 years; the first official introduction of paper was described in China by Tazi Lun. The word "paper" is derived from a Latin word papyrus (*Cyperus papyrus*). Papyrus is a thick, paper-like material produced from the pith of the *Cyperus papyrus* plant. Paper production has amplified globally and will continue to increase in the coming future and is one of the largest manufacturing sectors in the world. Paper is a mat or sheet-like structure and is used as raw material in various manufacturing processes, including wood (printing paper), cotton (filter paper and chromatography paper), jute, grass, bamboo, bagasse, straw, and hemp. Basically a thin sheet of paper is made by pressing the cellulose fibers [20]. Cellulose is the most abundant biopolymer on Earth and is the major source of paper. Cellulose paper is a homopolymer of glucose subunit. These

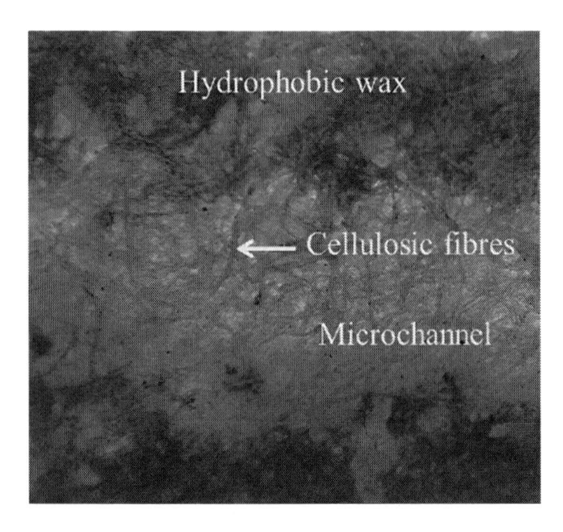

FIGURE 23.3 Demonstration of cellulosic fibers and patterning of microchannel through wax on paper.

subunits are attached with each other through glyosidic bonds. The hydroxyl groups of the cellulose chains make the paper hydrophilic with a negative charge. The cellulosic fibers are attached with each other through lignin. These cellulose fibers are separated by their manufacturing processes through mechanical pulping or chemical pulping. Filter paper which is generally used in every laboratory is made up of cellulose fiber. As shown in Figure 23.3, filter paper is made of cellulosic fibers creating hydrophilic microchannel by wax patterning. Along with adsorption of liquid, cellulose paper is flexible and hygroscopic in nature, and its flexibility depends upon the relative humidity and Young's modulus. Young's modulus of Whatman grade 1 qualitative filter paper ranges from 1.71 GPa (0%RH) to 0.46 GPa (90%RH). In fact, the rigidity of paper is a critical parameter of choosing paper as a substrate for MEMS devices.

23.4 SELECTION CRITERIA FOR THE PAPER AS SUBSTRATE MATERIAL FOR MICROFLUIDICS

Paper is an interesting and emerging material, and choosing paper according to our needs depends upon the paper manufacturing process. All papers are made up of cellulose but they differ from each other with respect to their material and chemical properties [21], as shown in Figure 23.4. In Figure 23.4, out of the known properties of commercial paper, relevant properties are shown in the context of microfluidic applications. Due to these major properties of paper such as pore size, porosity, stiffness, wet strength, chemical reactivity, and surface area, suppliers can change the properties of paper during manufacturing, due to which we can get different papers for different applications and can use paper as an analytical sensor in microfluidic devices.

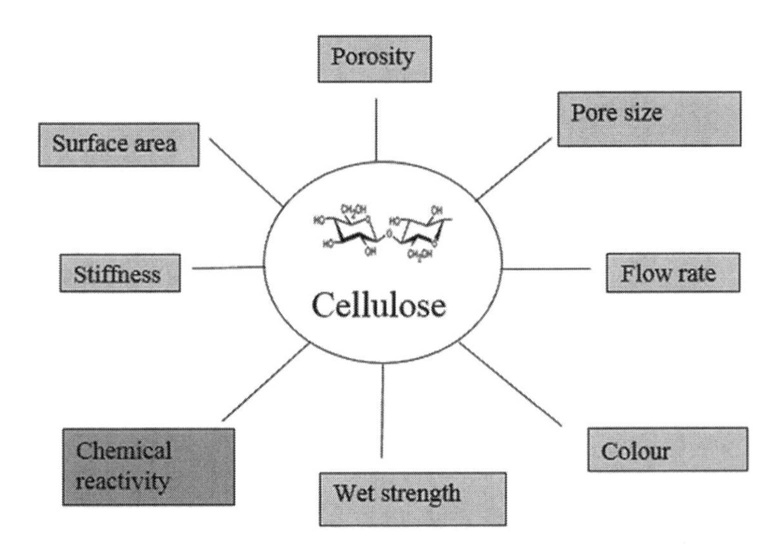

FIGURE 23.4 Critical parameters for preparing microfluidic paper.

23.4.1 Microfluidic Paper-Based Analytical Devices (μPADs)

The detection of disease or disease markers of particular disease on time is very rare. Proper detection technology may be accurate, robust, low cost, light weight, and simple to operate. The "ASSURED" criteria set by the World Health Organization means affordable, sensitive, specific, user-friendly, rapid and robust, equipment free, deliverable to end users. It may be possible if medicine/biotechnology moves on microfluidic concepts. Microfluidic devices use very small volume of liquid for the detection of a number of diseases from any biological samples (e.g., blood, urine, saliva, semen, sweat, etc.) in a very short reaction time in *in-vitro* diagnostics. Various materials, for example, silicon, glass, polymers, plastics, hydrogels, paper, etc., are being used for the fabrication of microfluidic devices. However, because of brittleness, cost, and lack of flexibility of silicon, glass, and polymers, paper is an emerging substrate having very difficult structural composition compared to traditional substrates like glass and silicon. Paper was first announced as a microfluidic analytical device by the Whitesides group in 2007. Paper is made up of cellulose fiber, and cellulose comprises hydrophilic hydroxyl group. Therefore, paper is porous and has a rough surface. Because of the above properties, paper adsorbs polar liquids such as water willingly, which are spread inside the paper through capillary action. The main task of paper microfluidics is to control the transport of the fluid of interest on paper. Therefore, microchannels are patterned on paper by blocking this flow on paper surface. The pore of paper can be hydrophobized or blocked physically, or there can be chemical modification during the paper manufacturing process. Materials such as wax [22], polystyrene, and photoresist [23] have accomplished this phenomenon. These materials form a hydrophobic channel to steer the liquid on a hydrophilic substrate, and can be printed on paper where desired microchannels are needed.

23.4.2 Types of Paper and Its Function

For the fabrication of μPADs, various types of paper can be used. For making microfluidic paper-based analytical devices (μPADs), the criteria of paper selection depend upon the fabrication method and the nature of analytes. Filter paper, chromatography paper, glossy paper, bioactive paper, and nitrocellulose membrane paper have been used for the fabrication of μPADs. Filter paper and chromatography paper (Whatman) are the most frequently used in laboratories among these various types of papers. Because of high pore size and uniform thickness of filter paper [24] and chromatography paper [25], their adsorption and retention properties are improved, which, in turn, improvs their wicking performance. Moreover, after the manufacturing process, both filter paper and chromatography paper carry out full bleaching, which eliminates almost all of the impurities from the paper matrix. Glossy paper and bioactive paper have also used as a substrate material for the fabrication of μPads. Both the papers are made up of cellulose fibers and inorganic fillers. Unlike traditional paper substrate, in bioactive paper, the surface properties may be altered by modifying the cellulose molecule with aldehyde and amide group to allow them to absorb more biomolecules. Due to the presence of some functional group on nitrocellulose membrane, it has been used as a substrate for μPADs. Nitrocellulose membranes have also been used in the DNA-RNA purification and in enzyme-linked immunosorbent assay.

23.4.3 Fluid Flow through Paper

Transport of fluid flow plays a vital role in microfluidic system and affects the results as well. Fluid flows in paper or any other porous materials through capillary force [26]. Capillary action is the ability of flow of liquid in a fine area from a micrometer scale without using any external force like gravitational force, for example, water between hairs of paint brush, in paper napkin, and in biological cell. It occurs between the intermolecular forces of liquid molecule and the surrounding solid and vapor surfaces. It means it occurs on the liquid–solid–vapor interface.

When a drop of liquid is put on a solid surface (paper), as shown in Figure 23.5, fluid flows in a paper substrate through capillary action (surface tension force) force. These three interfacial forces balance at three phases liquid, vapor, and solid surface contact line. Inside the liquid molecule, all the liquid molecules face a cohesive force with its neighbor liquid molecule and form an equilibrium; however, in the case of surface, between solid and liquid interface, half of the liquid molecules are present in the surface, so there is some need of energy or an adhesive force on the liquid–solid interface to require a shape of a molecule on this surface. This energy is called surface tension and surface free energy in the case of a liquid–vapor (LV) or solid–vapor (SV) interface, respectively. The additional energy or force present in the liquid–solid interface is less than the energy present in the inside molecules of liquid. So, liquid molecule do not expand their energy on the interface and cover the least surface area for the contact. These cohesive and adhesive forces of liquid and their surrounding surface regulates the wetting properties, contact angle, and meniscus forms in the surfaces. If the adhesive forces are greater than the cohesive

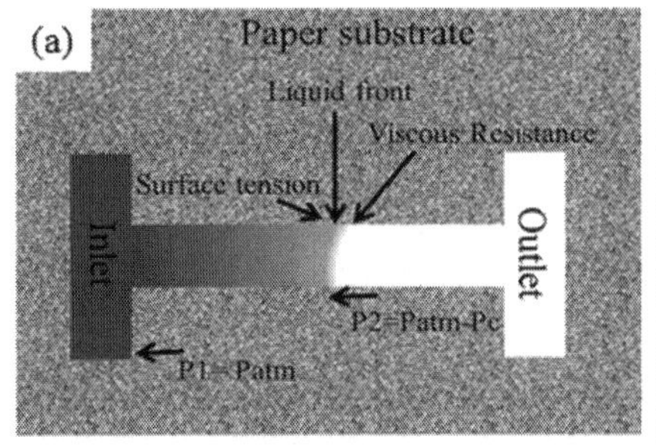

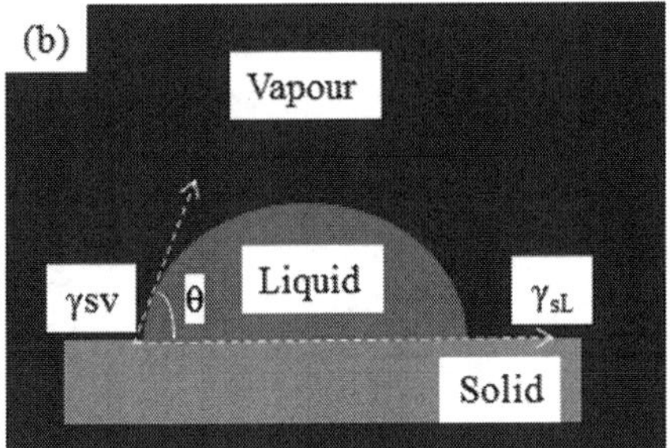

FIGURE 23.5 Transport of fluid in paper. (a) Schematic structure of fluid flow in a paper substrate. (b) Interfacial forces balance at three-phase surface contact line.

forces, liquid becomes flat and raised on the solid surface which is hydrophilic in nature. If a liquid fills in horizontal/vertical microchannel, the liquid–air interface connects the two walls of the microchannel. Because of the surface tension present on the interface, the shape of the liquid forms a curve meniscus as the surface area/volume ratio of liquid is low in that interface. This curve meniscus produces a negative pressure in the microchannel and liquid is pulled from the wetted to the non-wetted region.

23.5 CONCLUSION

In healthcare or point-of-care application, microfluidics is a developing technology for the unlimited possibilities and clinical diagnostics of diseases. The integration of microfluidics with biosensors can provide a portable and manageable microfluidics device, which can become a powerful tool for the analysis of clinical drug discovery, environmental monitoring, agricultural and food safety control, and security and defense. A wide variety of materials such as silicon, glass, and polymers are used for the fabrication of microfluidic devices; however, for the advancement of microfluidic devices, some critical properties such as biocompatibility, biodegradability, and flexibility are being used. Hence, paper is being seen as an attractive substrate for the advancement in flexible microfluidic devices in the future. Paper-based flexible microfluidic devices provide low cost, are easy to handle, are flexible, afford self-driven fluid flow through capillary action force, and simple fabrication technique without clean room. These paper-based flexible microfluidic devices can be applied as agriculture, water, food, and environmental markers.

REFERENCES

1. Whitesides, G. M. (2006). The origins and the future microfluidics. *Nature*, *442*(7101), 368–373.
2. Abgrall, P., & Gue, A. M. (2007). Lab-on-chip technologies: Making a microfluidic network and coupling it into a complete microsystem: A review. *Journal of Micromechanics and Microengineering*, *17*(5), R15.

3. Baret, J. C., Miller, O. J., Taly, V., Ryckelynck, M., El-Harrak, A., Frenz, L., Rick, C., Samuels, M. L., Hutchison, J. B., Agresti, J. J., Link, D. R., Weitzc, D, A., & Link, D. R. (2009). Fluorescence-activated droplet sorting (FADS): Efficient microfluidic cell sorting based on enzymatic activity. *Lab on a Chip*, *9*(13), 1850–1858.
4. Dutse, S. W., & Yusof, N. A. (2011). Microfluidics-based lab-on-chip systems in DNA-based biosensing: An overview. *Sensors*, *11*(6), 5754–5768.
5. Lagally, E. T., Simpson, P. C., & Mathies, R. A. (2000). Monolithic integrated microfluidic DNA amplification and capillary electrophoresis analysis system. *Sensors and Actuators B: Chemical*, *63*(3), 138–146.
6. Joyce, R., Yadav, S., Sharma, A. K., Panwar, D. K., Bhatia, R. R., Varghese, S., & Akhtar, J. (2016). Pattern transfer of microstructures between deeply etched surface for MEMS applications. *Materials Science in Semiconductor Processing*, *56*, 373–380.
7. Yadav, S., Joyce, R., Sharma, A. K., Sharma, H., Sharma, N. N., Varghese, S., & Akhtar, J. (2016, April). Process sequence optimization for digital microfluidic integration using EWOD technique. In *AIP Conference Proceedings* (Vol. 1724, No. 1, p. 020085). AIP Publishing LLC.
8. Kumar, M., Yadav, S., Kumar, A., Sharma, N. N., Akhtar, J., & Singh, K. (2019). MEMS impedance flow cytometry designs for effective manipulation of micro entities in health care applications. *Biosensors and Bioelectronics*, *142*(1), 111526.
9. Mazurczyk, R., Mansfield, C. D., & Lygan, M. (2013). Glass microstructure capping and bonding techniques. In Jenkins, G. & Mansfield, C. D. (Eds.), *Microfluidic Diagnostics* (pp. 141–151). Humana Press, Totowa, NJ.
10. Sia, S. K., & Whitesides, G. M. (2003). Microfluidic devices fabricated in pol(dimethylsiloxane) for biological studies. *Electrophoresis*, *24*(21), 3563–3576.
11. Kim, Y., Jang, G., Kim, D., Kim, J., & Lee, T. S. (2016). Fluorescence sensing of glucose using glucose oxidase incorporated into a fluorophore-containing PNIPAM hydrogel. *Polymer Chemistry*, *7*(10), 1907–1912.
12. Fallahi, H., Zhang, J., Phan, H. P., & Nguyen, N. T. (2019). Flexible microfluidics: fundamentals, recent developments, and applications. *Micromachines*, *10*(12), 830.
13. Martinez, A. W., Phillips, S. T., Whitesides, G. M., & Carrilho, E. (2010). Diagnostics for the developing world: microfluidic paper-based analytical devices. *Analytical Chemistry*, *82*, 3–10.
14. Gotoh, M., Mure, H., & Shirakawa, H. (2003). Biosensor. U.S. Patent No. 6,503,381Washington, DC: U.S. Patent and Trademark Office.
15. Bahadır, E. B., & Sezgintürk, M. K. (2015). Applications of commercial biosensors in clinical, food, environmental, and biothreat/biowarfare analyses. *Analytical biochemistry*, *478*, 107–120.
16. Claremont, D. J., Shaw, G. W., & Pickup, J. C. (1988, November). Biosensors for continuous in vivo glucose monitoring. In *Proceedings of the Annual International Conference of the IEEE Engineering in Medicine and Biology Society* (pp. 740–741). IEEE.
17. Koga, H., Kitaoka, T., & Isogai, A. (2015). Chemically-modified cellulose paper as a microstructured catalytic reactor. *Molecules*, *20*(1), 1495–1508.
18. Washburn, E. W. (1921). The dynamics of capillary flow. *Physical Review*, *17*(3), 273.
19. Darcy, H. (1856). *Les Fontaines Publiques de la Ville de Dijon (The Public Fountains of the City of Dijon)*, Dalmont, Paris. SAE Technical Paper Series. Society of Automotive Engineers.
20. Alava, M., & Niskanen, K. (2006). The physics of paper. *Reports on Progress in Physics*, *69*(3), 669.
21. Fernandes, S. C., Walz, J. A., Wilson, D. J., Brooks, J. C., & Mace, C. R. (2017). Beyond wicking: Expanding the role of patterned paper as the foundation for an analytical platform. *Analytical Chemistry*, *89*, 5654–5664.

22. Zhong, Z. W., Wang, Z. P., & Huang, G. X. D. (2012). Investigation of wax and paper materials for the fabrication of paper-based microfluidic devices. *Microsystem Technologies*, *18*(5), 649–659.
23. Martinez, A. W., Phillips, S. T., Butte, M. J., & Whitesides, G. M. (2007). Patterned paper as a platform for inexpensive, low-volume, portable bioassays. *Angewandte Chemie International Edition*, *46*(8), 1318–1320.
24. Songjaroen, T., Dungchai, W., Chailapakul, O., Henry, C. S., & Laiwattanapaisal, W. (2012). Blood separation on microfluidic paper-based analytical devices. *Lab on a Chip*, *12*(18), 3392–3398.
25. Martinez, A. W., Phillips, S. T., Wiley, B. J., Gupta, M., & Whitesides, G. M. (2008). FLASH: A rapid method for prototyping paper-based microfluidic devices. *Lab on a Chip*, *8*(12), 2146–2150.
26. Fu, E., Ramsey, S. A., Kauffman, P., Lutz, B., & Yager, P. (2011). Transport in two-dimensional paper networks. *Microfluidics and Nanofluidics*, *10*(1), 29–35.

Index

9780367564315